# POWER FAILURE

OTHER BOOKS BY WILLIAM D. COHAN

*The Last Tycoons:*
The Secret History of
Lazard Frères & Co.

*House of Cards:*
A Tale of Hubris and Wretched
Excess on Wall Street

*Money and Power:*
How Goldman Sachs
Came to Rule the World

*The Price of Silence:*
The Duke Lacrosse Scandal,
The Power of the Elite, and the
Corruption of Our Great Universities

*Why Wall Street Matters*

*Four Friends:*
Promising Lives Cut Short

# POWER FAILURE

*The Rise and Fall of*
*an American Icon*

WILLIAM D. COHAN

PORTFOLIO X PENGUIN

PORTFOLIO / PENGUIN
An imprint of Penguin Random House LLC
penguinrandomhouse.com

Most Portfolio books are available at a discount when purchased in quantity for sales
promotions or corporate use. Special editions, which include personalized covers, excerpts,
and corporate imprints, can be created when purchased in large quantities. For more information,
please call (212) 572-2232 or e-mail specialmarkets@penguinrandomhouse.com. Your local
bookstore can also assist with discounted bulk purchases using the Penguin Random House
corporate Business-to-Business program. For assistance in locating a participating retailer,
e-mail B2B@penguinrandomhouse.com.

LIBRARY OF CONGRESS CATALOGING-IN-PUBLICATION DATA

Names: Cohan, William D., author.
Title: Power failure : the rise and fall of an American icon / William D. Cohan.
Description: New York : Portfolio/Penguin, [2022] | Includes bibliographical references.
Identifiers: LCCN 2022030569 | ISBN 9780593084168 (hardcover) | ISBN 9780593084175 (ebook)
Subjects: LCSH: General Electric Company. | Electric industries—United States—History.
Classification: LCC HD9697.A3 U5335 2022 | DDC 338.7/62130973—dc23/eng/20220805
LC record available at https://lccn.loc.gov/2022030569

Printed in the United States of America
2nd Printing

Book design by Daniel Lagin

*To Deb, Teddy, and Quentin*

# Contents

### The Blind Men and the Elephant

It was six men of Indostan,
To learning much inclined,
Who went to see the Elephant
(Though all of them were blind),
That each by observation
Might satisfy his mind.

The *First* approached the Elephant,
And happening to fall
Against his broad and sturdy side,
At once began to bawl:
"God bless me!—but the Elephant
Is very like a wall!"

The *Second*, feeling of the tusk,
Cried, "Ho!—what have we here
So very round and smooth and sharp?
To me 't is mighty clear,
This wonder of an Elephant
Is very like a spear!"

The *Third* approached the animal,
And happening to take
The squirming trunk within his hands,
Thus boldly up and spake:
"I see," quoth he, "the Elephant
Is very like a snake!"

The *Fourth* reached out an eager hand,
And felt about the knee:
"What most this wondrous beast is like
Is mighty plain," quoth he;
"'T is clear enough the Elephant
Is very like a tree!"

The *Fifth*, who chanced to touch the ear,
Said: "E'en the blindest man
Can tell what this resembles most;
Deny the fact who can,
This marvel of an Elephant
Is very like a fan!"

The *Sixth* no sooner had begun
About the beast to grope,
Than, seizing on the swinging tail
That fell within his scope,
"I see," quoth he, "the Elephant
Is very like a rope!"

And so these men of Indostan
Disputed loud and long,
Each in his own opinion
Exceeding stiff and strong,
Though each was partly in the right,
And all were in the wrong!

—JOHN GODFREY SAXE (1872)

This was American capitalism. GE was America.

—U.S. TREASURY SECRETARY
HENRY M. PAULSON JR.

# POWER FAILURE

# SIASCONSET

The Nantucket Golf Club is so exclusive that there is no sign for it off Milestone Road, the six-mile artery that connects the quaint and historic town of Nantucket, where the ferries pull in after the thirty-mile trip from the mainland of Massachusetts, to Siasconset (often shortened to "Sconset"), the old fishing and artists' village on the far eastern edge of the island. If you make the mistake of turning onto the club's long driveway without a member, or an appointment to visit with one, you'll be stopped by security at the gatehouse a couple of hundred yards up the road and told to turn around.

The membership fee at the Nantucket Golf Club these days is $475,000 per person, more than double the $200,000 that investment banker Edmund Hajim corralled from about fifty initial members when he first conceived of the club in 1995 as an alternative to the nearby and infinitely snootier Sankaty Head Golf Club. At Sankaty Head, founded in 1923, the waiting list for membership is said to be decades, at best, although it's never been clear whether that's simply a ruse to keep out the unwanted, or whether the demand to get in justifies the waiting list. Sankaty Head secretes old-money charm and shabby gentility. Membership dues are much cheaper at Sankaty Head, assuming you can get in, and so it also includes among its ranks a disproportionate number of the year-round residents who would be unlikely to afford the pricier confines of the Nantucket Golf Club. And after Columbus Day weekend, the official end of the "season" on Nantucket, anyone can play a round of golf at Sankaty for about $25. But that's not even remotely the same as being a member. Some say Sankaty Head is changing; in recent years, a few Jews have been added as members and the cheap after-season access to the golf course seems to have disappeared.

Hajim, now eighty-six years old, got tired of waiting for Sankaty to let him in, despite his qualifications. A fine golfer, his Horatio Alger story reads like that of, well, Horatio Alger himself. After he was born in Los Angeles into a failing, dysfunctional family, his parents divorced when he was three and then his father kidnapped him and told him his mother had died in childbirth, even though she was very much alive. His father lost all his money in the Depression.

He overcame these setbacks. He received a scholarship to the University of Rochester and then went to Harvard Business School. Armed with his MBA, Hajim, of Syrian Jewish descent, headed to Wall Street, where he made a fortune big enough to give the University of Rochester $30 million in 2008, its largest single gift. And he has given full scholarships to more than two hundred students at a variety of colleges and universities across the country. He was chairman of the board of the Brunswick School, the tony private school in Greenwich, Connecticut, among the places he lives when not on Nantucket.

Sankaty Head rejected Hajim's bid to become a member, which was a shame as he had just built a nearly seven-thousand-square-foot house overlooking the Nantucket harbor. Since it seemed he wasn't going to be able to play golf on Nantucket, he briefly thought about leaving the island. But necessity being the mother of invention and all, rather than slink away, Hajim decided to get even and, along with golf-course genius Fred Green, began the nearly impossible task of build-ing, from scratch, a new golf club on Nantucket, just as they had in Vail and would later do at Queenwood Golf Club, in the London suburb of Surrey. He and Green negotiated to buy the three hundred or so acres of land they needed for $8 million from a branch of the Tristram Cof-fin family, among the original white settlers of Nantucket in and around 1660.

For the $20 million he needed to build the eighteen-hole links course, the clubhouse, and various other buildings, Hajim enlisted a group of other wealthy "new money" types—including Robert Green-hill, a former Morgan Stanley bigwig and the founder of his own invest-ment banking boutique, and Walter Forbes, the soon-to-be-convicted CEO of Cendant Corporation. Nantucket's decision to extend the run-way at its small airport in the 1960s had brought a wave of billionaires to the island, literally in their private jets. The new golf club, adjacent to Sankaty Head, opened officially in April 1998.

Like so many things on Nantucket, the club looks like it has been there forever. The clubhouse, a sprawling symphony of gray shingle-style architecture, sits atop a grassy knoll, giving it sweeping views of the otherwise flat but luscious terrain as it descends into the vast At-lantic Ocean. It could almost be mistaken for the Serengeti. The sun-sets from the outside terrace are breathtaking.

IT WAS HERE IN AUGUST 2018 THAT JACK WELCH, THE OCTOGENARIAN titan of American capitalism, invited me to lunch. We sat overlooking the ninth hole. Welch was the same age as Hajim but had not aged nearly as gracefully. At that time, he got around tentatively, with the help of either a cane or, as he called it, his "wagon," a three-wheeled, triangular walker. He'd had health problems for years, starting when he had a heart attack and quintuple bypass surgery in 1995. In 2009, he spent ninety-two days in a New York hospital battling a staph infection; he almost died. Never a physically imposing man, he now seemed even more gnome-like, a big head atop a shrinking body. But his mind remained razor sharp. And his personality remained a mix of infinite charm and biting candor.

During his nearly twenty years at the helm of the General Electric Company, now known simply as GE, Welch became a legend. Welch made GE the most valuable company in the world, in the same rarefied league now inhabited—depending on the day of the week—by the likes of Apple, Amazon, Google, and Microsoft, or a combination of them all. He also made GE perennially the world's most admired company and himself the world's most admired CEO, although there were a few years, early in his tenure, when he was deemed the most feared CEO, a label he doesn't quite understand, despite his willingness to be self-critical. At different moments in his career, he was referred to as Teflon Jack for his ability to sidestep blame for disasters on his watch. He was also referred to as Neutron Jack for a few years, after he gained a reputation for ruthless cost cutting, eliminating lots of people and expenses while leaving buildings standing. He hated that name, too.

By the time Welch retired from GE, on September 7, 2001, he was hailed as the CEO of the century. Welch had a chance to be one of the founders of Nantucket Golf Club but he demurred. Instead, he joined later, paying around $250,000 for the privilege.

In his day, Welch was nearly a scratch golfer, and he'd won several golf tournaments at the Nantucket Golf Club, although not its club championship. He had won the club championship twice at Sankaty Head, where of course he was also a member, and had been nearly forever. (No waiting time for Welch). For years, he had been a caddy growing up in Salem, Massachusetts, although he was a long way from that now.

Jack's house on Nantucket was around the corner from Sankaty Head, off of Hoicks Hollow Road. At the end of Hoicks Hollow are both the Sankaty Head Beach Club (members only) and the Atlantic Ocean, with access to a public beach. Like the members at Nantucket Golf

Club, the residents of Hoicks Hollow Road like their privacy. Bob Greenhill lives off Hoicks Hollow. Bob Wright, the former CEO of NBC and then of NBCUniversal when GE owned them and then, eventually, also a GE vice-chairman, lived on Hoicks Hollow, too. It's a funny road. Over the years, the sign for the street comes and goes; sometimes it's there, sometimes it's not. There is also a small cement marker with the name of the street stenciled on it. But it's often covered up with blue masking tape. If you know, you know.

During the glorious summer months on the island, Jack could be seen driving his boxy Mercedes convertible somewhat erratically between his sprawling house and his two golf clubs. You could always tell it was Jack, even though there were plenty of other Mercedes convertibles on Nantucket. The top was down and save for the appearance of a white baseball cap, it seemed like the car was driving itself.

Jack was already sitting at the table at the Nantucket Golf Club when I arrived. The staff was expecting me and I was ushered immediately to his table. He ordered the same lunch he always does, apparently: a small bowl of tomato soup, with a side of oyster crackers, and two chicken hot dogs. When he was GE's CEO, his lunch was always a turkey sandwich with lettuce, tomato, and mustard on whole wheat. He drank a diet Slice. He used to work while he ate. "And he's a fast eater," said Rosanne Badowski, his longtime assistant. Golf was one of the few constants in Jack's life. One way or another, there was always golf. You could learn a lot about someone playing golf, he liked to say. Matt Lauer, the now-disgraced former host of the *Today* show, recalled years ago the many times he played golf with Jack and how he had what Lauer described as "the mental wedge" when it came to the game. "He had a wicked sense of humor, razor sharp," he told the now-defunct *Talk* magazine. "He loves to get in your head." He remembered one time when Jack was on the opposing team in the foursome, Jack walked up to him as he was getting ready to tee off. It was an important moment in the match. "You know," Jack said to him, "you're really losing your hair. You better make the most of your career 'cause it's going." Lauer thought to himself, "You son of a bitch."

———

OUR LUNCH WAS THE DAY BEFORE THE START OF THE DELL TECHNOLO-gies Championship, the annual PGA event in Norton, Massachusetts. The famous golfer Phil Mickelson was at the Nantucket Golf Club for a practice round with some Wall Street bigwigs. At the next table over from us, Mickelson was having lunch with Bob Diamond, the former CEO of Barclays (the big British bank), who grew up on Nantucket, and with Paul Salem, one of the founding partners of Providence Equity Partners, a hugely successful private-equity firm. One after another, the three men came over to pay homage to Jack.

But Jack had more than golf chitchat on his agenda for our lunch. I was there to talk about what he had done to make GE the most valuable and most admired company in the world. But he had a few things he wanted to get off his chest first. What he really wanted to talk about was Jeff Immelt, his successor at GE. He seemed desperate to unburden himself about the guy and was already talking about him even before I could sit down. "I fucked up," he kept telling me, his squeaky Massa-chusetts patois getting crankier and crankier. "And I don't know why."

Jack had had no doubts about his decision to pick Immelt back in the day. But eighteen years later, Welch's perspective on Immelt had changed dramatically. He had utterly soured on him and the job he did as GE's CEO. He repeatedly told me that Immelt was nothing more than a "marketer" who had effectively sucked up to him and the GE board to win their collective approval in the competitive, world-famous succes-sion race. "He's full of shit," Jack said. "He's a bullshitter."

"But Jack," I asked, "didn't you choose Jeff?"

Yes, he conceded he had. "That's my burden that I have to live with," he continued. "But people have been hurt. Employees. People's pensions. Shareholders. It's bad." There were tears in his eyes. "I fucked up," he said again. "I fucked up." I knew Jack had been disappointed by Jeff's reign atop GE, but I had no idea of the depth of his disdain for his successor.

Jack actually seemed not to like Jeff one bit. Hated him, in fact, and had decided he was not a nice guy and kind of a prick. He wanted me to know. "Reminiscing about this pisses me off every time," Jack told me. He then, and not for the last time, engaged in a bit of spin control: he told me how I should think about the story of the rise and fall of GE, once the world's most respected company—and admired conglomerate—now in the process of being dismantled. "I think comparing me to Jeff is not the winning angle," he said. "The winning angle is: How did Jeff do with what he got? How did I do with what I got? Jeff had eight years of free run with my foundation"—meaning the version of GE that Jack had left him on Jeff's first day as CEO, September 7, 2001—"until he started fucking around. We were going beautifully. He was CEO of the year with those assets that I gave him. Or my team gave him. He was CEO of the year."

Jack wasn't finished, not by a long shot. "He got a ton of money and he got good businesses," he continued. "And he played with them. That's the story."

Needless to say, Jeff Immelt has a very different view of Jack and the company that Jack left him to manage. Jeff saw problems everywhere that Jack had either ignored or obfuscated, through a combination of his incomparable bluster and effusive charm that wowed investors and Wall Street for nearly twenty years. Jeff saw himself on a mission to fix Jack's mistakes and prepare GE for the twenty-first century, the digital century. "Following Welch was certainly no easy task," Jeff told me with considerable understatement in our first interview, in Cambridge, Massachusetts. He would have much more to say.

After lunch, Jack offered to drive me in his Jeep Cherokee the three miles or so back to my Sconset house. The supremely efficient and solicitous valet parking crew at the golf club had by then fetched

Jack's car, facing it out with the engine running and the doors open on both sides. I was a little disappointed we weren't in his Mercedes convertible. I got in the Jeep and put on my seat belt. Jack sat on his seat belt. Since he wasn't wearing it, the warning bell kept going off. I urged him to put his seat belt on, if for no other reason than to make the noise stop. But Jack didn't care about the ringing. Nor did he care about his own safety, it seemed. "I hate those things," he told me.

Off he drove. When he got to the left turn out of the Nantucket Golf Club, onto Milestone Road, he did something odd. Instead of keeping to the right side of Milestone Road, as other American drivers do, he decided to drive in the middle of the road, with the Cherokee straddling the yellow line. Needless to say, the drivers coming toward us on Milestone were freaking out. One after another, they all pulled off to the right onto the grassy edge of the street, giving Jack full clearance to continue driving down the middle of the road. He didn't seem to notice.

Amid thoughts of what it might be like to perish in a vehicle driven by the greatest CEO of the American century, it struck me that Jack's vehemence at lunch in pinning the failure of GE on his successor wasn't just a display of ego or of his bitter disappointment in watching his legacy evaporate in real time. It was a reminder of what GE once was and why it matters. GE was never just another company. At different times during its 130-year run, it had been a leader in technological innovation and entrepreneurial drive. It had been a leader in helping people buy GE products, and a variety of other things, using GE's money. The generation and distribution of electricity? GE. The lightbulb? GE. The jet engine? GE. The X-ray machine? GE. The world's first radio broadcast? GE. The first home television sets? GE. The first electric cars? GE. Beneath the hood, GE had also been a crucible of corporate leadership, producing one leading executive after another who was capable of managing massive, far-flung global businesses across a variety of disciplines—including, among many others, Boeing,

Honeywell, Allied-Signal, and Warner Bros. Discovery—and could do so while making a tidy profit. Although many people forget, if they ever knew, GE was also a financial powerhouse, larger and more profitable than many commercial banks, investment banks, private-equity firms, or hedge funds. GE embodied both the muscle of American business—entrepreneurial drive, inventiveness, financial legerdemain—and its weaknesses—unchecked egos, grandiosity, hubris, and corruption. The story of GE's glorious rise and distressing fall is not just the story of a power company or a jet engine company or a TV network or a finance behemoth. It's a cautionary tale about hype, hubris, blind ambition, and the limits of believing—and trying to live up continuously to—a flawed corporate mythology.

## Chapter One

# A CHILD OF
# TWO FATHERS

Ask most people about the origin of the General Electric Company in and around 1892, and you'll hear all about Thomas Alva Edison and his inventions—the carbon filament incandescent lamp, the dynamo, the phonograph, and the motion picture camera. But GE, and its extraordinary success after some initial financial hiccups, was actually more the doing of another restless entrepreneur, Charles Albert Coffin, whose visionary thinking and aggressive acquisitions drove the company forward in its early years. Coffin doesn't get nearly the accolades, or ink, of Edison, but it turned out that Coffin was by far the superior businessman, a gene Edison lacked. In any event, Coffin and Edison started out as competitors in the race to electrify America, but they soon joined forces and both of their powerful DNAs would become entwined in GE.

Born in December 1844, Charles Coffin grew up in Fairfield, Maine, north of Augusta, after his grandfather, a farmer, settled there at a time when the federal government was offering free land to people who would move to uninhabited wilds. Charles's grandfather was a minister of the Society of Friends, also known as the Quakers. After graduating from Bloomfield Academy in 1862, Charles moved to Lynn, Massachusetts, to live with an uncle so that he could attend a "commercial" school in Boston. That uncle was a partner in a shoe manufacturing company, Micajah C. Pratt and Company. When Pratt died in 1862, Coffin's uncle inherited the business and his nephew joined him there. At that time, Lynn was the center of the nation's burgeoning shoe industry, and Pratt's business was an important part of it. The industry was automating rapidly, with sophisticated machines replacing the artisans and craftsmen of an earlier era. Coffin quickly proved himself adept at shoe design. He then moved into sales and proved skilled at that task as well. He started making regular calls in the western United States. It turned out Coffin was a talented businessman. He "must have borne the hallmark of genius from the outset," John Broderick, a former colleague, wrote about Coffin.

In 1873, Pratt's shoe company was renamed Charles A. Coffin and Company for its leading executive. Coffin built a new plant by the railroad station in Lynn in order to snag customers as they were coming off the train, before they could venture farther into town to one of his competitors. The company thrived under Coffin and he soon began to consider other business opportunities. By the 1880s, new inventions like the telephone (1876) and the incandescent lightbulb (1880) were poised to create entirely new industries. There were fortunes to be made in high tech, and one such invention was sitting in Coffin's backyard.

The Lynn Grand Army Post, an armory in Coffin's hometown, was considering lighting its new building with one of the newfangled dynamo systems, a forerunner of the electrical generator. Dynamos were among the first commercially viable ways to generate the electricity used in manufacturing and, eventually, to light people's homes. Silas Barton, a local newspaper owner, and Henry Pevear, a local leather manufacturer, who were tasked with the project, noticed that the building had a rudimentary yet out-of-date six-light dynamo sitting in its basement. The name on the dynamo read "American Electric Co., New Britain, Connecticut."

Barton and Pevear set off for New Britain, where they hoped to meet with Elihu Thomson, a former Philadelphia high school teacher–turned–inventor, who had moved to Connecticut and founded American Electric. Thomson had partnered with Edwin Houston, his former physics teacher from Philadelphia's Central High School. Thomson and Houston together had built a number of electrical contraptions, including induction coils, an arc lamp, and a dynamo. Thomson showed the dynamo to a friend, who then invited a curious cousin to see a demonstration of how it worked. Thomson, then twenty-six, told the cousin he could build a better dynamo, "one that will run any number of lights you want." The cousin responded well to Thomson's idea: "Let's build a four-lighter. I'll stand the expense." Thomson installed his first souped-up dynamo in an all-night bakery. The next one was in a brewery. When the brewery later caught fire, one of the firefighters sent to douse the flames couldn't get over Thomson's dynamo. "What the dickens kind of a light is that?" he said. "You pour water on her and she won't go out."

But as is the case with start-ups (then and now), Thomson and Houston were in need of capital; their initial venture together was in the process of fizzling out. When they came to Lynn to install the armory's new generator, the newspaperman and the leather manufacturer put them together with Coffin to discuss the possibility of Coffin injecting fresh capital into American Electric Co. and moving it from Connecticut to Massachusetts. On February 12, 1883, the newly recapitalized American Electric Co. opened for business in Lynn, Massachusetts, under a new name, the Thomson-Houston Electric Company. Coffin had bought out the inventors' old investors and was now in the process of recapitalizing and retooling their power and lighting company for the future, under his control.

There was one small hitch. Unlike with the shoe business, Coffin now found himself leading a business without a market. Broadly speaking, there were few customers for an electrical power company in 1883. It's hard to imagine today, but there were no grids to deliver electricity to homes and businesses; there were few, if any, electric appliances and it was a serious challenge to convince consumers that electric light was preferable to whale oil or candles. What if the whole contraption exploded? Or went up in flames? Out of necessity, the Thomson-Houston sales force became proselytizers, fanning out across the country to share the powerful message of arc lamps and electric light. The company had to teach its engineers how to install and to operate the equipment. At times, it was a hard sell, in the same way that getting people to use the internet was not so easy at first.

This thirst for customers led to the creation of local power and light companies, backed by wealthy investors and supported by local governments eager to provide the new technology to their citizens. "Mr. Coffin and his associates set out to sell electricity," *The New York Times* reported. "Their main objective was to get electricity to the people. They began establishing power plants in every place possible, where people could make use of it simply by connecting up."

The company eventually moved into a new three-story building on Western Avenue in Lynn. The new building had so much space at first that Pevear, the leather manufacturer, wanted to use some of it to dry his animal skins. But the new electricity business was a big success, nearly immediately. Growth was swift. In 1884, the new venture was supplying electricity to five central stations with 365 arc lamps. A year later, there were thirty-one stations supplying 2,400 arc lamps. Business was booming in part because of a decision to use electricity to power streetcars. The horse-driven commuter systems had to go, and the electric streetcar system that the company built connecting Boston to Lynn, some ten miles away, helped put it on the map by 1888.

Coffin had an interesting approach to raising the capital he needed. Rather than approach friends and family, as was more typical, he preferred to put together a constellation of what he dubbed "men of large means." Some of his earliest investors read like a who's who of Boston Brahmins: Henry L. Higginson, an investment banker and the founder of the Boston Symphony Orchestra; S. Endicott Peabody, a merchant and father of the founder of Groton School (his great-grandfather, Joseph, one of the wealthiest men in America, made his fortune importing pepper from Sumatra); T. Jefferson Coolidge, a great-grandson of Thomas Jefferson; and George F. Gardner, another prominent Boston financier.

Soon enough, Coffin had sold off his family's shoe business to a company in Boston for $300,000 to focus his full energy on the high-risk but potentially lucrative electrical power business. He employed two tactics to fuel Thomson-Houston's growth, both of which would be familiar today—vendor financing and acquisitions. Thomson-Houston did not sell its equipment to the public but instead sold to small, poorly capitalized local electric lighting companies. While adding an element of risk to Coffin's business model—could customers afford to pay?—the strategy also allowed Coffin to achieve greater scale by being a wholesaler of electrical equipment rather than a retailer. But often Coffin's customers struggled financially. That's when Coffin decided that Thomson-Houston would provide what is now known as "vendor financing" to its customers, essentially allowing them to pay for the purchased equipment with a combination of some cash up front plus their debt or equity securities.

He also pursued an aggressive acquisition strategy, sensing that there would soon be a handful of power companies that dominated the industry, and that scale would be key to his company's success. Buying up rivals solved another burgeoning problem for Coffin: patent infringements. He had an ongoing feud of sorts with Charles Brush, the inventor of the new type of lightbulb known as a double carbon arc who had served a patent infringement notice on Coffin. To solve the litigation, Coffin bought Brush's company for more than $3 million.

When he wanted to enter the streetcar business, Coffin bought the failed Van Depoele Electric Manufacturing to get its patents, and then, for a stream of royalty payments, the Belgian-born electrical engineer and prolific inventor Charles Van Depoele himself. And in 1888, Coffin bought the majority of Fort Wayne Electric Light Company, which had been selling dynamos in the Midwest and South.

As often happens during fecund periods of innovation, two or three players battle it out for supremacy. And as Thomson-Houston grew and became more profitable, it increasingly found itself bumping up against the two industry titans: George Westinghouse, who had received his first patent at the age of nineteen and had the inventor Nikola Tesla on the payroll; and Thomas Edison, who in his lifetime would amass 1,093 U.S. patents. Coffin, who was once described as "a man born to command, yet who never issued orders," would soon find himself in a battle for market supremacy with Edison, while Westinghouse would remain a thorn in his side.

While Thomson-Houston had its genteel Boston backers, Edison had in his corner none other than J. P. Morgan, the titan of American finance and the senior partner of Drexel, Morgan & Co., and Henry Villard, the journalist–turned–railroad magnate. A lawyer who had been representing Edison on patent disputes related to the telegraph introduced Edison to J. P. Morgan to see if he might be interested in backing Edison's efforts to commercialize the incandescent lamp. He was. In 1878, with an investment of $300,000—most of which went to pay for equipment—Morgan, Edison, and small group of other financiers created the Edison Electric Light Company. Edison's immediate challenge was to create a filament for the electric lamp that would not burn up. By October 1879, Edison had the problem solved. In his lab in Menlo Park, New Jersey, a carbonized cotton filament burned for forty consecutive hours.

But of course it wasn't sufficient to create merely an electric light; Edison also needed to convince people of its efficacy and to create a mechanism by which electric current could be transmitted to their homes and businesses, so that the electric light could be used. Edison needed more space for this project. In 1880, he created the Edison Machine Works in New York City, to manufacture generators, and the Edison Lamp Works in nearby Newark, New Jersey, to manufacture lamps. Edison's lamp was little more than a glass bulb with most of the air taken out of it—nearly a vacuum—and endowed with a filament of carbon "for the reception of electric energy in the airless space," according to one description.

Edison's New York City office was at 65 Fifth Avenue. As his personal assistant in New York, Edison hired Samuel Insull, a young Englishman who had worked the previous two years as the secretary to his London agent. Edison and Insull, who would later become the controversial founder of the largest electric utility in Chicago, met for the first time in the back room of 65 Fifth Avenue on March 1, 1881, two weeks after Insull sailed on the SS *City of Chester* from Liverpool. Edison's office was spare, outfitted with only a few walnut rolltop desks. "Edison received me with great cordiality," Insull later remembered. "I think possibly he was a little disappointed at my being so young a man. I had only just turned 21 and had a very boyish appearance." Insull recalled in some detail what Edison was wearing at their first meeting—a "rather seedy" black diagonal Prince Albert coat, a white shirt "somewhat worse for the wear"—as well as his somewhat portly shape ("although by no means as stout as he has grown in recent years"). "What struck me above everything else was the wonderful intelligence and magnetism of his expression and the extreme brightness of his eyes," Insull said.

They immediately got down to business, including discussions of how much money Edison had in his bank accounts and what securities of European telecom companies were the most salable. Like Coffin, Edison had taken payments from customers in the form of their debt or equity securities. Edison wanted to make sure his incandescent lamp factory had enough money to operate. "He spoke with very great enthusiasm of the work he had before him [regarding] the development of his electric lighting system," Insull continued, "and his one idea seemed to be to raise all the money he possibly could, with the object of pouring it into the manufacturing side of the electric lighting business, and I remember how wonderfully impressed I was with him on this account, as I had just come from a circle of people in London who not only questioned the possibility of the success of Edison's invention, but often expressed doubt as to whether the work he did could be called an invention at all." They continued their conversations until around five in the morning. Instead of feeling exhausted by Edison, Insull seemed exhilarated, "feeling thoroughly imbued with the idea that I had met one of the great master minds of the world." He allowed that he "fell a victim" to Edison's "spell" during that first meeting.

The next day, Insull was off with Edison to Menlo Park, where for the first time he saw Edison's labs and the network of electric lighting he had created in the buildings that comprised his operation. He wrote of the visit that it was "unforgettable" and at ten o'clock at night he was so moved he returned to the Menlo Park railroad station and had the telegraph operator there send messages to his friends in London about the miracle of electric light he had just witnessed. "Menlo Park was naturally the Mecca of those who looked upon Edison as the great inventive hero of the time," Insull recalled years later. "It must always be looked at as the birthplace of the electric light and power industry."

The prize that Edison had his eyes on at the moment was lighting Lower Manhattan, or more explicitly the geographic area bounded by Wall Street on the south, Fulton Street on the north, and Pearl and Nassau Streets on the east and west, respectively. "We will make electricity so cheap that only the rich will burn candles," Edison told *The New York Herald* in 1879. The "most serious" task at the time was, according to Insull, the manufacture and installation of "underground conductors" in the area of Lower Manhattan where Edison wanted to build his electrical grid. Edison manufactured the conductors at the Electric Tube Factory, at 65 Washington Street, near the Hudson River. In 1882, Edison introduced the first electrical grid connected to his Pearl Street station in Lower Manhattan. The breakthrough made gas lighting nearly obsolete. On September 4, 1882, New York's high society turned out in droves to witness the first illumination of buildings in Lower Manhattan, thanks to Edison. Ironically, given what was to come in later decades, Edison backed off his further development of the phonograph to go all in on electricity. "Well, Sammy," he told his assistant, "they never will try to steal the phonograph; it is not of any commercial value and therefore no one will ever have any incentive to try to get it away from me." He doubled down on the lightbulb, ransacking the vegetable kingdom to find a substance that could create a long-lasting filament.

In 1887, Edison moved his operations to Schenectady, New York, near Albany. He simply had outgrown his operations in Manhattan and Brooklyn. Enticed by the combination of cheap labor and room to grow, Edison tried to buy an unused factory on ten acres of land outside Schenectady near the Erie Canal from the family that owned the Schenectady Locomotive Works. But his offer was rejected. To make the deal happen, Edison enlisted the help of the local citizens in Schenectady, who agreed to subscribe to a fund that bridged the financial gap between buyer and seller. At Edison's request, Insull moved to Schenectady, along with about 250 other employees, and he became the general manager of the operations. "Do it big, Sammy," Edison urged him. "Make it go." In five years, there would be more than six thousand men employed there.

Before making the move to Schenectady, Edison needed to get control of his disparate operations and clean up his keiretsu of businesses. Different companies had sprung up over the years on the backs of different inventions, some profitable, many not (including the Electric Tube Company, a forerunner of the modern-day subway that never got out of the station). Finally, in 1889, with the help of Henry Villard, J. P. Morgan, Deutsche Bank, and what is now Siemens, the big German conglomerate, Edison agreed to merge Edison Machine Works—comprising Edison Lamp Company, Electric Tube Company, and Bergmann & Company—with Edison Electric Light Company to form the Edison General Electric Company. Villard was elected the merged company's president. At the first annual meeting of the company, held at 44 Wall Street, Villard, Insull, and Edison, among others, were elected to the board of directors.

The business was capital intensive, and often capital was difficult to find. "The general business of the Edison General Electric Company is growing at a rate that is equally surprising and gratifying," Villard wrote to the bankers at Drexel, Morgan in March 1890. "Instead of one million, several millions are imperatively wanted to meet the current demands of the several manufacturing departments." Villard and Morgan began to look for a merger partner for Edison. Tension between Edison and Villard began almost immediately and only intensified when the railroad man broached the subject of Edison General Electric Company merging with Coffin's Thomson-Houston.

By 1891, both Edison General Electric and Thomson-Houston were relatively formidable enterprises. Edison's company had six thousand employees, $15 million in capital, and annual revenue of nearly $11 million and profit of $2 million. It had nearly four thousand customers and equipped 375 central stations, 180 street railways, and more than 2,200 street railway cars. Thomson-Houston had four thousand employees, one third fewer than Edison, but its revenue of $10.3 million was nearly the same. Thomson-Houston equipped 870 central stations, 204 street railways, and 2,760 street railway cars. It was also far more profitable than Edison, $2.7 million as opposed to $2 million, and needed much less capital—$10.3 million—to achieve the superior results. The combination of the two companies would create a power-house in New England and New York. But Edison insisted that his company should stay independent. "The new company must stand alone," the inventor wrote to J. P. Morgan, and then warned him, if "you make the coalition, my usefulness as an inventor is gone."

AS USUAL, THE MONEYMEN WON OUT. IN JANUARY 1892, VILLARD AND Coffin revived the merger negotiations that Edison had effectively killed some two years earlier. In early February, the news about the pending merger leaked to shareholders of the two companies, as well as to *The New York Times*. Villard confirmed the rumor and conceded the negotiations were "progressing rapidly" and a "conclusion satisfactory" to all parties would be "speedily reached." Villard admitted to the paper that he'd been working on the deal for two years and that it looked like it would be a merger of equals, more or less. "Ever since I organized the Edison General Electric Company, I have had the conviction that very great advantages to the stockholders" of both Edison and Thomson-Houston and to the public "could be secured" by a merger, "through greater economy of the management and the combination of the technical skill and experience which they comprise." He said a deal was almost reached the first time around "but failed for reasons which do not now any longer exist," an opaque reference, no doubt, to the fact that Edison had repeatedly nixed it and that two years later, he had less wherewithal to block.

There did seem to be at least one major casualty of the proposed merger: Villard. The *Times* reported that the word was he would be leaving Edison General—he would "cease to be active in the electrical field," is how the *Times* put it—and that the new president of the combined company would be Charles A. Coffin. Edison was not happy to hear that the merger was back, two years after his successful effort to kill it. "I have always regretted the abruptness with which I broke the news to Edison"—about the merger—"but I am not sure that a milder manner and less precipitate delivery would have cushioned the shock,"

Alfred Tate, his new personal secretary, observed years later. "I never before had seen him change color. His complexion naturally was pale, a clear healthy paleness, but following my announcement it turned as white as his collar."

"Send for Insull," Edison told Tate.

But it was true: the moneymen had cut a deal around Edison, who in the end had lost both financial and managerial control of the company he had founded. Of course, it wasn't the first time or the last time that a founder had lost control of his company. But given the nearly insatiable capital needs of the industry, and in an era before anyone had thought of dual-class stock and other protections for founders, it was probably inevitable that Edison would lose control. And despite the brave front he put up, it was also clear that Edison had moved on. The moneymen wanted more. Why not just create one big dominant company? The printed filings about the impending merger between Edison and Thomson-Houston circulated to shareholders—there was, of course, no Securities and Exchange Commission at the time—made clear that the shareholders of Thomson-Houston, and its management, would be the dominant forces in the merged company and on its board of directors. Every three shares of Thomson-Houston would be converted into five shares of the new General Electric, while every one share of Edison General would be converted into one share of the new General Electric.

It could have been worse for Edison General stockholders, based on the relative performance of the two companies—Thomson-Houston was much more profitable. That led to speculation in the press that Edison was more than just a little irritated by the proposed deal. The *New York Tribune* reported that in addition to feuding between Villard and Morgan, Edison also blamed Insull, his former private secretary, not only for letting the company's performance suffer but also for leaving "the Wizard," as Edison was known, "in a hopeless minority—as far as control was concerned." When the dust had settled on the creation of

Edison General, under Insull's watch, Edison himself had ended up with an ownership stake of less than 10 percent.

Edison, apparently, could not understand how that had happened. According to the paper, Edison also blamed Villard for "working contrary to his interests" in concocting the merger with Thomson-Houston, at which point Villard fingered Insull. "Were it not for your mismanagement," Villard said about him, "no consolidation would have been necessary." Insull apparently did not reply, and the *Tribune* reported that, in any event, Insull would be leaving the merged company. *The New York Times* also reported the controversy. "There is a sad lack of harmony among some of the principal men in the Edison Company," the *Times* reported on February 20. Edison and Insull lined up together against Villard. "President Villard is said to have replied with a good deal of heat that the way things had been going made a consolidation necessary," the *Times* continued. Given that Coffin was going to lead the merged company, the speculation was that *both* Villard and Insull would leave and that Edison would no longer serve on its board of directors. The *Times* wrote that the proposed merger was "not at all to [Edison's] liking" and with control of the new company "in other hands," the "gossip" on Wall Street was that Edison's "interest" in the company had been "greatly reduced."

Not so, Edison responded in an interview with the *Times* a day later, in an early example of corporate spin. He said that not only had he been the one to initiate the merger—although there seems little evidence of that—but also that he would greatly benefit from it. "I expect the consolidation will result financially to my advantage," he said. He was full of news. He said that both J. P. Morgan, the man, and Hamilton McKown Twombly, a wealthy Morgan banker from Boston who had built a grand estate in New Jersey and married into the Vanderbilt family, were conducting the negotiations "on a friendly and satisfactory basis" to get the merger done. He said the proposed name of the combined company was likely to be the General Electric Company, removing references to any of the three founders—Edison, Thomson, and Houston—although that had not yet been agreed.

He also announced that Villard was out and made clear that his relationship with Villard was strained, as was Villard's with Morgan. "He made the negotiations when the Edison General Electric Company was formed," Edison said of Villard, "and I suppose he made some money from the transaction." He also said he expected Sam Insull to join him once again on his new ventures and that there was nothing to the rumors that their relationship was somehow soured by recent events. "We are on the best of terms now," Edison said. "He is a valuable man." (In the end, General Electric offered Insull a job as a vice president, but he declined it and moved to Chicago. He later came to some disgrace during the Great Depression.)

But in the end, Edison made it clear he was leaving electricity and electrical machines behind. "I cannot waste my time," he said, "over electric-lighting matters, for they are old." In 1893, Edison sold his stake in GE for a reported $1.5 million (around $430 million in today's dollars). He was over the whole electricity thing. "I have a lot more new material on which to work," he continued. He said all he cared about General Electric was to get "as large dividends as possible from such stock as I hold. I am not businessman enough to spend my time at that end of the concern."

Squabbling and disagreements aside, on April 15, 1892, the General Electric Company was incorporated in New York City as a holding company of sorts for the two companies, one operating in Schenectady, New York, the other operating in Lynn, Massachusetts (both of which factories remain open), with Schenectady being the corporate headquarters at that time. General Electric also acquired the international operations of Thomson-Houston. "It may be likened to a huge tree," John Broderick wrote of the newly created company, "the growth of which is traceable, not only to its own roots, but to a great variety of scions that from time to time have been grafted upon it."

The company began its operational business life on June 1. The market capitalization of the new company was around $35 million, with Thomson-Houston's shareholders controlling a little more than 50 percent of the General Electric stock and its most important management positions. The theory behind the merger made sense (as theories behind mergers often do). The creation of General Electric stemmed from three imperatives: a desire to eliminate competition; a desire to have first dibs on the capital needed to sustain a capital-intensive business; and a desire by both management and investors to run a more effective organization. All three factors led to the inevitable combination of Thomson-Houston and Edison General.

There was little doubt about who would be calling the shots at General Electric. Edison had a spot on the company's new eleven-member board of directors, along with J. P. Morgan and his partner, Hamilton McKown Twombly, and then a group of Coffin's financial backers from Boston, including Henry Higginson and Thomas Jefferson Coolidge. But from the outset, it was definitely Coffin's show. Villard was gone. Insull, Edison's right-hand man, was gone. And soon Edison would be, too.

*Chapter Two*

# TWO CRISES

In General Electric's first annual report, issued in April 1893, Coffin acknowledged the strain of trying to merge some parts of the two operating companies. "The difficulties inherent in such a reorganization were many and serious," he wrote. But that was Coffin's only acknowledgment that General Electric had had a rough start. The rest of the seven-page report made things sound pretty rosy. In its first eight months, GE generated revenue of nearly $12 million and profit of nearly $3 million, an enviable 25 percent margin. The company paid out nearly two thirds of its profit to its 3,272 stockholders. The business was predominantly comprised of six thousand customers providing local electric lighting and operating electric streetcars. Some 1,277 "central station" electric companies were using General Electric equipment, supplying 2.5 million incandescent lights and about 110,000 arc lamps. "The growth of these companies has been phenomenal," Coffin wrote. He also noted the improvements in the size of the machines that General Electric was being hired to manufacture, vastly increasing the electricity they could generate. Its railway car business was also growing fast, with 4,927 miles of road in operation as of February 1893, more than double the amount of the year before. The number of railway cars in operation more than tripled, and the company was also making electric locomotives for use in the mining industry, with two locomotives of 1,600 horsepower each then under construction. In the Schenectady factory, GE made the equipment for the six-mile electric elevated railway used at the 1893 World's Fair in Chicago. Coffin had a "holy faith" in electricity, he once said, and if he hadn't, "we never should have dared the things we did."

But Coffin's early optimism about the company would dissipate quickly. At the very moment that he published General Electric's first annual report, the worst financial crisis in the republic's history—known as the Panic of 1893—had begun. "I saw beggars and bums everywhere on the street," the journalist Upton Sinclair, who covered the panic, said years later. "I saw women and children weeping, sitting on furniture, which have been dumped on the pavement." Speculation in railroads was at the heart of the panic. "Railroads were over-built, and many companies continued growth by taking over competitors, endangering their own stability," according to one contemporaneous account. But there were other causes, too. Silver from mines began to flood the market. Farmers, particularly in the Midwest, suffered a series of droughts that left them short of cash to pay their debts, driving down the value of their land. In February 1893, the Philadelphia and Reading Railroad failed, raising concern about the health of other large railroads and leading to a crash in the stock market on May 5, following the collapse of the National Cordage Company, one of the most widely traded stocks on the exchange. In a matter of weeks, Cordage's stock had declined precipitously by 75 percent from $75 a share to $18.75 a share, before the company eventually filed for bankruptcy.

Interest rates spiked on short-term loans, as fear seeped into the market. On July 25, the Erie Railroad failed. Banks around the country, especially those in New York, suspended depositors' access to their own money. The Treasury was running out of the gold needed to exchange for paper money on demand, so silver was substituted. Further panic ensued while more than fifteen thousand companies and five hundred banks failed. Unemployment hit nearly 20 percent. Life savings disappeared. Many Americans lost their homes when they could no longer afford to pay their mortgages.

General Electric nearly went down the tubes, too, barely a year into its existence. New business fell precipitously in May and June, and GE had to borrow to meet its payroll, at exorbitant interest rates. Orders for its products declined 75 percent. Two thirds of its eight-thousand-person workforce was dismissed and there was hardly enough work to do for those who remained. As of July, the company had $1.3 million in cash and $10 million of long-term debt, an obligation compounded by the fact that it had provided vendor financing to many of its customers who were struggling to pay their bills.

Wall Street lost faith in General Electric and thought it was headed for bankruptcy. "During the summer of 1893, even old and strong customers were obliged to ask for leniency in paying their accounts and notes," Coffin wrote in the 1894 annual report. "Under these circumstances your [c]ompany found itself with its own obligations to meet, but unable at that time to collect the money with which to meet them." Bankruptcy loomed. The company's bankers, many of whom were also on its board of directors, kept Coffin on a very short leash. An executive committee of the board was set up by the directors to override his decisions, if that became necessary. "There were months that seemed like scalding centuries," Coffin said years later.

As he had done before when Thomson-Houston was facing financial difficulties, Coffin came up with a seemingly clever plan to raise cash and to avoid financial catastrophe. He decided GE would sell, at a substantial loss, the stocks and bonds that it owned in various local companies. These securities had a face value of $16.2 million at the time, and GE marked them at $9.2 million on its balance sheet, or 57 cents on the dollar. Coffin figured in the midst of the panic, these securities would go for, at best, 25 cents on the dollar. He proposed valuing $12 million of the best of them at a price of 33 cents on the dollar and offering them for sale to the company's shareholders; they could buy them at that price once the panic had abated. In the meantime, a trustee would manage the securities and Wall Street bankers would underwrite their sale while also immediately advancing GE a much-needed $4 million. Coffin's Boston bankers, Lee, Higginson & Company, signed on to his plan immediately.

But the deciding vote on the GE board belonged to J. P. Morgan himself. During the summer of 1893, he was in Maine, virtually unreachable. Coffin decided to take a train up north to go see the banker and find out what his vote would be. As he was entering the door to see him, Morgan's messenger was going out of it with the news that Morgan had accepted Coffin's plan "without reservation." The cash infusion from the fire sale of the securities was critical and helped GE restore confidence with investors.

Coffin was a changed man after helping GE to survive the Panic of 1893. Keeping GE far from financial danger thereafter became an obsession with him. "We had a close shave," he said when it was over. "We were headed for the poorhouse, and the prospect was anything but pleasant. Hereafter we shall cherish no illusions. Our aim, as we strive for expansion, shall be to build up plenty of reserve strength and to maintain it." According to John Broderick, who worked for Coffin at GE, Coffin's "sensations" after the Panic of 1893 "were like those of a commander whose ship has reached port after having encountered a terrific storm and been dangerously close to the rocks. His experience had been a nerve-racking one and he never forgot it. Its effect upon his judgment was permanent."

He observed that when it came to the General Electric Company, "financial writers" have a "custom" of describing how conservatively the executives valued the company's assets. "That conservatism was a consequence of the panic of 1893," he continued, "during which a spade was called a spade, and little or no recognition was given to contingent or potential values of any kind. Having kept the enterprise of which he was the chief steward from being totally wrecked, Coffin resolved to make it invulnerable, so far as this was humanly possible." (Somewhat famously, Coffin also coined the phrase when valuing assets on a company's balance sheet, "Use cost or market value, whichever is the lower," an accounting aphorism that has stood the test of time.)

At the end of January 1894, GE had debt of $6.75 million and cash in the bank of around half a million dollars. GE vastly curtailed its practice of extending credit to its customers and of taking back equity in them in exchange for access to GE's products. If you wanted a generator, it had to be paid for in either cash or "short credit to desirable customers," according to the GE annual report. GE decided it would take less profit for the time being, and its "surplus" went from a positive $1 million to a negative $12 million. The negative surplus continued for some fifteen years.

By the middle of the following year, business improved. Factories were once again humming six days a week. Coffin noted in the January 1894 annual report that the last two months had seen "a gradual improvement." The streetcar business struggled, as did the arc lighting business. The best performer during this difficult period was the incandescent lighting business, which did relatively well, in large part because its customers were more conservatively run and were able to pay for the GE products. A year later, the company had made material strides in reducing its debt load by paying off its floating-rate debt, by funding its working capital needs out of cash flow, and by buying back $1.25 million of its long-term debt at around 89 cents on the dollar.

Notwithstanding the fact that GE had managed to stave off bankruptcy, the Panic of 1893 touched off a serious economic catastrophe across the country, which would remain in a depression for the next five years. And it would take GE just as long to pull out of its malaise. Sales remained stagnant at $12 million annually. In the meantime, as he had often done, Coffin was willing to take chances on new technology. He brought Charles G. Curtis, a New York City engineer, to Schenectady to try his hand at what eventually became the precursor to today's steam turbine—a business that GE would dominate for decades. In exchange for the right to purchase patent rights, Coffin allowed Curtis to use the GE facilities while he tried, and often failed, to perfect his invention. It was a tough slog, but with the help of William LeRoy Emmet, an engineer in the lighting department, and another thirty-six experienced mechanics, they succeeded in designing a five-thousand-kilowatt turbine. Coffin offered the new machine, which was one tenth the size of what it replaced and could generate far more power, to Samuel Insull, Edison's former right-hand man, who was then president of Chicago's Commonwealth Edison Company and who was staking his reputation on GE's new steam turbine. Coffin decreed that the new turbine needed to be working by March 7, 1900, in order to be shipped to Insull. The team finished three days early. But the machine still had to be installed. A wave of GE construction workers moved to Chicago for the complex assembly at the new generating station that Insull had ordered built on Fisk Street. For months things went wrong. Whenever this new turbine generator ran well, GE's stock increased in value, and when the turbine failed, GE's stock fell. Three months later, though, the new steam turbine was installed and was running flawlessly. Orders for the new machines streamed into GE.

By 1898, GE's revenues had finally begun to grow again, reaching $22 million for the previous year. New business opportunities were everywhere, including providing electricity to the Manhattan subway system and to the ones serving the outer boroughs. There was the contract to electrify Grand Central and one to do the same for the Manhattan Elevated Railway, for more than $1 million. That's a deal that Coffin sealed himself by going to the railway's offices and personally making the pitch for the business.

He also had made a concerted effort to clean up the risk on GE's balance sheet, clearing the way for the company to pay stock dividends. Under Coffin's leadership, GE had paid off its long-term debt and was no longer factoring—selling off at a discount—its receivables. It had also not borrowed any money. Customers paid cash for everything, or else weren't able to buy from GE. And the securities of its customers that GE was selling were now achieving prices above par after years of GE's having been forced to sell them at a discount.

By April 1900, the *Times* reported that GE seemed intent on world domination after it agreed to buy the U.S. business of the big German electrical manufacturer Siemens-Halske Company, with which it had been associated from the outset of its corporate life. Upon completion of the acquisition, the *Times* reported, GE "will in all probability control all of the leading electrical vehicle companies in America." The funny thing was that Levy Mayer, GE's general counsel, impolitically agreed with the paper's assertion. "It is to be understood distinctly," Mayer said, "that the purpose of the General Electric company is to absorb as many companies as possible under the law." And Coffin made sure to welcome the new entrants to the GE family. "We may be entertaining an angel unawares," he liked to say. In 1905, GE bought Allis-Chalmers, a major competitor in the electrical and power equipment businesses. And the company was busy working on the "machinery" that would allow an electric train to travel at speeds of 125 miles per hour "and revolutionize the railway transportation of the world" and that, "if successful," would mean the trip from Boston to New York by high-speed train could be done in two hours (an accomplishment that still seems fanciful today).

But before world domination, GE first had to survive another financial crisis—the Panic of 1907. Just as the company was finally hitting its stride following the previous financial crisis, and the markets were booming, the seeds of a new financial crisis had been sown. There was an enormous demand for capital, but since the money supply was pegged to gold, and gold mining production had not kept pace with the demand for money, there was growing concern that financial trouble was lurking around the corner. The supply-and-demand equations were out of whack. Such was the concern in early 1906 that Jacob Schiff, the senior partner at Kuhn, Loeb & Co.—a rival of J. P. Morgan & Co.— warned his Wall Street brethren, "If the currency conditions of this country are not changed materially . . . you will have such a panic . . . as will make all previous panics look like child's play."

He would be right, and the ensuing crisis would again showcase J. P. Morgan's skills as a savior of the nation's financial system—it was perhaps his finest moment—while also highlighting the general danger posed by having no prudential regulatory authority to step in when panic ensued. Recall that this was an era without financial regulations. There was no Federal Reserve Board or system of Federal Reserve banks. There was no Securities and Exchange Commission. There were very few laws indeed governing financial transactions and disclosures. There literally was a Wild West feel to Wall Street, even though it was still mostly located in Lower Manhattan.

Unlike so many others, though, GE had built a bit of a fortress balance sheet in the years leading up to the Panic of 1907, and it weathered the crisis with relative ease. Earlier in the year, before the panic had set in, Coffin decided to use the company's good fortune to sell $13 million of long-term bonds, at 5 percent interest, in case the markets should turn, which of course occurred in the second half of 1907. Orders in the second half of the year fell 23 percent. If they could get access to the capital markets, businesses were borrowing money at rates of between 10 percent and 12 percent, far higher than what GE had just accomplished. "But although GE factories slowed up, there was no financial trouble," Charles Wilson, who was one of Coffin's successors, explained, "and Coffin even helped out some of his utility customers with loans from the bond issue, resulting in more good will and more orders in the years that followed."

An important but underappreciated part of how Coffin was able to rescue GE in 1893 and to keep it above the fray in 1907 had to do with his willingness to engage in what we would today call "financial engineering." He was able to recognize what assets GE had on its balance sheet that could be quickly turned into cash, even at a severe discount, in order to raise the money needed to pay down some debt while restoring confidence among equity investors that the company was taking necessary steps to keep itself out of bankruptcy. Pretty clever.

Coffin first concocted the idea when he was running Thomson-Houston. In 1890—two years before the Edison merger—he organized the United Electric Securities Company to raise capital for Thomson-Houston in the face of a growing reluctance by banks to provide the money Coffin needed. He contributed to United Electric the securities Thomson-Houston had been receiving from its utility customers in exchange for the products Coffin was selling them. Capitalized with these securities—some of which were obviously worth more than others, depending on the customers' ability to pay them off—United Electric was able to raise debt using the securities as collateral. The proceeds of United Electric's debt issuance were then transferred to Thomson-Houston, which owned 100 percent of the stock of United Electric. Based in Boston, United Electric survived the creation of General Electric and continued to serve the purpose that Coffin had intended: to occasionally sell to the public the securities GE often took from its customers. In 1904, Coffin created another wholly owned subsidiary, the Electrical Securities Corporation, to do in New York City what United Electric was doing in Boston. But Coffin was still not done. In 1905, he created the Electric Bond and Share Company as another GE subsidiary. Coffin's idea this time was to create a way to help GE's customers—among them the smaller, regional utilities that were buying GE's power equipment—to get the capital they needed at more favorable rates. In effect, Coffin had devised a plan to finance GE's customers using GE's bigger balance sheet. The Electric Bond and Share Company helped GE's customers get access to equity-like capital and to get better pricing on their more senior debt; in many ways it was a precursor to junk bonds. (Eventually, GE spun off EBASCO, as it was later known, to its shareholders. But it was eventually liquidated after the federal government won a lawsuit against it.) Although Coffin's precise creativity in GE's early years—collecting the financial stakes in its customers into a separate entity and selling them at a considerable discount to GE's shareholders in return for much-needed cash—would fall by the wayside, the intellectual inspiration behind his steps—saving the company at all costs—would become a leitmotif of GE's next hundred years.

In 1913, having steered GE successfully through two major financial crises while also making the company a powerful force in industrial America, Coffin relinquished his title as GE's president. Coffin remained chairman of the GE board for another nine years, retiring fully from the company in May 1922. He then established a private office for himself at 120 Broadway, around the corner from Wall Street. He remained a director of GE until his death, after a week's illness, in July 1926. At his death, he was one of America's richest men.

A month after he retired as the chairman of the GE board, Coffin gave a rare interview to a reporter from the New York *World*. It was a masterful combination of a practical assessment of what electricity had done for America since its inception forty years earlier and prescience about where more abundant and cheaper electricity would take the country. "Take electricity out of New York," he mused, "and one of the items that comes out with it is about nine-tenths of her population." New Yorkers loved the telephone and the electric light. "The skyscraper would be impossible without electricity," he continued. "Not only is it dependent upon the electric light and the electric elevator but the very processes of steel manufacture, which made modern construction possible, depend at many different stages upon electrical machinery. The modern apartment house is unthinkable without electricity."

He predicted—correctly—that electricity's next frontier was the American countryside. (He anticipated the passage of the Rural Electrification Act by eleven years.) Bringing electricity to rural America would transform it by bringing meaningful work, in the form of factories that were closer to raw materials. Instead of shipping those goods to the cities, where they could be turned into finished products, electricity would allow manufacturing plants to be built in rural areas, too. "Electricity made the big city," he said, "now it may be expected to unmake it." All the advantages of the city "will soon be possible for the farm," he continued, "without having to put up with the unendurable indignities of city life. . . . Electricity is not only the cleanest and most efficient servant that mankind ever had but it is also the cheapest. It works for less than a coolie's wage, and its wages are going down every day while its efficiency is being constantly increased. In addition to this it does its own traveling, at the rate of 186,000 miles a second, and doesn't have to be transported."

He also anticipated the perquisites of the farm-to-table and other environmentally friendly movements. "In the country, where our food grows, is the best of all places to eat it," he said. "Sending it to the city costs much more than getting it out of the ground, and it has lost a lot of flavor by the time it has reached the ultimate consumer." He believed that the "next big step in industry" would be the development of the "small, electrically driven factory" near to where the raw materials are found. He predicted that industry would move out of the Northeast in order to be closer to raw material. "Cloth will more and more be manufactured where the cotton is grown," he said. "The flour mills will leave the cities and go back to the farm." A new wave of population migration out of the cities would follow, he predicted. "[T]he country life of the future will not be the dull, forbidding, solitary thing which the concentration of industry has made now."

Finally, in his way, Coffin also anticipated the internet. "There will be ample communication with the outside world," he predicted. "The radio has just burst upon us; there is no telling where its influence will reach. Moving pictures may be broadcast. It may not be necessary to come to New York to see and hear the newest Broadway drama, possibly all that will be necessary will be to adjust the switchboard in your own home in Maine or Washington or New Mexico, and take in a first-night performance as full as if you had an orchestra seat." He knew he was on a bit of an ideological ledge. "This is dreaming of course, for no one can prophesy with any accuracy concerning such matters," he concluded. "Of one thing only, I am certain. That is that we are just in the beginning of an electric age."

Coffin died, in Locust Valley, New York, after a short illness. He made a point of reviewing GE's financial statements until his dying day. Titans of industry attended his funeral. Charles Dawes, Calvin Coolidge's vice president, said Coffin's death was "a loss to the entire country" and added that "his undaunted courage in the early development of the electric power business of the nation, his vision, and the confidence and kindliness with which he assisted its pioneers, made him a unique figure in American industry." He left an estate worth $6 million, and he also left behind a legacy of moral fortitude.

The executives whom he chose for key leadership positions after his departure included Edwin Rice, his longtime colleague and Thomson's former high school student. Rice was an executive who wanted to hear the bad news. "Let's know what the worst is," he said, "then we can seek for the best." Coffin also found Gerard Swope, a graduate of the Massachusetts Institute of Technology and the son of German Jewish immigrants from Saint Louis, and named him the first president of GE's international business, the International General Electric Company. In 1922, Swope succeeded Rice as the company's president, and a young lawyer from Boston, Owen Young, became the chairman of GE's board. Soon enough, GE's revenue increased to $337 million, with some eighty thousand employees and factories across the United States.

## Chapter Three

# THE BATTLE OF
# THE AIRWAVES

The outbreak of World War I opened up entirely new business lines for GE—including radio broadcasting. GE's executives played an important role in the war effort, and its inventors were at the service of the military. W. R. Whitney, the head of GE's research lab, served on the Naval Consulting Board. When the army wanted a lighter searchlight to help spot bombing airplanes, GE created it. When the navy asked for a submarine detector, GE built one in ten weeks and was soon churning them out. In December 1917, three GE engineers joined a group of American naval men on a British trawler, located a German submarine using the new GE technology, and destroyed the vessel. The list of GE's contributions to the war effort was long and varied: special submarine motors; compasses for airplanes; winch drives for balloons; enough telephone switchboards to fill a sidewalk a hundred miles long; and bomb-release mechanisms. GE's executives were not mere businessmen any longer. They were beginning to take roles in the government and laying the groundwork for what President Dwight D. Eisenhower would call the military-industrial complex. The innovations from the war years might have been initially pursued for the military, but often the technology would have civilian applications, applications that could potentially spawn entirely new sources of revenue for GE.

To navigate GE through the war years, the company turned to Owen D. Young, who led GE with great success. Born October 27, 1874, in Van Hornesville, in the Mohawk Valley of upstate New York, Young was the first person in five generations of his family to venture beyond the local school; the Youngs were farmers and their mission was to preserve and protect the farm. When he was twelve, Owen was sent to East Springfield Academy, seven miles away from Van Hornesville. His father would take him there on Monday morning and retrieve him on Friday night. Two years later, age fourteen, Owen Young graduated top in his class. He wanted to be a lawyer, a thought cemented in him during the spring of 1887 when he spent some time back at the homestead working on the farm on a break from the academy. It was very hot and dusty. Planting the corn was then "back breaking" work, he observed in a later speech. Meanwhile, an uncle was involved in a lawsuit in Cooperstown, sixteen miles south. The uncle wanted young Owen to testify in court on his behalf. "Well, I would testify to anything to get a day off," he said. The courthouse was crowded but cool and shady, and he couldn't quite believe people made a living that way. "[T]his is the life for me," he continued. "From that time on I never swerved from my desire to study law." His plan was to study for the Regents Exam and go to Cornell. But he was only fifteen years old; one had to be seventeen to take the exam.

Young ended up at St. Lawrence University, eighty miles north in Canton, New York. St. Lawrence was looking to expand its student body, then about a hundred students. Word of Young had reached the campus after one of its theological students had seen Young conduct a Sunday-school class at the local Van Hornesville Universalist Church and recommended him as a candidate for enrollment to the university's president. The Youngs didn't have the money for the full tuition, but the university president himself agreed to lend the family the balance. That sealed the deal. Young was off to St. Lawrence (and later became one of the school's biggest benefactors).

From upstate New York, Young set his sights on Boston. He wanted to go to Harvard Law School, and had the qualifications, save one. He could barely pay the tuition and made clear on his application that he would also need to take a job—he hoped to be a tutor—to get the extra money he needed to pay the tuition bills. But Harvard said no. "The course was so heavy that they felt that a man should give it all his time," wrote his biographer Ida Tarbell, the famed muckraking journalist. The snub, and its absurdity, was one Young would not forget, even when thirty years later Harvard gave him an honorary degree.

Instead of Harvard, in September 1894, Young enrolled at the law school at Boston University. It was glad to have him. The law school dean arranged for Young to tutor a wealthy man living thirty miles away in Haverhill, Massachusetts, and promised that Young could graduate in two years instead of three, saving him a year's tuition. He graduated cum laude in June 1896 and was selected by his peers to be a class speaker.

He took, and passed, the Massachusetts bar exam over the summer and began in the fall a $10-a-week job as a clerk in the law office of Charles Tyler, an older BU law school graduate who had a thriving practice in real estate and corporate law. (A job he had been offered working for a Boston judge dissolved after the judge's sudden death.) He also secured a second job as a lecturer at the law school. He found himself, in the fall of 1896, working on any number of corporate reorganizations and refinancings necessitated by the Panic of 1893 and its aftermath. He was also engaged in helping small public utilities across the country figure out a way to get back on their feet. And it was in the course of that work that he started to intersect with General Electric, not only one of the largest suppliers of electrical equipment in the country but also a creditor of many of the small utility companies that Young was advising.

Tyler, Young's boss, made clear to Young early on that he had no intention of having any partners. But such were Young's legal skills, as well as the high regard in which his clients—and in particular Stone & Webster, the Boston-based engineering firm—held him that against his wishes Tyler had little choice. Stone & Webster sealed the deal by telling Tyler that it would take its considerable business from Tyler's firm if he did not make Young a partner. In January 1907, the "card of the new firm, Tyler & Young was sent out," Tarbell wrote. With the real money he started making for the first time in his life, Young bought his parents the biggest home in Van Hornesville. He kept the family's dairy farm outside of town and started building up its herd. He bought a thirteen-room house high on a hill overlooking Lexington, Massachusetts.

While working on the corporate reorganization of the Dallas Southern Traction Company, on behalf of Stone & Webster, Young came to the attention of the GE brass, and then of Charles Coffin, its CEO. Dallas Southern was considering whether to branch out into Fort Worth, then a small city, only to discover that the GE subsidiary Bond and Share was thinking of doing the same thing. "It was not long before the two interests were locked in a bitter struggle and were carrying their troubles to court," Tarbell wrote. That's when Young's legal skills made an impression on the GE attorneys, who at first did not rank him "as a formidable opponent" and "scrutinized curiously his lank, lean figure as he sat slouched down in his chair" in the courtroom. But through the course of the trial, the GE team began to change its view of Young, chiefly because of his "penetrating discernment and his clear-cut effective arguments."

Then fate intervened. In April 1912, Hinsdale Parsons, GE's general counsel, died instantly in a car accident near Albany. He was forty-five years old and had been at the company for eighteen years. GE needed a new general counsel. Alerted to Young's talent, Coffin consulted his board members from Boston and confirmed for himself that Young had an excellent reputation. Hearing that Young and his wife were visiting New York City, Coffin invited the lawyer to GE's office at 120 Broadway and offered him Parsons's job as head of the company's legal department. He accepted the job on the spot, even though Coffin offered him less money than he had been making as a partner in his own law firm. Going to GE was Young's chance to abandon the drudgery of his firm's real estate practice. "It was an escape from the medieval," he said, "a chance to operate in what was still the no man's land of law." What he did, from the outset, was to professionalize the general counsel's office—by organizing the lawsuits, by studying the existing contracts, by working to settle disputes rather than just litigating them, and by taking a fact-based approach to the law. He also led the way in trying to be less confrontational with GE's unionized workforce and in creating what we think of now as a "human resources" department.

Young also presided over GE's entry into the modern radio business. In 1901, the ability to communicate across the ocean, from ship to shore, had been introduced. One of the leading corporate practitioners of the new communications technology was the British Marconi Company and its publicly traded U.S. counterpart, American Marconi. Messages could be sent across the Atlantic but were often garbled. "What was needed, said those who knew the science, was something which would give a high frequency current capable of steady reporting," Tarbell observed. In 1915, a GE scientist, Dr. Ernst Alexanderson, armed with a high-frequency alternator that he had developed in Schenectady, asked American Marconi to give his discovery a practical test at its factory in New Brunswick, New Jersey. Alexanderson's alternator was such a success that Guglielmo Marconi himself asked Young to meet him at Holland House, a hotel at Thirtieth Street and Fifth

Avenue, to share with him the news that the British Marconi Company, in England, wanted to buy the exclusive rights to Alexanderson's alternator. At the time of that meeting with Marconi, Young knew little about radio and less—nothing, in fact—about what Alexanderson had created in GE's research center. What Young did realize, though, was that a deal with Marconi for the alternator would mean millions of dollars of new revenue for GE. He wanted to make the deal.

But the war, and national security concerns related to it, made the deal utterly untenable. The U.S. Navy understandably wanted control of the nation's radio-related assets, including local radio stations, which had started to appear across the landscape, and of Alexanderson's alternator. The navy blocked GE's negotiations with British Marconi. Instead, Alexanderson designed and built a powerful radio transmitting system—two hundred kilowatts—that he installed in New Brunswick for the federal government's exclusive use during the war.

After the war, British Marconi tried again to get Alexanderson's alternator. In March 1919, just as Young was considering whether to lay off GE employees engaged in the manufacture of the alternator because the war was over, Sidney Steadman, the general counsel of British Marconi, and Edward Nally, the general manager of American Marconi, arrived at Young's New York office. British Marconi had prospered during the war and "was now embarked upon the ambitious project of attempting to dominate wireless communication throughout the world," Gleason Archer wrote in his *History of Radio to 1926*. "The company needed the Alexanderson Alternator and needed it very much." Steadman came calling with a big order for GE: twenty-four alternators at a price of $127,000 per alternator, a $3 million deal. Young countered with a royalty deal. Marconi countered that it would buy the machines outright, plus another $1 million in lieu of Young's proposed royalty deal. But GE was still hoping to get a royalty payment from Marconi based on each word broadcast using the alternator.

As the negotiations between GE and Marconi were heading toward a successful conclusion, the navy sent GE a letter asking for Alexanderson to make a report about an alternator that the navy was using on Long Island. In response, on March 29, 1919, Young wrote to Franklin Roosevelt, then the acting secretary of the navy while Josephus Daniels tended to peace talks to resolve the war, that GE was more than willing to help the navy again, as it had in the past, but that it also had to consider its own commercial needs. In fact, Young wrote to Roosevelt, GE was about to conclude a deal to sell the valuable alternators, along with other radio equipment, to both the American and British Marconi companies. "This letter produced immediate reactions in Washington," Archer wrote. Roosevelt turned the matter over to a deputy, Stanford Hooper—known as the "father of naval radio"—who quickly deemed the proposed sale of Alexanderson's alternator to the British an existential threat. "When I heard about this impending deal," Hooper wrote in *Radio Broadcast* magazine in June 1922, "I became convinced that the whole future of American radio communications was involved" and that the government's "established radio policy" would "fail utterly" if British Marconi got its mitts on the alternator.

Wasting no time after receiving Young's letter, Hooper mobilized the government's power to block the sale of Alexanderson's alternator to the British and launched a rather remarkable and audacious plan to keep the valuable technology in American control. Hooper and his team demanded, via "long distance phone," according to Hooper, to have a conference with Young and the GE board of directors where "the entire matter" could be "laid before them from the Navy's point of view." On April 4, the day after the call, Roosevelt wrote to Young inviting him and the GE board to Washington "due to the various ramifications of this subject." But Admiral William H. G. Bullard, the new director of naval communications, decided the matter could not wait a week. He and Hooper headed to New York City the next day, April 5, to see Young. "The date," Tarbell wrote, "in the history of the radio which takes about the same place as the Fourth of July in the history of the country." The men carried with them a message from President Wilson, then also in Paris for the peace talks, that, according to Admiral Bullard, the president was "deeply concerned over the matter of checkmating British domination of wireless" and had given Bullard "special instructions" with regard to America "keeping control" of the Alexanderson alternator, according to Hooper's 1922 account in *Radio Broadcast.*

According to an interview Young gave to Gleason Archer in 1937, Bullard told Young that Wilson was "deeply impressed" by the fact that Europeans had been able to listen to his "Fourteen Points" speech when it was broadcast, using the Alexanderson alternator from New Brunswick, in January 1918. Admiral Bullard told Young he had been in the Balkans and "found school children learning the Fourteen Points as they would learn their catechism." Bullard said that Wilson had asked him to undertake the job of "mobilizing the resources of the nation in radio" to prevent "our international neighbors" from getting control of it. "The whole picture puzzle had to be put together as a whole in order to get an effective national instrument," he told Young. Young later testified, "On that day, for the first time, radio made an indelible impression on my mind. Here was something so important that the President of the United States, in the midst of the peace negotiations had sent one of his admirals from Paris to New York to talk about it."

On April 8, Bullard and Hooper came to New York City and met with a group of GE executives and board members. The two navy men asked Young not to sell the alternator to either British or American Marconi because it would "fix in British hands a substantial monopoly of world communication," given that the British also controlled cable communications. Their request was not particularly welcome news around the GE executive suite. "There were no other prospective customers for that kind of apparatus," Young told Archer. GE had made a major investment of labor and of capital in developing the alternator, and without customers, the production of the alternator would be closed down, at a great cost to the company and its shareholders. The GE executives argued there were basically no other customers for the alternator and that even if GE refused to sell it to the British, they could get their hands on the Poulsen arc transmitter, a competitive device that could also likely get the job done for Marconi. But Admiral Bullard's lengthy appeal to their sense of patriotism won the day.

GE agreed not to sell to Marconi, and to make up for the lost sales, Bullard and Hooper, the navy men, had a solution for GE: it should form a new American company, under the auspices of GE, to control the various radio assets and patents that had been developed during the past decade or so and to serve as a bulwark against the British. At first, Young's response was to deflect: Didn't they realize that GE was an electrical device manufacturing company, not a wireless communications company, and that it had no interest in becoming one? Bullard responded that Young was being shortsighted, that a huge opportunity was staring him in the face. GE could not only manufacture the communications equipment but could also sell it to a ready-made customer that it controlled. It was a gift, Admiral Bullard insisted. "The argument could not fail to capture the imagination of such a far-seeing man as Owen Young," Archer noted. GE succumbed to the navy's arm-twisting.

But there were a ridiculous number of obstacles to be overcome. American Marconi was the dominant communications company at the time. For GE to start a competitor from scratch would be foolish at best. American Marconi had to be brought under GE's tent for Wilson's plan to work. American Marconi had to be made an offer it couldn't refuse. The postwar power of the U.S. government was brought to bear. It played hardball. Since federal officials had decided that it was in the national interest to make sure that no foreign companies were allowed to dominate the country's burgeoning wireless communications network, American Marconi would have to accede to the wishes of the federal government or try to figure out a way to survive in America without access to either the Alexanderson alternator or the Poulsen arc, both of which were made in America. "Without one or the other of these great machines," Archer wrote, "the Marconi stations would be at a great disadvantage." There was also the problem of who controlled the patents on the various pieces of equipment needed to make a radio transmitter. During the war, for national security purposes, the U.S. government took control of the various patents, which before then had been owned by different companies—GE, AT&T, and Westinghouse, among others—that made the component parts. "The Westinghouse Company, the American Tel. & Tel. Company, the United Fruit Company, and the General Electric Company all had patents but nobody had patents enough to make a system," Young said years later when remembering the postwar landscape. "And so there was a complete stalemate. Radio could not go ahead at all, and the patent mobilization was not for the purpose of creating a monopoly, but it was to relieve the restraints upon the art, so that somebody could create and develop radio."

Out of the stalemate was born, inside GE, the legendary Radio Corporation of America, which would unleash years of American technological ingenuity and the creation of what we now know to be radio and television broadcasting. "It was an amazing thing, which we had not contemplated at all," Young remembered. After his initial hesitation, Young soon saw the opportunity clearly for GE. After working the politicians in Washington—in particular Senator Henry Cabot Lodge, the powerful chairman of the Senate Foreign Relations Committee, who was a fierce opponent of government ownership of corporate assets—Young approached Edward Nally, his counterpart at American Marconi, to see if a deal could be made for GE to buy the assets of the company. In approaching Nally, Young had a few weapons: he knew that the U.S. government wasn't going to let American

Marconi buy the equipment it needed and that the government had cast a wary eye across the Atlantic at British Marconi, which still owned about 25 percent of American Marconi.

Young told Nally if American Marconi was sold to GE, American Marconi's executives would be the ones running the new subsidiary—Radio Corporation of America—because GE had no expertise in the radio business, only in the manufacture of a component part of the business. GE wanted only to be able to keep selling the parts to its new subsidiary. "It was not only good diplomacy but sound business as well to offer to the responsible officials of the American Marconi Company employment of the same nature with the new corporation should the consolidation be effective," Archer observed. Nally, at American Marconi, saw the wisdom in Young's proposal and said he would go along with it, as long as British Marconi agreed to sell its roughly 25 percent stake in American Marconi. Nally realized American Marconi was "doomed" if it tried to go it alone after the war and that GE "offered the only means of salvation."

In June 1919, Young's representative and Nally set sail for London to try to persuade the executives at British Marconi to sell their stake in American Marconi to GE. It was an arduous negotiation. And was all about who would begin to control the post–World War I world: the British or the Americans. The spoils were nothing less than the ability to communicate instantly across borders. Up until the Americans entered World War I, the British controlled international communications through the Marconi companies and they controlled the cables running between the two companies. Engineers at GE had changed that calculus.

The fight was about power and who would own the vital and still emerging radio technology. Since Alexanderson had created the crucial component at GE, it made sense for GE to become the dominant force in creating and running RCA. But that didn't make the negotiations with the British any easier. The Americans were challenging

Britain's long domination of global communications in a particularly aggressive fashion. "All this would require time for reflection and debate," Archer wrote. "Weeks passed in weary negotiations." GE eased the pain for British Marconi by giving it the right to use the Alexanderson alternator for twenty-five years, plus any other relevant new inventions, and vice versa. That was salve that helped heal Britain's wounded pride. On September 4, 1919, *The New York Times* broke the news that GE was negotiating to buy American Marconi, and in particular the 25 percent or so stake in the company owned by British Marconi. Details were sketchy but the paper made clear that the new company would be backed by American capital "so that control will rest absolutely in this country." The next day, British Marconi agreed to sell its 364,826 shares in American Marconi to GE, a purchase of more than $2 million.

On October 17, 1919, RCA was incorporated in Delaware as a wholly owned subsidiary of GE. A month later, on November 20, RCA merged with American Marconi in a deal valued at $9.5 million. "The way was now paved for the carrying into effect of the ambitious plans on which Owen D. Young had been working for months past, by which he sought to consolidate the wireless facilities of the nation—to put together the 'jigsaw puzzle' of conflicting interests in this great field," Archer wrote. Nally became the first president of RCA; Young became chairman of its eight-member board of directors, four of whom were GE executives and four of whom had been on the board of the now-defunct American Marconi. In January 1920, Nally invited Wilson to appoint a navy officer to the RCA board of directors. At Nally's urging, Wilson, who had recently suffered an incapacitating cerebral hemorrhage, chose Admiral Bullard, who served on the board for the next eleven years. Nally wrote to Wilson that Bullard would have the "right of discussion and presentation of the Government's views and interest concerning matters coming before the directors and stockholders."

A month later, the federal government returned radio stations to their owners. On March 1, 1920, RCA transmitted its first commercial messages between New Brunswick, New Jersey, and England, at a fixed cost of 17 cents per word, undercutting by 32 percent the cable rate of 25 cents per word. In 1922, GE went on the air with its first radio station, in Schenectady, and the words "This is station WGY"— the call letters standing for wireless, General Electric and Schenectady. At its start, RCA was the largest radio communications company in America. It not only manufactured radio equipment but also controlled the nation's first network of radio broadcasters, known as the National Broadcasting Corporation, or NBC.

At the time of the creation of RCA, David Sarnoff was American Marconi's "commercial manager." He would come to have the same job at RCA and play a pivotal role in its future, eventually rising to become the president of RCA. He was also the founder of NBC. Sarnoff, the Jewish Orthodox son of Russian immigrants, had joined American Marconi on William Street, in Lower Manhattan, as a messenger and office boy in 1906. He was fifteen years old and had previously been a messenger at *The New York Herald*, making $5 a week plus what he made delivering newspapers. He eventually saved up enough money to buy a telegraph machine, learned Morse code, and exchanged messages with a colleague at the newspaper. When Sarnoff confided he might like to be in the radio business, the man suggested he try to get a job at American Marconi. "The idea of flinging messages in a code through space appealed to Sarnoff," the *Times* later reported, "and he applied for a job as an operator." Instead, he was offered a job as an office boy for $5.50 per week. In Siasconset, Massachusetts—the same village on the eastern edge of Nantucket near where Jack Welch would later own a house—there was one "lonely" Marconi station, the *Times* reported, and at age seventeen, Sarnoff moved there in 1908 as an operator for $60 per month. The station opened in 1901, the first of its kind in the United States to transmit and to receive messages to and from marine vessels; it was also used to broadcast the results of New

York baseball games to the island's summer residents, many of whom, like Sarnoff, hailed from New York City. At the station, there was a library stocked with technical books about radio. "There was little to do outside working hours but read the books," the *Times* reported. "But that was not sufficient inducement to make it easy to keep men contented there." Sarnoff read the books, became a full-fledged operator, and got a raise of $10 a month. But, sufficiently bored, he took a pay cut, back again to $60 per month, to move back to New York City.

At RCA, Sarnoff's career grew in tandem with radio's utility. Over the next decade it would evolve from a way of communicating wirelessly with ships off the coast to communicating across oceans to finding its way into American homes, movie studios, movie theaters, and the phonograph industry. None of this would have been possible, Sarnoff believed, without Owen Young liberating the nascent technology from patent restraints and organizing it into a business. As it did with electricity, GE would need to create a new industry, a business for the new technology of radio, creating programming and building radios for home use. This was no easy feat. "The development of the radio tube, for example, the very heart of the modern radio set required over thirty separate and distinct patents," Sarnoff explained in testimony before the U.S. Senate. By the end of 1920, he continued, RCA had "demonstrated that a program of entertainment, education and information could be regularly broadcast to the home by radio." He said RCA, under GE's ownership, helped to develop new industries in radio broadcasting and television broadcasting—NBC was started by RCA, and at the time of its launch, RCA owned 50 percent of NBC, GE owned 30 percent of NBC, and Westinghouse owned the remaining 20 percent—but no company could have accomplished this alone, not even RCA or GE. In 1921, the retail value of radios sold in the United States was $5 million; seven years later it was $650 million. In 1921, the radio audience in the U.S. was 75,000 people; at the beginning of 1929, it was 40 million.

Three years later, in 1932, the RCA bonanza was over, at least for GE. The troubles started in May 1930. B. J. Grigsby, the chairman of the Grigsby-Grunow Company, a Chicago manufacturer of radio tubes and radio sets, had since 1928 believed that GE's control of RCA had put his company at a competitive disadvantage because it could not get access, at any price, to RCA's patents. He brought his concerns to Congress, which in turn encouraged the Justice Department to bring an antitrust suit against GE, and others, in order to dissolve the ownership structure of RCA. By then, RCA had a 20 percent market share in radio receiving sets and a 40 percent market share in radio tubes.

For his part, Young had always supported some form of regulation for the radio industry. "I have no doubt that effective regulation can be established," he testified before the U.S. Senate in 1932, "fair alike to the people rendering the services and to the people served." In the end, Young chose compromise rather than continued litigation. He agreed to a consent decree that required GE (and Westinghouse) to spin off half of its ownership stake in RCA to its shareholders within three months (with the balance being spun off within three years), to make management changes atop the company, and to "alter" its relationship to the four thousand or so patents that RCA controlled. Young could remain RCA's board chairman for another five months. David Sarnoff, by then RCA's president, informed RCA's stockholders that RCA could—and would—operate as an independent company. To pay off the balance of the $18 million of debt that RCA owed both GE and Westinghouse, RCA agreed to transfer to GE the RCA building, a stunning art deco spire at Fifty-First Street and Lexington Avenue in Manhattan, and to issue new debt in the amount of $4.25 million to GE and Westinghouse. The balance of $8 million or so in debt would be canceled as part of the settlement agreement. Also as part of the settlement, GE and Westinghouse agreed not to manufacture "radio devices" for the next thirty months, the time RCA needed to get geared up to be an independent company.

In his congressional testimony about the end of GE's ownership of RCA, Owen Young was particularly unsentimental. "In 1919," he said, "after the close of the great war, the General Electric Company, at the suggestion of the President of the United States, undertook to secure the cooperation of all American concerns interested in wireless in the creation of a unified organization to develop and protect the communication interest of the United States throughout the world." He then explained that fourteen years later, per the consent decree, GE's ownership stake in RCA, 5.2 million shares, was spun off to GE's shareholders in the form of a stock dividend with a value of $26.5 million. He was unemotional about the outcome.

He told the senators about what had made GE successful during its first forty-seven years. The key, he noted, was the particularly American ideal of the power and importance of "individual leadership" in developing new industries and new companies within the new industries. Although the reference was likely lost on the senators, Young then invoked, with some hyperbole, Ralph Waldo Emerson in *Self-Reliance*. "Truly it may be said that the General Electric Company was and is only the lengthened shadow of Edison and Thomson in the field of science and of Charles A. Coffin in the field of business," he said. "It was their energy and capacity, functioning largely without governmental restraint, that enabled them to build in half a century one of the most spectacular industries in the world."

*Chapter Four*

# THE INCREDIBLE
# ELECTRICAL CONSPIRACY

The respect for GE's technology and management prowess was such that, in September 1942, President Roosevelt seconded GE's president, Charles Edward Wilson—known as "Electric" Charlie Wilson so as not to be confused with Charles Erwin Wilson, the CEO of General Motors, who was known as "Engine" Charlie Wilson—to Washington to serve as the executive vice-chairman of the War Production Board, which commandeered the nation's manufacturing resources in the war effort. The Allied victory in World War II redounded to GE's benefit. At its peak, in fact, GE not only was considered a leader in scientific and financial innovation but also was revered as a crucible of corporate leadership. In the 1950s, as management became an academic discipline taught at business schools, Ralph Jarron Cordiner found himself GE's newly minted CEO and a leader who was keen to test out theories about how to turn managers into entrepreneurs—especially in a company that was growing into a far-flung conglomerate.

By the time Cordiner was closing in on the president's job, succeeding Wilson, GE had become one of the largest and most respected corporations in the country. It manufactured roughly 200,000 different products in more than 170 plants in thirty-one states and also in several foreign countries. It had laboratories where more than twenty thousand scientists and engineers worked on cutting-edge technologies. It employed some 250,000 people, more than the population of all but forty cities in the United States, and, according to *Fortune*, was "often the social center" of dozens of "company towns" where "the G.E." is "more important than city hall" and where company schools had more students enrolled—32,000—than most U.S. universities. It had net income of more than $150 million on revenue of more than $3 billion.

Ever since he had joined GE in 1928, Cordiner had bristled at its bureaucracy and vowed that if he ever got a chance to run it, he would create a system that rewarded ambitious men like him and remove the layers of fat in the company that he found sclerotic. Born in 1900—eight years after Charles Coffin created General Electric—on his parents' 1,280-acre wheat farm near Walla Walla, Washington, Cordiner worked his way through Whitman College, a small liberal arts school in his hometown, by doing odd jobs and by selling wooden-paddle washing machines on behalf of the Pacific Power & Light Company. When he graduated from Whitman, he went to work for Pacific Power. He was a star salesman and, as a result, came to the attention of a GE subsidiary, the Edison General Electric Appliance Company, which lured him away. He was a man in a hurry. At twenty-eight, he was named head of the northeast office of the GE subsidiary and then, two years later, vaulted over sixteen others to become head of the Pacific Coast office. Then he moved back to GE's new merchandising department in Bridgeport, Connecticut, to "get closer to the hub of the corporate wheel." By the time he was thirty-eight, he was running GE's Bridgeport office. On his watch, sales in the electric heating division of GE increased 60 percent. His boss during those years was Wilson, who had left school at age twelve to work as a stock boy at the Sprague Electrical Works, which GE had acquired.

While Wilson was serving his country, Cordiner left GE to become the president of Schick, Inc., a struggling manufacturer of razors and blades that he helped resuscitate. GE's bureaucracy frustrated him. "He was outspokenly sour on G.E.," T. K. Quinn, a former GE vice president and the author of *I Quit Monster Business*, said about Cordiner. "He referred to the layers of fat in the company that smothered talent like his. The fat was anybody above him." In 1943, Cordiner returned to GE as the personal assistant to Gerard Swope, who had come out of retirement to lead GE in Wilson's absence. When Wilson returned to GE after the war and resumed his position at the top of the company, he enlisted Cordiner in what turned out to be a five-year project to reimagine GE for what would surely be a postwar era of nearly unfettered growth. In December 1950, Wilson returned to Washington to help Harry Truman manage the Korean War, just as he helped Roosevelt, his predecessor, manage World War II.

Wilson presented only one candidate to the GE board to succeed him: Cordiner. "In this big company, don't we have another man?" asked GE director Sidney Weinberg, the senior partner at Goldman Sachs, who had worked with Wilson at the War Production Board.

"Why do you need more than one?" Wilson replied.

Cordiner wasted little time in laying out his plan for massive decentralization as a way to empower a greater number of GE executives to run their own businesses. He split the behemoth that GE had become into twenty-seven independent divisions comprising 110 small companies, each the perfect size "for one man to get his arms around." Each leader would run his own show, would make day-to-day decisions, would propose his (rarely her) own budget and marketing strategies, would be able to approve capital expenditures up to $200,000, and would basically be "freed up," as Cordiner liked to say, to make long-term strategic decisions about the direction of the business. But with authority and autonomy, Cordiner believed, came accountability and responsibility for higher performance.

Cordiner didn't mince words in explaining to *Time* magazine why he restructured GE: "When I took over in 1951, I told lots of people immediately that this company was not going to be a sinecure for mediocrity. The old G.E. had a reputation as a good and complacent place to work if you kept your nose clean. I wanted to get rid of that idea and create more risk and opportunity."

It was one of largest and most comprehensive reorganizations in corporate history. "American business spends too much time on thinking about this month, this year," he continued. "It ought to spend more time preparing for 15 to 20 years from now—the next business generation." Gone were "assistants, coordinators and committees," *Time* reported. And the reason? "G.E. has no place for committees as decision-making bodies," Cordiner said. "A committee moves at the speed of its least informed member and too often is used as a way of sharing irresponsibility."

Soon after taking over, Cordiner implemented his plan. He called together the top fifty GE executives and laid down the law. He had long made clear that he did not intend to run GE the way his predecessors had run it, in a centralized, autocratic fashion. (Swope used to give people four minutes to make their case.) "So shocked were G.E.'s executives by the swiftness of his move," *Time* reported, "that G.E. Board Chairman Philip A. Reed got to his feet to relieve the tension. He told the story appropriate to the moment, of the little girl who began making a picture of God. Said her mother: 'Honey, God is a spirit and no one knows what he looks like.' Replied the little girl: 'They will when I finish.'" Cordiner could be brutal with those executives who misbehaved. "If you did something wrong," Sidney Weinberg said, "Cordiner would send for you and tell you, you were through. That's all there would be to it."

In December 1955, *Fortune* magazine tried to assess how Cordiner's "overhaul" of GE was going. The article focused on Cordiner's struggles getting the GE rank and file to turn around the battleship and to accept his new strategic plan, while also promoting his lofty goals for the future of the company. "Six billion dollars by 1963" was Cordiner's mantra, and one that he did not achieve. In 1955, GE's revenue was barely half that.

Cordiner's plan was so radical—and so upsetting to the GE rank and file—that it was nearly stillborn. Until then, GE had been managed, according to *Fortune*, by "a central cluster of autocrats, with one superb autocrat placed grandly at the top." It was Gerard Swope, who ran GE after Owen D. Young and before Charles Wilson, who first realized something had to change. Swope had regretfully observed that many of his "big and fundamentally sound ideas often failed to pan out in practice." For instance, GE's appliance division—the makers of refrigerators and ranges, among other things—couldn't seem to become a market leader or to be particularly profitable. The problem, or so it seemed at the time, was that engineers were designing these products, not marketers or style leaders. The people making the appliances were too far removed from the people selling the appliances, let alone the ones buying the appliances. Something had to change. That change began to happen when Cordiner returned to GE from Schick as Swope's assistant while Swope was filling in for Wilson. "Wilson and Swope," Cordiner explained, "freed me up to think about the future."

He landed on the plan of pushing as much authority as possible down to the operating levels, empowering division chiefs and the general managers of business units to make the important decisions, with directional oversight from the corporate headquarters on Lexington Avenue. Cordiner field-tested his ideas in the late 1940s at several of GE's affiliated companies—at Hotpoint (an appliance manufacturer), at Telechron (a clock manufacturer), and at Trumbull (a manufacturer of electric parts)—with the intention of rolling them out company-wide in 1951 before Wilson's scheduled retirement. Wilson, the avuncular and familiar old hand, could introduce the changes more easily than could a new leader. But fate intervened in December 1950 when President

Truman again summoned Wilson to Washington to lead the effort to produce goods for the Korean War.

As GE's new leader, Cordiner faced an immediate dilemma: whether or not to roll out his decentralization plan, and when. He knew it would be controversial and unpleasant and disruptive. But he decided not to wait. He got the board's approval during his first board meeting as CEO in January 1951. The next day, he met with twenty-two direct reports at the Waldorf Astoria and then took to the road to announce the "momentous news" to six thousand GE executives at six regional meetings.

It's not hard to see why the company nearly rejected Cordiner and his disruptive plan. For the first time in GE's history, individual business managers were given the responsibility to make their own decisions, empowering and frightening them at the same time. There was pushback. Cordiner created twenty decentralized divisions and shifted around some two thousand men into new management assignments around the country, causing many of them to have to uproot their families and lives for some new management theory. People were moved from Bridgeport to Louisville or from Schenectady to New York City or from Pittsfield, Massachusetts, to Rome, Georgia. "Many people not transferred thought they may be at any minute," *Fortune* reported. Men in engineering, manufacturing, or purchasing who were used to giving orders to operating executives were told to stop. No one could tell the operating executives what to do or how to do it anymore. They could consult but not make final decisions. "Many of these men felt, understandably, that their personal careers in General Electric had been swept away," *Fortune* reporter William B. Harris observed. Large staffs and assistants were eliminated. "An assistant," Cordiner said, "is really a prop for a manager who can't get his own work done. A coordinator is an official with a vested interest in keeping two or more people apart."

In the new world of GE, there were now one hundred independent operating units, combined into twenty-two divisions that were further consolidated into four operating groups and one distribution group. The four operating groups were comprised of large electrical components, industrial products and lamps, appliances and electronics, and atomic energy and defense products. The latter group, which comprised about 15 percent of GE revenues, included Hanford, the large Atomic Energy Commission project in Washington State that employed nine thousand people (and would soon enough cause GE another round of headaches and scandal). There was also a so-called Atomic Power Equipment Department, a manufacturer of nuclear power plants that had already sold one, for $45 million, to Commonwealth Edison in Chicago. "A share in General Electric might be called a share in an investment trust engaged in manufacturing," *Fortune* wrote.

To oversee the sprawling operation, Cordiner created the Office of the President, a fancy name for a group of top GE executives who continued to pull the strings from high above Lexington Avenue. He also created what he called the Services Division—researchers, accountants, controllers, treasurers, community relations, PR—that served the operating divisions. The members of the Office of the President had only secretaries—no staff—and spent much of their time on the road, thinking about what the future General Electric would look like, much as Wilson and Swope had allowed Cordiner to do a decade earlier.

One practical example of how the Office of the President worked was what became known as Appliance Park, in Louisville. It was the largest major-appliance plant in the world when it opened in 1952, at a cost of about $200 million. Cordiner and his lieutenants decided in the early 1950s that GE either had to get out of the business of manufacturing major appliances or get serious about doing it better. GE got serious and consolidated seven manufacturing facilities, spread out across the country, into the one massive Louisville facility. Cordiner's critics soon dubbed Appliance Park "Cordiner's Folly" because they figured it would sit half empty for years.

Instead, it filled right up, driven by demand for GE's better-designed and better-functioning appliances. GE's refrigerators had top market share or were second to GM's Frigidaire division. GE was number one in ranges. GE's washing machines were second only to those made by Sears. GE dishwashers and clothes dryers were also top industry performers. By the end of 1956, GE's appliance division was for the first time generating respectable profits. A similar success occurred in GE's Weathertron division, which made heat pumps, described as a breakthrough "gadget" designed to help keep homes at set temperatures despite the outside weather. Other innovations were in the works, too, including an "electric oven," an early precursor to the microwave, and manmade "synthetic diamonds" that could cut extremely hard metals with the precision and ease of real diamonds but at a fraction of the cost.

GE was also working hard to develop better relations with its labor force, and in particular its unions, and was busy studying what characteristics made for a successful and admirable manager. "He must be a teacher of men," Cordiner explained. "If he can teach, he can lead and select men. A manager must also be a crisp boss, willing to tell his men the bad as well as the good, yet still be so compatible men will want to work with his team." It was during Cordiner's reign that GE created its world-famous research and training facility north of New York City, along the Hudson River in Crotonville, New York. "Just think," Cordiner said of Crotonville, "how wonderfully mobile our management will be when this thing works out." All of these wonderful innovations and strategies added up to a vision that Cordiner hoped would lead to $6 billion in revenue by 1963.

It all might have worked out that way for Cordiner and for GE, but for the pressure that his new organization placed on executives to hit quarterly and annual numbers and the illegal steps that some of them took to try to impress the boss. What Cordiner had failed to account for in his management theories designed to unleash an entrepreneurial spirit across GE was the temptation that some of his key executives might feel to collude with their counterparts at other power companies, and break the law in the process. Aware of the clubby and collegial nature of the power business in the 1950s, Cordiner would reaffirm a central tenet of Wilson's leadership as he made the rounds selling his vision for a decentralized GE. He told any executives who would listen—frankly it's remarkable that he even felt the need to do it—that GE employees should not collude with businessmen from other companies to set prices on similar products. Cordiner signed on to a Wilson manifesto, known as "instruction 2.35," on compliance by GE employees with antitrust laws, first issued by Wilson in 1946 and then again in 1948 and 1950: "It has been and is the policy of this company to conform strictly to the antitrust laws. . . . Special care should be taken that any proposed action is in conformity with the law as presently interpreted." Complying with this simple directive had, apparently, been a problem in the previous decade around GE, with some thirteen antitrust lawsuits filed against the company, including for price fixing and "patent pooling," when two or more parties get together and decide who can license a patent.

GE, being one of the nation's most successful corporations, had also become a bit of a battleground for two divergent and competing strains of economic thought. One, endorsed by Cordiner, was more of the Adam Smith variety: Why descend to collusion when GE had superior people and superior technology? Why not go for the competitive jugular and push for every advantage? That's why Cordiner had fired three GE employees, in 1949, for engaging in price fixing—*pour enourager les autres*, as Voltaire wrote in his famous novel, *Candide*. "Every company and every industry—yes, and every country—that is operated on the basis of cartel systems is liquidating its present strength and future opportunities," Cordiner said. The opposing argument swirling around GE in those days was that surely upstanding gentlemen could figure out a way to agree on prices that wouldn't hurt customers too much and would make life much easier for the manufacturers of complex products. Obviously, these kinds of agreements violated antitrust laws. "Sure, collusion was illegal," one GE veteran told *Fortune*, "but it wasn't unethical."

Clarence Burke joined GE's heavy-equipment business in 1926, from the Georgia Institute of Technology. Burke's group contributed about 25 percent of GE's revenue by then and was considered the backbone of the company. It was also a hotbed of GE executives who believed in collusion, despite the repeated admonitions against such behavior from Lexington Avenue.

In June 1945, Burke had moved to Pittsfield, Massachusetts, as a sales manager for distributional transformers, a component used by electrical utilities. Within a month, H. L. "Buster" Brown, the division head, told Burke he needed to attend a convention of the National Electrical Manufacturers Association, or NEMA, in Pittsburgh. There were about forty men from the industry there, as well as representatives of NEMA and the Office of Price Administration, a federal price-control agency established after the U.S. entered World War II. Once the OPA and NEMA representatives left the conference, the businessmen got down to the real business, that of colluding. Burke became a regular at these monthly meetings. Soon enough, instead of discussing how best to lobby the government about OPA, the men started discussing setting prices among themselves. They agreed upon prices "to the penny," Burke later revealed. Brown, Burke's boss, assured him that these agreements did not violate GE's antitrust directive.

In 1946, the Justice Department started an investigation into price fixing—resulting at GE in three firings by Cordiner, then GE's president, in 1949—and the word went out to the GE rank and file, from the GE lawyers, to cool the price talk. Sales managers, such as Burke and Brown, were barred from attending NEMA conferences, and they were replaced by GE engineers at these events. The GE salesmen referred to this development as being placed "behind the iron curtain." The new policy lasted nine months. During that time, Wilson issued one of his periodic antitrust directives. And then, suddenly, the iron curtain was lifted. "Word came down to start contacting competitors again," Burke recalled. He remembered that it officially came from Brown, but he suspected Brown was being ordered around by his bosses. "I think the competitive situation was forcing them to do something," he continued, "and there were a lot of old-timers who thought collusion was the best way to solve the problems." The men started meeting together in hotel rooms and were careful not to reveal in their expense reports what they were doing. Instead of showing on their expense reports where they actually went to these meetings, they would write the names of other cities equidistant from Pittsfield. The conspiracy continued in this fashion for the next few years, Burke remembered. What about the antitrust directive they had initialed, indicating they understood not to collude? "When anybody raised a question about that," he said, "they would be told it doesn't apply now."

In September 1951, Clarence Burke got a new job at GE, selling industrial circuit breakers. The job was open for Burke to fill because the previous executive had resigned from GE on principle, given that he had signed Wilson's antitrust directive and felt guilty about violating it. Burke got the new job in Philadelphia because he was considered an important part of the "conspirational wheel." "They knew I was adept at this sort of thing," he said. "I was glad to get the promotion. I had no objections." His boss in circuit breakers was Henry Erben, a very senior GE executive who, Burke recalled, made it known he had cleared with Cordiner the conversations with competitors.

Burke's first assignment was to get to know executives at Westinghouse, Allis-Chalmers (again back as an independent company, following GE's ownership of it), and Federal Pacific and to introduce them to his team at GE. Part of his assignment was perpetuating the conspiracy, all with Erben's approval. "Erben's theory had been live and let live," Burke said. "Contact the competitors." That "theory" trickled down throughout the division. The men who felt uncomfortable participating in the conspiracy would be replaced. "Not replace them for that reason of course," Burke continued. "We would have said the man isn't broad enough for this job, he hasn't grown into it yet."

At GE, the price-fixing conspiracy in the circuit-breaker division lasted from 1951 to 1958 and, according to the Justice Department, involved some $650 million of revenue, or around $75 million per year, most of which related to sales of electrical equipment to private utilities. The idea was to rotate the business among the four manufacturers of circuit breakers on a fixed percentage basis: GE would get 45 percent, Westinghouse 35 percent, Allis-Chalmers 10 percent, and Federal Pacific 10 percent. Every two weeks or so, the executives would meet to decide whose turn it was next, based on how much business each company had been doing lately. Also to be determined was what price would be the lowest bid for the job.

There was a conspiracy argot: The attendance list was known as the "Christmas-card list" and meetings were known as "choir practice." Each company had a number, to be used instead of its name. (GE was 1; Westinghouse 2; Allis-Chalmers 3; and Federal Pacific 7.) Only first names were used. A telephone call to a conspirator at home would sound something like "Bob, what is 7's bid?" At the hotel meetings, it was essential not to register with the name of one's employer or to be seen having breakfast with a competitor. "The GE men observed two additional precautions," *Fortune* wrote. "Never to be the ones who kept the records and never to tell GE lawyers anything."

What kept the conspiracy humming was a combination of the vicissitudes of the utility business and, at least at GE, the constant pressure from Cordiner that division executives make their numbers. The electric-power business was a "feast or famine business," Burke explained. "At one time everybody was loaded with orders, and ever since they wanted to stay that way. When utilities decide they need more generating capacity, they start buying and we have three years of good business—and then three years of bad." (The boom-and-bust cycle in GE's power generation business would plague the company for years.) Then there was the pressure from the corner office. "All we got from Lexington Avenue," Burke said, "was 'get your percentage of available business up, the General Electric Co. is slipping.'" Cordiner's new organizational structure had created a pressure-cooker environment. "Cordiner's asset is stretching men," said Harold Smiddy, a GE vice president. "He can push them and he did."

Burke, for one, felt the pressure. "We did feel that this was the only way to reach part of our goals as managers," he said. "Each year we had to budget for more profit as a percent of net sales, as well as for a larger percentage of available business." His boss would not approve a budget unless it was a "reach budget," he said. And, at least according to Burke, that meant "we couldn't accomplish a greater percent of net profit without getting together with competitors. Part of the pressure was the will to get ahead and the desire to have the good will of the man above you. He had only to get the approval of the man above him to replace you, and if you wouldn't cooperate he could find lots of other faults to get you out."

But the conspiracy grew complicated. The instinct to compete is a powerful one, and once-collegial coconspirators began to wonder what would happen if, say, one company were to lower its prices a bit and win more business away from its competitors. That's what Burke experienced in 1953 and 1954, after he had again been promoted to be the general manager of GE's high-voltage switchgear business—an annual $25 million revenue division. In Cordiner's decentralized GE, Burke was solely responsible for hitting that $25 million figure. Worse was that his corporate collaborators weren't adhering to the rules of the game. "We at GE were being made suckers," he said. "On every job someone would cut our throat; we lost confidence in the group." GE got out. The company boycotted the cartel through most of 1954, the same year that Cordiner distributed his directive about not flouting the antitrust laws.

But 1954 was a bad year at GE. Company-wide revenue fell around $175 million from the year before—the first sales slump in Cordiner's tenure. Profit margins had also contracted. Cordiner was not pleased. General managers were told that their prices were not competitive. A last straw of sorts seemed to come when, toward the end of 1954, GE lost a major order for turbines to Westinghouse, at a considerable discount. Top GE executives decided that could not stand. The next contract GE must win.

It turned out to be a $5 million opportunity for transformers and switchgear with Ebasco, a onetime GE subsidiary. The bidding was fierce, as were the discounts offered. Ultimately Ebasco got a deal at 40 percent below the list price. Soon enough, the industry suffered through the so-called white sale of 1954 and 1955, a period of rampant discounting by retailers on appliances as large as 45 percent off book price. Since the products being delivered took a long time to make, the discounts flowed through the industry income statements for years.

With the "white sale" in full bloom, Burke's boss, George Burens, a steadfast defender of price competition, succumbed to the pressure, too. He called Burke into his office one day and told him the "old cartel was going to be cranked up again" and that, in fact, Burens was going to make the calls himself. Soon enough, Burke and Burens "trotted off to mix in a little conspiracy with a little golf" in Bedford Springs, Pennsylvania, with some top salesmen at Westinghouse. They agreed GE and Westinghouse alone could not set prices. The Westinghouse folks agreed to contact Allis-Chalmers, as well as another smaller company, ITE Circuit Breaker. There was a follow-up meeting in January 1956 and then another golf outing, with plenty of alcohol, all followed by monthly memoranda listing pending jobs, whether sealed or open bids, and what the listed price of the equipment needed to be.

The group would then meet some more to agree on the price and "to forestall any chiseling" from the price they had just agreed. In various hotel suites across the country, there were nine of these meetings in 1956. Everything was going quite well on the collusion front until the first months of 1957.

That was when McGregor Smith, the chairman of Florida Power & Light, "lit the fuse that blew them up," *Fortune* reported. Even though it was a bit down in the weeds, McGregor Smith liked to negotiate equipment contracts for his company. But in 1957 the Florida Power executive received few, if any, discounts from the manufacturers, despite his efforts. On one Monday, Smith agreed to transformer purchases from a consortium of manufacturers, including GE and Westinghouse. The next day, Clarence Burke, the GE executive, got worried reports that Westinghouse had approached Smith and offered to sell him $1 million worth of circuit breakers with a 4 percent discount that it proposed to bury in its transformer order.

The cartel was cracking. GE verified the discount and quickly matched Westinghouse's offer. Smith then agreed to split the circuit-breaker order between GE and Westinghouse. But the Westinghouse salesman was so peeved by Smith's decision, and told him to his face, that Smith cut a new deal for the circuit breakers to buy them all from GE. A new price war was on. Westinghouse offered Baltimore Gas & Electric a 5 percent discount on switchgear and circuit breakers; Allis-Chalmers gave Potomac Electric 12 percent off. A week later, Westinghouse gave Atlantic City Electric a 20 percent discount. And it got worse. That winter, during the new "white sale," prices fell dramatically, to as much as 60 percent off list price. That seemed like the end of the cartel—in circuit breakers.

But it was still thriving in other parts of GE's decentralized corporate sprawl since, according to *Fortune*, "each general manager of a division or department took a strictly personal view of his participation in any cartel." At least seven other GE divisions were involved in the price-fixing conspiracy: among them industrial controls; power transformers; and turbine generators. In the fall of 1957, at the Barclay Hotel, near GE's headquarters in Manhattan, the conspirators agreed to give GE "position" in the sale of a 500,000-kilowatt turbine for the Tennessee Valley Authority. (With some humor, apparently, the hotel placed small ads in *The New York Times* boasting that "antitrust-corporation secrets are best discussed in the privacy of an executive suite at the Barclays. It's convenient, attractive and financially practical.")

In 1958, against long odds, the price-fixing conspiracy resumed once again in GE's circuit-breaker and switchgear division, then still run by George Burens, with Burke's help. Supposedly, the two men were dead set against rejoining the conspiracy. The pressure was taking a toll. Burke's colleagues noticed his physical deterioration. He went from being a "hail fellow well met" kind of guy before the pressure placed on him to deliver short-term profits, *Fortune* reported, to a guy who "simply shrank into himself," where "everything got to be cold turkey with him, no warmth at all." At an electric utility conference that fall, held in California once every four years, the beleaguered Burens ran into Fischer Black, the amiable editor of *Electrical World*. (It was Black's son, also named Fischer Black, would go on to become famous as one of three men who created the mathematical model known as Black-Scholes—still used to this day to price options and other financial derivatives—and as a partner at Goldman Sachs.) Black told Burens that others in the industry were angry at GE generally and Burens in particular because he was still refusing to rejoin the conspiracy. Black offered to set up a meeting with other industry executives, if Burens would attend. He told Black he would. The meeting took place on October 8, 1958, at the Astor Hotel. Like a scene out of *The Godfather*, executives from GE, Westinghouse, ITE Circuit Breaker, and Federal Pacific met and spoke generally not only about past recriminations but also about possibilities for the future. Black, who paid for the suite, encouraged everyone to order lunch. They agreed to meet again, on November 9, at the Traymore Hotel in Atlantic City. The circuit-breaker executives quickly reached agreement that they would stick to list prices and end the excessive discounting, which had lowered prices as much as 55 percent in recent months. (They then bickered for some ten hours about how to carve up the percentages among them for the sealed-bid business, in order to accommodate several new entrants into the market.) Burens ran the conspiracy for the next three months during a series of meetings in hotels around New York. Prices increased. And in January 1959, much to his excitement, Burens was promoted to run GE's lamp division. It didn't take long for Lewis Burger, Burens's replacement, to join the conspiracy that Burens had left him.

In May 1959, Cordiner testified before a Senate subcommittee after the Tennessee Valley Authority complained of identical secret prices it received for electrical equipment on three bid contracts. He defended GE against the implication that somehow the bidding had been rigged. He noted that in 1958, the assistant attorney general in charge of the Antitrust Division "cited the General Electric Company as the 'number one example' of companies which have made earnest efforts to live up to the antitrust laws."

Human nature being what it is and all, Cordiner probably felt confident in his theory of the case. After all, Gerhard Gesell, GE's outside counsel at Covington & Burling (and later a distinguished federal judge), had investigated the allegations of price fixing inside GE by combing through reams of pricing data and determined that nothing was amiss, and that in fact GE was just the target of yet another government witch hunt. Cordiner also got reassurance from Gerard Swope Jr., the son of the former GE CEO and a GE internal attorney, who went to Pittsfield, Massachusetts, to talk to executives and asked them point-blank if they had been fixing prices with competitors. When no one spoke up, Swope Jr. concluded that nothing was amiss.

But the GE executives were terribly mistaken, perhaps blinded by a belief that Cordiner's decentralized management approach would defy human nature. In June 1959, a grand jury had been convened in Philadelphia to investigate "pricing policies" among the manufacturers of electrical equipment. Then, in September and October, word came to the top GE executives that the names of GE general managers had been shared with the grand jury because they had participated in pricing discussions with competitors. When confronted by Cordiner's internal investigators, some of the named GE managers finally fessed up to what they had been doing. "Most of the guilty lied," *Fortune* reported, "gambling that the exposures would not go any further than they had." Still, Cordiner "felt obligated" to share what he learned had been going on with the GE board of directors. The company had been implicated in at least four cartels "and an immense number of questions had to be answered," *Fortune* reported.

But in fact the conspiracy extended beyond four cartels and was widespread throughout the company. In November 1959, one general manager told GE's general counsel to find a way to cut a deal with the government because "this thing would go right across the board." But the attorney "just laughed at me," he said, and then told him, "You're an isolated case—only you fellows would be stupid enough to do it." When it was discovered that there were actually nineteen cartels, spread across the company, accounting for more than 10 percent of GE's revenue, "the company found itself in the ludicrous position of continuing to proclaim its innocence while its executives were being implicated by platoons," *Fortune* reported. The first federal grand jury was examining conspiracies in insulators, switchgear, circuit breakers, and other products. A second federal grand jury was looking into transformers and industrial controls. The dam was breaking. Defense attorneys were seeking immunity deals with the Justice Department in exchange for testimony.

On January 5, 1960, after the Justice Department's subpoenas started flying but before anyone had been indicted, Cordiner gathered the company's leaders together at the Homestead Resort in Hot Springs, Virginia, where he once again tried to lay down the law on collusion. (The resort had, ironically, also been the site of some of the samizdat meetings to fix prices.) Cordiner was trying to impress upon his top managers the severity of the situation. "The first item on the agenda is one to which I attach the greatest importance," he said. "The subject of business ethics in a competitive enterprise system. We in General Electric believe in and benefit from vigorous competition." But, he said, the system would remain "free and competitive" only if businessmen and women were "capable of the self-discipline required" to keep from cheating. He warned that without greater self-discipline, government would regulate business more assiduously, effectively killing the golden goose.

He said those who conspired to fix prices would be disciplined through demotion, pay cuts, or firings. Beyond that, he insisted that such acts were simply not good business, given GE's competitive advantages. "If you considered compliance with the antitrust laws on only the basis of business opportunities there is nothing, in my opinion, that is less intelligent than an attempt to have price restrictions with your competitors who do not have [our] modern facilities, or have this company's organization or public acceptance on which to sell company values to customers," he continued. "Every company and every industry—yes, and every country—that is operating on a basis of cartel systems is liquidating its present strength and future opportunities."

He reminded the managers that GE had 275,000 employees, some 50,000 vendors and suppliers, and more than 400,000 shareholders. "All of us must conduct ourselves in the interest of these larger groups and the customers and the public," he continued. "What has occurred is to be deeply regretted, and I fear its blot upon the company's good name will live with us for many years because our detractors will publicize it, repeat it, and attempt to have the public interpret it as typical of the acts of all General Electric people." He exhorted the troops not to lose faith because of the "inexcusable acts of a few of our associates."

The indictments started being handed up in February 1960. On February 16 and 17, the federal grand jury in Philadelphia charged forty companies and eighteen individuals with fixing prices in the market for electrical products. Then, on June 22, 1960, a federal grand jury in the Eastern District of Pennsylvania handed up another round of indictments. Half of the twelve men indicted were GE employees, including Lewis Burger; his boss, George Burens; and Burens's boss, Arthur Vinson, a GE vice president and a very senior executive of the company. For at least the years 1958 and 1959, the federal government charged that these men, and others, had "engaged in a combination and conspiracy in unreasonable restraint" of interstate trade, in violation of the 1890 Sherman Act. GE pleaded not guilty.

After trying for months to make the case against Vinson, in early December 1960, the feds dropped the case against him. The federal prosecutors also dropped the suggestion that any of Cordiner; Robert Paxton, the new GE president; or the GE board of directors "had knowledge of the conspiracies . . . nor that any of these men personally authorized or ordered any commission of any of the acts charged." With Vinson suddenly in the clear, GE quickly pleaded guilty to the "major indictments" against it and pleaded nolo contendere, with the government's consent, to the minor charges against it. That paved the way for the GE executives, among the others, to be in the courtroom on February 6, 1961, to hear Judge J. Cullen Ganey tell them, "What is really at stake here is the survival of the kind of economy under which this country has grown great, the free-enterprise system."

Unlike many of the other company conspirators, and to its considerable credit, GE fired the fourteen employees who pleaded guilty before Judge Ganey. *Fortune*'s Richard Austin Smith spent some time interviewing the various executives involved in the conspiracy following their guilty pleas, only to discover that many of them believed they had become the "fall guys" for American business. "They protested that they should no more be held up to blame than many other American businessman, for conspiracy is just as much 'a way of life' in other fields as it was in electrical equipment. 'Why pick on us?' was the attitude. 'Look at some of those other fellows.'" Smith then was given a little editorial license to sum up his reporting. "The problem for American business does not start and stop with the scofflaws of the electrical industry or with anti-trust," he concluded. "Much was made of the fact that G.E. operated under a system of disjointed authority, and this was one reason it got in trouble. A more significant factor, the disjointment of morals, is something for American executives to think about in all aspects of their relations with their companies, each other, and the community."

A few weeks later, Cordiner appeared at a crowded $7-a-plate dinner meeting of the New York Society of Security Analysts to discuss the guilty pleas. "We don't think anybody's been damaged," he said. He explained how he had spoken with more than twenty government officials and officers of utility companies and "I've yet to encounter the first man who said, 'Cordiner, we've got a complaint, we've been damaged. We intend to resist. It will be a neat problem to prove damages.'"

Cordiner took the same defiant tone at the 1961 shareholders' meeting, held in April in Syracuse. "It has been said by some," he said, "that I, as chairman and chief executive officer, either knew of these violations and condoned them or that I was derelict in not knowing of them." Neither was true, he continued. "We were diligent in the light of the facts as we then knew them," he said. His arrogance did not go down well with certain of the shareholders. One, Wilma Soss, described as "an inveterate needler" at such meetings, said Cordiner was "an embarrassment to the company" and should resign. "I am your employer," she said. "You are not an act of God."

James Carey, the president of GE's largest union, quoted from a speech given two weeks earlier by Henry Ford II, a GE board member, who said the company needed to clean house. "We say amen and Godspeed," Carey said, much to the embarrassment of Ford, who was seated onstage. But Soss and Carey were in the minority and, at various times, other members of the audience told Carey to "shut up" and said that he should be tossed from the meeting. At one point, many of the 2,700 GE shareholders at the meeting gave Cordiner a standing ovation and passed, with 98 percent approval, a new slate of directors. A proposal was also defeated that would have required GE to conduct an independent investigation into the price-fixing scandal and whether Cordiner knew about it or participated in it. After the meeting, Cordiner was reelected the board chairman and GE's CEO.

But the scandal took its toll on Cordiner, and his moral authority. The day before the April 1961 shareholders' meeting in Syracuse, George Burens, the former head of GE's switchgear division, testified before the Senate Judiciary Committee that Cordiner, Paxton, and Vinson all knew about the price-fixing conspiracy as early as 1952. Burens told the senators about the meeting in Philadelphia between Burke, Vinson, and him. And the whole thing was plastered on the front page of *The New York Times*.

In June, Cordiner testified before the Senate that he was "not proud of" GE's record on antitrust violations for the "last fifty years" and that the price-fixing scandal had been "a humbling experience" for him. But he did not agree with Tennessee senator Estes Kefauver, the chairman of the Senate panel, that the scandal was a "shabby record amounting to a corporate disgrace." He told Kefauver GE had made "too many contributions to the nation's welfare over the years" to be described as a "national disgrace" and so he could not "concede we deserve such an indictment."

Cordiner's status at the helm of GE was again on the front page of *The New York Times* in August 1961 about how he was still the company's CEO but had named two new executives as part of his top management team. He tried to talk about how the two new executives would fit into GE's complicated corporate structure, but the reporters focused instead on Cordiner's future at the company. He reiterated that his future at GE was a matter for the board of directors, not him. He noted that he was sixty-one and a half years old and the mandatory retirement age was sixty-five. "I am not trying to be evasive," he said. "I have no contract. Nobody here does. It's just that I don't think I have any longer a free choice to go down to my ranch"—six hundred acres near Tampa, Florida—"and count cows like I had planned several years ago." In other words, Cordiner planned to stick around.

Part of the reason Cordiner survived the scandal was the support he enjoyed from GE's shareholders. And there was good reason. Under Cordiner, GE's revenues and profits were soaring. It was clear that, somehow, his decentralization plan was working, even if it had led to a price-fixing scandal exacerbated by ongoing demands for financial performance with minimal central oversight. In 1961, GE's revenue was nearly $4.5 billion, a 6 percent increase and an all-time high. Revenue in 1960 was higher than revenue in 1959, which was higher than revenue in 1958—both of which were also record years for GE revenue. Net income, $242 million, also a record, was 21 percent higher than it had been in 1960. The GE stock was also at an all-time high. In other words, despite the magnitude of the scandal, Cordiner was getting the job done and shareholders were appreciative. "It appeared obvious from these results," the *Times* reported in January 1962, "that General Electric, the nation's largest electrical manufacturer, had overcome the effects of the Government's successful prosecution of the company on price-rigging charges."

## Chapter Five

# JACK FROM PLASTICS

Jack Welch joined GE in 1960, just as the worst aspects of its role in the price-fixing scandal were beginning to trickle out publicly. He seemed not to have noticed the controversy or to have cared much about it at the time. But according to Ben Heineman, GE's longtime general counsel, the price-fixing scandal "influenced his view of the importance of GE being a high integrity company."

Much of Welch's biography is well known, of course, given how famous he became as GE's longtime CEO and the popularity of *Jack: Straight from the Gut*, the autobiography he wrote (with John A. Byrne, a writer at *Fortune* and then *BusinessWeek*). The tales and anecdotes from Jack's youth have taken on a high gloss, whether apocryphal or not.

Born in November 1935, an only child of Catholic parents—"It was a miracle that I showed up," he said—Jack grew up in Salem, Massachusetts, on the so-called north shore of Massachusetts, just southwest of the grittier Gloucester and west of the tonier Marblehead. It was a working-class town with a decent-sized tourist trade, thanks to its elegant housing stock, the infamous witch trials at the end of the seventeenth century, and the House of the Seven Gables, a large mansion built in 1668 (before the witch trials) by wealthy merchant John Turner and made famous in Nathaniel Hawthorne's 1851 novel of the same name.

Welch was a classic overachiever. His father, John, was a conductor on the Boston & Maine Railroad, the train line that runs from Boston to Newburyport, Massachusetts, and was also head of the local union. The train passed through the same ten stops along the coastline day after day, including one in Lynn, where GE had an aircraft engine plant. "He brought home the newspapers from the big shots who rode on his train from Boston to Marblehead," Jack said. They weren't particularly close. "He had crazy shifts because he had to be in in the morning and he had to take some workers home at night from Boston," he continued. "I didn't spend a lot of time with him. But he was generous to me. He was good to me. But he wasn't a big influence in my life. He was distant." He summed up the relationship: "He was there all the time. But he wasn't *there*." They lived in a modest six-room stone house across from a valve-manufacturing factory. His father was convinced it was a great location; it would be quiet over the weekend since the plant wouldn't be operating.

Jack's mother, Grace, a housewife from Salem, was a different matter entirely. She was descended from Irish immigrants who came to the United States around 1810. Jack adored his mother. And his mother adored him, and not only because his parents had been trying for years to start a family—a big Irish Catholic family—but ended up with only one child: Jack. Grace was in her early forties when she had Jack. "That makes the sun set on you," he said. She never went to high school. But, Jack said, "she was smart as a tack." She did the taxes for everyone in the neighborhood. "She was my best friend," he told Irishamerica.com in 2001. "So, if things didn't go right, I'd talk to her about them, talk to her about my girlfriends, talk to her about everything. She was my buddy, my manager, my critic. She was everything." He often wondered why his mother didn't do better in school and why she and her family didn't progress further in America than they had. "They'd been here like three generations," he said of his mother's family. "And she was the smartest one."

Even though he was a small kid, he was an athlete. He played on the basketball team, the baseball team, and the hockey team. His mother never made him feel diminutive. "I never felt short in my life," he said. "That's what my mother did for me." She made him feel both big and important. "She'd kick my butt when I didn't do well and she praised me to the hills," he said. His career in team sports seems to have peaked when he was twelve, before he and his peers hit puberty. But his interest in sports had less to do with his own physical prowess— of which he admitted there was little—but rather more to do with enthusiasm for team sports, leadership, and, of course, winning. "Winning is good," he said.

Sports, especially baseball, taught him something about what he called "differentiation," the fact that some people were better at some things than others. He and his friends would go to the local Salem playground—called the Pit—and choose up sides for a game. They would pick players by tossing up the bat and then alternate choices depending on where you grabbed the bat. The best athletes played shortstop or pitcher. The last guy picked would go to right field. When he was young, the bigger kids would always stick him in right field. "I didn't want to be there," he said. "Desperately." As he got older and a little bigger, he was the best player on the team. He pitched. "That's the way we were brought up," he said. "I never stopped in seventy years."

Jack also had a pronounced stutter when he was child. He would sometimes make fun of himself when he was midstutter. But his mother had another idea about that, too. "She convinced me that my brain was so big and smart that my tongue couldn't keep up with it," he recalled. "She told me that a thousand times." (But it took a while for Jack to overcome the stutter. Jack liked tuna fish sandwiches and sometimes, even as an adult, when he ordered a tuna fish sandwich, he would be brought *two* tuna fish sandwiches by mistake.) She also made sure he did well in school, at both Salem Elementary and Salem High. "She made me a good student," he told me. "She was smart as hell."

Grace also tried to keep Jack's emotions in check. It wasn't easy. One time, at the final hockey game of his senior year, against rival Beverly at the hockey rink in Lynn, the score was tied at two after regulation. Jack had scored both goals for Salem. The game went into overtime, but shortly thereafter, Jack's team lost. It was the team's seventh loss in a row. Jack was pissed. He threw his hockey stick in anger, then skated to get the stick and marched off to the locker room.

Next thing he knew, his mother was in the locker room, too. She bounded right up to him, oblivious to the fact that the guys around her were in various states of undress. She grabbed him by the jersey in front of everyone. "You punk," she yelled at him. "If you don't know how to lose, you'll never know how to win. If you don't know this, you don't belong anywhere." He paused for a moment, recalling the memory. "She was a powerhouse," he said. "I loved her beyond comprehension."

She was also the family disciplinarian. She would discipline Jack when he played hooky from church or altar boy practice. One time when he skipped the practice to play hockey with friends on a supposedly frozen pond near his Salem home, he fell through the ice and got soaking wet. (He was fortunate to have not suffered a worse fate.) He decided he needed to dry his clothes before he went home. He stripped down, built a fire, and then put the wet clothes above it. But when he got home, his mother, of course, smelled the smoke on his dry clothes. "Ducking altar boy practice was a big deal to someone who hung a crucifix on the wall, prayed the rosary, and considered Father James Cronin, the longtime pastor of our church, a saint," he wrote in his memoir. She "whacked" him with one of the nearly dry shoes he had been wearing. He described himself as "an incredibly serious, believing Catholic" growing up.

She was more lenient another time when he lost it on what turned out to be his final time caddying at the Kernwood Country Club, in Salem. Jack had started caddying at Kernwood when he was ten years old with the encouragement of his father, who would overhear the successful businessmen on the train talk about their golf games. He figured his son should learn the game, too, and perhaps it would be useful to him in whatever business endeavor he ended up pursuing. Playing golf might also help Jack escape Salem, his father thought. The pay stank—$1.50 for carrying one bag eighteen holes, with minimal if any tipping—and the work was laborious and hard, especially on hot summer days. But as a caddy, Jack learned how to play golf well, a skill that helped him make his way up the GE corporate ladder. He always said golf was an important element of success at GE.

On what turned out to be his last day at Kernwood after eight years—"which was probably a little too long for my own good," he said—Jack was caddying for "one of the stingiest members," a guy who owned a clothing store on Salem's Main Street in an era when owning that kind of business, even in a small town, made you relatively wealthy and important. Kernwood was founded in 1914, and Jewish families predominantly comprised its membership, in the same way that the Nantucket Golf Club—where Jack and I first met for lunch—was established, as Jews were not allowed at many country clubs across New England and had to make their own.

On this particular day, Jack got assigned to caddy for the Salem store owner after doing his best to try to avoid it. On the sixth hole came trouble, for both Jack and his patron. The hole is a longish three-hundred-yard par four (depending on where you tee off from), with a pond about a hundred yards from the start. The guy topped his drive; his ball ended up in the pond, about ten feet in. He asked Jack to go in and retrieve the ball in the mud. Jack refused. "When he insisted, I told him to go to hell," he told me during our lunch at the Nantucket Golf Club. "I tossed his clubs into the water, told him to get his ball and clubs himself and ran off the course." That was his last time caddying at Kernwood. He told me the story with great pride. But it cost him his share of the coveted—by caddies anyway—Francis Ouimet college scholarship.

Jack also lost out on his opportunity for a four-year ROTC scholarship he was hoping to get. He and two high school friends had passed the naval exam, the first step in trying to secure the four-year scholarship. Jack's father persuaded state representatives to write letters of support on Jack's behalf. He had "a battery" of interviews. His two friends—one went to Tufts, the other to Columbia—got the coveted scholarships. Jack didn't. He was hoping to go to either Dartmouth or Columbia. But he couldn't attend without the scholarship; his family didn't have that kind of money. He was plenty disappointed. But he wasn't much of a student at that time anyway. "No one would have accused me of being brilliant," he said of his high school years. (In his office, when he was CEO, Jack had a sign on his desk that read "Where Is It Written That To Be Successful In Business You Have To Be Smart?")

Instead of the Ivy League, Jack applied to the University of Massachusetts in Amherst, a state school where the tuition was $50 per semester. For under $1,000 he could be a college graduate, something only one other member of the Welch family—a cousin—had accomplished. He didn't hear he had gotten into UMass until June, a few days before he graduated from Salem High. He figured he had been on the wait list without knowing it.

Jack had never been away from home before for any extended period of time, and even though Amherst was only about a hundred miles from Salem, he was desperately unhappy. "It was awful," he said. He was homesick. Three weeks in, his mother showed up on his doorstep and read him the riot act. It was a repeat of the scene she made when he was upset about losing a hockey game. They were walking out on the quad when his mother turned to him while pointing out some of the other kids. "She said, 'See that kid there? You're smarter than him.

Why are you worried about him? See that one there? They're not having these problems. Get off your horse, or get on your horse and get going, and you're better than all these kids here.' She went down, pointed right at them." His "anxieties," as he described them, went away in a week's time. "She's a total hero," he told me about his mother. "I'm going to cry if I keep going."

He was captain of the UMass hockey team, which wasn't much in those days. The university didn't even have its own rink then. (They played at Amherst College.) He was in a fraternity, Phi Sigma Kappa, drank a lot of beer, played poker, and went to parties. In other words, he was a typical 1950s college undergraduate. He decided to study chemical engineering at UMass, of all things, and found that he was a natural. He was good at math. The equations made sense to him. He "fell in love" with chemistry, he said. "I was really good at it. I just really understood it. Don't ask me why." He had a 3.7 GPA his freshman year and was on the dean's list every semester of his four years. "I was the best student UMass ever had at that time" in chemical engineering. (Actually, he was one of the two best students in chemical engineering at UMass at that time.) During the college summers, Jack took jobs as a chemical engineer at Sun Oil, near Swarthmore, Pennsylvania, and at Columbia Southern (now PPG Industries) in Ohio.

When Jack graduated from UMass, in 1957, he had many corporate job opportunities but decided instead to take a graduate-school fellowship at the University of Illinois in Champaign, to get his master's degree in chemical engineering. "I picked Illinois because they had more Nobel Prize winners there," he said. Within two weeks, though, of arriving in Illinois, he found himself in a little trouble.

By then, he had met a girl, and on a Saturday-night date, they were getting busy in a VW Bug that Jack's father had bought him after he graduated from UMass. Suddenly, the flashlight of a campus police officer was shining inside the little car. Such was American life in the late 1950s, especially in the Midwest, that any form of premarital sex was a major infraction. Jack was suddenly at risk of losing his fellowship and getting tossed out of the university. He had a meeting with the university's provost, no less, to discuss the matter. He had to think fast. He quickly made an ally of the department head—the fellow was impressed that one of the graduate students in chemical engineering was not just a nerd—and begged him to intercede on his behalf with the provost to give him another chance. The plea worked. Jack still had to meet with the provost, but he got off with a warning.

The plan had been for Jack to spend one year at Illinois, get his master's degree, and find a job somewhere in corporate America. But 1958 was a recession year. In 1957, he'd had his pick of jobs. But a year later things had changed. He had two job offers: one from an oil refinery near Tulsa, Oklahoma, and one from Ethyl Corporation, in Baton Rouge. On the flight down to Baton Rouge to visit Ethyl, he had an epiphany. The stewardess asked him if he wanted a drink and addressed him as "Mr. Welch." Meanwhile, Jack's colleague sitting next to him was also asked if he wanted a drink. The stewardess addressed him as "Dr. Gaertner," since he was a doctoral candidate. Jack thought being referred to by the "doctor" honorific sounded better than the "mister" label, so he decided to stay at the university for a few more years and get his PhD in chemical engineering. "I thought, 'Shit, all I have to do is stay here two more years and they'll call me Dr. Welch on the plane,'" he said. It meant long hours in the lab, working on chemical experiments while keeping meticulous notes and records.

It helped that the job market was tough in those years and that Jack liked and admired his professors. His PhD thesis was about dropwise condensation in nuclear power plants. "I thought it was the most important thing in my life," he told me. I asked him to translate into English what his thesis was about. "How you make steam condense on a pipe to transfer heat," he said. "The smaller the drops, the more area is exposed, and the better the heat transfer." There were equations, of course, and high-speed film of the water drops. He spent hours in the lab developing film. "You had to work your way through problems to which you didn't know the answers," he said. He got his doctorate in three years. "I got out of Illinois faster than anybody," he said. "I was damned if I was going to stay there." He graduated in 1960.

As was often the case, especially back then, Jack was pushed to use his newly furnished doctorate to get a teaching job, one that might put him on a track to be a professor. He liked teaching and was good at it. But he resented being shoved into teaching and being made to feel bad because he wanted to get rich, or rich enough. "You're shamed" into wanting to teach, he said, and you were somehow "dirty" if you were interested in a corporate job or making money. But he wasn't ashamed in the least about wanting to get a big paycheck. "Oh, Jesus, was it important? I wanted to make $30,000 by the time I was thirty," he told me. Why? I wondered. "Because I never had any," he said. "I had my nose against the glass about money. I never had any. I caddied and didn't have any money." (He was a talented teacher, though. And loved to mentor future GE leaders. That's why he was a regular at Crotonville and spent millions refurbishing it as a serious center for the development of senior GE executives.)

Rather than get stuck on a teaching track, Jack decided he wanted a job that would combine working in a lab and interacting with people. "I truly love people," he said. "Maybe you don't know that, but I do." He decided a career in business would be the best way to combine his interests in research, people, and making money. "I could bullshit, okay?" he said. He was convinced that being a chemical engineer was a perfect background for a businessman, since "there are no finite answers to many questions." They were both ways to figure out a "darker shade of gray," since there were "rarely black and white answers. More often than not, business is smell, feel, and touch as much as or more than numbers. If we wait for the perfect answer, the world will pass us by." He interviewed at both Syracuse and West Virginia Universities but quickly abandoned that path. (There was no question of military service, even though he was in the ROTC at UMass. "I missed them all," he said of the American wars when he was a young man.)

Jack had several job offers and decided that two of them made sense: one from Exxon, in a development laboratory in Brownsville, Texas; the other from General Electric, in a new chemical development operation in Pittsfield, Massachusetts, across the state from where he grew up. He didn't know much about the company, other than that Pittsfield was in Massachusetts and that GE was willing to pay him $10,500 a year, more than his other offers. He took the job at GE.

In Pittsfield, Jack met with Dan Fox, the scientist at GE often credited with discovering Lexan, the ubiquitous plastic resin that GE started selling in 1957. (A German scientist also developed the substance at about the same time as did Fox.) Fox was also working on developing a new thermoplastic product—polyphenylene oxide, or PPO—that could withstand high temperatures and that could potentially replace copper pipes and stainless-steel medical instruments. Jack's job would be to get PPO out of the lab, into production, and for sale in the market. GE's plastics division was seen as something of a backwater within the company—it had only three hundred employees at that time—and there seemed to be a general concern that it could be closed or sold off by corporate at any time. The prevailing gallows humor was summed up by the joke "What's a Plastics Division optimist? Answer: Someone who actually brings his lunch to work." But Jack didn't see it that way. "The job was perfect," he said. He was part of a six-person "chemical development operation."

One of his first tasks was to figure out how to put a plastic coating on the electrical wires that GE also manufactured. The plastic coating had just been invented at the GE research lab in Schenectady; the invention was then transferred to Jack's little group in Pittsfield to commercialize. The plastic pellets came from the lab in small quantities of little chips. Jack's job was to figure out how to make more of the material, to scale up its manufacture and try to create applications for it.

He was the first PPO employee outside the research lab. "How to commercialize it," he said. "How to process it. But the first thing was to learn how to make enough of it." The first employee he hired, Gus Kanalas, was his lab technician. "We brought him home," Jack said. "He became my friend."

By then, Jack's home was a rental apartment in Pittsfield, heat not included. "The landlord told us to get blankets," he told me. The "us" was Jack and Carolyn Osburn, his bride of a few months. They had met at the University of Illinois, where Carolyn was getting her master's degree in English literature, thanks to a $1,500-a-year fellowship. She had graduated with honors from Marietta College, in Ohio, where she was class president. Jack "first spotted" Carolyn at the Catholic church on campus during Lent while she was doing the Stations of the Cross. They were later introduced by a mutual friend at a downtown bar. Their first date, in January 1959, was at a University of Illinois basketball game. They were always together after that and were engaged five months later. On November 21, 1959, two days after Jack's twenty-fourth birthday, they were married in Arlington Heights, Illinois, Carolyn's hometown outside of Chicago, and spent their honeymoon driving in Jack's Volkswagen across the country and in Canada as he interviewed for jobs. Carolyn taught for a while in Pittsfield but soon became a full-time mother. "She got pregnant right away," Jack said, "and she got pregnant right away, and she got pregnant again right away."

From the outset, Jack's hugely competitive spirit was on full display in Pittsfield. Peter Foss saw that firsthand on the golf courses around Pittsfield, where he and Jack played regularly. Foss, who grew up in Pittsfield, met Jack when he used to come into the pharmacies that Foss worked in growing up and then eventually owned. For Jack, golf "was a way to compete with people," Foss said. "And he loved to win." They played often. "He was a fun partner," Foss continued. "It wasn't as fun to compete against him, because Jack was always nudging you and making comments. He liked to really dig at you going around the golf course. And he said, 'You're too easy on those guys. You've got to start needling them.'" They would socialize together, too. "We had a lot of fun," he remembered. "I had dinner at his house. He'd have these informal staff meetings sometimes after golf. And I'd sit and listen to how he'd do things. It was quite an education." Years later, in 1977, right before Jack moved to Fairfield, Connecticut, to try to compete for the top job at GE, he persuaded Foss to join GE. "He said to me one day, as only he could do, 'Yeah. You gotta get a real fucking job someday, someplace,'" Foss recalled with glee. He wondered what Jack was talking about.

"You need to run these popcorn stands," Jack told him.

"Well?" Foss replied.

"You should come to GE," Jack said. "You'd love it. You know, you run your own business. You have personal pressures. You have to make a payroll up every Saturday. But you don't have to do that at GE."

Jack talked Foss into it. He went to work for GE Plastics. "When I joined the company, he put me in an area where not many people knew I even knew him," he said. "I was just a new guy that got hired. And so I didn't have to put up with people going, 'Oh, yeah. It was Jack.' That went on for years until one day he called me at work." (Ironically, Foss also worked with Jeff Immelt at Plastics as Immelt was climbing the GE corporate ladder; they remain close friends and own homes near each other on Kiawah Island, off the coast of South Carolina.)

In Pittsfield, Jack was on a mission to commercialize Dan Fox's little bags of plastic pellets. "Welch was literally a one-man engineering operation," Fox told *The New York Times* in December 1980. "If things weren't moving fast enough for him, he moved them." After his first year on the job, his boss gave him a $1,000 raise for a job well done. At first, he was thrilled. After all, a 10 percent raise was not insignificant, and it made a big difference to Jack and his growing family. But Jack couldn't resist comparing notes with the employees in his unit. Much to his consternation, he discovered the other five guys also got

$1,000 raises. "I went apeshit," he told me. He thought to himself, "I'm better than those fucking guys and you give me the same you gave them?" His boss, Burt Coplan, was unsympathetic.

Jack resolved to quit. He began looking at the help wanted ads in the back of *The Wall Street Journal* and *Chemical Week*. Within a month he had found a new job at International Minerals and Chemicals. "I quit the next day and moved my family to Chicago, my wife back home," he said. Coplan was happy Jack quit. "I was crowding him a bit," Jack said. And Jack pretty much hated Coplan. In 2002, he told MSNBC's Chris Matthews about his first year at GE, "The first year, you come home and you hate your boss and you hate the environment, and your boss is a bum and you're smarter than your boss."

But his boss's boss, a man named Reuben Gutoff whose office was on Lexington Avenue in New York City, wasn't so keen on Jack's departure. In the previous weeks, Jack had impressed Gutoff with a presentation he had made with a new idea for how to use the plastic pellets and how they might fit in a market dominated by Dow and DuPont. His five-year-business-plan presentation to Gutoff "blew away" the ones done by his colleagues. "That came out of my ass," Jack said. He did more than Gutoff expected of him as a way of "getting out of the pile," he said. A less charitable view was that Jack had figured out, and mastered, early on, the art of corporate politics. In any event, his sucking up to Gutoff seemed to have worked. An hour or so before Jack's Pittsfield going-away party, a month or so after he had resolved to quit, Gutoff asked to see Jack alone. He came up to Pittsfield from New York City.

He invited Jack and Carolyn to dinner at the Yellow Aster, in Pittsfield. "Trust me," Gutoff told Jack. "As long as I am here, you are going to get a shot to operate with the best of the big company and the worst part of it pushed aside."

"Well, you are on trial," Jack replied.

"I'm glad to be on trial," Gutoff said. "To try to keep you here is important."

He told Jack he wanted him to stay at GE and offered to give him a $4,000 raise, bringing his compensation to $15,000 a year. "He gave me a sales pitch," Jack said. Gutoff promised Jack he would no longer have to report to Coplan and that he would have the "leadership role" in commercializing PPO. Even though the job in Chicago paid a bit more, Jack decided to stay. "It was one of my better marketing jobs in life," recalled Gutoff. "But then he said to me—and this is vintage Jack—'I'm still going to have the [going-away] party because I like parties, and besides, I think they have some little presents for me.'"

It was a career-making decision. "All of a sudden I had a rabbi," Jack said. "This guy loved me. He was a slick operator. I learned from him." The decision to stick around GE reinforced in his mind the importance of "differentiation," he said, the old lesson he had learned on the baseball field that different people have different skills and capabilities and need to be rewarded accordingly. Subliminally, or not, what the early experience at GE taught Jack was that he was that rare corporate executive who could play by his own rules and somehow succeed. That's what he proceeded to do. At seemingly every turn, his successes were celebrated, as were his failures. His bad behavior was overlooked. For reasons having to do with his unbridled ambition, his business acumen, his charm, his personality, his intelligence, and his powerful political skills, he was able to turn his misfortunes, or his mistakes, into corporate victories that would likely have flushed out those with lesser political skills. For the next seventeen years, Jack Welch ran Pittsfield as if it were his own fiefdom. He had his own private jet. He ran the business out of office space in a downtown Hilton hotel. He pretty much did his own thing, making a huge success of what became known as GE Plastics.

One of his first nicknames at GE was Teflon Jack. He proved it soon after he cut his deal with Gutoff to stay, free of annoying supervision, when the PPO factory in Pittsfield literally blew up one spring day in 1963. Jack was twenty-eight years old. He survived the explosion both physically and politically. "It was one of the most frightening experiences of my life," he remembered. He was in his office across the street from the plant when the explosion occurred, blowing the roof off the building and the windows out of the top floor. The damage was significant. "As the boss, I was clearly at fault," he continued. He had been working with scientists "bubbling oxygen through a highly voluble solution" in a large tank when an "unexplainable spark" ignited the solution, leading to the explosion that blew the top off the tank and the roof off the building. It was a disaster, mitigated only by the fact that somehow no one was seriously injured or killed.

The next day, Jack was summoned to Bridgeport, Connecticut, to explain to Charlie Reed, the group executive, what had happened. Reed was Gutoff's boss. While Gutoff was also at the meeting, he was not in a position to protect Jack on this one. "I was prepared for the worst," Jack recalled. Once again, the GE gods shined on Jack. Being responsible for an explosion at a chemical plant was a fireable offense, or certainly should have been. But Reed was sympathetic. Like Jack, he was a chemical engineer, with a PhD from MIT, where he had also been a professor. He took a clinical, rather than an emotional, approach to the incident. "Charlie understood what could happen when you were working at high temperatures with volatile materials," Jack continued. Reed understood that sometimes things go wrong in pursuit of innovation. At least no one had been killed or seriously injured. Jack's project would continue, as would his career at GE.

Reed's reaction impressed Jack. Instead of attacking Jack when he had erred, Reed sought to restore his confidence, to put him back on the path to success. That was good for Jack and good for GE. It's a lesson he tried to remember as he continued his rapid ascent up the corporate ladder. "Don't get me wrong," he wrote in his memoir. "I enjoy challenging a person's ideas. No one loves a good and passionately fought argument more than I do. This isn't about being tough-minded and straightforward. That's the job. But so is sensing when to hug and when to kick."

By 1964, Jack and his team were starting to make real progress in the development and commercialization of PPO. He managed to convince Charlie Reed of the progress. Reed, in turn, persuaded the GE board to approve $10 million for a new plant to manufacture PPO on a large scale. Jack explained, "We got the money on the basis that we had a breakthrough plastic that was a step beyond" Lexan, GE's other plastic product, which was already changing the industry. The first thought was to build the new PPO plant near the Lexan plant in Indiana. But Jack didn't want to move to Indiana. He wanted to stay in the Northeast. One weekend, he dragged his wife and three kids along to Selkirk, New York, about an hour west of Pittsfield. He picked a 450-acre site that once belonged to the New York Central Railroad and that had access to the Hudson River. Jack admitted he wanted the location out of selfishness—he wanted to stay put in Pittsfield and to run his own show. He won out, again, despite the skepticism of other GE executives who wondered why the plant shouldn't be in Indiana, next to the Lexan plant.

And then he won the promotion to become the general manager of the Selkirk plant, after the then manager got a promotion to corporate headquarters. "Life appeared to be perfect," he recalled. The only regret he fessed up to having at the time was that he couldn't share his burgeoning success with his parents, especially his beloved mother. On January 25, 1965, she died of a heart attack. She was sixty-six years old. Jack was devastated. He had seen it coming for years. She had suffered her first heart attack when he was a UMass undergraduate, and when he heard the news, he was so concerned he literally started running the 110 miles between Amherst and Salem until he was able to hitch a ride home. She spent three weeks in the hospital before she recovered. Three years later came her second heart attack and another period of recovery. Another three years passed before her third and final—and fatal—heart attack. His parents were in Florida at the time, thanks to the $1,000 that Jack had given them from his GE bonus to allow them the extravagance. "Thank God I did it," he remembered. "One of my life's regrets is not being able to give her all of the things I could if she were alive today." Upon hearing the news from his father that his mother was in a Fort Lauderdale hospital, he immediately flew from Pittsfield and went straight to her room. "She was in bad shape, weak and frail," he remembered. She asked him to wash her back. He obliged. That night, he and his father returned to the motel to sleep. They never saw her alive again. His father and aunt traveled with his mother's body by train back to Salem. Jack drove his father's car, spending a few fitful hours in a North Carolina motel trying to get some rest. Mostly he just cried. "I cried for at least four or five days steady," he told me. He was angry at God for taking his beloved mother away from him.

Fifteen months later, his father was also dead. Jack was on a business trip in Europe when he got the news that his father was in the hospital, the result, essentially, of eating too much salty food, causing his body to retain an excessive amount of fluid. His mother used to keep his father's diet in check when she was alive. But without her, he resorted to eating whatever he liked, including the foods that he knew he should avoid. In a replay of the previous year, Jack rushed to the hospital. But by the time he made it to his father's bedside, he had died. He was seventy-one years old.

Jack's parents are buried together in the cemetery of the St. Thomas the Apostle church in Salem. Orphaned at thirty-one years old, Jack was "thrown for a loop," he said. "My mother and father were gone and I was feeling awfully sorry for myself." Fortunately, he said, he had his immediate family: Carolyn and their three children, with a fourth, Mark, coming two years later. "She was a real rock for me," he wrote of Carolyn.

Jack wasn't sure he would be viewed around GE as ready for the promotion to be the general manager of the Selkirk plant, but that didn't stop him from taking his appeal directly to Gutoff, his rabbi. He was utterly brazen. After a team dinner at a restaurant in Selkirk with Gutoff, he climbed into the front seat of Gutoff's Volkswagen and made his case. Gutoff argued that Jack didn't know anything about marketing, and selling PPO was all about marketing. Jack was relentless. He stayed in Gutoff's car for more than an hour. Another man likely would have been fired. He didn't get an answer from Gutoff that night. But Jack was certain he had been persuasive. For the next week, Jack peppered Gutoff with calls and justifications. Somehow, it worked. Gutoff gave Jack the job and told him, "You'd better deliver."

Jack certainly was driven to succeed. One story goes that an older fellow working in the next office over from Jack was having a heart attack. Jack and another coworker tried hard to revive him. "They're on their hands and knees trying to pump desperately to bring him back, and they failed," explained someone familiar with the story. "He died right in front of them. The story that was told after that was that Welch just slapped his hands together and said, 'He's gone. Let's get back to work.'"

Nearly immediately upon getting his new job, Jack was faced with an existential problem: As PPO aged, it got brittle and cracked under the high temperatures it was designed to withstand. PPO wasn't going to be much use as a replacement for copper tubing or pipes if it cracked at high temperatures. Worse, the GE scientists had no idea why it was happening. "I had lobbied myself into a potentially career-killing challenge," Jack explained. He remembered standing outside in Selkirk on a cold winter day staring at a huge hole in the ground where the $10 million plant was being built and having no idea what to do. It took six months of tinkering with the chemistry of PPO on a daily basis until the GE chemists, led by Dan Fox (the man who recruited Jack to GE in the first place), figured out a solution to the problem by adding rubber and polystyrene to the manufacturing process. Called Noryl, the product that Jack helped to develop and bring to market generated $1 million in revenue in 1966 and nearly $24 million in revenue five years later. It was a $50 million business by 1973. (Eventually, the product was generating sales of more than $1 billion per year.)

Or at least that was Jack's version of how the PPO cracking problem was solved. Another version, according to journalist Christopher Byron, had it that Eric Cizek, an "obscure" GE chemical engineer who joined the Pittsfield team in August 1961—around the time that Jack was thinking of quitting the company—had grown disenchanted with how frenetic his colleagues, including Jack, had become in trying to solve the problem. Instead of joining the fray and the manic behavior, Cizek went about his business on a daily basis trying to solve the problem. "Cizek would laboriously test out his ideas and write down the results in lab notebooks, which he would fill from front to back, then place in his desk, one after the next," according to Byron.

One thing led to another for Cizek, the loner, and in 1964 he was encouraged to look for a new job. He soon left GE and headed to Ford. According to Byron, Jack and his team spent another three years and "more millions of dollars" to try to fix the PPO problem before someone had the bright idea to look in Cizek's desk, found his notebooks, and discovered that he had actually solved the problem before he left GE. "Since no one would listen to him," Byron wrote, "he simply wrote it down in his lab book and put on a longer and grumpier face than ever when he was around the office."

One day in 1968, Cizek got a letter from GE at his home in Ann Arbor informing him what had been found in his notebooks and asking him to sign over the rights to the discovery. He agreed, giving GE the rights to Noryl. Cizek did get a few consolation prizes, though. GE gave him a paperweight and flew him and his wife to New York City for a Broadway show. Also, on May 14, 1968, Cizek was awarded the patent—number 3,383,435—for a "blend of polyphenylene ether and a styrene resin," essentially what became Noryl, which, Byron noted, "went on to launch Welch in his ride to the top of GE." (A 2000 book, *Unlikely Victory: How General Electric Succeeded in the Chemical Industry* by Jerome Coe, confirmed the PPO problem that Cizek and a team of others, including Dan Fox, helped to solve and that Cizek got a patent for his invention.)

Having solved the PPO crisis, and having made Noryl a success, Jack continued his GE career on its upward trajectory. "He was a kind of Billy the Kid, the enfant terrible whose exploits became even more legendary because he broke the rules and never seemed to get caught," the journalist Thomas F. O'Boyle wrote in his hard-nosed book about him and GE, *At Any Cost.* He won raise after raise and promotion after promotion. Everything seemed to be a cause for a celebration in Pittsfield. He bought a new car—a Pontiac LeMans—and a new house on Cambridge Avenue in Pittsfield. "When I got a three-thousand-dollar bonus in 1964," he explained, "I threw a party for all the employees at the new home." He celebrated when a customer bought $500 worth of Noryl. He celebrated when ten new customers bought Noryl. He was the prince of Pittsfield. "We had our own thing and I started hiring more guys," he told me. "I had a band of Indians that I loved. We drank and we played and did all the things that young guys do when they've got some money and living in a small town." In his memoir, Jack was equally cryptic about what he and his "band of Indians" had been doing in and around Selkirk, New York, and Pittsfield, Massachusetts. He mentioned only "frequent parties at local bars" and "behavior that wasn't the norm."

It turned out Jack was understating the extent of the revelry, as one is prone to do in a memoir or in conversations with a journalist. But Byron, in his well-researched, albeit snarky, *Testosterone, Inc: Tales of CEOs Gone Wild*, pulled back the curtain on some of the more boisterous, if not downright offensive, aspects of Jack's behavior during those days. It bears repeating that Jack's antics occurred while his wife,

Carolyn, was home in Pittsfield caring for their four children, two boys and two girls, born between 1962 and 1968.

When they were not working hard to expand the market for Noryl—including building plants overseas in the Netherlands and in Japan—Jack and his Indians liked to have a good time, often at the expense of men not in his circle and of women who, well, just happened to be around. Byron told the story of how Jack treated Larry Burkinshaw, a colleague who took years to win Jack's favor. Passing him in the corridor one day, Jack said, loudly enough to be heard by others, "Pick up his skirt and kick him in the cunt." At a dinner party at the home of one of his Plastics colleagues, he "goosed" the wife of the host in front of the other guests. There were Fitzy and Rex, two of Jack's closest confidants and, like him, Irish Catholics who loved partying. "The workdays in Pittsfield all ended the same way," Byron reported, "with Jack rounding up the Jack Pack and heading for town." There was Five Chairs, on the edge of Pittsfield, a "kind of mirrors-and-velvet pickup joint," Byron wrote, and the Bubble, closer to the Selkirk plant, which was a combination bowling alley and year-round swimming pool. Not far from Jack's new plant in Selkirk, there was the Feura Bush Tavern, which the Jack Pack called the Bush, as in "Anybody up for some bush?" One night at the Bubble, Jack and Fitzy arrived around 10:00 p.m. to find the dance floor already hopping with the latest Motown songs. Jack wasted little time "zeroing in on his target, an attractive young woman who appeared to be alone," according to Byron, "and prepared to hit her with the irresistible Welch charm."

Within minutes, amid the wildly dancing bodies, Fitzy realized that Jack was fighting, in the middle of the scrum, with the husband of the woman he was trying to pick up. The fight moved across the dance floor, out the door, and into the parking lot. "But it was what happened next, as Fitzy rushed to the door along with the group from Engineering, that quickly became part of the Legend of Jack Welch,"

Byron wrote. Jack was, apparently, running around the parking lot "screaming for someone to call the police," while the woman's husband was running after him with a beer bottle and yelling, "I'll get you, you little prick."

It was a different time. The "boys will be boys" mentality was still reigning supreme in American culture, to say nothing about the corridors of power in corporate America. There was no social media. There was no internet. While the gossip columnist for the local paper may have known about Jack's extracurricular antics, that was not the kind of stuff that made it into one of her columns, not, that was, if you wanted to keep relationships on an even keel with one of the most powerful local employers. There was no #MeToo movement. Back then, there was simply no accountability for that kind of behavior; in fact, it was more or less celebrated as a macho credential. Needless to say, Jack's career was not the least bit thwarted by his carousing. All that seemed to matter were the financial results his business posted year after year, and they continued to be good and kept getting better. His trajectory was ever upward.

In the summer of 1968, Jack was promoted to the manager of the Plastics operation. Apparently that made him worth writing about in GE's internal corporate magazine, *Monogram*, a corporate staple since 1923. When the writer came to Pittsfield to interview Jack, he referred to him as *Dr.* Welch, finally granting Jack the honorific he had craved back when he was still a graduate student. Jack would have none of it. "I don't make house calls, so call me Jack," he responded, prompting the reporter to include the line in the *Monogram* article. He was revved up. His colleagues were a "turned-on bunch," Jack said, who—gulp— generated their own "electricity." Under his leadership at Plastics, he proclaimed (correctly), revenue grew more in his first year than in the previous ten. "There's gold here, and we were lucky enough to come along and dig the mine," he said. He predicted the plastics division would continue to break revenue and profit records. (In his memoir, he expressed incredulity at his audacity. "Those who read the article must have damn near choked," he wrote.)

But as troubling as was Noryl for Jack and his team to perfect, Lexan was like a dream come true. "It was clear as glass and tough as steel," he wrote. "It was flame resistant and lightweight." And better still, he was its shepherd. Customers loved the stuff. Boeing used four thousand pounds of it on its 747 jets, instead of heavier, less durable materials. Jack ramped up the marketing of Lexan. He hired Hall of Fame pitcher Bob Gibson—even though he won three games in the 1967 World Series against Jack's beloved Red Sox—to throw fastballs at a sheet of Lexan in a television commercial to prove it wouldn't break. He green-lighted an ad of a bull in a Lexan china shop, where of course nothing broke because everything was made of the durable plastic. He directed much of the Lexan advertising toward Detroit, where he was hoping that the Big Three automakers would use Lexan instead of steel—and in just about every other part—on their cars. He recruited another famous pitcher, the Detroit Tigers' Denny McLain—the last Major League pitcher to win thirty games—to throw pitches at him while he cowered behind a sheet of Lexan in the parking lot of GE's Detroit office. The sales of Jack's division exceeded even his boastful predictions.

In 1971, Jack was promoted again to head up GE's worldwide chemicals business, its man-made industrial diamond business, and its global metallurgical operations. It was a major promotion to head up what was then a $400-million-revenue business. But, being Jack, he was so unconventional that GE's HR department felt obliged to note the risks of his promotion in a memo. "Despite his many strengths," according to the July 1971 document, "Jack has a number of significant limitations. On the plus side, he has a driving motivation to grow a business, natural entrepreneurial instincts, creativeness and aggressiveness, is a natural leader and organizer, and has a high degree of technical competence. On the other hand, he is somewhat arrogant, reacts (or overreacts) emotionally—particularly to criticism—gets too personally involved in the details of his business, tends to overrely on his quick mind and intuition rather than on solid homework and staff assistance in getting into and out of complex situations, and has something of an 'anti-establishment' attitude toward General Electric activities outside his own sphere." HR blamed Jack's "youthfulness and lack of maturity" for his "limitations." It did not block the promotion, and his new rabbi, Herm Weiss, supported him. "Now I had to figure out how to run a whole portfolio of materials businesses, including carbide cutting tools, industrial diamonds, insulating materials, and electro-materials products—and do it all with very different people," Jack recalled.

In March 1972, he was promoted to be a GE vice president, meriting his first mention in *The New York Times* (along with two other executives, never to be heard from again, at least at GE). He had once again assessed the competition and figured he could best them. After delivering his July 1971 review to him, the GE HR department, Jack told me, thought he was "too ambitious" and "too arrogant" and added for good measure "too well spoken," "too candid," and "too obnoxious." But his bosses loved him, and his ambition. "The energy that radiated from him was like nothing the employees at GE had ever seen before," Byron wrote. "Here he was, barely half-way into his thirties, and he was opening production plants wherever one looked, creating what amounted to an entire shadow-GE on what seemed to be little more than his own say-so." Thomas Morton, the former managing editor of *The Berkshire Eagle,* who covered GE's Pittsfield activities for more than thirty years, said of Jack and his time in Pittsfield, "Jack Welch was a man of great ambition. He always played to win, whether in golf or in business."

## Chapter Six

# THE BENIGN CYCLE
# OF POWER

In January 1973, Reginald Jones—known to one and all as Reg, with a hard *g*—succeeded Fred Borch as GE's CEO. Borch had followed Cordiner as CEO, in October 1963, after Cordiner finally saw the writing on the wall and retired two years early. At the time Borch took over from Cordiner, GE's annual revenue was $4.8 billion and its net income was $266 million. Thanks to Cordiner's decentralization policy, Borch found himself in charge of 112 operating units. He was a native of Brooklyn, where his father was an electrical engineer with Brooklyn Edison (now part of Consolidated Edison). In 1931, Borch graduated from Case Western Reserve, in Cleveland, and then joined GE's light-bulb division, which was also based there. Over a long career at GE, he worked in marketing, operations, and consumer products. At the time of his promotion to CEO, the most *The New York Times* could say about Borch was that he lived in Darien, Connecticut, was married with two children, and was "an enthusiastic golfer."

During Borch's time at the top of GE, sales doubled to $10 billion. The company's profit more than doubled to $530 million. He made big investments in the manufacture of jet engines—GE was the leader in supplying jet engines—as well as in computers and nuclear power, both of which GE has since abandoned. Borch had tried to bring order to the unwieldy organization he was left by Cordiner. He was also widely credited with creating what has become the Business Roundtable, one of the most powerful corporate lobbying organizations in the country. Borch retired unexpectedly from GE at the end of 1972, three years ahead of the mandatory retirement age of sixty-five. His wife had wanted him to retire early, while he was still in good health. At the time, he said he wanted to "work on his 16 handicap" at his home in Naples, Florida. But since no one voluntarily retired from being the CEO of GE, it was obvious that the GE board had lost faith in Borch, despite the company's financial performance under his leadership.

When GE announced, in June 1972, that Borch was retiring, it also announced that Reg Jones would succeed him. Jones, born in July 1917, grew up an only child in a row house in Stoke-on-Trent, smack in the middle of the English countryside. His father was a foreman in a local steel mill. His mother pushed to go to the United States in search of a better life, and when Jones was eight, the family moved near Trenton, New Jersey, where both of his parents found work at the Acme Rubber Manufacturing Co. His father was an electrician's helper; his mother sorted and packed the rubber rings that helped to seal Mason jars. Jones liked to tell people, "I am English. I am too damn poor to be British." Reg Jones graduated from the Wharton School at the University of Pennsylvania, which was where he met his wife, Grace Cole. He joined GE's business-training course in 1939, and between 1942 and 1950 he was a traveling auditor, even though by his own admission he had taken only five hours of accounting at Wharton. "I'd never consider myself the epitome of a bookkeeper's bookkeeper," he once said.

After posts in the supply and construction division, Jones was named VP of finance in May 1958, based in the corporate headquarters at 570 Lexington Avenue. "I've really spent more time in general management than in finance," he told the *Times*. He wanted to refuse the job because he liked being an operating executive, but he found that he also liked finance. "Reg made suggestions that were far beyond what you would expect from a financial man," a retired GE vice-chairman told *The New York Times Magazine*. "Although he might be talking to others about spending money, he'd chime right in on marketing or technical matters. In time, he got to be considered a good back-up man on anything... but he was the best financial man I ever met."

Jones and two other GE executives worked on a task force to study GE's role in the computer industry. Jones had been one of the GE executives who had recommended that GE get into the computer business despite its being dominated already by IBM. By then, GE had lost several hundred million dollars trying to compete. They concluded that GE would pay "a terrible price" to continue to be in the business, especially because of the investment required on research and development. The experience was a bit of a black eye for Jones and GE. Observed the *Times*, "G.E. did not escape the barbs of critics who wondered why a company that had successfully ventured into aerospace, nuclear energy and pollution-control systems couldn't hack it in computers." But Jones's bosses were happy. He recouped $240 million from selling the computer business to Honeywell in a tax-efficient way. "Jones's sale of the computer business was masterful," the former vice-chairman told the *Times*. "He helped G.E. and he even helped Honeywell."

Borch then asked Jones to set up a long-range strategic planning group so that GE executives would have some idea "in advance" which businesses to keep and which businesses to divest. He concluded that GE's future lay in plastics—Jack's division—medical care, transportation, and power generation, where, he said, "we can allocate our very vital resources in a more intelligent manner for greater growth." He was particularly enthusiastic about the potential in medical devices and services, where GE was an early pioneer of the X-ray machine. He expected X-ray technicians to speed up their medical diagnoses through the use of computer equipment.

He conceded that determining what businesses GE should be in at the executive level was a break with both Cordiner's and Borch's decentralized approach, which "sharpened our operating abilities" but failed miserably from a long-term-planning perspective. Under Jones's concept, GE would use centralized planning tools—computers, for instance—to closely monitor the financial performance of the various divisions. According to Jones, what had happened as the company grew and diversified its product offerings, and then adopted Cordiner's decentralized approach to management, was that the CEOs of the various business units—hundreds of them—would come across a possible acquisition. The CEO would then get input about the wisdom of doing the deal from a variety of internal GE centralized specialists—the manufacturing specialist, the finance specialist, the engineering specialist, the marketing specialist—each of whom would give the CEO a memo about the proposed deal from his perspective "but none taking an overall business point of view," Jones explained. "None of them asked things like whether the proposition fit with the mission of the General Electric Company, or whether we had the necessary resources—human, physical, or financial—given all the other problems and opportunities facing the corporation as a whole."

This was no longer tenable. In 1970, Borch created a "corporate executive staff" and asked three top executives, Jones included, to come up with a system whereby GE could decide which deals to do and which business lines to pursue, not from a tactical standpoint but rather from a strategic, long-range-planning perspective. At the time, Jones said, GE had three big business lines that required huge amounts of capital: the jet engine business; the nuclear power business, "which was a real problem in those days"; and the computer business, "which was already absorbing just about every spare cent General Electric had." Borch put Jones and his two colleagues on the project full time. They launched three studies of the three business units and quickly decided to exit the computer business, to revamp the nuclear power business, and to invest heavily in the commercial jet aircraft business, which, Jones recalled, "led us up the road to one of our greatest successes."

That road began after the United States entered World War I, in 1917. As part of the war effort, the federal government was looking to improve the performance of the nation's fledgling air force by figuring out how to boost the power of its planes at higher altitudes. The precursor of NASA asked GE's engineers, as well as those of another company, to see who could devise a solution first. It was the first military aircraft engine competition in the nation's history. It was also a secret.

The problem that Sanford Moss, a GE engineer, was trying to solve was the fact that the Liberty aircraft engines—used widely during the war—performed well at low altitudes, but at higher altitudes the lack of oxygen in the air caused the engines' output to drop, often in half, in the same way that it gets harder for hikers to breathe as they go up a mountain. Above an altitude of around fifteen thousand feet, the planes' performance dropped dramatically.

What to do? At the time, of course, GE's business was primarily focused on building gas turbines for power plants to generate electricity. The federal government was hoping to enlist GE, and Moss, whose specialty was gas turbines, to help solve the problem and to make the military's planes more effective at higher altitudes. Moss and his team built a 350-horsepower, "turbocharged" Liberty aircraft engine, which drew energy from the engine's exhaust gases. In June 1918, GE's steam turbine business, in East Lynn, Massachusetts, shipped the prototype of Moss's design to the War Department. "It was a teenage marriage: an untried turbocharger wed to an engine that one year earlier had been just a glimmer in the War Production Board's eye," according to *Air & Space Magazine*.

Moss quickly realized the prototype engine had to be tested at altitude to know if it would work. He came up with the idea of taking the engine to the top of Pikes Peak, a 14,100-foot mountain in Colorado that had the twin benefits of being at high altitude and having a twenty-eight-mile access road to the top. In September 1918, Moss and his team drove a converted Packard with the engine mounted on the back to the Pikes Peak summit. (Part of this expedition was later memorialized in GE television commercials.) They were there for a month and completed twenty-five tests. The results were impressive, with the engine's power mostly holding up at altitude, giving confidence to GE and the military that the engine would work. This was the moment GE entered the airplane engine business. Moss's engine changed everything—both for the country and for GE. "The first plane powered by a turbo-supercharged Liberty engine was a Le Pere biplane," according to a GE Reports publication. "It took off for the first time on July 12, 1919, and later scored a record of reaching 137 mph at 18,400 feet, compared with 90 mph without a supercharger." The successful Le Pere test flight led to the creation of GE Aviation.

GE Aviation was responsible for many aviation firsts: the creation of the first jet engine, in 1942, and the first turbojet engines that allowed jets to fly three times the speed of sound; innovations in private-jet engine aviation and in engines used on helicopters. One of the first private jets—the first Learjet 23—was powered by a pair of GE jet engines that were originally designed for a fighter jet and that could cruise at five hundred miles per hour. Frank Sinatra bought a Learjet 23 and used it to shuttle his Rat Pack friends between Los Angeles and Las Vegas. GE was one of the largest suppliers of engines for military jets and the commercial airline industry. (The business remains the envy of the world. The division received orders for $26 billion worth of jet engines in 2021 and its backlog totaled $303 billion.)

The directive to focus on the jet engine business was a natural extension of GE's electrical-power-generation origins—a jet engine and a power turbine are more similar than you might think. As for what to do about the other two hundred GE businesses, Borch and Jones hired both McKinsey and the Boston Consulting Group to make recommendations. They decided to group GE's two hundred businesses together according to those that operated on similar criteria, for instance, similar customers, similar competitors, similar missions. The first cut brought GE businesses down to fifty. From there, Jones structured the company into six divisions. Instead of the BCG four-box matrix then all the rage—developed by a BCG partner in 1968—GE created a nine-box matrix, consultant jargon used, theoretically, to help CEOs to decide which businesses to keep and which to sell. The BCG approach pigeonholed businesses as "cash cows," "stars," "dogs," and "question marks" and recommended that corporate executives act based on those categorizations by keeping the cash cows and the stars and jettisoning the dogs and possibly the question marks; GE took a broader approach to the labeling. It became clear that some of GE's businesses were keepers and would require major investment to allow for growth; other businesses would be sold, merged, or otherwise divested. "In between were a lot of businesses that had to be approached with selectivity, and we graded those on the nine-box matrix," Jones said.

He was convinced that GE's strategic planning efforts were much more inclusive than the "services" apparatus—the internal domain experts that opined on deals—that existed during Cordiner's decentralization years. Back then, "services" was seen, he said, "as the glue" holding the company together because it had a "right to look" at any business at GE, judge for itself how that business was doing, and report back to the top executives. It seemed like a good idea, Jones said, but flaws became apparent. "The operating people would tend to go into their clam shell and become very, very wary of the so-called 'help' from services." The beauty of strategic planning, Jones believed, was that "everybody was party to working out where they wanted to take a given business." The business managers could share their hopes and dreams. "They always felt they were able to go right to the top to have their innings and get an audience," Jones said. "Then, when we would suggest to them that certain changes in the plan were in order, they tended to accept the changes more readily. And it helped that we were wiser, older heads saying things like 'When we look at this from a company standpoint, it becomes apparent that we now have to complement this other business in its activities,' or 'We have to draw back here a little bit, because we don't quite have the funds to do that.'"

By the end of the 1970s, GE had been transformed from a company where 80 percent of the net income derived from selling electrical equipment and supplies to one where that business represented 47 percent of net income. And not because it was any less profitable but rather because GE had diversified its business into the rapidly growing commercial aircraft jet engine business and locomotives as well as into various materials, such as man-made diamonds and engineered plastics, that grew to 27 percent of GE's profits at the end of the decade, from 6 percent at the beginning of the decade.

Outside the walls of GE, Jones had also transformed Borch's idea of the Business Roundtable into a powerhouse organization, with some two hundred corporate CEOs showing up in Washington on a regular basis. "We recognized that labor spoke with a monolithic voice in Washington and had enormous clout," he said. "We recognized that special-interest groups were growing up across the nation, and that they had tremendous impact on our legislators. They were saying, quite frankly, 'We hold you accountable for our special interests—if you're going to get our vote, you're going to mind our special interest.' Whether it was consumer groups, environmentalists, or other factions, they were there—in force—and they were heard. Business, on the other hand, was so often after the fact, not being causative in its approach."

Jones liked to expound on the topic of GE, its origins, and its future strategic direction. "From its founding," Jones liked to say, "the

General Electric Company was built on what we call 'The Benign Cycle of Power.'... We created all of these products that could produce electricity, the turbines and the generators. Then we had to produce all the transmission equipment to get that power out of the station, all the distribution equipment to get it to the homes of the people in the nation and the industries and so forth in the country and then we had to produce all the equipment that was going to use that electricity and the more we produced in appliances and motors, the more we had the opportunity to produce more turbines and this was the so-called 'Benign Cycle of Power.'" (This was Jones's subtle refinement of what his predecessor, Ralph Cordiner, called "the benign circle of electric power.")

Oftentimes, Jones continued, the Benign Cycle of Power produced other business opportunities, especially in technology, that GE opted to pursue even though they didn't necessarily fit into the Benign Cycle. For instance, GE needed electric insulators, which took the company "haltingly," Jones said, into the chemical business in order to develop the insulators it needed. When the U.S. Air Force needed superchargers for its planes, it turned to GE to adapt the turbines it manufactured. "They said to us, 'Well, you know, a supercharger is so close to a turbine, why don't you build them for us?'" This turn of events, Jones said, led directly to GE getting into the jet engine business. "There was a case," he continued, "of our technology taking us into an entirely new business not related directly to the Benign Cycle of Power, although, we're now taking those aircraft engine turbines and using them to produce electricity." He explained how in the old days, the company's strategic planning revolved around exploiting the Benign Cycle of Power. But with Jones's endorsement, GE was starting to exploit the new technologies that it was developing as businesses in their own right. "It's a very conscious change from the General Electric Company that we had for so many years to the General Electric Company that we envision for the future," he said.

Plastics was a big part of that future. And Plastics, of course, was

Jack Welch, as ambitious a man as there ever was. He wrote on his annual self-evaluation form, as early as 1973, that he wanted to be the CEO of GE. He wrote it without irony. He had been getting promoted rapidly and he had not seen the negative HR memo (which he said would have caused him a meltdown had he seen it). "I was blunt and candid and, some thought, rude," he once wrote. "My language could be coarse and impolitic. I didn't like sitting and listening to canned presentations or reading reports, preferring one-on-one conversations where I expected managers to know their businesses and to have answers." Whatever he was doing worked. In his evaluation of Jack from 1973, Gutoff, his rabbi, noted that Jack's ambition was to be "Chief Executive Officer of General Electric Company" and noted that technically he was "very strong" and had "entrepreneurial drive." He was also "bright," was "quick," and "sets high standards for himself and demands same of people who work for him." Gutoff put Jack's 1972 performance as "outstanding," and the list of his stellar attributes nearly ran off the page: "extremely imaginative and creative," likes to operate "outside the dots," while also being "maze bright." He was also very "profit & results oriented" and "disdains diversions that detract from business accomplishments." He had "great marketing skills" and was a "strong leader." Gutoff once compared Jack to basketball superstars. "It's sort of the Michael Jordan kind of thing, or the Dr. J thing," he said. "When they do moves around the basket that are unscripted, they just sort of come out of their genius."

*Chapter Seven*

# REG AND JACK

Reg and Jack couldn't have been more different, as different as "chalk and cheese," *Forbes* once observed. Jones was the ultimate buttoned-down, chain-smoking, patrician corporate CEO, despite having grown up in Trenton. "He was regal," explained one former GE executive. "Jones had just an aura about him. I remember being in a room and when he walked in, it was like the king walked in." Where Jones was reserved, Jack was gregarious. Jones was tall—six foot four—while Jack was short—five foot eight on a good day. "Reg was a formal man," Jack told me. "He wore a tie and jacket on Saturdays." For Jack to move up, he would need to figure out a way to bridge the divide between their two very different personalities.

Around GE, going to see Reg Jones was like going to see the president in the Oval Office. Going to see Jack was like going to see a fraternity brother at a tailgate party. Jones also had the spectacular misfortune of being GE's CEO at a particularly difficult time in America: the cultural revolution that started in the mid-1960s called into question the basic tenets of American capitalism, and global events—the Yom Kippur War, oil embargoes, the stock market collapse—didn't help. Some of this rubbed off on Jones, who seemed unable to make the same dynamic moves as the CEO who had gotten him the top job. It appeared to others, including Jack, that Jones's interests lay elsewhere. It wasn't that he didn't care about GE; it's just that he cared more about policy, and influencing policy in Washington—tax policy, depreciation rules, capital formation—than he cared about the operations of GE. He used his platform as CEO of GE to get the policies he wanted in Washington. "He was the most admired CEO in the day," Jack said. "In those days, they became statesmen."

Jack didn't care about Washington, or about policy. Jack cared about operational and financial results: hitting his numbers, streamlining operations, exploiting new opportunities, being the toughest competitor. Jack was all about flouting convention, breaking the rules, delivering results, and having fun. It was at an otherwise boring and predictable budget and planning session—the inevitable hockey-stick projections for both revenue and earnings—early on in Jones's tenure at the helm, that Jack cut through the chaff. He was a pro at distinguishing himself with these kinds of presentations, having salvaged his GE career from the outset by appeasing the corporate suits on Lexington Avenue. His presentation to Jones about the expected performance of the plastics business was laid out in fine detail by product line on a quarter-to-quarter basis. He juxtaposed his supply forecasts with those for consumer demand, detailing the new products Plastics had planned and what markets would be excited about them. It was a corporate, bureaucratic tour de force. Jones was wowed. "This is really something," he thought to himself as he read Jack's presentation before the actual meeting. Jones was even more impressed by Jack during the meeting itself, as he answered question after question thoughtfully and with aplomb, as if it were his PhD orals. After the meeting, Jones pulled aside Jack's boss, Herman Weiss, a GE vice-chairman. By then, Weiss had been at GE more than thirty years, having worked his way up from being a craftsman in the lamp division in Pitney, Ohio, to become one of three vice-chairmen, in charge of corporate administration and business development. Jack and Herm got along great. "Herm Weiss was a guy's guy," Jack told me. "All-American football player. He and I were brothers, simpatico." After the meeting, Jones told Weiss, "Herm, this Welch fellow is someone we've got to be bringing along." Jones thought Jack should come to corporate headquarters—then still at 570 Lexington Avenue in New York City but soon to move to a new campus setting in Fairfield, Connecticut—and see "some more of GE than what's up there in Pittsfield."

At about that time, in June 1973, Reuben Gutoff, one of Jack's early mentors and the man who kept him at GE when he had wanted to quit early on, got promoted to be head of corporate development at GE. Jack took his job as a group executive, expanding his purview beyond plastics and metallurgy, in Pittsfield, to include medical systems, based in Milwaukee; appliance components, in Fort Wayne, Indiana; and electronic components, in Syracuse. It was a big deal, especially since he was only thirty-six years old. He would be managing 46,000 people and a portfolio of businesses that generated more than $2 billion in annual revenue. One negative: the new job required Jack to move to corporate headquarters in New York City, just as Jones wanted him to do. In an effort to play the team game, Jack went down to Lexington Avenue and then selected an office, from the architectural drawings, at the new Fairfield headquarters—carefully counting the ceiling tiles to make sure that he got a big one. He picked out his office furniture.

But that was a big head fake. Jack wanted the promotion but did not want to move to Fairfield. He did not want anything to do with Lexington Avenue. "I hated corporate headquarters," he told me. "A bunch of stiffs." He met with Weiss at 570 Lex and pleaded his case. By then, Jack had his fiefdom and his family—four young children and his wife—and a well-established hatred of the corporate bureaucracy. He told Weiss he didn't want to move. He was happy in Pittsfield, as was his family. He also hated the idea of moving to headquarters and being one of twenty guys milling around Jones and the three vice-chairmen. "I was free as a bird up there," he continued. "I had leased a plane. I was operating out of a hotel. I was outside of the mainstream, and I wanted to stay there. I was happy as hell." He wrote about being "scared stiff" to discuss this measure of corporate disobedience with a vice-chairman. But he was determined to stay put. He told Weiss he would be spending much of his time on a plane anyway, so it didn't really matter in which city he started his day. He also promised never to miss, or be late to, a meeting of the divisional heads in Fairfield, even if it meant getting up at 5:00 a.m. and driving the 110 miles between Pittsfield and Fairfield.

Once again, Jack had worked his magic. Weiss agreed to both give him the promotion and let him stay in Pittsfield. "I practically jumped up and kissed him," Jack wrote. "I got out of his office in a hurry, before he could change his mind or report his decision to Reg, who I was sure wanted me at headquarters." Jones was certainly befuddled by Jack. "He's a big fish in a small pond up there," Jones thought. "But he's going to have to make up his mind."

At first, Jack wondered if perhaps he had made a mistake by turning down the Fairfield move. But Jack was not much for second thoughts once a decision had been made. Instead, he basically set up his own corporate headquarters of sorts in downtown Pittsfield, in offices on the second floor of the Berkshire Hilton, and had a staff of five people working for him. He regularly chartered a Citation jet to take him around the country. "Over the next five years," he said, "I kept my promise to Herm to never miss a meeting." He kept on delivering the numbers for Fairfield, whether he was actually there in person or not. And when you get right down to it, in corporate America that's the bottom line.

He had also learned how to serve up plenty of corporate pabulum, another job requirement, even though he was not one to toe the line if he could avoid it. "I learned to support high-growth businesses like medical and plastics and how to squeeze everything out of slow-growth operations," he wrote about those years. "It was a great set of experiences." Jack endorsed the development of what became known as CAT—computer-assisted tomography—scanners that enabled doctors to detect and then to diagnose internal injuries. The machines were expensive—about $1 million each—but soon GE was flooded with orders for them, yet another part of Jack's success story.

During the summer of 1976, Jones gave Jack a particularly tough— and particularly unwanted—assignment: the point man on trying to reach an agreement with the state of New York on how to handle the fact that for years GE had been dumping PCBs—polychlorinated biphenyls—into the Hudson River in upstate New York. For some thirty years, ending in 1977, GE used PCBs in its capacitor manufacturing plants in both Hudson Falls and Fort Edward, New York. During this time, according to a December 2018 New York State report, the two GE plants "discharged" PCB oils both directly and indirectly into the Hudson. Since World War II, PCBs had been used in lubricants, paper coatings, hydraulic fluids, plasticizers, paints, inks, and adhesives. Because of their inertness, fire resistance, and ability to hold an electrical charge, PCBs were thought to be well suited to the manufacture of capacitors and generators.

Monsanto, the big chemical company (which is now part of Bayer), had started manufacturing and selling PCBs in the 1930s, without causing much, if any, concern. "They didn't smell, lie in slicks atop the water or turn fish belly up," Peter Hellman wrote in *The New York Times Magazine* about PCBs. "They simply sink to the bottom." He also pointed out how "extraordinarily useful" the chemical was to industry. Questions about the consequences of PCB use started to arise in 1964 when a Swedish scientist found the chemical in eagles and in plankton—the top of the food chain to the bottom of the food chain— even though the chemical was not made in Sweden. He found the chemical could attach itself to airborne dust particles, raindrops, and snowflakes and could travel long distances. His discovery seemed little more than a scientific curiosity.

But then things got more serious in 1968 at a rice factory in western Japan. A machine using PCBs sprang a leak and contaminated rice oil that was used for cooking. Thousands of people became ill, with swollen eyelids, darkening skin pigmentation, and painful joints, among other ailments. More disturbing, pregnant mothers who had been exposed to PCBs in passing gave birth to children with dark skin pigmentation, slower tooth development, and below-average weight. Years later, the incidence of liver cancer in the region was as high as fifteen times the normal rate. As the use of the chemical became more widespread, other physical anomalies began to show up: Minks failed to reproduce after eating fish in the Great Lakes that contained PCBs. Rats fed PCBs developed tumors. Monkeys fed the chemical developed physical problems similar to what the Japanese had experienced. In the face of these developments Monsanto began to cut back on the manufacture of the chemicals and to limit how they were used.

GE continued to be Monsanto's biggest customer for PCBs, buying around 4,800 tons a year in 1970, most of which was used at the two capacitor plants two hundred miles up the Hudson. With the permission of both federal and state authorities, GE continued to discharge thirty pounds of PCBs a day in the Hudson. Some New York State environmental scientists began sounding the alarm, according to *The New York Times*, but too often their superiors were indifferent, afraid of offending GE and of potentially losing 1,200 upstate jobs.

The effect of GE's chemical discharge into the Hudson was plenty controversial. The *Times* reported the river had about 500,000 pounds of PCBs, which bind themselves to particulates in the water and also to sediment on the river floor. The state began an enforcement action against GE in the early 1970s, severely limiting PCB discharges into the Hudson and requiring that a water treatment facility be built near the Fort Edward plant. In 1973, a dam near the Fort Edward plant that contained some of the chemical was removed. "Because of this removal," according to the 2018 state report, "PCB-laden sediments once captured behind the dam were remobilized and deposited downstream." In August 1974, the EPA decided to test the Hudson for PCBs, just below the two GE plants. The EPA scientists found record amounts of PCBs wherever they looked; it hadn't been a secret, of course, that GE was dumping the chemical in the river. Their report sat around for a year without much notice until an EPA administrator in Duluth, Minnesota, read it and sent a message of concern to another EPA administrator in New York. That administrator sent it on to the new head of the state's Department of Environmental Conservation, Ogden Reid, who wondered why he was just hearing about the problem for the first time, and from an EPA administrator in Minnesota, no less. Reid discovered some of his scientists knew about the PCB discharges; others chose not to do much about them.

Reid warned the public not to eat fish from the Hudson and closed the river to most forms of commercial fishing. He ordered a hearing for November 1975. "A parade of state witnesses testified that PCBs are toxic to a wide range of living creatures, including man," Peter Hellman reported in *The New York Times Magazine*. It was also made clear that the two GE plants were the culprits—again no surprise—for contaminating the Hudson with the PCBs. In its defense at the hearing, GE argued that it had federal and state permission to dump the PCBs into the river. In a 1976 front-page article, the *Times* was unequivocal about the harmful effects of the chemical. "PCB's have caused cancer in laboratory animals and a variety of illnesses among General Electric workers who have been in contact with the substance for a long period of time," it claimed. "The toxic chemical was discovered recently in the milk of more than 30 women living in Michigan. In Japan, more than 1,000 people became ill from PCBs that had leaked into rice oil used for cooking."

In Hellman's longer piece a few months later, he introduced readers to Luke Hart, the plant manager of the two GE factories in Hudson Falls and Fort Edward where the PCBs were once heavily used. He held up a vial filled with "brilliantly" clear, viscous fluid. "This is the stuff that's caused the fuss," he said. "This is PCBs." Hellman described PCBs as "a nearly indestructible and highly dangerous group of industrial compounds that for many years seemed to be responsible for nothing but good works." GE used PCBs to make its capacitors run more efficiently. But according to Hellman, "the effect of PCBs is, at worst, fatal and, at best, unknown."

Jack had a very different view about the toxicity of PCBs, and even years later disagreement about the health effects of the chemical could still get him plenty agitated. "I fought those to the death," he told me about the PCB battles with the state and environmentalists. He said he hired a doctor from Mount Sinai Hospital in New York to run tests on all of GE's employees who worked in the two plants, and "none of them had cancer, none of them." For good measure, he added, "I knew as a chemist, they're not dangerous. It's bullshit. . . . I have a PhD in chemistry, I know chemistry, and they're wrong and I wasn't going to cave." As he did with me, he gave an interviewer, in 1997, a full dose of his controversial views. "There's no question that nobody who polluted in the '50s and '60s knew a thing about the fact that they were polluting," he said. "I was a Ph.D. chemical engineer from a great school. I came to Pittsfield, Massachusetts, and bathed in Phenol in those days. We never had one course from '57 to '60 dealing with the environment. It wasn't part of the chemicals we did. We worked in it, we bathed in it. We did things. I can see trucks driving out of my plant now with waste chemicals that I sold to a waste company. I don't know where the hell they went. Now we say how bad they were, those polluters. It was not mean."

In September 1976, Jack cut the deal of a lifetime with the state of New York. In a negotiated settlement with the state Department of

Environmental Conservation, GE agreed to pay $3 million "toward cleansing" the Hudson of PCBs, the *Times* reported on its front page, and also agreed to pay another $1 million to research the effects of PCBs. The state also agreed to spend $3 million for the cleanup. The next day, deep inside the paper, the *Times* ran a picture of Welch signing the agreement with Peter Berle, the state's new environmental commissioner. By then, Reid had resigned, caught between a desire to solve the problem and the state's political leadership not wanting to piss off GE. The *Times* article reiterated Jack's argument that while GE had been dumping the PCBs into the Hudson for about twenty-five years, the company had "requested and obtained" state and federal permits to do so legally. The agreement called for GE and the state to work together "in attempts" to cleanse a fifty-mile stretch of the Hudson from Fort Edward to Albany. The Columbia University law professor who conducted eleven days of testimony and hearings about the effects of the chemical in the Hudson declared the settlement "an effective precedent for dealing with situations of joint culpability." Reg Jones told *The New York Times* about Jack after he cut the deal with the state of New York, "Mr. Welch grew up in our materials business where the problems of clean air and clean water are ever present. He addressed them first." The agreement Jack negotiated was another notch in his belt and seemed too good to be true. And it was.

While Jack focused on getting an agreement with the state of New York about cleaning up the Hudson, Jones was busy putting the finishing touches on the largest acquisition in corporate history—GE's December 1976 $2.2 billion purchase of Utah International, a major international mining company with interests in copper, coking coal, uranium, and iron ore, among other minerals, including gold, silver, oil, natural gas, and rhenium, a member of the platinum family used in no-lead gasoline. Utah International, which started in 1900 as a construction company that helped to build the nation's railroads, had been transformed into a mining company by its CEO, Ed Littlefield.

Utah was immensely profitable, having increased its earnings four-teenfold between 1966 and 1976, when its revenue was $1 billion and its net income was $181 million. Jones said the acquisition, though large, would be immediately accretive to GE's earnings. (Jack hated the deal, although he didn't object to it publicly. Nor had Jones asked him for his views.)

Jones saw the acquisition as an important way for GE to accom-plish three strategic goals: to have an important hedge against infla-tion, which was then spiraling nearly out of control, with interest rates hovering around 15 percent; to diversify internationally; and to diver-sify industrially. He told financial analysts that the Utah deal gave GE its first entrée into natural resources. "This is an area of enormous po-tential for future earnings growth," he said, "because the world is indus-trializing and demand for fuel and raw material is increasing rapidly." Utah provided GE shareholders with a "valuable" hedge against infla-tion, he said, because there was a finite supply of these materials "and so their value keeps rising over the long term, offsetting the effects of inflation," which he described as "one of the most pervasive problems facing business today."

In many ways, the deal was an inside job. Littlefield had been on the GE board for twelve years prior to the acquisition. And there were plenty of other directors who sat on the boards of both GE and Utah. Discussions about an acquisition of Utah by GE began in the late spring of 1975 when Littlefield was in New York City for a GE board meeting. Ever since GE moved to Fairfield, Jones had made a habit of coming to New York the night before board meetings and to have dinner with whichever directors were in town. Over dinner at the Waldorf Astoria Hotel, around the corner from the GE building on Lexington Avenue, Littlefield told Jones he had been giving thought to how he could di-versify Utah for its shareholders. "He recognized that their very suc-cess was creating a real problem for them in that this very substantial growth in earnings came with a very substantial risk and that was there was no diversification essentially in that company," he said. One way for Littlefield to diversify Utah for its shareholders was to merge with a company that was already diversified and that could afford the $2 billion price tag. That meant, according to Jones, "the first few companies on Fortune's 500 list," namely "auto manufacturers, oil companies or the General Electric Company."

Jones was immediately interested. For the first few months, the conversations between Jones and Littlefield were private chats, just to think through whether the acquisition was "a truly feasible proposition" or not. At the end of the summer of 1975, the two men brought in "our top financial man" and did a deep dive into the numbers. "The overall scope ... of this merger was so large that it took us a few more months to digest all the relevant statistics and to gain familiarity that would enable us to reach a reasonably informed opinion," Jones said. That fall, Jones went to Queensland, Australia, to look at Utah's coking coal mines. The real purpose of Jones's visit was kept secret, of course, so the cover story was that he was in Australia to visit GE's Equipment Motors Control. By coincidence, Jones arrived in Australia on the same November 1975 day that the Labor government lost power. At that moment, the U.S. had no ambassador. Soon after Jones landed in Sydney, there was a lunch in his honor, to which the press was invited—he was the CEO of GE, after all—and the rumor circulated that Jones was in Australia to check out the place prior to becoming the new U.S. ambassador. The local press ran the ambassador story. "Well," Jones said, "it turned out to be a perfect cover. We visited the mines and left Australia without comment."

On December 15, working in tandem, both Jones and Littlefield presented the acquisition idea to their respective boards of directors. The GE board met in New York; the Utah board met in San Francisco. "They were most enthusiastic," Jones said, and authorized him and his team to dig deeper into Utah's books and records and to negotiate a merger agreement. A year later, after months of negotiation, digging, and antitrust reviews (GE agreed to put Utah's uranium-mining business into a separate corporation controlled by independent trustees, to allay concerns that GE might have some "unfair advantage" in the market by packaging together the sale of uranium with its nuclear power plants), GE held a shareholder meeting at the Shakespeare Theater in Stratford, Connecticut, to vote on the proposed merger. It was approved overwhelmingly. Jones noted that some people likened the protracted merger process to *The Perils of Pauline*. But he preferred to think of it as *A Midsummer Night's Dream* that "went on to become a 'Winter's Tale' and there were moments of 'Tempest,'" and when the "Justice Department first responded and turned us down it looked as though it was 'Love's Labor Lost,'" but "finally," on December 15, it was *All's Well That Ends Well*. The largest acquisition in American corporate history closed five days later.

Jones used the occasion of the acquisition of Utah to promote his vision for what he called "the New GE." In a presentation to security analysts at the Pierre Hotel on December 16, 1976, he gave a twelve-page presentation on the subject that concluded with the thoughts that the company had been transformed in the previous decade into one "less dependent on traditional business lines"; one "more widely diversified" into "fast-growing" services, materials and new equipment businesses; one that was more international in scope; one less susceptible to inflation and business cycles; one with "greater containment" and "diversification" of risks and that was also more productive and efficient and better able to finance future growth, thanks to its AAA credit rating and a net debt-to-equity ratio "in the middle teens"; and, no surprise, one with a management team that was the "most sophisticated" in the company's history. "It's a new General Electric," he concluded, "but its financial objective remains unchanged: high and sustained earnings growth."

Five months later, talking to a reporter from *Fortune*, Jones cited Harvard sociologist Daniel Bell's writing about the "post-industrial society" and the growth of service businesses and noted how that had become GE's credo, too. "Product services, repair services, financial services," he said. "What we found is the earnings of these businesses in the last few years have been growing at a compounded annual growth rate of 17 percent. Considerably faster, in other words, than the overall company results. . . . We have decided that over these last few years that this is an opportunity that we've got to be sure we catch and catch at the right time." He also couldn't resist sharing what he told his top four hundred managers every year at the three-day off-site at the Belleview Biltmore in Belleair, Florida: "I just remind these fellows . . . you're not only stewards for all these tremendous physical resources of this company, but you're stewards for its human resources. You're really stewards for the economic lives of your employees and the economic life of an employee is a big chunk of his total life. It conditions, really, his total life. Don't ever forget it."

Toward the end of 1977, Reg Jones called Jack to have the Talk. Jones wanted to see Jack in Fairfield, urgently. It was the call Jack was both hoping to get and dreading to receive. Jack was in Fairfield the next morning. "You're obviously very ambitious, Jack, and you want to move up in the company," Jack recalled Reg telling him. "You've done very well, but with a small slice of this business, and as you know we are diverse way beyond just the Plastics and Materials part with which you've been associated." That was the carrot. "If you aspire to a higher level, you've got to come to Fairfield to work at some other businesses of this company," Jones continued. "You have to understand the breadth and diversification of General Electric."

Jack's recollection of the meeting with Jones was similar but included the fact that the CEO offered him the job as head of GE's consumer products businesses, making Jack one of six direct reports to him. "The gig is up. You have to come to Fairfield or else I can't pick you," Jack told me that Jones told him. "You're done. . . . You can't be a big fish in a small pond anymore. If you want to be considered for bigger things, you're going to have to come here." Jack packed his bags. Reg Jones knew what Jack could do with a growth business; now he wanted to see what he could do with consumer products, a business that was going nowhere.

The six Welches moved to New Canaan, Connecticut. Two of the children were in high school and two were younger still. Before leaving Pittsfield with the family's packed Buick station wagon, Jack stopped off at the office of a local real estate agent and completed the purchase of five acres of land on a mountain in the Berkshires. "For some reason, it made me feel better," he said. His $300,000-a-year salary bought him a much smaller house in Connecticut than it had in Pittsfield. He was disappointed. "I downsized," he said. And he missed his Pittsfield posse. "I was sucking my thumb like I did the first four months of college," he told me.

Suddenly Jack found himself in a very unfamiliar place: in a vicious competition with five other men, all of whom hoped to succeed Jones. There was Robert "Bob" Frederick, then head of corporate planning and development; Stanley Gault, then head of the industrial products division; Tom Vanderslice, a former Fulbright Scholar who ran the power systems division; Ed Hood, who had run GE's jet engine business and was then head of technical products and services; and John Burlingame, who was head of GE's computer division and then head of employee relations before heading up the international division. All six men had been lifelong GE employees and had worked their way to the top of GE's byzantine corporate structure—twenty-nine levels between the bottom and Reg Jones at the top. "I was excited by the possibilities," Jack said, "but apprehensive about whether the Pittsfield act would play in the Fairfield bureaucracy."

For a competitive guy, Jack said he hated the daily grind of the competition. It was essentially a three-year bake-off. "I was in a parking lot with a bunch of other guys I wasn't particularly fond of," he said. They had to have lunch with Jones, or with each other, nearly every day they were in town. He said he felt like he was always "walking on eggshells" in the corridors of the new headquarters. Worse, he told me, "guys were kissing ass" and "we were looking at each other." He also felt alone, especially since his first rabbi, Gutoff, had left the company in 1975, and his second rabbi, Herm Weiss, had died from lung cancer the next year. He spent the first four months living out of a suitcase at the Stamford Marriott while his house was being renovated. He was a long way from the Berkshire Hilton.

In the patented GE way, Jack's new job was very different from the one he had spent years perfecting in Pittsfield. This, of course, was part of the competition that Jones had architected. Who among these various GE lifers could learn about a new GE business and run it successfully? That was the test that Jack found himself confronting. His assignment was to run GE's consumer products business, a division of around $4.2 billion in revenue—20 percent of the company's overall total—that comprised GE's best-known products: major appliances, air conditioners, lights and lightbulbs, small housewares and audio products, as well as televisions, and radio and TV stations.

Jack had taken over the job from Stanley Gault, with whom he'd had a long-standing feud after Gault's refusal to use Noryl as a housing for GE air conditioners. But the betting seemed to be that Jack was put in the pole position, since Wilson, Cordiner, and Borch had all run the consumer products division before getting the top job. The division had 45,000 employees and was not making any money. To make matters worse for Jack, or so he thought, one of his immediate bosses, a vice-chairman named Walter Dance, favored his protégé, Gault, for Jones's job. And he made no secret of it. The other GE vice-chairman, Jack Parker, favored *his* two direct reports, Hood and Burlingame. "It was the first time in my seventeen years at GE that I had a boss who was not rooting for me," Jack said. And then, sounding like any other employee in corporate America, he continued, "There is probably nothing worse in business than to work for a boss who doesn't want you to win. This can happen anywhere, at any level—and probably happens more often than we think. Until I came to work for Dance, I had never had it happen to me."

He resolved to try to ignore the internal politics—as nearly impossible as it was—and to focus on turning around the consumer products businesses. He fired 25,000 people, cutting the workforce by more than half, including firing the father of one of his son's classmates at high school. He made the business profitable—with plenty of collateral damage. There was the time, on a visit to GE's lightbulb headquarters in Cleveland, that Jack reportedly shoved a lightbulb in the face of the top man and asked, "They can make these things in Hungary for half of what we charge! Any idea why?" He then sent the Hungarian bulb to be analyzed in a GE lab and discovered that the two bulbs were identical, except for the fact that the Hungarian bulb cost much less. That bit of news, plus the fact that the Cleveland headquarters had a full-time barber, a dentist, and a full-size swimming pool for executives, made Jack nuts. "It was all the ammo Welch needed: Bye-bye department head," Byron wrote.

Jack also had an aversion to fat people. Once, when visiting the small-appliance operation in Bridgeport, up the road from Fairfield, he asked the plant manager how much he weighed, causing an assistant to run off and bring back one of GE's new digital scales. Jack stepped on and off the scale a few times, only to discover that his recorded weight was different each time. At first he was unhappy that the plant manager was overweight. Then he was unhappy because the scale didn't work. "What the fuck do I pay you for?" was one of Jack's favorite expressions. But Jack didn't care what people thought of him, or said he didn't. "I was just tearing it up," he told me. "I got in there and there were sleepy businesses. I fired a lot of guys. I didn't give a fuck. . . . I wanted to make the company great."

A hidden jewel in Jack's new portfolio—one that much of the GE brass couldn't care less about—was General Electric Credit Corporation, a "popcorn stand" as Jack described it when he took it over in 1978, that had been started in 1932, at the height of the Great Depression, as GE Contract Corp. At a time when Americans could barely put food on the table, they were in no mood to buy new GE refrigerators or GE stoves. GE tried to solve that problem the old-fashioned, American way: by offering them a way to get the new appliance now while allowing them to pay for it later. GE Credit lent consumers the money to buy the new refrigerator, charged them interest, and let them pay it back over time. Reg Jones was pleased with GE Credit but only in a utilitarian way, as a method to sell more GE products that people couldn't otherwise afford.

Jack, on the other hand, *loved* GE Credit. He was determined to grow it. "I always commented on how easy it was," he told me. "Try having a union, try bending metal, try thousand-degree-temperature-firing jet engines." He also figured out that he could leverage GE's AAA credit rating as a way for GE to borrow money very cheaply in the unsecured short-term commercial-paper markets—the benefit of its stellar credit rating—and then lend the money out to consumers, and others, at much higher rates, pocketing the difference as profit. ("Commercial paper" refers to short-term IOUs issued by companies with superior credit, allowing them to borrow money inexpensively and without providing collateral. It's a market that rewards big and powerful corporations with access to cheap money.) GE was a huge issuer of commercial paper. In short order, Jack turned GE Credit into a moneymaking machine. It was pure financial alchemy. "What I saw in 1978 was an immense opportunity—not just the benefit you get on a balance sheet, but the additional leverage you get by putting together two raw materials: money and brains," he wrote.

In September 1978, *Forbes* did a feature on Jack, the future prospects for the businesses he was running, and his own prospects for succeeding Jones. "At 42, John Welch runs a $4-billion business—and he has a clear shot at a much bigger job," the magazine reported. He was still "John" then in the business press. He was described as "intense," "positive-thinking," and the man who had successfully turned around GE's seemingly moribund consumer products division—largely by firing half the workforce—into a business that on its own would have been the thirtieth-most-profitable industrial company in the country and that accounted for 30 percent of GE's profits. He was portrayed as the man to beat but also the man whose business lines were most vulnerable to rising interest rates, inflation, and a slowdown in the economy and especially in housing starts.

If Jack was the least bit worried, he made sure to keep it to himself. He budgeted for an increase in earnings in his division for 1979, despite the looming economic woes, and said he thought it was "realistic." He was especially confident about the lighting division and its array of new products—a more efficient fluorescent lamp, a new generation of industrial lamps, and a new camera flashbulb. He thought that GE Credit's "relatively new" two insurance businesses, Puritan Life and Puritan Casualty, could earn more than $10 million in profit in 1978. "GECC is even promoting home loans and second mortgages," *Forbes* reported. Jack also had the reins of GE's "profitable and steadily growing" broadcasting businesses, including eight radio stations, three TV stations, and a dozen small cable companies. He wanted to build the business, especially cable, although the "competition for properties is intense," he said. Other positives for Jack's succession prospects included a fast-growing heat-pump business—already a $500 million–a–year business expected to double in three years and in which GE had a 40 percent market share—GE's microwave oven business, and its food-processor business. (He was worried about GE's money-losing color TV business and was trying to get a joint venture approved with Hitachi, in Japan.) "Batter Up!" *Forbes* declared.

To try to whittle down the field of his possible successors, Jones engaged them all in what he called "the airplane interviews," in which he asked the rather asinine questions of whom each of the candidates would choose to be the next CEO of GE if he were killed in an airplane crash, then, if he somehow survived the crash, whom he would choose to work with as a GE vice-chairman. Borch had done the same thing to Jones, who was returning the favor in an expanded way. "I sat down with each of the seven or eight candidates," Jones said. "They didn't know the purpose of the meeting, and I made sure they didn't tell the others, so that everybody came in surprised. You call a fellow in, close the door, get out your pipe, and try to get him relaxed."

Jack's turn with Jones came in January 1979. He hated the questions (and the process) and instinctively looked for a way to emerge from the hypothetical wreckage unscathed. Jones wouldn't let him. You're dead, he kept telling Jack. Finally, Jack allowed that he would pick Ed Hood to run GE—"He's thoughtful and smart," Jack said—with Tom Vanderslice, from power systems, to be his number two. "Tom's decisive and tough as nails," he continued. Much to Jack's annoyance, the interviews went on for months, with executives Jones was considering seriously and with those he wasn't. After the first round of "airplane interviews," nobody picked Jack to lead GE. Most picked Gault; several picked Hood.

In June 1979, Jack was back in the hypothetical plane with Jones. But this time, Jack managed to survive but Jones didn't. Whom would Jack choose as GE's leader in this unlikely scenario? "That's better," Jack replied. "I'm the guy." Asked who his team would be, he picked Hood and Burlingame, whom he said he respected because he was comfortable in his own skin and was highly intelligent and technically proficient. In this second round of interviews—the one where the prospective CEO candidate walked away unscathed—each man picked himself as CEO and then—as to whom they would choose as the number two—Jack emerged among the top vote getters, one fewer than Gault and tied with Hood.

During the interviews, Jones was stoic and inscrutable. He conducted four "airplane interviews" of each candidate. "He never gave us a hint if we were in good shape or not," Jack remembered. "At times, he could appear remote and hard to connect with." According to *Passing the Baton*, a 1987 book about CEO succession by Richard Vancil, a Harvard Business School professor, "by the end of the third round it was apparent that [Jones] was asking them to define feasible teams. So, the contenders began talking to each other, and some decided they would rather be a vice-chairman on a winning team than simply another loser. The effect was a severe politicization of the organization. By mid-summer 1979 the lid was ready to blow." While Jack's gut feeling was that he was a serious contender, his head wasn't so sure. He and Jones were so different, after all—the gritty Irish kid from Massachusetts and the patrician-seeming Brit.

But the more Jack thought about it, the more he came to believe he and Jones weren't all that different really. Jones may have been portrayed as a "courtly statesmen" and as a presidential adviser who looked like "an industrious church deacon." But Jack came to understand that Jones, like him, was not from a privileged background. He was a self-made man, Jack said, "from working-class roots who worked and fought hard for what he achieved." Over time, Jack came to realize that he and Jones were far more alike than anyone understood; this insight gave him an increasing amount of comfort as the succession race unfolded. Despite opposition to him and his candidacy from Jones's vice-chairmen, Jack came to believe, as he spent more and more time in Fairfield, that Jones preferred him to the others.

That intuition became more apparent over time. Jack took some comfort from the fact that he was prominently featured in a major April 1978 *New York Times* article about how Jones and GE were rewriting the rules about corporate planning and strategy. But for Jack, the article would have been prima facie evidence of corporate pabulum, circa late 1970s. Jack, then forty-two and described as "flamboyant," was unabashed about the importance of careful strategic planning. "The idea is to get out front and get on the trends," he told the *Times*. "Quick follow is a very successful strategy for a company like us." He spoke about being hip to such cutting-edge products as food processors and smoke alarms. The combination of the introduction of new products with GE's "reputation for quality" as well as its distribution and service networks got Jack pumped. "But add to that being first now and then," he said, "and think of the size of the hit." The more Jack thought about the top job, the more he wanted it. "Man, did I want it," he told me. "Oh, man . . . I knew I could make a better company. I knew I could get a lot of the bullshit out."

In retrospect, two decisions that Jack made while he was in Fairfield competing for Jones's job were as important as anything else in his career. The first involved his decision to pursue—and then to abandon reluctantly—an acquisition of Cox Communications' cable and broadcasting businesses, known as Cox Broadcasting Corporation. Tucked inside Jack's growing—and increasingly profitable—consumer products empire was not only the neglected (and underperforming) GE Credit business but also a small television and cable business. Although Jones didn't think much of it, GE was a pioneer in the burgeoning cable television business and owned twelve systems with some 170,000 subscribers. GE also owned three local television stations—in Schenectady, Denver, and Nashville—and five FM and three AM radio stations. Working closely with Norman Blake Jr., his business development man, Jack decided GE should buy the publicly traded Cox Broadcasting, which owned forty-eight cable television systems (with 500,000 subscribers), five TV stations, and twelve radio stations. The wealthy Cox family was concerned that the newly emerging cable business was going to hurt its TV business. The Coxes wanted to sell Cox Broadcasting in a tax-free transaction. Only a few companies at that time were big enough to do a deal: once again it was GE and several oil companies.

In the summer of 1978—during the middle of the succession battle—Jack persuaded the GE board of directors to do the Cox deal.

Jack knew it was risky to him personally, but he didn't think it was risky professionally. He and Blake were convinced that cable and local television were growth businesses despite the regulations and competition. The businesses were "on the verge of breaking out," Jack remembered. Jones and the board agreed. "I wanted cable in a big way," Jack told me. "I saw cable as a big opportunity. And it was." The companies started talking about the deal in September. "It was as close to a mutual get-together as you can have," Clifford Kirtland Jr., the president of Cox Broadcasting, told *The Washington Post*. "There was a recognition on our part that we needed to do something to relieve the cross-ownership pressure" of the Cox family's control of both Cox Broadcasting and Cox Enterprises, which owned newspapers in several of the same cities where Cox Broadcasting owned TV and radio stations—a regulatory no-no at that time.

In early October 1978, the two companies announced they were entering into negotiations to try to conclude a deal to combine these two parts of their businesses. GE was to pay for Cox with GE shares, offering 1.3 GE shares for each Cox share, or about $467 million. It was the largest media deal up to that time, making a nice pairing of sorts with the Utah International deal, then still the largest merger ever. The deal "came as a complete surprise," *The Washington Post* reported, to the research analysts who covered GE, despite the fact that GE had started RCA and had "a long history in telecommunications," the *Post* continued, "being responsible for some of the key technical developments that made radio possible." But, the paper noted, GE's interest in the industry had waned, unlike that of rival Westinghouse, "which has a strong and profitable broadcasting division." If the deal were to close, the two Atlanta sisters—Barbara Cox Anthony and Anne Cox Chambers—would have owned about 1.5 percent of GE, making them the largest individual shareholders, with stock valued at around $250 million. A December 1978 *Forbes* article praised GE's dealmaking prowess—for both the Utah and Cox deals—at a time when the Justice

Department's antitrust division was questioning the wisdom of such mergers and suggesting "bigness is bad." The magazine quoted Jones, "A growing number of economists rebut the myth that there are no advantages for economies of scale."

With what looked like an agreement with Cox—and while awaiting the Federal Communications Commission's approval of the combination—Jack decided to get a jump on the integration process. He asked Bob Wright, a young lawyer he had come to respect in the plastics business in Pittsfield, to go down to Atlanta and begin working closely with the management of Cox Broadcasting on combining the two businesses. He thought the combination of Wright's salesmanship and legal background "would be perfect" for the cable industry. And Wright wanted to go. He moved to Atlanta in January 1980.

Wright was, in many ways, a younger version of Jack, although he was far more refined. He was also an only child and was raised Catholic on Long Island. After attending the all-boys Chaminade High School in Hempstead—among his classmates were Lou Gerstner, a future CEO of IBM, and the actor Brian Dennehy—Wright was off to Holy Cross, in Worcester, Massachusetts, and then to the University of Virginia Law School. He got married in law school and his wife got pregnant. Then he was drafted into the army for the Vietnam War. He ended up in an Army Reserve unit that was assigned to civil affairs in Utica, New York, and could have been called for duty in Vietnam at any time. He had interviewed with GE when he was in law school. GE persuaded him to become a staff lawyer for the plastics division, in Pittsfield, using the logic that it was sufficiently close to Utica that if he needed to be there for the five years he was on service duty, he could make the trip quickly.

Wright stayed in Pittsfield for eighteen months. He then took a job clerking for a federal judge in New Jersey. "It was spectacular, just spectacular," he told me about his clerkship. "We had the major gangster cases that were the first RICO cases." He then practiced law in

New Jersey for a short time. But GE kept calling him. He really wanted to see the world and he thought working for GE would allow him to do that. "I was like a sailor," he said. Instead, GE said, "Go to Pittsfield again. Do you remember Jack Welch?" He said he knew of Jack but didn't really know him. "Now he's really a big shot," he was told. "You can come up and be his lawyer." He agreed. "That sounds pretty exciting," he said. He was Jack's lawyer for fourteen months. Then, seemingly out of the blue, Jack suggested that Wright quit the legal stuff and become a business executive. "I never looked back," Wright said.

In February 1979, GE and Cox announced they had finally signed "a definitive agreement" for Cox Broadcasting to merge with GE in a tax-free stock deal. There were also provisions added to the agreement that would pay Cox shareholders more money if the deal did not close by September, which was likely given the slow FCC approval process. In July, the Cox shareholders approved the deal. But as the FCC's review of the combination dragged on and the prospects for the cable business generally improved, the Cox family began to feel it had sold the business too cheaply. That idea was reinforced by the downward movement in GE's stock price at the time, which meant that—unless the deal was renegotiated—Cox shareholders were looking at a sale price closer to $400 million, rather than $470 million. In August, in the face of public concern that the Cox stations were not sufficiently diverse, GE agreed to hire more minorities and to include more "minority affairs programming" at Cox's Atlanta TV station, a move that *The New York Times* characterized as solving "a major obstacle" to closing the merger.

Jack and Norm Blake were struggling with what to do. "With the price escalating every time Norm and I met with the Cox people, I was coming to the conclusion the deal couldn't get done at any price," Jack remembered. "The Cox family had changed their mind about selling and was using the price as a way to end it. Losing this big deal in the politically charged atmosphere of a succession race was a potential disaster. With all that was riding on it—the acquisition and my own future at GE—we wanted the deal badly." Jack kept doing the analysis and agonizing with Blake. Finally, during the summer of 1979, Jack told Jones he had decided to walk away from the deal. He had concluded that whatever GE would offer the Cox family would never be enough.

Jones asked Jack to tell the board of his decision at the next meeting, scheduled for Saint Louis. "Now I not only had to bare my soul to Reg," Jack said, "I also had to face all the company's directors." He thought for sure he was going to have to "eat crow" with the board over the failure of a deal he "had been touting so hard for over a year." He said he didn't know what to expect. At the board meeting the next morning, Jack explained what had happened with the Cox deal. He still wanted to buy the company but was not going to be undisciplined with regard to price. He figured walking away from only the second big acquisition in the company's history would show him to be financially disciplined, which was about the best outcome he could hope for them to think under the circumstances. There was a bigger risk in that they might have concluded he was a poor dealmaker and, worse, that he was no longer in a position to run the company. Some board members wondered why not just keep raising the price. "Cox kept moving the goalpost," he told me. He told the board, "Continuing to chase it wasn't right for GE." He thought the meeting went well. "I was hoping the directors would overlook my inability to close the deal and appreciate that I faced a tough call," he remembered. "I had little idea what they really thought." Later that afternoon, on the golf course—where Jack was by far the best golfer, at nearly a scratch handicap—he took comfort from the body language, and the jokes, of the directors in his foursome.

But despite what appeared to be the death knell for the Cox deal, it would not go away. The two sides kept negotiating for months. Wright remembered that GE was offering around $400 million and the Cox family wanted $600 million. He thought their request was justified. "The cable thing was in every magazine and newspaper and the values were going up dramatically," he recalled. Finally, in April 1980, the two companies announced the deal was being abandoned. What had been a $467 million offer was now worth $414 million as the GE stock price continued to fall. *The New York Times* reported, without mentioning Jack at all, that GE had offered Cox another $125 million "to sweeten the deal" but Cox rejected the additional compensation. "This offer proved insufficient to satisfy Cox's demands," according to a GE press release. "We cannot justify the transaction on the terms requested by Cox." According to one report, Cox wanted $637 million for the company; GE had offered $570 million. But still the deal was not dead. In the *Times* article, GE indicated it would be open to further negotiations with Cox, despite the public pronouncements. Finally, the bidding stopped at $575 million. "Obviously, after having worked and planned for the concept for the past 18 months, my associates and I wanted Cox and wanted it badly," Jack later wrote in his June 1980 self-assessment. "I recognize there were no winners on this one. But from what I can gather, the termination in no way tarnished the reputation of the Company."

There was some collateral damage: Bob Wright, whom Jack had seconded to Atlanta to try to get the deal done and integrated, ended up staying and working for Cox. At the request of the CEO of Cox Communications, Wright became the president of Cox's cable business. "By then, I was hooked on the TV explosion," he remembered. "How could I resist?" He and his wife, Suzanne, built a big new house in Atlanta—"All of a sudden I had money," he told me—and moved his three children into private school. Cable, he said, "was the best, most exciting business opportunity I saw on the horizon, and I had already seen a few good ones during my time at GE." For the next three years, he successfully acquired cable franchises across the country, totaling around one million subscribers. He had the chance to buy 20 percent of ESPN and

49 percent of CNN—for a pittance—but both potential acquisitions made the Cox family nervous, and it ended up passing.

Still, Wright was having a blast. "The Cox people gave me enormous license," he told me. "I was able to do things that I could have never done at GE. Now I had chances to run two businesses with GE before this, so it wasn't my first rodeo. I hired people. They gave me great support. They got a little afraid when I started to win these [cable franchise] awards, because the cost is enormous. But I felt like a different person." He said he felt like an entrepreneur and was getting paid "a lot more money" and "they were quality people." He told Jack, "I'm going to stick this out. This is a chance of a lifetime, to be in a media company which is growing dramatically, in a business that's growing dramatically. I get it." Jack told him he could come back to GE whenever he wanted. He was happy for Wright. "It was a good deal for him," Jack told me. (It also turned out to be a good deal for the two Cox sisters; at the time of her death, in January 2020, at age one hundred, Anne Cox Chambers was worth $17 billion. Her sister predeceased her by thirteen years and was worth $12 billion.)

If Jack's willingness to walk away from the Cox deal at a crucial moment in the succession process revealed a dealmaking discipline rare in most business executives—who often get seduced into thinking that doing big deals enhances their wonderfulness—equally meaningful to Jack's success, especially in retrospect, was the insight he had into how GE Credit could be transformed, with a little focus and wisdom, into a major contributor to GE's bottom line. "Of all the businesses I was given as a sector executive in 1977, none seemed more promising to me than GE Credit Corp.," he observed. "Like plastics, it was well out of the mainstream, and like plastics, I sensed it was filled with growth potential." The business had been ignored. "It was the orphan child in a manufacturing company," Jack wrote. Since the Great Depression, there hadn't been much to it, aside from helping customers buy GE stoves and refrigerators as well as the home furniture that stores were also selling. Things got a little more exciting at GE Credit in the 1960s when it started financing purchases of Yellow Iron, heavy construction equipment made by Caterpillar. By the end of the decade, GE was leasing other equipment, too. A decade later, still small, GE Credit had further branched out into providing second mortgages and financing for commercial real estate, manufactured homes, and private-label credit cards. Not having much of a background in finance, Jack didn't understand the business particularly well—at least at the beginning—but his gut told him it had the potential to be big. In many ways, GE's financing businesses, especially under Jack's command, were the precursor of what Mike Milken created decades later and what became known as the high-yield or "junk bond" market. As Milken would do, GE had figured out there was lots of money to be made lending to the less well-heeled. "At GE Capital, I made heroes," Jack told me. "I loved it. To play with money was so much better than bending metal." To prove his point, Jack liked to share with people the fact that per-person profit at GE Credit in 1977 was around $9,500, while at his plastics business the same metric was closer to $2,100. Not a bad insight for a chemical engineer.

He set out to learn finance and wasn't afraid to ask dumb questions, or what must have seemed like dumb questions to executives steeped in a world filled with jargon. But when Jack got answers he didn't like or couldn't understand or that weren't clear enough, he didn't offer many second chances. He was on a mission to upgrade the quality of the leadership, as well as the rank and file, at GE Credit. John Stanger was running GE Credit when the business came into Jack's portfolio. A 1947 Harvard graduate, Stanger joined GE Credit right out of college. He was vice president for commercial and industrial finance for thirteen years, until 1975, when he took over as the CEO. He is credited with having the insight that GE could make piles of money in its finance business by arbitraging its AAA credit rating and borrowing in the commercial-paper market.

Instead of impulsively firing Stanger, Jack surrounded him with more and more talent. One of those uncut gems was Larry Bossidy, who was born in Pittsfield—of all places—and joined GE as a finance trainee after graduating from Colgate in 1957. He had worked at the family shoe store growing up and harbored dreams of becoming a major league pitcher, after a stellar college career. But his mother nixed that opportunity by barring a baseball scout from entering the family home. Jack first met Bossidy in 1968 when he showed up in Pittsfield to audit the books of the plastics division that Jack was running. They met again ten years later, in Hawaii, at a GE Credit management meeting. They found themselves playing a raucous game of late-night Ping-Pong. "As if our lives depended on it," Jack recalled. It was time to catch a plane back to New York. But Jack didn't want to leave the game. He had found a like-minded soul in Bossidy. They became lifelong friends, and Jack did much to promote Bossidy's career. "I was excited by this guy who was so full of life and so competitive," he said. He thought Bossidy was just the kind of guy he needed to elevate into the top ranks of GE Credit, at Stanger's side. But as their conversations continued, Bossidy told Jack he was going to leave GE, to take a job at Lone Star Cement. He hated the GE bureaucracy. "This place is driving me nuts," Bossidy told him. But Jack implored him to stay, in a repeat of what Gutoff had done with Jack years before. "You're just what we need," he told him. "This is going to be a different place." Bossidy stayed at GE. A year later, in 1979, with Stanger's support, Jack promoted Bossidy to chief operating officer of GE Credit. Led by Jack's enthusiasm for the business of making money from money, Jack and Bossidy, plus a new, highly talented supporting cast of characters, would transform the renamed GE Capital from a business making $67 million in 1977 with fewer than 7,000 employees to one, by 2000, making $5.2 billion in earnings with 89,000 employees. It wouldn't have happened without the impetus and vision of Jack Welch.

John Myers, who spent thirty-seven years at GE, ran its pension fund for years, and sat on the GE Capital board, explained to me the keys to the success of GE Capital: GE's AAA credit rating, allowing it to borrow money very cheaply. "Banks weren't even rated AAA at that time," he said. "We borrowed money cheaper than anyone could." GE could also use GE Capital to reduce the taxes GE would otherwise pay on earnings from its very profitable industrial businesses. Here's how that worked: If, say, an airline bought a new jet, it would have an asset that would depreciate over time, and the airline could use the depreciation to reduce its taxable income. But, at that time anyway, most airlines didn't make much money, if any, so the value of the depreciation deductions was of little use to them. But to GE, the depreciation—the tax deductions—would be very valuable as a way to reduce GE's pretax income and therefore to pay less in taxes. With that logic, GE Capital would buy the jets, lease them to the airlines at commercially attractive rates, and then use the depreciation on the jets to reduce the pretax income at GE.

GE Capital soon branched into many other lines of business—buying accounts receivable, financing leveraged buyouts, financing the purchase of manufactured homes, financing big real estate purchases, and on and on. "We put it all under an 8-to-1 leverage, and it was shown to Wall Street as just one big business," Myers said. "We never had to have the transparency of what real estate was doing, what LBO lending was doing, what accounts payable, what distressed, whatever. It was all going up."

In GE's 1978 annual report—Jack's first as a "sector executive" and published in the spring of 1979—there is a picture of a young, smiling, vibrant executive with nearly a full head of neatly combed hair, extolling the success of the consumer products and services division. He was in charge of a hodgepodge of businesses—lighting products, major appliances, air conditioners, small housewares (toaster ovens, smoke alarms, curling irons), the broadcasting businesses (then still in the midst of trying to acquire Cox), and GE Credit. The revenues of the division were up 15 percent; the earnings were up 17 percent, to $377 million. The division earnings were some 30 percent of GE's overall earnings. GE Credit contributed $77 million of net earnings, or 6 percent of the company's overall earnings. Its earnings had increased 15 percent from 1977, thanks to the decision to expand its financial services beyond GE products to the point where it had "extensive interests" in financing a wide range of commercial and industrial businesses across the country.

As he was taking real and valuable steps to improve the financial performance of his new portfolio of businesses, Jack also found himself, of course, in the middle of a very competitive race to succeed Jones. He was the outsider. But he did have one consistent accomplishment that transcended office politics. "Reg was a numbers guy and he loved my performance, and I was just tearing it up," he told me. "I got in there and there were sleepy businesses. I fired a lot of guys. I didn't give a fuck. I wasn't trying to win a contest with them. I wanted to make the company great. Reg was a straight shooter, and he had a liking for me. I knew he always liked me. I trusted him. I was the most different than him that you can imagine. I liked him, he liked me, and I felt it."

But he wasn't sure at all. The "hall gossip," according to Jack, had Al Way, the GE chief financial officer, as the favorite. Like Jones, he was a numbers guy. He also worked with him nearly every day. He advised Jones on the acquisition of Utah International, the biggest ever. And Way helped Jones merge GE's computer business with Honeywell's— also perceived initially as a smart move. Jack believed in himself and in his connection to Jones. But he still had his doubts. And with good reason. He was an iconoclast in what was predominantly a rigid corporate culture. The two vice-chairmen were against him. As was the HR department.

In late 1974, well before the "airplane interviews" with the finalists, Jones asked the "executive manpower staff" to prepare a list of potential successors. "The initial pool of 96 quickly melted down to ten," Jones said. Jack's name wasn't on it. "At that point, he was only a 37-year-old division manager," Jones explained. But Jack could never figure out why Roy Johnson, the head of HR, kept him off such lists. Maybe it had to do with comments in his personnel file: "Not on best candidate list despite past operating success. Emerging issue is over-whelming results focus. Intimidating subordinate relationships. Seeds of company stewardship concerns. Present business adversity will severely test. Watching closely." Still, he was considered worthy of "intensive development," which Jack took to mean he was still in consideration for "bigger jobs." His doubts remained, though. Always the target of headhunter calls as a senior GE executive, around this time Jack de-cided to take one, from Gerry Roche at Heidrick & Struggles: Allied Chemical Corporation needed a new CEO. Was Jack interested? He flirted with the idea. But opted not to pursue it. His "lack of self-confidence was rearing its ugly head," he wrote in his memoir.

Jones decided to retire in 1980 and created a "personal and pri-vate" road map for succession and over time brought the GE board of directors into his thinking. The board became "very deeply and inti-mately involved" in the succession process, he explained. By then, Jack was on the short list of successors, and his reading of the political landscape in his favor proved justified. On August 2, 1979, after a board outing at the Blind Brook Club, in Rye, New York, Jones told his two vice-chairmen that he had narrowed the CEO selection process down to three men: Jack, then forty-four, as well as Hood, then fifty, and Burlingame, then fifty-eight. It was just as Jack had diagrammed the play during his second "airplane interview." The remaining candi-dates could either keep their jobs or leave the company. The two exist-ing vice-chairmen would also have to retire. Jones told the two men he would be asking the board to ratify his plan the next morning. The two vice-chairmen—Parker and Dance—were against Jack still being in consideration to succeed Jones and laid out their objections at the board meeting. But, no surprise, Jones prevailed; the board vote was unanimous, despite lingering support for Stanley Gault.

After the board meeting, Jack Parker asked Jack to stop by his office. "I want you to hear this from me and not anyone else," he told Jack. "I didn't support you, and I don't think you're the right guy to run GE. I don't want you to screw this company up." Deadpanned Jack, "I admired his candor, even though I disagreed with his assessment." According to Vancil, the Harvard professor, the "compromise" of naming three vice-chairmen was done "to maintain the race for some period while the board became more comfortable with Welch as the next CEO. The rationale for the compromise was that Welch was a risky maverick." Overnight, it had essentially become a three-man race, although Burlingame was a long shot given his age and the mandatory retirement age of sixty-five.

The three men jettisoned in August quickly found gainful employment at other companies, initiating a pattern that would come to epitomize the idea that GE was a management machine and even those passed over for top jobs were great candidates to fill other corporate openings. Tom Vanderslice became president of General Telephone and Electronics Corporation (later known as GTE and now part of Verizon); Alva Way became president of American Express; and Stanley Gault became the CEO of Rubbermaid, which is now part of Newell Corporation. (After Rubbermaid, he became CEO of Goodyear.) Robert Frederick became president of RCA.

In the face of rampant inflation, Jack's division kept performing beautifully in 1979. Some people say about Jack that he was the luckiest corporate executive who ever lived. His timing always seemed to be perfect. Revenue increased 12 percent and net earnings were up 6 percent. GE Credit's profits were up 17 percent to $90 million, while continuing to provide 6 percent of GE's overall earnings. To help grow its assets, GE Credit purchased a financial services operation in Hawaii, started providing leasing on imported cars, and entered into a joint venture with Toshiba Credit Corporation to provide consumer financing in Japan. It was also the year that Jack and his marketing team created the advertising slogan "GE brings good things to life," which GE hoped would "strengthen the public consciousness" of the brand and would set "the tone" for GE's commitment to consumers.

Part of the way Jack made his numbers was by aggressively slashing costs. During a visit to Louisville, the home of GE's major-appliance business, Jack more or less terrorized a group of GE executives who showed him a new GE dishwasher. "What the fuck are you showing me this for?" he demanded to know. "I haven't been in a kitchen for ten years!" It was a rhetorical question. "I know what you're doing," he continued. "You brought that thing out here so you can say, 'See this new machine? Isn't purple a great color?' Then if the machine doesn't sell you can say that I agreed purple was a great color so you're off the hook. Well, fuck that, I'm not going to say it."

As it became clear Jones's tenure at GE was coming to an end, the paeans to him started to appear in the press. In September 1979, Isadore Barmash, writing in *The New York Times Magazine*, declared the GE CEO "America's Most Influential Jones." Barmash's piece started with Jones emerging from an "absurdly long black limousine" at the White House and being greeted by President Jimmy Carter himself, not by a presidential aide as would be the norm. It was one of his many trips to the White House to see Carter. He was also a frequent guest at Camp David, the presidential retreat in Maryland. Carter reportedly offered Jones two senior cabinet-level posts—chairman of the Federal Reserve Board and director of the Office of Management and Budget—but he turned down Carter, repeatedly, out of his obligation to stay at GE to "complete unfinished work" at the company—namely succession.

Jones would go on to be remembered as a paragon of the patrician, understated CEO. His routine as GE's CEO was to leave his relatively modest colonial home in Greenwich, arrive at his corner office in Fairfield at 7:15 a.m., "with a collar that sometimes seems a bit loose," the

*Times* reported, and then start getting antsy fifteen minutes later. Stories about Jones's modesty still make the rounds occasionally. In a May 2020 article in *The New Yorker* about the changing political mores in Greenwich, Connecticut, journalist Evan Osnos, who grew up in the town, cited Jones's "modest brick" home and then interviewed his daughter, Grace Vineyard. She recalled, "He asked my mom, 'Do you want anything more?' And she said, 'Why would we want anything more?'"

## Chapter Eight

# THE BRASS RING

In June 1980, Jones asked Jack to write him a memorandum about why he deserved to be CEO. Jack hated the assignment and said so. "You know how I felt about this assessment," he wrote to Mr. R. H. Jones, "but having said that, here is my best effort. It may be more about Welch than we both need or want to know." The note was classic Jack: self-deprecating, charming, ambitious, and deadly serious. His eight-page letter started by touting his accomplishments as a member of the Office of the CEO. Who knew that Jack had shepherded GE's involvement in Epcot—where GE's "vision of the 80's will be portrayed to a wide audience"—in Mexico, where "multi-Sector coordination" will "significantly improve" GE's prospects and in something called the World Iron Project, "where a Sector thrust was extended to a Company-wide opportunity." Better known was Jack's role in trying to solve GE's PCB and asbestos problems, "where I've pressed for broad Company involvement." He also sought to improve GE's "manpower" and "pressed the system" to preserve management continuity and excellence. As examples, he cited the reorganization he implemented at

Plastics and the promotion of Larry Bossidy at GE Credit, although he did not name him specifically.

Jack saw these accomplishments as examples of his leadership capabilities. In noting that he had held executive positions at "all levels of the organization," he said he had "vigorously advocated" for solving problems. As a result, he wrote, "the people with whom I have been associated have worked harder, enjoyed it more, although not always initially, and in the end, gain increased self-respect from accomplishing more than they previously thought possible." Launching into some serious corporate-speak, he continued that "the initiation of action to stimulate business advancements and accelerate human resource development and growth in an environment marked by success and excitement" have been "the constant" of his GE career. He viewed his toughness as an asset. "While I have and will continue to demand that people attain high performance standards, I have at the same time provided numerous 'leapfrog' opportunities to promising employees and have helped to create an atmosphere which attracts talented and ambitious men and women."

Jack did make space for his professional challenges—but managed to turn them into examples of perseverance against the odds or of his own good judgment. It was the 1980 equivalent of writing on a 360 review that your biggest fault was that perhaps you worked too hard and cared too much about your job. He had problems with Robert Kurtz, a GE senior vice president and, like Jack, a member of the nine-member GE "corporate policy board." Kurtz was head of corporate production and operating services. In other words, after Jones, Jack, Burlingame, and Hood, Kurtz was one of the company's most senior executives. Jack's relationship with Kurtz was, he wrote, "disappointing." He believed Kurtz could "do more" for GE and then "hopefully feel better about himself." He admitted he could not "stimulate" Kurtz "intellectually" because "he consistently treats our relationship as superior/subordinate, and hence we have been unable to engage in peer type discussions." Jack wrote that Kurtz was responding to some of the "operations challenges" that Jack had pushed on him. "There has been progress," he wrote. "However, overall, I have not been able to strike a sympathetic chord. I would like to. I haven't given up and I won't." (Kurtz would be gone from GE within two years.)

Jack also didn't shy away from discussing the Cox debacle. How could he, considering it was his strategic initiative that had fallen flat, and very publicly no less? He managed to turn it into a positive in his self-evaluation form. He wrote that before GE had "broken off" the Cox negotiations, he had "tested" Wall Street's reaction to the deal. It was unenthusiastic. "Their response was not only disappointing, it was a telling point," he wrote. "When General Electric is making a $600 million acquisition in a highly attractive, highly visible industry and the impact of the acquisition, because of our size and complexity, is not considered significant, it reinforces the view that what we have to sell as an enterprise to the equity investor is consistent, above-average earnings growth throughout the economic cycle. Our size may dictate that as the only option. The discipline to balance the short-term and long-range is the absolute of such a strategy." (His focus on having GE achieve "above-average earnings growth" and convincing Wall Street that he could do it quarter after quarter would become a Jack Welch signature move.)

He wrapped up his "paper," as he called it, with a combination of self-confidence and obsequiousness that left little doubt that he both wanted and felt he deserved the job. "In the end," he wrote, "when one looks at an executive and where he is going, there are at least two questions—how far along is he, and how much is there in him. I believe I have proven the capability to perform in the current assignment"—as vice-chairman—"and more importantly, I believe I have the capacity to reach for more." But he ended with a flourish. "Obviously, in looking at any of these issues, there is a great distance between where you are and the three of us," he concluded. "However, I feel I have the intellectual capacity, breadth, discipline, and, most of all, the leadership to get there."

Soon after he sent his memo, he started getting "positive vibes" about getting Jones's job. Being a gregarious extrovert, Jack thought nothing of pumping GE's human resources professionals about where he stood in the pecking order. At a party at his home during the summer, he "pinned" Dave Orselet, his human resources partner, against his refrigerator to try to discover what he knew about the horse race. "Bless his heart," Jack later wrote, "Dave would never tell me I was the front-runner, but he reluctantly divulged just enough information to make me feel upbeat about the eventual outcome." He got another positive sign in September 1980 when board member Ed Littlefield, one of the largest GE shareholders, invited Jack to be his golf partner at the Cypress Point Member Guest Tournament in Northern California. The club, in Pebble Beach, long had a reputation for exclusivity. As the comedian Bob Hope once quipped, "Cypress Point is such a beautiful place, but is also very exclusive. They had a membership drive last month. They drove out forty members." Of course, by then, Jack was used to exclusive golf clubs. The invitation to Jack wasn't that startling. "I was the best golfer in the group by far," Jack told me. Jack had never played Cypress Point before and he took it as a positive sign that Littlefield, who was close to Jones, would choose Jack as his playing partner. It was a shotgun tournament, meaning that groups of golfers started on different holes at the same time. On Jack's second hole of the day—a par three—he shot a hole in one. It was his first hole in one in thirty years of playing golf. "That sure made it easy to meet everyone," he quipped.

A month later, the GE board met in Saint Louis. That's where and when Jack believed he got the job. He wasn't part of any deliberations, of course, even though he was on the board. There was no interview with the board. But during the golf rounds at the Old Warson Country

Club, Jack again began to pick up on the positive body language of his fellow board members. Jones was also giving him a good vibe. "I could tell from October on," he told me. "Reg was cozier. Because I was always the odd man out, I was getting no signals. Reg a couple of times winked at me the right way, you know? I mean, I felt it." Unbeknownst to Jack, on November 20, at a dinner meeting of the board, Jones had recommended Jack as his successor; the board unanimously agreed to the selection. But then Jones gave the board a month to think about its decision and to raise any questions.

There were none. After the next board meeting in December, there was a dinner. At one point, Jones looked at Jack and gave him a thumbs-up. But Jack was always leery of Jones and didn't quite trust him. On December 15, Jones walked into Jack's office in Fairfield and hugged Jack and told him he had the job. He told Jack about the November 20 dinner meeting where the board had unanimously approved his selection. Jones told Jack that Hood and Burlingame would stay on as Jack's vice-chairmen—the soundness of their working relationship already well established, thanks to the airplane interviews—and that Jones would stay on until April 1 to help Jack with the transition. Jack would be formally named CEO and board chairman on December 19. He told a colleague after he got the word, "I want a revolution" at GE.

Afterward, Jack and Jones held a press conference at 570 Lexington Avenue. "I think I'm the most happy man in America today," he said, "and I'm certainly the most fortunate." He said he was pleased to continue to work with Hood and Burlingame. "We all came to the party as good friends," he said. "It's my goal to create an atmosphere of sharing both the good and the bad, where they can grow and I can grow and we can continue to have a good time." He said he expected that 1981 would be "a wonderful year" for GE and that while he believed the Japanese consumer appliance manufacturers would be tough competitors, GE would be equal to the competition. For his part, Jones said that while he believed the company had just put in place the "right" management team, he allowed that Jack should have "complete freedom" to choose his own "organization structure."

Asked years later to recall the general reaction to the announcement, Jack told me, "I'd say forty percent cheered, twenty percent didn't know, and forty percent died. Hated the fucking thing. Hated the idea that Jack Welch had won. This guy who gave the finger to the traditions." But Jack once recalled the best piece of advice he ever got came from Paul Austin, the former chairman of Coca-Cola, who came up to him after one of his earliest GE board meetings, where he had been uncharacteristically quiet, and told him, "Jack, don't forget who you are and how you got here."

The reaction to Jack's selection, at least publicly, was nearly euphoric. *The Wall Street Journal* said that GE had "replaced a legend with a live wire." That night, Jack threw himself a party at his home in Fairfield. Stanger, from GE Capital, was there. Bossidy was there. "He's my guy," Bossidy said about Jack. "I love the guy." Paolo Fresco, GE's suave head of its international business, was there too. He had come to see Jack earlier in the day, after the announcement, to tender his resignation. He told Jack that he hadn't supported him for CEO. He had supported Burlingame. He gave Jack a letter of resignation. "Tear that up," Jack told Fresco. "I need you. I love you and you're going to be great. And he was."

Jack was feeling great, too. "On top of the world," he said. And not the least bit nervous. "Not in the slightest," he told me. "Don't forget, I was an insider. I watched it all. I wasn't a rookie. The guy on the outside has a huge disadvantage. A change agent from the inside is what you need." Jack was taking over a company with a AAA credit rating, with $2.2 billion in cash on its balance sheet and a 19.5 percent return on equity—a financial and industrial powerhouse.

On February 24, Jones—still in charge—decided to throw a party for Jack at the Helmsley Palace Hotel, on Madison Avenue in New York. The hotel had been built atop the historic mansion that was once owned by Henry Villard, the venture capitalist who was the CEO of Edison General Electric before the merger that created GE. Jones wanted to introduce Jack to other important CEOs and to transfer his own relationships with them to Jack. "It was a terrific bash," Jack later wrote. "I had a ball. Everyone was relaxed, and nearly everyone drank a little too much, except Reg, who wanted to make sure I was introduced to every one of the 50 or 60 guests. He wanted to get me off to a perfect start." Later in the evening, Jones asked Jack to make some remarks. Unbeknownst to Jack, apparently Jones thought he had slurred some of his words. He'd had a few "extra vodkas," he said, and was slapping his golfing buddies on the back. He had a great time.

The next morning, Jones made a beeline to Jack's office. He was very angry. "I've never been so humiliated in my life," he told Jack. "You embarrassed me and the company." Jack said he "felt like shit" after their conversation. He thought maybe he would be fired even before he had started as GE's CEO. "I was stunned," he wrote. "I'd had a fantastic time and thought it was a great party." He was despondent for the next four hours. His mind was racing from one emotion to another. *Had he let Jones down? Was Jones being too much of a stick-in-the-mud? Would he be fired?*

"I felt sorry for myself because maybe I hadn't made the great impression I thought I had," he continued. "I couldn't believe our guests didn't have a good time. They just couldn't have faked it that much. I'd been to enough parties to know a good one." Just before noon, Jones came back into his office and apologized. In the interim, he had received more than twenty calls from the guests telling Jones it was the best party they had ever attended. "All I'm hearing are good reports about you and the party," he told Jack. "They liked you. I just misread the evening."

He told Jones, "Reg, that's who I am."

"I'm sorry," Jones replied. "I never saw that. They all saw it, but I never saw it."

"I'll do my best," Jack said, "but I'm me and that's who I am, I like people."

JACK COULD NOT WAIT TO GET GOING. JONES FINALLY CEDED CONTROL to Jack on April Fool's Day 1981. And Jack was confident his selection as CEO would transform the company. "If you pick a CEO, you're picking the fate of a company," he told me. "They end up being an important factor in people's livelihoods." He wasted little time blowing up the GE sacred cows. Bossidy remembered how he and Jack went off to Florida soon after Jack took over to begin to figure out how GE needed to be changed. "I knew it was going to be difficult," he said, "because it wasn't changing a company in duress, it was changing a company that was successful." One of Jack's first targets was something called the Elfun Society, an organization of current and retired GE employees "dedicated to the encouragement of cooperation, fraternity and good fellowship and to the betterment" of the community. Among other things, the organization published a four-volume hagiographic history of the company. The name was a contraction of Electrical Funds, a mutual fund in which Elfun Society members could invest. The society's motto was *Semper Paratus*, always ready. It was for GE's white-collar executives, and becoming part of it was considered a "rite of passage" into the company's management ranks. Jack hated it. "It was an organization of suck-ups," he said. He especially hated the "superficial congeniality" that Elfun represented, and not only at Elfun. He hated the "superficial congeniality" he perceived throughout the company. He wasn't going to put up with that anymore.

In the fall of 1981, the Elfun Society asked Jack to speak to its annual leadership conference of about a hundred men, at Longshore Country Club in Westport, Connecticut. They thought they were going to get a pep talk; instead Jack gave them a punch in the gut. "Tonight, I'd like to be candid," he told them, "and I'll start by letting you reflect on the fact that I have serious reservations about your organization." He wasn't done. "I can't find any value to what you're doing," he said. "You're a hierarchical social and political club. I'm not going to tell you what you should do or be. It's your job to figure out a role that makes sense for you and GE." That was the organization's death knell, at least in that form. He tried to ameliorate the blow by hanging around the bar for an hour after his speech. But, he remembered, "no one was in the mood for cheering up." Within weeks, with Jack's encouragement, the Elfun Society had been transformed into a group of some 42,000 community volunteers, trying to do good deeds rather than trying to curry favor with other GE executives. No one was excluded from Elfun at GE. The only requirement was a desire to give something back. Members built parks, playgrounds, and libraries. "Elfun's self-engineered turnaround became a very important symbol," Jack said. "It was just what I was looking for."

At around the same time, Jack began to take steps to raise the public profile of GE Credit, the fast-growing division that he had fallen in love with four years earlier. In October 1981, *New York Times* business reporter Leslie Wayne wrote about GE Credit. Wayne, a graduate of Columbia Business School, had wanted to shed light on a profit center inside a company that the average reader thought of as a power and appliance company. She described it as "one of the invisible giants of finance," which owned the nation's largest fleet of tankers, was the largest lessor of industrial equipment, and had financed such businesses as Atari, a video game manufacturer. Like Ford Motor Credit and General Motors Acceptance Corporation, GE Credit had learned how to make money from money, in addition to making money from building cars and jet engines. At the time, GE Credit had $9.3 billion of assets, most of which were secured loans to small businesses—such as a brass manufacturer or the owner of a fleet of trucks—that most commercial banks would reject. Wayne described how GE Credit was into leveraged leasing, real estate loans, second mortgages, commercial real estate financing, and mortgage insurance. She didn't mention it, but GE Credit was also, by then, financing leveraged buyouts—companies bought using as little equity and as much debt as possible—and investing in venture capital. The company also owned a life insurance business "but, to date, its contribution to net earnings has been modest," Wayne wrote.

But the article also made clear that some of those banks were getting miffed that GE was becoming a fierce competitor. Bankers argued that GE had a competitive advantage of being less regulated. But John Stanger, the head of GE Credit, countered that argument, explaining that his business was leveraged only seven to one, whereas banks were leveraged twenty to one. "Banks operate with outrageous leverage," Stanger told Wayne. "We have to operate without the lender of last resort"—the Federal Reserve—"and at substantially lower leverage. Banks choose to ignore that unfair competitive advantage versus commercial and industrial finance companies." Wayne came up with the idea of writing about GE Credit on her own and cold-called the company to get its cooperation. "GE was interesting because people, then, thought of it as a maker of refrigerators, appliances and airplane engines," she told me in an email. "No one really thought of it as a finance company. So the idea was that this place people thought of for appliances had this whole other, and very big, side to it."

Two months later, Jack made his Wall Street debut, appearing before a group of research analysts at the Pierre Hotel, on Fifth Avenue. He may have misread the room. There was not a single earnings-related number mentioned. Instead, Jack summarized Carl von Clausewitz, the nineteenth-century Prussian general and military theorist, and his final, unfinished book, *On War.* "Men could not reduce strategy to a formula," he told the research analysts. "Detailed planning necessarily failed, due to inevitable frictions encountered: chance events, imperfections in execution, and the independent will of the opposition. Instead, the human elements were paramount: leadership, morale, and the almost instinctive savvy of the best generals." There were risks in Jack's approach. "We don't know what the hell he's talking about," one research analyst was overheard to say.

Jack forged ahead, tying von Clausewitz to General Electric—not the easiest thing to do—and declared that in the 1980s, which he predicted would be a slow-growth decade, the only way for companies to survive was to "insist upon being number one or number two" in every business they chose to be in. The corollary to his thesis was to wonder if it still made sense to be in businesses where it was not possible to be among the industry's market leaders. He cited the famous question posed by the management consultant Peter Drucker: "If you weren't already in the business, would you enter it today?" If the answer was no, then came the second question: "What are you going to do about it?" He told the assembled crowd that the companies that hung on to their losers in 1981 wouldn't be around in 1990. That would not be GE, he insisted. The reaction in the room made it clear that this crowd thought they were getting more hot air than substance. He knew he had to do better. "Wall Street had listened," he wrote, "and Wall Street yawned." GE's stock was up 12 cents. "I was probably lucky it didn't drop," he said.

# Chapter Nine

# GROWING PAINS

When Jack took over as CEO, GE had a market capitalization of $12 billion, and by the end of 1981, it had $28 billion in revenue, $1.6 billion in profits, $1.5 billion in cash on its balance sheet, and a AAA credit rating. It was the largest diversified industrial company in the country. It had recorded twenty-six straight quarters of improved earnings, through several recessions. "But all these good things have not brought GE's stock to life," *Fortune* magazine quipped, playing on Jack's new advertising slogan. GE stock, which had traded at twenty-two times earnings a decade earlier, was now trading at a mere eight times estimated 1981 earnings. Jack, the magazine continued, "would obviously love to have Wall Street, not just Madison Avenue humming" the new slogan.

One lingering problem for GE's stock price was GE's manic relationship with the computer industry. On the one hand, as *Fortune* pointed out, Jones put himself on the fast track to the corner office by selling off GE's computer business to Honeywell "and breaking even on the deal." But the sale put GE out of the semiconductor game, hurting its own technological development and its productivity. "GE was far behind in electronics," *Fortune* observed.

For a company that had played an integral role in the invention and implementation of electricity, jet propulsion, and MRIs, being on the back foot in the burgeoning computer industry was a marked departure from GE's inventive past. Quietly, or as quietly as it could, GE attempted to rectify the situation, spending $235 million on buying Intersil, a world leader in the manufacture of integrated circuits, another $170 million on Calma, the fourth-largest manufacturer of computer-aided-design equipment, and $50 million on four software start-ups. Computer-aided-design systems, like those produced by Calma, were able to cut production times in half and would, it was hoped, form the basis of a transformation of the factory floor. Intersil, which provided "state-of-the-art semiconductor technologies," was GE's second foray into the world of integrated circuits but the first in twenty years, after selling that business off to Honeywell in 1970. Jack had disagreed with Reg Jones's decision to sell GE's computer division to Honeywell, although he conceded it may have been the right decision at the time. He always thought that GE was going to give IBM some tough competition in computers and that his predecessors didn't give the division, and its leaders, enough of a chance to succeed.

That sort of miss would not be happening at Jack's GE. Take, for example, GE's first $20 million foray into reimagining the lightbulb. In 1979, under Jack's supervision, a small group of GE scientists and engineers in the lighting division set out to create the halide arc lamp, known as the Halarc. The idea was to create a bulb that would shine like a 150-watt bulb, use only one third as much electricity, and last five times longer than a regular bulb. Bulbs would cost $10 each and be sold directly to consumers. GE expected the lifetime savings from the purchase of one bulb to be $40 in a typical American city and as much as $75 in New York City (where everything was just more expensive). Given the second energy crisis in a decade, the Halarc bulb seemed like the right product at the right time.

There followed months of design failures and technological challenges. The final product worked only in a lamp and not in an overhead lighting fixture. But after two years, in 1981, it was ready for testing by consumers. There were two ten-month studies, one in Des Moines, where the price per bulb was $10, and one in Salt Lake City, where the price per bulb was $15. The bulbs bombed. But Jack didn't want the men who tried and failed to bring Halarc to market to have their careers tarnished. Jack promoted Richard Kashnow, Halarc's team leader, to run the quartz and chemical products division, and gave him a bonus. The seventy members of the Halarc team were given a VCR "instead of exile," *Forbes* reported. Jack got the GE board to approve a multimillion-dollar discretionary fund for rewards. "It's really the heart of the issue," Jack told *Forbes* at the time. "We have to get people to trust that they can take a swing and not succeed. In big corporations, the tendency is not to reward the good try." When the decision was made in 1983 to wind down the Halarc project, there were parties, dinners, and ceremonies, some of which Jack attended. (Eventually, GE's Halarc bulb made it to market, with some success; original bulbs from the early 1980s can be had on eBay for around $200 each.)

Jack kept taking risks. GE pumped $130 million into buying three microelectronics manufacturing plants and was also making robots in joint ventures with Japanese and Italian companies. The question was why. While not precisely clear, the answer seemed to lie in Jack's desire to create the "Factory of the Future," where networked computers tell machines what to do and when to do it. Creating the next generation of factories became a "holy grail" inside Jack's GE. But it failed. Jack's vision for "Factories of the Future" proved "ineffective," according to Paul Street, a former McKinsey consultant who was a senior GE Capital executive.

In the meantime, while there were still people in the GE factories, Jack made sure to be there, too. "I was everywhere," he told me. "I walked factory floors everywhere. The union guy has a problem with his wife, I'd get on a plane and take her to the hospital. I'd do that naturally. I cared about every one of them." He'd be walking on a beach in Marblehead, Massachusetts, near where he grew up, and someone would come up to him and say hello. He'd reply, "Hi, Bill. How's Mary?" People couldn't believe he would remember the names of near strangers and then be able to ask about their families. "They knew I had their back," he said. "They trusted me." I tried to explore with Jack why his personal touch mattered to his employees. "I get in their skin," he said, "and make them do things they never thought they could do.... I didn't become the captain of the hockey team because I was the best skater.... My passion for what I'm doing rubs off."

One of the paradoxes of his early years as CEO was his ability not only to connect with GE employees one-on-one but also to eliminate whole swaths of jobs when the bottom line required it, just as he had done when he was working his way up the ladder. One of Jack's passions was to eliminate as much of GE's bureaucracy as possible. "I wanted to slim down the layers," he told me by way of explanation. "I mean, GE had layers of people that did nothing, nothing." He was determined to make decision making fast and decisive.

The media quickly picked up on what Jack was doing and, not surprisingly, described it in less charitable terms than did Jack. "They call him 'Neutron Jack,'" *Newsweek* explained in a July 1982 article about how the recession was resulting in an increasing number of layoffs among white-collar employees. Jack had, by then, already chopped seven thousand white-collar jobs, or 4 percent of the salaried employees, with more layoffs coming in the second half of the year. "With so many people being let go," one fired executive told the magazine, "Jack's like the neutron bomb. Soon only the buildings may remain." In his first nine months, he fired forty thousand employees, and another sixty thousand during his first two years. "The early days of his tenure were brutal," Denis Nayden, a longtime GE executive who eventually served as CEO of GE Capital, told me. "He was a brutal bastard," until he one day woke up and decided, "You know, I've got to do a better job of nurturing the people and developing relationships and really growing the intellectual capital of this company. And then his whole management style changed."

Until that C-suite chrysalis turned into a butterfly, Jack was ruthless. Jack didn't like the "neutron" moniker, his head of human relations, Bob Nelson, told me, "but I thought it was accurate." Nelson remembered that in Louisville, for instance, the home of the major-appliance business, "there were crowds of people down there pushing paper." Nelson "did the analysis" that enabled Jack "to order big reductions in staffing." The six big factories and the warehouses at Appliance Park were left. Ten thousand people, or half the people who once worked there, were let go. "The buildings were still standing," Nelson said, "and there were a lot less people." Jack fired every strategic-planning specialist that Jones had hired and nurtured. Jack wanted more "candor" and more "constructive conflict" from the top GE executives. He decided instead to arrange for regular meetings with the heads of the business units, along with Burlingame and Hood, his vice-chairmen, "one on three, two on three, in a small room." It was intense. "I was a workaholic," he told me. "I called guys like me 'hoary-handed sons of toil.' We'd have personnel meetings that would go ten hours." As Jack said, "I have more energy than most people."

As filled with trepidation as these meetings made the top GE executives feel, the real fear was in the GE factories and offices when Jack would show up. That's when "local managers would panic," Byron reported. "All fat people were told to stay out of sight." A plant manager in Cincinnati kept around a list of deceased former GE employees who had worked in the factory in case Jack showed up and asked to see progress in job reductions. At the plastics plant that Jack had built in

Selkirk, New York, the manager was sufficiently worried about an up-coming visit from Jack that he told Joseph Jett, a newly hired engineer from MIT—and the plant's first Black scientist—that he should stay out of sight during the visit, since he did not know "whether Welch felt the same about blacks as he was said to feel about fat people," Byron continued. (Jett remained at GE for another three years before leaving to attend Harvard Business School; he would return to a GE subsidiary a few years after graduating and then be the person chiefly responsible for a financial calamity.)

One ironic twist of Jack's reign of terror was that the economy of his supposedly beloved Pittsfield—the place he got his start and the place he didn't want to leave even if it meant climbing the corporate ladder—was nearly destroyed. "In round after round, the layoffs came," Byron wrote, "beginning with the closing of the city's transformer factory, and the decimation of the ordnance operation, until 80 percent of GE's entire workforce in the city had disappeared." A similar level of decimation occurred in other Hudson Valley cities, including Schenectady—once the location of the corporate headquarters and the home of GE's crop of Nobel Prize–winning scientists—and Waterford, New York, once the home of GE's silicone products division. (Much of the decimation that GE's departure wrought in these communities has been captured in the melancholy photographs of Gregory Crewdson, who lives near Pittsfield.)

Jack also moved quickly to reorganize GE, removing the structure that Jones had implemented in 1977 when he set up the succession competition. Two new "sectors"—Information Services and Technical Systems—were created. Larry Bossidy ran the fast-growing "services" business, including GE Credit and GE Information Services, a computer time-sharing business that GE had taken full control of a few years after the sale of the rest of the computer business to Honeywell, and the Engineering Materials Group, which included Jack's old plastics division, the silicone manufacturing business, and various other metallurgical businesses. The other new division, Technical Systems, run by James Baker, was the home of GE's effort to create the Factory of the Future and included a variety of GE's microelectronics businesses, as well as its aerospace business, its mobile telecommunications business, and its medical systems business.

GE Healthcare was the world's leading manufacturer of diagnostic imaging equipment, thanks to the decision, in the early 1970s, to convert an X-ray machine into a CAT scanner. In the early 1980s, GE started manufacturing nuclear magnetic resonance, or NMR, imaging machines, which used powerful magnets to produce cross-sectional photos of body tissue, without X-rays, surgery, or pain, and could also look through solid mass, an improvement over a CAT scan. The CAT scan machines cost roughly $1 million each; the NMR machines, which were similar to MRI machines, cost roughly $2 million each. Despite the price tags, they were in high demand.

Jack saw these two business divisions as the fast-growing future of GE—one a hybrid of financial and computing divisions; the other a big bet on medical devices and research and development. These new divisions would not be spared the same yardstick as the existing ones. Per Jack's dictum, GE would need to be number one or number two in each business segment. Jack spelled out this strategy in his shareholder letter in February 1982, his first as GE's CEO and board chairman. A new GE, oriented around these themes, he wrote, would form the basis of an "operational atmosphere" in which "people will dare to try new things" and "where their own creativity and drive will determine how far and how fast they grow." He wanted GE to become a company "more high-spirited, more adaptable and more agile than companies a fraction of our size." As he had also alluded to from time to time, Jack also wanted GE to grow its earnings reliably quarter after quarter, to the point where Wall Street could count on that earnings growth and reward GE with a higher multiple of earnings. Jack Welch was not going to be satisfied with a stock price equal to eight times GE's earnings, not by a long shot. As was becoming increasingly fashionable on Wall Street during the 1980s, if he told research analysts GE was going to earn a certain amount of money in a given quarter, come hell or high water GE was going to earn that amount of money.

Dennis Dammerman, GE's CFO, said that what Jack was doing to GE was essential. "When I started with the company, it was an incredibly hierarchal and formal place," he recalled. "The chance you would ever see your boss's boss's boss was just beyond imagination." The layers upon layers had to be eliminated. Documents no longer needed twenty signatures when one would suffice. "You had to get rid of some of that," Dammerman said. Aside from continuing to generate steady profits—a prerequisite for everything, in Jack's mind—he wanted to turn GE's "bottom-line wizards" into "world-class competitors." He wanted "to take the strengths of a large company and act with the agility of a small company." Gregory Liemandt, the head of GE's new Information Services business (known as GEISCO), told *The Wall Street Journal* what Jack would always demand to know of him: "How big" can GEISCO get and "how fast" can GEISCO grow? Liemandt said Jack had made him feel like an owner of the business, rather than just a caretaker. (Receiving GE stock options probably didn't hurt—another Jack innovation for GE execs.)

To achieve this vision of being number one or number two in all the different sectors of GE's business would require a new sort of manager. Integral to transforming the company would be an indoctrination camp for its best, brightest, and future leaders at the newly refurbished Crotonville, run by James Baughman, a former Harvard professor. Jack had badgered the Crotonville town government into granting him the right to land a GE helicopter near the facility—a right Jack insisted on having, or else.

While Jack was CEO, getting an invite to Crotonville was an honor. But it was also known as "Scrotumville" because of the many men who were emasculated in "the Pit," the 110-seat amphitheater where Jack would do his proselytizing and that was named after his old play-ground in Salem. Starting in 1981, thousands of GE "hi-pos"—high potentials, in GE speak—went through Crotonville to learn from Jack. To Jack, Crotonville was like "a coffee pot. . . . It doesn't just percolate. It gives off aromas that draw people from all over the company." One senior executive told me that Crotonville was the "glue" that "made" GE managers. "GE is like the Army," he said. "I was always stunned by the huge organization, with a lot of what I would think of as non–Ivy Leaguers. Jack didn't like Ivy Leaguers. They would work their way up and there would be a filtering process."

Jack thought his idea of being number one or number two in a business was a simple message. Either lead or get out of the way. But it turned out to be a lot more difficult to implement his strategy across GE's massive workforce or its forty-two business units than it was to talk about it in a room full of Wall Street analysts at the Pierre Hotel. In January 1983, he was having dinner with Carolyn at Gates, a res-taurant in New Canaan. He was trying to explain his "vision" for GE to her and started writing with a felt-tip pen on the napkin that had been under his drink. "I often drive people crazy by sketching my thoughts out on paper anytime, anyplace," he remembered. He drew three interlocking circles—Jack's version of a Venn diagram but with nothing inside the overlapping areas—each meant to illustrate GE's areas of focus: its "core" businesses (major appliances, lighting, tur-bines, transportation, and motors); its "high technology" businesses (medical, aerospace, aircraft engines, materials, and industrial elec-tronics); and its "services" businesses (financial, information, con-struction and engineering, and nuclear). Any business outside the three circles, he told his wife, he would "fix, sell, or close." Outside his circles were businesses such as small appliances, television stations, central air-conditioning, large transformers, television and audio de-vices, and mobile radio. "The chart really hit home," he said.

It certainly did. Those GE employees who worked in Jack's keepers felt "a certain sense of security and pride," he recalled. But those outside his rings were upset and immediately felt marginalized. "Am I in a leper colony?" Jack remembered that some of them wondered. "That's not what I joined GE to become." Union leaders complained. The leaders of cities and towns where businesses to be divested were located did, too.

To implement what he had mapped out on his cocktail napkin, Jack turned GE into a mergers and acquisitions machine, buying and selling companies with abandon in a never-ending effort to construct the perfect portfolio of high-performing, high-profit assets. This was an ambition that Jack, at least at first, didn't broadcast widely. To run this M&A machine, in December 1982 Jack hired Michael Carpenter. Carpenter was a thirty-five-year-old Brit with an arrogant demeanor who had been working as a consultant at Boston Consulting Group for a decade after graduating from Harvard Business School in 1973. He joined GE as vice president in charge of corporate business development. He had never done any consulting for GE but got to know Jack over the years. Whatever he had gleaned from his experience almost buying Cox's media assets had whetted Jack's appetite for the deal business. Doing deals was nothing particularly new at GE. "Always," Jack told me. "It's part of the DNA of the company."

Carpenter started work at GE in February 1983, a month after Jack had shared his vision with his wife. Carpenter wasted little time executing his mandate. During the next few years GE would divest itself of 118 businesses, generating proceeds of more than $3.5 billion. There was the highly publicized sale of GE's small-housewares business to Black & Decker for $300 million and the low-key sale of Family Financial Services, a Stamford, Connecticut–based originator and purchaser of second mortgages with a loan portfolio of $560 million, for $600 million to the Philadelphia Savings Fund Society. He also sold GE's central-air-conditioning business to Trane for $135 million in cash and announced his intention to sell all eight of GE's radio stations and two of its TV stations. Jack considered these deals "peanuts" financially but of great importance culturally inside the company. He thought the sale of the air-conditioner business, in particular, sent a needed message. There were three plants and 2,300 employees in the unit. "It wasn't very profitable," Jack remembered. It had a market share of 10 percent, puny by Jack's standards. He hated the business. "I felt it had no control over its destiny," he wrote years later. He hated that the air-conditioning units were sold through distributors who cared little about them or how they were installed and maintained, leaving GE with the blame and criticism when things went wrong, as they inevitably did. "This was a flawed business," he continued. But the people in Louisville who made the air conditioners took pride in them and were shocked when the business was sold to Trane. "It really shook up Louisville," he said.

He did not feel their pain. Quite the contrary. The correctness of selling the business to the market leader was reinforced when, a month after the close, he spoke with Stan Gorski, who ran the business when it was part of GE and then ran it at Trane. "Jack, I love it here," Gorski said. "When I get up in the morning and come to work, my boss is thinking about air-conditioning all day. He loves air-conditioning. He thinks it's wonderful." That was quite the contrast with Jack's view of the business. With Jack, it was all about customer complaints and the collapsing profit margins. "You hated air-conditioning," Gorski continued. "Jack, today we're all winners and we all feel it. In Louisville, I was the orphan."

Jack's response? "Stan, you've made my day," he said, before hanging up and concluding that he had made the right decision not only for Stan and his 2,300 employees but also for GE, which used the $135 million proceeds "to help pay to restructure other businesses," Jack said. This would be his template for the remainder of his tenure, or so he claimed: sell businesses that were not leaders in their industries and use the proceeds to shore up the businesses he wanted to keep and grow. Of course, it didn't always work out for Jack as he planned.

Jack's most radical early move was his decision, in January 1983, to sell Utah International, the commodities mining company, for $2.4 billion to Broken Hill Proprietary Company, Ltd., a large Australian steel and natural resource company. Utah International had been Reg Jones's baby and the largest M&A deal ever when he completed it seven years earlier for $2.2 billion. Jack didn't like the company, and commodities with their unpredictable price swings didn't fit into his vision for a company that would be able to deliver steady earnings each quarter.

## Chapter Ten

# THE NEW JACK PACK

With the divesting of Utah International and GE's small-appliances division to Black & Decker, Jack had shown that he was capable of dumping businesses he deemed losers. But he could not be everywhere. He needed a team of like-minded and fiercely loyal lieutenants to bring his vision for a new GE to life. Mike Carpenter, his mergers and acquisitions guy, was just one member of the new team that Jack was assembling around him. There were, of course, the loyal lieutenants he had elevated or kept around as vice-chairmen, including Larry Bossidy and Paolo Fresco.

There was also the return of Bob Wright. Jack had sent him down to Atlanta to close the Cox deal, and Wright ended up catching the cable bug and becoming the president of Cox's cable business. In May 1983, Wright returned to GE, for the third time, from Cox Cable to run GE's housewares and audio division, both of which, of course, were outside Jack's three-circle future for the company. Wright had spent three years at Cox and by the end had started reporting to William Schwartz, the CEO of Cox Communications, the parent company of Cox Cable. Their relationship wasn't great. "He was a smart guy, but he was a micromanager," Wright told me. "He used to call me up for the damnedest things. He was always nervous that he didn't know enough about the cable business, and he was afraid of it."

Jack called Wright about coming back to GE while Wright was feeling increasingly peeved by the oversight from his boss at Cox. Jack had an opening in the housewares division he needed to fill. "Listen, I know you're doing a good thing down there, but I've got some stuff here that I think you really should take a look at," Jack told him. "I have a housewares and audio electronics business, two of the oldest businesses at GE, and we put a lot of money into them and I'm not sure that I made the right call. I'd like you to come back up here to run these businesses, and then tell me over time whether you think we should sell them or keep them, because I'm really nervous about how much money I've invested. And by the way, if you do that and you do that well, I would like you to go to GE Capital."

That was not a promise from Jack, Wright explained to me. There was no obligation or contract. Just Jack's ability to persuade and cajole. Jack said he always "meshed well" with Wright and remembered they lived "three houses down" from each other in Pittsfield back in the day. "Smart," Jack said of Wright. "High IQ, flexible mind." They also lived near one another on Nantucket. Wright moved to Bridgeport to run housewares. But if Utah International exited the company on cat feet, the excising of the housewares business caused a major ruckus. Jack got a bunch of letters from irritated employees.

"How could we be GE and not make irons and toasters?" read one.

"What kind of a person are you?" another employee asked. "If you'll do this, it's clear you'll do anything."

Recalled Jack, "If e-mail had existed, every server in the company would have been clogged up." He knew he had hit a nerve and had angered the GE rank and file. The problem was, Jack continued, "There was a lot more to come, a helluva lot more."

Dennis Dammerman, known around GE as "Triple D" (his middle

initial was *D*), was another one of the people Jack plucked from obscu-
rity inside the company, elevating him to become GE's CFO. Jack had
summarily dismissed Tom Thorsen, Dammerman's predecessor, despite having a relationship with Thorsen that dated back to Plastics.
The son of an electrician from Grand Mound, Iowa—population 619—
Dammerman joined GE in 1967 after graduating from the University
of Dubuque, in Iowa. He started in the highly regarded Financial Management Program, or FMP as it was known, at GE Credit, in Louisville. After Jones selected Jack to rotate out of the plastics division to
head up the sector that included major appliances, lighting, and GE
Credit, Jack wanted to find two young analysts to help him understand
the two larger businesses he didn't understand. He chose Bill Nelson,
who had been at major appliances, and Dammerman, who had been at
GE Credit.

In his job interview, Jack asked Dammerman a brutal question, in
front of his boss: What would he tell Jack to determine whether his
boss was doing a good job or not? Calmly, Dammerman answered the
question—focus on the big assets, like airplanes and oil refineries, not
loan volume—while his boss glared at him. Jack was impressed. "He
demonstrated incredible smarts," Jack recalled of Dammerman during that period. "He could slice and dice the smallest details of the
appliance business one day, then analyze the most complex deal at GE
Capital the next." Jack also believed Dammerman could spot talent,
one of the key aspects to becoming a better, even great, executive. In
1981, Dammerman returned to GE Credit to run the real estate financial services division. Three years later, in March 1984, Jack called
Dammerman at his desk at GE Credit and arranged to meet him at
Gates, the restaurant in New Canaan, for dinner that night. He also
swore Dammerman to secrecy. When Jack arrived at the restaurant,
Dammerman was already at the bar. Jack sat down right next to him
and got to the point. Was Dammerman okay with Jack going to the
board and recommending that Dammerman be GE's new CFO? No

surprise, Dammerman agreed, and then began asking a bunch of questions.

The Dammerman appointment shocked the GE rank and file. Dammerman was thirty-eight years old and the youngest CFO in GE's history, and therefore a nice parallel to Jack, the youngest CEO. Dammerman had never made a presentation to the GE board. Most people had never heard of him. It was nearly a total gut-instinct promotion. "I wanted him to lead his own revolution," Jack said. Dammerman didn't think he was qualified, or ready, for the job. But Jack disagreed, and that's all that mattered. "I knew he could do it and I was completely committed to help him," Jack said.

People who worked with Dammerman as he climbed the corporate ladder were struck by his ongoing modesty. Norman Liu, a graduate of Yale and Harvard Business School, made it far within GE, and for a long time served as CEO of GE Capital Aviation Services, known as GECAS, one of the world's largest aviation financing companies. He remembered that Dammerman never forgot about his Iowa roots and always treated him respectfully. "Norm, I'm a little dude," Dammerman would tell him. "Norm, I grew up in Iowa. They used to drop me in these little sheep pits on a little bungee cord or something. I had, like, this barbershop shearer. Oh my God, as the little guy, just go shave all this shit out, and they would drag me out of this thing." He would allow Liu to join him on the corporate helicopter flights between New York City and Fairfield County, saving Liu the headache of commuting. On the flight, he would take the time to ask Liu about his family life. "No senior guys at GE treat you like that," Liu told me. "He was just great." But he could be tough, too, when he had to be. "Dammerman was a complete asshole," remembered a longtime underling. "He was the hardest guy in the world to work for. I worked for him for fifteen, twenty years. The guy was gruff, rough, really incredibly challenging, but I got along great with him."

Jack complicated life for his new CFO within months, though, when he wrote Dammerman a three-page memo in May outlining his thoughts for how the job should be disrupted. "There is no place that quantum change is needed more than in Finance," Jack wrote. "Finance is not an institution—it has to be . . . the driving force behind making General Electric 'the most competitive enterprise on earth.' . . . Everything done in the past is open for question—question not criticism." He wrote the memo to Dammerman, he told me, as a "rallying cry to Dennis to shake up the finance organization and get out of the 'bean counting' business and get it into operations."

Where Dammerman, Fresco, and Bossidy were all GE lifers whom Jack took a shine to and promoted, he reached outside GE to fill another important position, his new general counsel. For that important role, in 1987, Jack selected Ben Heineman, who was then the managing partner at Sidley & Austin and a well-respected constitutional lawyer. Some eight years earlier, Nick Lemann, writing in *The New Republic*, described Heineman as the "possessor of the best résumé in America." And it was true. Heineman was the son of another Ben Heineman, a prominent Chicago attorney-turned-businessman. After a successful proxy fight, Ben Sr. became the CEO of North Western Railway and over time turned it into the conglomerate Northwest Industries. He was also a close confidant of Lyndon Johnson, who offered him a variety of cabinet positions—all of which he turned down. "More than anyone, he understood what we were trying to do," Joseph Califano, Johnson's chief domestic aide, told *The New York Times* about Heineman's father. "He was quite a selfless person. He had no personal agenda. He told it like it was, which is very hard, and the most important thing you can do for a president or one of his top aides, like me, because people are usually fawning all over you."

After Harvard, a Rhodes Scholarship at Oxford, Yale Law School (where he was editor of the *Yale Law Journal*), a Supreme Court clerkship, and a stint as a reporter for the *Chicago Sun-Times*, Heineman Jr. worked for the Center for Law and Public Policy, in Washington, DC, as a public-interest lawyer defending the mentally disabled. When Califano became Jimmy Carter's secretary of health, education, and welfare, he recruited Heineman to be his executive assistant.

Heineman met Jack in 1987 after being introduced by Zoë Baird, who was then working in GE's legal department. Baird knew Heineman from the Carter administration, where she had a few different jobs, including associate counsel to the president of the United States. (She would go on to infamy after being nominated to be President Clinton's first attorney general, only to later withdraw her name from consideration after telling Clinton she had hired illegal immigrants but had not paid their social security taxes.) "Ben seemed a bizarre choice to head our legal staff," Jack remembered. Heineman thought so, too. "Remember, I'm a constitutional lawyer," he told Jack. "I'm not a corporate lawyer. I'm not a New York lawyer." Jack didn't care. "You'll hire good lawyers," he told him. "That's what I want you to do." Jack wanted Heineman to build a legal department to rival what Dammerman was building in finance and Carpenter was building in business development. He paid Heineman well—$2 million a year in the early days—and allowed him to rebuild the department, as long as he hired the best.

"This was a classic case of As hiring As," Jack said. Heineman hired attorneys such as John Samuels, a former partner at Dewey Ballantine, to head the tax department. He hired Brackett Denniston, the former legal counsel to Massachusetts governor Bill Weld, to run the litigation department. By the time Heineman finished, he had created a group of more than four hundred well-paid attorneys, as big as any law firm in the country and, many people thought, among the best. "I loved Jack," Heineman told me. "I think he was phenomenal. I worked with presidents, Supreme Court justices, cabinet secretaries. The guy was phenomenal."

By the end of 1983, GE had generated $27 billion of revenue and $2 billion in profit. GE made everything from refrigerators and lightbulbs to jet engines and turbines, while also financing LBOs and investing in venture capital deals. The problem, as Jack correctly identified on the cocktail napkin, was the difficulty of succeeding in so many businesses at the same time. And GE was failing at that: GE's sixteen major businesses, where it was a top player, accounted for 92 percent of its profits and 87 percent of its revenues; another twenty business lines, where it wasn't particularly competitive, resulted in $54 million in profit on revenue of $3.5 billion.

That's why Jack was pushing for the elimination of the businesses from the GE portfolio that weren't industry leaders. Out were the profitable audio unit, the household appliance business, the radio business, a couple of television stations, an air-conditioning business, and, of course, Utah International. The question was: What was GE going to do with its $5 billion cash pile? But there was also the fact that with interest rates so high—short-term Treasury securities were then yielding around 10 percent annually—GE could make plenty of money sitting on the cash, doing nothing, and sticking to financial discipline on any acquisitions. GE had considered some six thousand potential acquisition candidates, but Mike Carpenter, the head of business development, deemed prices too high and passed on the purchase of Hughes Helicopters, for instance, allowing McDonnell Douglas to buy it for more than $450 million. Along with being in no rush and refusing to overpay, GE's acquisition strategy seemed to be that any deals had to fit in with the existing dominant business lines. The ceramics business that GE bought from 3M in 1983 helped both the lighting and jet engine businesses, while a tank-car-leasing business complemented well what was going on at GE Credit.

In 1984, Jack made his first big strategic move. He decided to buy Employers Reinsurance Corporation, known as ERC, from Texaco, for $1.08 billion in cash. ERC had been part of Getty, the diversified oil company that Texaco had bought two months earlier after a long and controversial battle, with the promise of selling Getty's noncore assets.

ERC, based in Overland Park, Kansas, had been around since 1914 and was in the business of reinsuring the risk taken by other property and casualty insurance companies. In 1983, ERC generated revenue of $787 million and net income of $71 million. Initially, GE had agreed to pay $1.1 billion for ERC. But when the ERC deal team showed up in Fairfield for a dinner to celebrate the deal, it informed Jack that ERC was going to miss its projected numbers. Jack was not pleased. He wanted a price adjustment. He tracked down John Weinberg, the senior partner at Goldman Sachs who had represented GE in the ERC acquisition, on the golf course in Augusta, Georgia. Jack told Weinberg to call Jack McKinley, the CEO of Texaco, and negotiate a lower price. McKinley was "a gentleman," Jack allowed, and gave GE a discount of $25 million, bringing the ERC price to $1.075 billion. (For most of the rest of Jack's tenure as CEO, ERC was a home run. Under GE's control, ERC earnings peaked at nearly $800 million in 1998, an eightfold improvement.)

Nineteen eighty-four was not without its bumps: In July, Jack got an honor he didn't want or appreciate: America's toughest boss. *Fortune* magazine named him the country's most difficult CEO and again referred to him as Neutron Jack, a name he still disliked. He blew away the competition; he received more than twice as many nominations as William Klopman, the CEO of Burlington Industries, the runner-up. To win, Jack had to prove he was "autocratic, ruthless, grueling and intimidating."

Jack also found himself tainted, for the first time (but not the last), by the aroma of scandal. In March 1985, a federal grand jury in Philadelphia indicted GE on 108 counts of falsifying claims and lying on work related to a nuclear warhead system. The indictment alleged that between January 1980 and April 1983, GE managers Roy Baessler and Joseph Calabria falsified time cards and, as a result, overcharged the government "at least $800,000" for work on the missile system. Baessler, who had since left GE, and Calabria were also indicted on two counts each of lying to a federal grand jury.

Three years before the grand jury indictments, John Michael Gravitt, a machinist foreman at GE's military and commercial jet engine plant in Cincinnati, stood up in the middle of an "assertiveness-training session," of all things, and told his thirty colleagues that he believed GE had been cheating the government by falsifying time cards related to the government defense contracts. "The time cards going into supervisors' offices aren't the same ones coming out," he said at the meeting, according to his June 23, 1988, testimony before a House of Representatives subcommittee. He said another foreman at the meeting tried to quiet him by grabbing his arm and pushing him back into his seat. "Shut up," he told Gravitt, "you're going to get fired." The training class was halted abruptly, but not before a bunch of other foremen stood up and voiced their agreement with Gravitt. "That's when I realized how big the cheating was," he said.

Gravitt worked at GE's Evendale, Ohio, plant for three years from June 1980 to June 1983. In spring 1983, Gravitt was told he would be laid off from his $35,000-a-year job at GE "due to a so-called lack of work." He and his wife, who also worked at GE as a machinist, began documenting the fraud. In October 1984, he filed a lawsuit against GE, arguing that the company had violated the 1863 False Claims Act. Gravitt's allegations led to the 1985 indictments. According to *The New York Times*, which published the story on the front page, GE appeared to be "by far" the biggest Defense Department contractor ever criminally charged in connection with a government contract. If convicted on all charges, GE would be fined around $1 million and Baessler and Calabria could end up spending ten years in prison. Of course, with $28 billion in revenue and some $2.25 billion in profit, the fine would be akin to a parking ticket to GE. But GE also had $4.5 billion in military contracts with the government that could be at risk.

GE denied wrongdoing. "It is alleged that incorrect charges were entered on employee timecards submitted five years ago," a company spokesman told the *Times*. "Involved are 100 timecards out of approximately 100,000 timecards." Two days later, the air force suspended GE from participating in any new contracts with the Defense Department, for which it made engines for the B-1 bomber and the F-16 fighters. GE called the suspension "highly unusual and disproportionate to the issues under dispute." The air force also demanded that GE repay $168 million—real money—that it claimed were "excess profits" from the manufacture of engine parts for the government during a six-year period.

The company's bravado evaporated on the eve of the May 14 trial. Unexpectedly, GE agreed to plead guilty to defrauding the air force and was fined $1.04 million and ordered to repay the $800,000 it had overcharged the government. It was front-page news, the most important story in *The New York Times* that day. GE decided to plead guilty—after insisting it had done nothing wrong—after one of the two men indicted, Roy Baessler, admitted to federal prosecutors that he had falsely billed the government. He also implicated "higher management" in the scheme.

While the denouement was swift, it was of little consequence to GE's bottom line. GE's indictment was a decade or so before such a thing could so badly shake the confidence of investors, customers, and employees as to effectively sound a company's death knell, as it did for companies such as Enron, WorldCom, and Arthur Andersen.

The billing fraud was the third time (at least) in a generation that GE had been involved in scandal. There was the price-fixing scheme of the 1950s that culminated in the jailing of three senior GE executives, the forced resignation of another twelve, and the early retirement of Robert Paxton, the company's president, amid accusations that he had been aware of the fiasco. Then, in 1973, Hoyt P. Steele, a vice president for international sales in GE's power business, was charged with authorizing the payment of a $1.25 million bribe to a Puerto Rican official to try to obtain a contract for a $93 million power plant project on the island. According to the July 5, 1985, *Wall Street Journal*, Steele, then sixty-four years old, who was paid $185,000 per year and was nearing GE's mandatory retirement age, did not benefit personally from the bribe; he did it to try to make sure GE won the contract, since GE had lost an earlier deal in Mexico for a similar design and was anxious to develop a prototype that could be sold to other developing countries. In his testimony at his 1981 trial, Steele said he was afraid that rival Westinghouse would win the contract. "I authorized a payment, yes; not a bribe," he testified. His objective, he said, "was to avoid being cheated out of an order that belonged to us." In the end, after a three-week trial, both GE and Steele were convicted of mail fraud, wire fraud, conspiracy, and interstate or foreign travel in aid of racketeering. Steele faced up to thirty years in prison. But the convictions were reversed on appeal, and three counts were ordered retried. Prosecutors settled for a no-contest plea from GE to a fraud charge and dropped charges against Steele.

The third strike—the 1985 billing scandal—prompted some in the media to wonder if GE—supposedly a pillar of American corporate rectitude—had lost its way. The Justice Department briefly considered ordering that GE be broken up as a way to try to stop the bad behavior. Jack seemed determined to minimize such breaches in the future. He stepped up the teaching of ethics to everyone passing through Crotonville. He appointed a corporate ombudsman and authorized him to investigate any allegation of wrongdoing. He thickened the GE employee policy manuals with new sections on how to properly account for time spent working on government contracts and for the rules around making gifts to government employees. That meant no more free lunches. Time cards contained bold new warnings that falsifying records was a crime. He shared a video message with GE's 330,000 employees that there would be no tolerance for cutting corners in pursuit of achieving corporate financial and operational goals. He encouraged whistleblowers to come forward. "Whistleblowing, speaking out, telling it like it is, is part of what we're after," he said in the video. "I can assure anybody who wants to talk about abuses of integrity in this company that they will be welcomed."

Despite his entreaties to GE's employees to behave, neither scandal nor failure seemed to touch the man in the corner office. Jack was well on his way to perfecting his skills at building and protecting his public image. On the one hand, GE's soaring financial performance made it easy. Despite the tumult in the capital markets in the wake of the October 1987 stock market crash, GE kept steamrolling. By nearly every financial measure, GE was a juggernaut. Revenue for 1988 topped $50 billion for the first time; net income was $3.4 billion. Earnings per share grew 17 percent, and return on equity was 19.4 percent, up a full percentage point from the previous year. Operating margins were higher than the year before and Jack expected them to increase further still.

On the other hand, it was during this time that Jack developed close, collegial relationships with the journalists who covered GE and set the agenda for the business press. He was also not above the practice of what came to be known as "catch and kill," trying to stop problematic stories from ever seeing the light of day. One telling illustration of his penchant for controlling his image came when a major problem occurred with a new type of rotary compressor in certain models of GE refrigerators. This fiasco threatened GE's entire refrigerator business, and it set Jack on a collision course with a *Wall Street Journal* reporter named Thomas F. O'Boyle, who had broken the story of the faulty compressors.

Jack quickly realized the new compressor was an engineering disaster and that every one of them would fail eventually. Lawsuits alleged that the flawed refrigerators caused fire, food poisoning, injuries, even heart infections that led to a stroke. Roger Schipke, GE's head of major appliances, quickly ordered the replacement of millions of compressors in a variety of new GE refrigerators. In 1988, GE took a charge of $450 million to account for the cost of the damage to its appliances and to its reputation for engineering excellence.

Jack made no mention of the disastrous compressor problem in his subsequent memoir *Jack: Straight from the Gut*, and according to O'Boyle, his only mea culpa occurred when, during the filming of an employee video, he asked that the camera be turned off and then said, "On that one, we nearly broke the company." In his 1988 shareholder letter, Jack shared none of the details that O'Boyle uncovered. He wrote that there was "no safety issue" involved with the compressor fiasco and that every faulty compressor would be replaced at a "substantial" cost, although, he wrote, "we still had a record [financial] performance" for the year. His hope was "to come out of this situation with our reputation for customer support and satisfaction not only intact—but if anything—enhanced."

Born in Newark, New Jersey, O'Boyle grew up in Morris County. He used to take school trips to West Orange, to the Edison Invention Factory. He'd get home and build a voice recorder out of aluminum foil and a cylinder. "I was a geeky kid," he told me. "I found that stuff thrilling." He graduated from Northwestern's Medill School of Journalism in 1979, spent a year at *Crain's Chicago Business* as it was starting up, and then moved on to the *Pittsburgh Post-Gazette* as a business reporter for the next sixteen months.

In September 1981, the *Journal* hired him to cover the Rust Belt. He wrote about the steel and the aluminum industries. When he attended Bethlehem Steel Corporation annual meetings, he would see Jack's predecessor, Reg Jones, who was a Bethlehem Steel board member. He started to get a sense of the kinds of changes Jack was making at GE and how different he was from Jones. By then, Jack was already known as Neutron Jack. "A lot of people were upset because Jack was really slashing and burning pretty heavy then," O'Boyle told me. "They were laying off people left and right. And in my reporting, I later found that he was getting these letters from widows and orphans, basically, 'After so many years, you're eliminating my husband's job, he's been so loyal to GE for all these years.' And that's when someone in GE told me that he had commanded and stricken the word 'loyalty' from the GE playbook, because it was bugging him. It was an offensive thing to him. To his credit, he did have enough of a conscience and soul that it did bother him." In July 1985, O'Boyle wrote a piece for the *Journal* about the deterioration of loyalty across corporate America, what with the increase in mergers and acquisitions, corporate raiders, LBOs, and the drive for higher profits. O'Boyle said that piece was "really my first intersection with Welch and what he was doing." From January 1986 until November 1989, O'Boyle was the *Journal*'s correspondent in Germany, covering German politics as well as Germany's heavy industries.

In November 1989, his bosses asked him to return to the United States and spend more time covering the industrial businesses of GE, many of which were in the Midwest—jet engines in Cincinnati, major appliances in Louisville, lighting in Cleveland, among others. O'Boyle's immediate boss at the *Journal*, Carol Hymowitz, also had a hunch that there was more to the GE compressor story than anyone was saying. She was right. "I spent about—oh, gosh—six or nine months looking into it and found out through the footnotes and everything that it really had been a complete fiasco," O'Boyle remembered.

In his *Journal* article, O'Boyle quoted Roger Schipke, the former division chief, who had since moved on to Ryland, a home builder, describing the failed compressors as "your worst nightmare come true. I don't even want to think about it anymore." Jack declined O'Boyle's invitation to be interviewed. But, O'Boyle wrote, the new compressor "flopped so badly" that GE had to take a $450 million pretax charge in 1988 and that it had voluntarily replaced 1.1 million defective compressors. He wrote about how GE engineers were determined to adapt a compressor from GE air conditioners for a new updated line of GE refrigerators, despite others at GE thinking it was a bad idea and that the compressors should be imported from Japan or Italy. (Jack, of course, was part of the group that was determined not to rely on foreign manufacturing, especially from Japan.) O'Boyle explained how the compressors that GE used in the refrigerators were nearly identical to the compressors used in the GE air conditioners, with one exception: powdered metal, instead of hardened steel, which was less expensive and more easily molded to fit the need. "It turned out to be disastrous," O'Boyle wrote. Problems with the refrigerators, which sold well, started showing up in July 1987—first in Philadelphia, then in Puerto Rico. Six months later, the GE engineers still couldn't figure out what was wrong. Schipke decided to pull the plug. One GE engineer told O'Boyle the decision literally made him sick. "How do you think the engineers at Thiokol felt when the Challenger blew up?" Schipke asked.

O'Boyle began to get "the true picture" of GE during his reporting on the refrigerator debacle. He decided he wanted to write a book about the company. He wrote to his literary agent, Edward Novak, that a book about GE had all the elements of a Shakespearean tragedy: "power, glory, scandal and hubris." There was also a "charismatic, innovative and powerful CEO; a corporation with brand-name recognition; a colorful cast of characters doing flat-out bad things; plenty of angry sources both inside and out of the company who are willing to spill their guts; and a sense of a faltering giant being brought down by corruption that emanates from the top on down." There was more. Jack was "responsible for a series of stunningly inept blunders," was leading a "corporate culture that has tolerated and even promoted recklessness, lawlessness and the persistent waste of corporate assets." Jack was "combative to the point of viciousness" and was "a brilliant but abrasive boss" with a "voracious appetite for detail, a photographic memory and the stamina to work most subordinates under the table." In the proposal, O'Boyle quoted a former college roommate who remembered that Jack wasn't "blessed with a lot of grace or athletic ability" but still "trounced people by trying harder," and he quoted a former employee, Don Awbrey, about Jack, "He delivers, but at what cost?" Jack, Awbrey said, was "totally ruthless" and created "an aura of fear at GE" and didn't "engender trust and integrity."

He also claimed that GE's board of directors "censured" Jack for "persistent womanizing" and that his "illicit sexual behavior ruined his first marriage, which ended in a messy divorce." There were the stories about how when Jack would visit the GE medical system business in Milwaukee, he would demand to be set up with hookers while in town. "That shit wouldn't fly, obviously, today, but this is the 1980s and there were a lot of things that happened in the past that wouldn't fly today," one longtime GE observer told me. "It wasn't #MeToo then; it was all for Jack." O'Boyle felt he could relate to Jack and get inside his head, as they were both Irish Catholic guys from hardscrabble backgrounds. "I get the Welch thing," he told me. "Every day you have to prove yourself. That's how I am, and that's how he was. And part of the proving yourself is not just—and this is where he had genius—it's not just what you do, it's selling what you do."

O'Boyle proposed to write a 100,000-word manuscript and have it finished a year from the signing of a contract. He planned to take a sabbatical from the *Journal*, assuming the publisher's advance was sufficient. It was a compelling—and counterintuitive—proposal. After all, Jack was being generally lionized at this point—a lionization that would continue for years—and a book about GE's underside, that ran counter to the prevailing narrative, was likely to find a home. In short order, Jonathan Segal, the vice president and senior editor at Knopf, bought O'Boyle's proposed book for an advance of $375,000. "At the time, that was a whole heap of money," O'Boyle told me. His boss, Carol Hymowitz, who legally witnessed O'Boyle's signature on the contract, told him, "You're going to be a millionaire!"

Somehow, a few weeks later, Jack had got hold of O'Boyle's book proposal. According to O'Boyle, Jack was beyond furious, tore up the proposal, and tossed the pieces across the room. "Who is this fucker O'Boyle and what does he think he's doing?" Jack screamed. Jack called Paul Steiger, the *Journal*'s managing editor, to complain about O'Boyle. Steiger called Hymowitz. Hymowitz called O'Boyle. "Watch out," she told him. "Welch is beyond ballistic on this one." That was putting it mildly.

On July 7, 1992, Dan Webb—a litigator at Winston & Strawn—wrote O'Boyle a threatening eight-page letter on behalf of Jack and GE. Webb was also a former U.S. Attorney. It was the start of a six-year battle between Jack and O'Boyle designed to intimidate him and to influence the book he was writing. Sources try to do that to journalists all the time, of course. That is the nature of the beast. What made this lengthy interchange so creepy was that, in an age before email and social media, GE had gotten O'Boyle's book *proposal*—he had not even written a word of the book—and engaged the services of one of the nation's top litigators to try to thwart him because Jack and GE did not like the direction in which O'Boyle seemed to be heading.

Webb also knew that O'Boyle had received a book contract and that he was taking a leave of absence from *The Wall Street Journal*. "Our clients, of course, respect your right to write a book about them," Webb began. "Per GE's usual practice regarding the media, they'll respond to any reasonable inquiry from you to help insure accuracy and balance." But, Webb continued, the book proposal was "replete with defamatory, false and biased statements which violate both the law and journalistic standards. Our clients strongly object to such statements and ask that you immediately retract the libelous statements and observe traditional journalistic ethics by fairly finding the facts before reaching your conclusions." Webb found it "incredible" that O'Boyle could write a book about GE despite only having written one story about the company—about the refrigerator compressors—and without having talked to any of GE's "senior officials." Webb claimed that O'Boyle had sold the book on a "lurid—and wholly unsubstantiated—'decline and fall' thesis" about a company that "is, in fact, one of the most successful in America." Webb's lengthy letter was both nasty and chilling. He made sure to copy Paul Steiger, the *Journal*'s managing editor; Alberto Vitale, the CEO of Random House, the publisher of O'Boyle's forthcoming book; and Sonny Mehta, the editor-in-chief of Knopf.

During the next six years, as O'Boyle worked on his book, he said he had received a dozen nasty letters from Webb. (Webb did not respond to my repeated requests for comment about his letters to O'Boyle.) He also received a pernicious and threatening letter from Walter Wriston, GE's lead independent director and the former CEO of Citibank. Thanks largely to these ongoing efforts to discredit his work, in June 1997 O'Boyle found himself in a bit of a dustup with Jonathan Segal, his editor at Knopf, and Elise Solomon, a lawyer at Random House, Knopf's parent company. "If there was a dust-up," Segal emailed me, "it couldn't have been a very bad one, because I stuck with the book, when I think it's possible other editors and publishing houses might not have, and Tom stuck with us." Jack and his lawyer had raised many questions about the accuracy of the assertions made in O'Boyle's manuscript—without having read the pages, of course— forcing O'Boyle to defend his reporting to his publisher. "I am aware of your concerns and will seek to answer each and every one of them," O'Boyle wrote to them, "down to the smallest scintilla of doubt on your part. I have no desire to publish anything that could be deemed defamatory or libelous." He was rightly concerned that doing so would damage his reputation and expose him to financial liability. "At the same time," he continued, "I refuse to be bullied just because my subjects, a wealthy corporation and a powerful CEO, do not like what is being said about them. . . . If we are indeed people of integrity, we have to be guided by what we believe to be right rather than bow to the path of least torment." All of this was highly unusual, of course, and resulted from Jack obtaining the purloined copy of O'Boyle's original book proposal—apparently from one of O'Boyle's former *Wall Street*

*Journal* colleagues—and then hiring a top-flight former prosecutor to make O'Boyle's life difficult.

In an interview, O'Boyle said Jack tried to "suppress" his book. "On some level, even though he made my life hell on earth, I don't blame him, maybe," he said. "But to me, it was—when I talk about the guy being a lizard, this is what I'm talking about." He said Jack was "very good" at making friends and very skilled at spin. "Jack knew how to play the dimensions of publicity better than anybody," O'Boyle said. "There will never be another Jack Welch. I don't think there could be, because the world has changed so much with social media and the ability to get alternate information and the fact that there's always going to be a detractor out there. You can't control the narrative the way you could then. It speaks to a time when the narrative could be controlled."

Thanks to Jack and his legal pressure, the final manuscript was "triple vetted" by two different sets of attorneys and an in-house counsel. "They were really, really worried, because GE threatened to sue," O'Boyle said. He hired Roz Litman, a First Amendment attorney in Pittsburgh, to help him as his interests started diverging from his publisher's. "I think [she] may have saved the project," he said. "She was a real tiger. She went after them as hard as they went after me and gave me the space that I needed to write a manuscript."

When the book was published, in November 1998, neither O'Boyle nor Knopf heard a peep from GE, Jack, or Webb. The years of legal intimidation had all been bluster. "GE didn't do anything," O'Boyle explained. He said Knopf had "big, big plans" for the book but after a harsh *New York Times* review, written by Roger Lowenstein, a former *Journal* reporter who had written a bestselling book about Warren Buffett, Knopf shelved them. Lowenstein wrote that O'Boyle presented Jack as a "thin-skinned monarch who has destroyed a company and a culture in pursuit of short-term profit." He praised O'Boyle's "muckraking" ability and for excavating a number of scandals that plagued

Jack's tenure at GE. But then he criticized O'Boyle for failing to have "equal resolve at getting both sides of the story." Forget that Jack wouldn't talk to him and that GE had tried to intimidate him for years. There was also a critical review in *BusinessWeek* written by John Byrne, soon to be Jack's ghostwriter. "Much of O'Boyle's material makes for some entertaining, but not very enlightening reading," Byrne wrote. "However, the entertainment value comes at the expense of the author— for in the end, O'Boyle hacks not at Welch's credibility but his own."

To be sure, O'Boyle got several laudatory reviews and warm personal letters that took some of the sting out of Lowenstein's review. But the *Times* being the *Times*, Lowenstein's review put the hurt on his book sales. *The Wall Street Journal* ignored the book entirely, even though O'Boyle had been a star reporter at the paper. There was no review and no excerpt. He said he thought it might have been because Paul Steiger, then the *Journal*'s managing editor, "was close to Welch." He did get a handwritten note from Norm Pearlstine, his former boss at the *Journal*. Pearlstine wrote that he had read O'Boyle's book and had read Lowenstein's review and that he agreed with the review. "As Roger told me, reading your book was like reading a biography of Babe Ruth that covered everything except his home runs," he wrote. "Much of our disconnect, I am sure reflects a true difference of opinion about the role of the publicly held corporation and its CEO. I also think Welch's unwillingness to spend time with you deprived you of needed insights into the man. I have had the opportunity to spend a fair amount of time with him over the past decade and my impressions are obviously very different than yours."

Knopf "shut it all down" after the *Times* review, O'Boyle said. "It was kind of a kick in the teeth, especially after you'd worked on something for six years." The experience of writing *At Any Cost* changed O'Boyle. "This was such a tough experience and tough circumstance for me as a writer and a reporter that it really did bring me closer to Christ and to the Lord," he said, a remark that was a bit of a shock, especially from a hard-boiled former *Journal* reporter. He now works as the director of communications at the Beverly Heights Presbyterian Church in Mount Lebanon, Pennsylvania. "I would say it is a major contributor to why I am working at a church today," he continued. "It became the most adversarial project and frankly, on some level, frightening. I had to go to a place where I could find faith in something larger than myself, and that faith was found in Christ. It changed my life. It truly changed my life. It was one of the seminal events. When I look back on my life, it was one of the seminal events in my spiritual journey. You feel so threatened and so anxious about a situation that you don't really have anyplace else to turn, and I turned to God. I prayed every day for four years about that project. I had a lot revealed to me through it about how the Lord is revealed to you in times of tribulation and trial. You have to lose your life to find it. . . . I truly thank Jack Welch, because he helped to bring me to Christ."

## Chapter Eleven

# A TALE OF TWO ACQUISITIONS

By the mid-1980s, Wall Street—doing deals, trading stocks and bonds—began to loom larger and larger in the American imagination, and GE (or at least its CEO) was not immune to its magnetic pull. This was an era, incredibly, when investment bankers attained a rock star–like status in New York City and in well-heeled commuter towns like Jack's hometown of New Canaan and nearby Greenwich. Dealmaking and its practitioners were now the stars of Hollywood movies such as Oliver Stone's *Wall Street*. Tom Wolfe's 1987 novel *The Bonfire of the Vanities* had as its protagonist a bond trader. And people ate it up.

Against this riotous canvas, GE was busy building up its own internal financial prowess while also pursuing two supposedly transformational deals—one for RCA, the mini conglomerate that owned NBC (and that GE had once owned), and another for Kidder Peabody, the venerable investment bank—deals that would showcase the strengths and weaknesses of Jack's tenure as CEO. By then, the three major television networks were in play. Back before the internet and streaming services, these were crown-jewel assets. They were where the vast majority of Americans got their news and where they turned to be entertained by regularly scheduled programs. Owning a television network conferred a certain cachet.

Given that GE still had $3.2 billion in cash on its balance sheet, even after the $1.1 billion acquisition of Employers Reinsurance, the inevitable question persisted: What would Jack buy next? Jack badly wanted a TV network, especially after his unrequited dalliance with Cox. In April 1985, there were rumors that GE had talked to CBS about buying the broadcast network in the event that someone made a hostile bid for the company. (Acting as a "white knight" for a company facing a hostile takeover was becoming a thing on Wall Street at the time, and GE was more than willing to play that role.) According to *The New York Times*, GE had agreed to pay nearly $4.5 billion for CBS in the event that it became the target of an unwanted bid, in particular from Ted Turner, the founder of Turner Broadcasting. Both CBS and GE denied the rumor.

Despite the public denials about GE's interest in CBS, Jack told me on several occasions that he tried desperately to buy CBS. "It came down to the wire," he said. He wanted to buy a business, like a television network, that foreigners could not own because of the rules against foreign ownership. He was most worried about the Japanese, especially how formidable they were as competitors in the manufacture of televisions. A television network, he told me, would give GE a protected source of cash flow, or at least protected from foreign competition. He said GE pursued CBS "quietly." He had dinner with Tom Wyman, the CBS CEO, at GE headquarters in Fairfield to discuss GE being CBS's "white knight." But in the end, in 1985, Wyman decided to sell the network to Larry Tisch, the billionaire investor who already owned about 10 percent of the company.

Complicating Jack's ambition to buy a television network was the fact that, also in 1985, Capital Cities Communications agreed to buy ABC for $3.5 billion. That left Jack with an obvious choice: to buy back for GE the RCA Corporation, which the Justice Department had forced

GE to divest in 1933. Wouldn't it be a touching bit of corporate irony if Jack were able to buy back RCA and make it all look like GE had never owned it in the first place? Coincident with Jack's growing interest in buying RCA, in August 1985, a low-level lawyer at RCA had called his counterpart at GE and began the process of ending the fifty-two-year consent decree that prevented GE from buying *any* RCA stock. Two months later, the Justice Department eliminated the consent decree, paving the way for GE to consider buying RCA outright. "Talk about dumb luck," Jack said. "None of us had a clue the consent decree existed." RCA still owned NBC, which had been a perennial third-place broadcaster, although its fortunes had started to improve with the success of *The Cosby Show* and the hiring of Grant Tinker to run the business.

At the time, RCA was an unwieldy conglomerate, a hodgepodge of disparate businesses, a mini version of GE itself. In 1984, NBC accounted for 23 percent of RCA's revenue of $10.1 billion and 35 percent of its pretax operating profit. In addition to NBC, RCA also owned car-rental company Hertz, the Random House book publishing business, a television manufacturing business, and a defense contractor. Incredibly, there was also CIT Financial, a growing financial institution, Coronet Carpets, and the Banquet Foods Corporation, a leading supplier of frozen foods, particularly of breaded chicken. André Meyer and Felix Rohatyn, two legendary investment bankers at Lazard, were often the ones who were called when RCA wanted to do deals. In 1966, Rohatyn advised David Sarnoff, then still the CEO of RCA, on its acquisition of Hertz. (Lazard received a $750,000 fee, then one of the largest ever.) In 1970, Rohatyn also sold Banquet Foods, as unlikely as that might have seemed, to RCA. Peter Lewis, also a Lazard partner, spoke with Rohatyn after he said he was going to sell Banquet to RCA. "Felix, I know you can sell anything to anybody but how in the world are you going to sell Banquet Foods, a frozen food company, to RCA?" Lewis once told me. "They're not in that business." Replied Rohatyn, who had advised many a conglomerate on their sometimes crazy acquisitions, "Nuts. That's why they're going to buy it."

Jack told people that Thornton Bradshaw, the RCA board chairman, was worried that RCA would get gobbled up and then split up by a private-equity firm. He decided that rather than have KKR or Forstmann Little buy RCA, he would do the buying and the splitting. "We had done some of the earliest private equity at GE Capital in 1984 and 1985, and Jack knew all about it and Bradshaw did, too," Bob Wright explained to me. "He just felt they were going to attack him piece by piece. . . . He'd rather see GE break it up than some private-equity firm."

By the fall of 1985, Lazard's Rohatyn had caught wind of Jack's interest in CBS. At about the same time, RCA had started selling some assets. In 1984, it sold CIT to Manufacturers Hanover Bank for $1.5 billion, and in June 1985, it sold Hertz to UAL Corp., the parent of United Airlines, for $587 million. An astute banker, Rohatyn had befriended Bradshaw as he had Sarnoff, and they began discussing RCA's strategic options. By paring down RCA's business, Bradshaw figured RCA was becoming a more attractive acquisition candidate. Rohatyn had done nothing to discourage Bradshaw in this thinking. Among those rumored on Wall Street to be interested in RCA were Ford Motor Co., United Technologies, and General Dynamics. The wealthy Bass family of Fort Worth, Texas, had accumulated RCA shares for "investment purposes." Donald Trump also owned a stake in RCA, as did the Bendix Corporation. KKR, the private-equity firm, offered to work with management to take the company private in a leveraged buyout. Marty Lipton, a senior partner at Wachtell Lipton and one of the country's leading M&A attorneys, also advised RCA. RCA did what it could to try to thwart hostile raiders eager to get their hands on its $2 billion of cash and NBC: it put in a "poison pill"—Lipton's specialty—making it more difficult for a raider to get control of the company, and it also staggered the terms of the board members, making it harder for activist investors to get control of the board. Failed overtures were made to

Disney about buying NBC and to MCA, the powerful Hollywood movie and television conglomerate, about buying the whole company. RCA was in play.

Bradshaw had joined RCA from ARCO, a big oil company, four years earlier. He found his successor in Bob Frederick, one of the GE executives whom Jack had bested for the chance to succeed Reg Jones. In March 1985, RCA named Frederick CEO, with Bradshaw remaining as chairman. *Time* described Frederick as "a decisive, demanding administrator who can often be found running on a treadmill at a Manhattan health club by 7 a.m." On the wall of his office, Frederick had the photographs of his eleven predecessors. The first picture was of Owen Young. Four of the pictures had gone up on the wall in the previous nine years, an indicator of the company's management turmoil before Bradshaw came along to try to stabilize things. "If all goes as the RCA powers hope," the *Times* reported, "the 13th picture will not be nailed up for another six years," when Frederick reached RCA's mandatory retirement age of sixty-five.

But it was not to be. "Poor Bob," Jack told me. In early November 1985, Rohatyn called Jack "out of the blue," he said, and asked if he would like to meet Bradshaw. That wasn't really what happened. A *New York Times Magazine* article from September 1987 reported that, *actually,* Jack had asked Rohatyn to set up the meeting with Bradshaw. A *Time* magazine cover story from 1985 quoted Jack as saying he had placed an "early morning" call to Rohatyn and asked, "Can you arrange for me to meet Brad?" as Bradshaw was known around town. Rohatyn was impressed. "You don't see the chairmen of some of the very big old-line companies making decisions that way," he said. "His grasp of numbers is superb. It is superb."

GE had been Bradshaw's first choice, so karma had a role in the deal, too. Rohatyn and Bradshaw were especially concerned about confidentiality. If the deal leaked, RCA would terminate discussions. "Leaking just upsets a lot of people," Rohatyn said. "It upsets the markets. It upsets the employees. And if you announce that you're in a merger discussion and it fails, then you've gotten two things wrong." Jack told Rohatyn that only he and Jack's "chief planning guy" would know about the discussions, which was almost true.

On November 6, Jack and Bradshaw met at Rohatyn's elegant eighth-floor apartment at 810 Fifth Avenue, overlooking Central Park, for a drink. On the wall were a 1729 Canaletto painting of the Grand Canal in Venice, looking toward the Rialto Bridge, and a 1738 Bernardo Bellotto of the Campo and the Church of Santa Maria Formosa, also in Venice. Bradshaw, fresh from an RCA board meeting where he and Frederick had presented RCA's strategic plan, was dressed in a tuxedo; he had a formal dinner at the Navy League to go to after the meeting and had to give a talk. Jack reported that there was no mention of a specific deal that night but there was good chemistry between the two men. They both wanted to be important business leaders. They both were nervous about the competitive threat coming from Asia. "I discovered that we thought almost alike," Jack told *Time*, "and when you meet people with the same philosophical bent and you both see global markets and you can both agree, you move."

They spoke for less than an hour and didn't make an immediate plan for a follow-up visit. Bradshaw left Jack with two thoughts: there could be no leaks about a possible deal, and Jack would need to make the first move. Even Bob Frederick, the RCA president, didn't know about the meeting at Rohatyn's apartment. Jack was smitten. He named the project "Island" and put together a team comprising Dammerman, his CFO, and Carpenter, his M&A guy, to learn everything they could about RCA from public information. By that point in Jack's tenure, GE had looked at an astounding three thousand companies as possible acquisitions. But 90 percent of them, as Carpenter had said, didn't make sense: wrong price, wrong fit, wrong culture. But RCA seemed like a good fit. Carpenter prepared a sixty-page briefing book for Jack. Only fifty other companies had merited the briefing book.

On the day before Thanksgiving, Jack brought the team together, along with Bossidy, to "wallow" in the pros and cons of the deal. Asked what he meant by "wallowing," Jack once told me that meant to make sure the staff and the team had a chance to speak their minds. "Take time to wrestle acquisitions, divestitures, labor negotiations, etc. to the ground," he said. "Business is full of paradoxes. Wallowing is encouraging people to hang out in the paradoxes and sort them out." For four hours before the Thanksgiving holiday, the GE men hashed out whether to do the deal. They all loved and wanted NBC, of course, which, thanks to Grant Tinker, was soon gushing cash. But there was also a fit between the two companies' defense businesses: GE made land-based radar systems; RCA's were seaborne. GE made military satellites; RCA made civilian satellites. They each had a semiconductor business and they both made television sets. There was also a sense that there was a unique window from an antitrust point of view, because the Reagan administration seemed to be letting more and more deals get done without an antitrust challenge. The GE team valued the broadcast business at $3.5 billion, based on its previous work analyzing CBS. "If we could stomach paying about $2.5 billion for everything else, then the deal could be a home run," Jack said.

Jack urged everyone to think the matter over some more during the Thanksgiving holiday. He challenged them to answer the question "Ten years from now, would you rather be in appliances or in network television?" Jack and his family went to their home in North Palm Beach. "On Monday morning," he said, "to a person, we all came back fired up to go." The numbers worked, he said, and many of RCA's other businesses fit with GE's. "It was a go," Jack said. On the same day, December 2, John Weinberg at Goldman Sachs called Rohatyn to set up a second meeting. As that was happening, Jack called Rohatyn and told him that "at the right price," GE was interested in buying RCA.

Jack and Bradshaw met alone on Thursday, December 5, at the Dorset Hotel, where RCA had a duplex suite that Bradshaw had been using while his apartment was being redecorated. "I'd like to buy your company," Jack told him. "Our companies are a perfect fit." Jack said he would offer RCA shareholders $61 a share in cash, a 28 percent premium to where the RCA stock had been trading. He also made clear he was flexible on the price. Bradshaw, "in his professorial way," Jack recalled, let him know the offer was insufficient—no surprise—and so they parted ways. They agreed to pursue the deal but disagreed on the price GE would pay.

On Friday, Jack said, "things got a little hairy." It turned out that it was time for Bradshaw, the RCA chairman, to tell Frederick, the RCA CEO, that he had been discussing selling the company to GE. Frederick had been in Los Angeles for a business meeting, and the 6:30 a.m. call from Bradshaw, to tell him the news, woke him up. Frederick felt betrayed. He immediately flew back to New York City and began to try to organize board members to block the sale. He had two days to try to persuade them. The RCA board was meeting on Sunday to approve the GE deal, or not.

It was a long shot for Frederick, but not impossible. His chief disadvantage, being new, was that he did not have the depth of relationships with his fellow board members that Bradshaw did. He found sympathetic ears in John Petty, the chairman and CEO of Marine Midland Bank; in Donald Smiley, the retired CEO of R.H. Macy & Co.; and in Robert Cizik, the CEO of Cooper Industries. Petty told Bradshaw he was irritated that Bradshaw had failed to inform the board about his discussions with Jack, especially about the November meeting at Rohatyn's apartment. He was also peeved that Bradshaw had not discussed with Jack what was going to happen to RCA's top management in a sale to GE. He wondered why Bradshaw had decided not to auction the company, which might well have resulted in a higher price for RCA and would have given board members legal cover in the event of shareholder lawsuits alleging that the company was sold too cheaply. "Bradshaw was unmoved," according to the *Times*.

Petty tried rallying a few of his fellow board members to his cause from an empty office in Rockefeller Center. But he struck out: they were hard to find or didn't respond to his calls. (It was the era before email and cell phones.) Defeated, Petty flew to Moscow, where he had previously agreed to serve on a panel. Frederick, with the most to lose (including his job), could have made a bigger stink. But he wasn't one to rock the boat sufficiently to get done what had to be done. "If Frederick had been more aggressive, there'd still be an RCA," one former RCA executive told the *Times*.

By the time of the Sunday board meeting, at the offices of Wachtell Lipton on West Fifty-Second Street, the decision to sell RCA was a fait accompli. The eleven-member board voted nine to one to pursue a deal with GE, with Frederick being the lone dissenter. (Petty, in Moscow, did not vote.) Bradshaw called Jack with the "good news," he recalled, but also explained that the price was still too low. Over the next few days, Jack, Bradshaw, Rohatyn, and Weinberg hammered out a deal at GE's suite in the Waldorf Astoria Hotel on Park Avenue. Bradshaw wanted $67 a share for the company; Jack had offered $65. Instead of splitting the difference—$66 per share—Jack decided to go further. He offered Bradshaw $66.50 a share, in cash, "probably 50 cents more than he expected," Jack remembered, and then explained why: "I always tried to leave some goodwill on the table when the seller's ongoing involvement was important to the company's success."

At 6:00 p.m. on December 11, the GE board approved the RCA deal. An hour later, by unanimous vote, the RCA board approved the sale to GE. (By this time, Frederick and eight other RCA executives had been given four-year employment agreements. If he stayed, Frederick was to receive his $2 million annual salary plus $4.45 million for his RCA stock options.) Jack and his team left the Wachtell offices after the vote and headed over to 570 Lexington—the former RCA headquarters building that GE obtained in 1933 as part of the agreement to spin off the company—and broke out the champagne to celebrate.

Jack, Bossidy, Dammerman, and Carpenter "were like kids," Jack remembered, adding how excited they were to look out "through the fog" to Rockefeller Center three blocks away and to see the RCA name atop 30 Rockefeller Center. "We could hardly wait to get the GE logo up there," he said. From the first meeting at Felix Rohatyn's apartment to announcement, the deal—the largest nonoil merger in history—had taken thirty-six days and sealed Jack's reputation as one of his generation's savviest dealmakers. "The acquisition of RCA is the boldest move ever for the company," *Time* reported.

The $6.3 billion acquisition was front-page news. *The New York Times* described how Rohatyn had been a frequent breakfast partner of Jack's and how at one of those breakfasts Jack had asked Rohatyn to set up the November meeting with Bradshaw. The paper also reported how Rohatyn had left New York City on the weekend of RCA's Sunday board meeting to return to Paris to visit his ailing mother. It was noted that he had taken the Concorde both ways to make sure he got back in time. (This was the kind of publicity that Rohatyn had long relished and that Lazard could not buy. That and a fee rumored to be $16 million made the deal a bonanza for the firm.) There was the interesting detail that Jack had wanted to secure an option for GE to buy two television stations from RCA—in New York and Los Angeles—as a way to discourage another bidder from coming along and offering a higher price than GE did for RCA. But RCA rejected that request and instead gave GE an option to buy 28 million shares of RCA stock at $53 a share, in effect giving GE a big breakup fee—worth hundreds of millions of dollars—if someone else came along.

The deal was a triumph for Jack. At the press conference the next day, he was jubilant. "This is going to be one dynamite company," he said. Both he and Bradshaw portrayed the deal as an American bulwark against foreign competition, specifically from Japan, Korea, and China. But there was plenty to do before the deal closed. There was the February 1986 RCA shareholders' meeting in New York City. The Justice

Department had to approve the deal to make sure it didn't violate antitrust laws. (GE still owned a television station in Denver.) The Federal Communications Commission also had to approve the transfer of NBC and its more than two hundred affiliated local television stations to GE. RCA also directly owned five TV stations and eight radio stations. Getting the necessary approvals was looking like a nine-month process. But Jack wasn't worried. GE had already reached out to officials in Washington.

He was right not to worry. GE easily won the approvals it needed for the deal. Many thought Jack had gotten the deal of the century by buying RCA for a cheap price. Two years later, *The New York Times Magazine* wondered, "Did RCA have to be sold?"

Jack was not concerned about such intellectual arguments. He knew he had pulled off a major coup with the RCA acquisition. "Foreigners couldn't own a network, so Japan couldn't buy the network," he told me. "I had a cash flow and a moat"—protection against possible competitors—"so immediately I wanted to buy a network. Not out of any strategy, other than to have cash. Cash is king. If you have cash, you own the world. No matter how big the company, if you don't have the levers to pull to change, you're dead. It's just a matter of time. I wanted cash and I wanted a moat, because it turns out nobody had a moat. We were getting killed in television sets by the Japanese. Everyone was going down the drain because of the Japanese. I was going to have a barrel of cash coming from NBC every day, and I could maneuver GE Capital and other things from that cash. It's a solid strategy, and no Japanese."

A bigger problem for Jack was who would run NBC. He wanted Grant Tinker to stay. Thanks to prime-time shows such as *Cheers, The Cosby Show,* and *The Golden Girls,* Tinker had made NBC number one in a few short years and improved the network's cash flow to $333 million in 1985, from $48 million when he took over in 1981. But his five-year contract was up in July 1986 and he had been commuting to New

York from his home in California. At sixty, he was tired of the traveling and wanted to return to California. Jack tried to persuade him to
change his mind. Over dinner in New York, "I offered him, what was
for me, an ocean of money," Jack said. But Tinker turned him down—
something that did not happen to Jack very often. "There was nothing
I could do to get him to reconsider," Jack continued.

The Wall Street Journal speculated on four NBC insiders who could
succeed Tinker, not including Brandon Tartikoff, the thirty-seven-year-
old NBC programming whiz, who the newspaper said was a "long shot"
because he wanted to stay in his role in charge of programming. But
Jack had his own succession plan for NBC and had it "from day one":
his old friend Bob Wright, who was running GE Credit after GE had
sold the housewares division to Black & Decker. "Bob was the perfect
fit," Jack said. "He had a feel for the business, was familiar with GE,
and was a close ally as we moved into strange territory." Jack gave
Wright the job in August 1986. "The media asked, 'How could this GE
guy run a network?'" Jack recalled.

As Wright was settling into his new role, Jack set about dismantling the pieces of RCA that didn't fit his best-in-class agenda for GE.
By the fall of 1987, he had mostly succeeded, cementing the RCA deal
as one of his best deals, if not his very best. He sold RCA Records to
Bertelsmann AG, a German media conglomerate. TVs went to Thomson, which was owned by the French government; Radio to Westwood
One. The carpet business and a variety of other businesses were spun
off into a separate company. The David Sarnoff Research Center, in
Princeton, New Jersey, was given to SRI International, along with a
promise of $250 million worth of contracts over five years. GE realized
around $1.3 billion from the sale of the RCA businesses, not including
the additional $1 billion it reaped from the deal with Thomson.

Jack's deal with Thomson was particularly brilliant and would pay
dividends to GE for decades. His idea was to swap a highly competitive
business he hated—the manufacture of televisions—for a business he

coveted—medical equipment. Whereas once RCA and GE were leading players in the manufacture of consumer televisions, by the late 1980s, the Japanese manufacturers had dusted the Americans. They could make TVs better, cheaper, and more profitably than anyone else. Jack may have hated the business, but he wanted to keep his options open on the small chance that the GE/RCA combination might give him an unexpected boost in TVs. In early 1987, Jack decided to allocate $20 million to modernize the big RCA television-manufacturing plant in Bloomington, Indiana, and to resume domestic production. In July 1987, *Fortune* wrote about how GE and Jack were relying on management consultant Ira Magaziner, then thirty-nine, whose insights into how American manufacturing might better compete with the Japanese had become Jack's "bible" for the TV business.

It was all a big head fake. A month earlier, in June, Jack was in Paris entertaining customers at the French Open along with Paolo Fresco, a vice-chairman. NBC was broadcasting the tournament; being Jack's guest was a hot ticket. Alain Gomez, the chairman of Thomson, the big France-owned electronics company that was once the French subsidiary of the Thomson-Houston Electric Company, visited with Jack in the GE box. The next day, Jack and Fresco went to see Gomez. Jack had a plan. He coveted Thomson's medical imaging business, CGR, which Jack described as "very weak." He wanted it anyway. He wanted to combine it with GE's medical equipment business—comprising X-ray equipment, CT scanners, and MRIs—which was number one in the U.S. GE had no medical equipment business in France because the French government blocked competitors from entering the market. Jack figured if he owned CGR the French government would have to allow GE into the country.

Gomez quickly let Jack know he did not want to sell the business. Jack was unbowed. Magic Marker in hand, he offered to swap one GE business after another to Gomez in exchange for CGR. Gomez had no interest in GE's semiconductor business. But when Jack suggested swapping GE's TV manufacturing business, Gomez "liked that idea immensely," Jack said. "Alain saw the trade as a way to unload his losing medical business and overnight become the No. 1 producer of television sets in the world." Gomez agreed to work with Fresco to see if the swap could happen. Back on the street, Jack grabbed Fresco by the arm. "Holy shit," he said. "I think he really wants to do this."

Excitement aside, the proposed swap was not the easiest to engineer. The French government's ownership of Thomson complicated things. As did the fact that GE's consumer electronics business had $3.5 billion of sales in 1986 and 31,000 employees while Thomson's medical equipment business had revenue of about $770 million. But despite the challenges, by July 23, Jack had pulled it off. The two companies had agreed to swap GE's TV manufacturing business for Thomson's medical equipment business. Thomson also agreed to pay GE $1 billion in cash and transfer a portfolio of patents that, according to Jack, generated $100 million a year of after-tax profit, for fifteen years. The agreement made Thomson the largest manufacturer of television sets in the world and tripled GE's market share of medical equipment in Europe, to 15 percent. *Time* started referring to Jack as "Jumping Jack Flash" because "he leaps on a good deal when he sees one." The sale of the business came as a bit of a "surprise" to the man running GE's television business, Richard Miller, who was the only senior RCA executive whom Jack retained in GE's senior management. Miller met Gomez in Paris in the first week of July. They hit it off and Gomez asked Miller to keep running the business under Thomson's ownership. He agreed.

IN 1987, THE RCA DEAL ADDED AN UNEXPECTED 14 CENTS PER SHARE to GE's earnings—Jack had assumed the deal would be dilutive to GE's earnings in its first year. He also used the proceeds of the asset sales to pay down GE's debt, further preserving GE's AAA credit rating. And despite NBC having higher earnings during 1987 than RCA as a conglomerate had recorded during its entire history, Jack was determined to streamline the broadcast network. In a closed-door session for NBC executives in March 1987, Jack "sent a chill through the network," the *Times* reported. Yes, he was proud of what NBC was accomplishing, but, he reportedly said, a bunch of NBC executives were "turkeys" who would soon be looking for work. In fact, more than eight thousand RCA employees were fired, another three thousand retired, and some thirty thousand went with the businesses that were sold. "In July 1987, the red RCA sign high above Rockefeller Center passed a milestone," the *Times* reported. "It was just 50 years since the Rockefeller interests, eager to have General Sarnoff occupy their flagship building, permitted RCA to become the only company in the complex with an illuminated sign. It sits there still, though RCA is nothing but three initials inside G.E." Soon enough, those three letters would be gone, replaced atop 30 Rock with the two letters of the company that had bought it and dismembered it.

IN APRIL 1986, WELL BEFORE THE RCA DEAL HAD CLOSED, JACK'S EGO got in the way of his strategic judgment on a very different sort of acquisition. Jack had been trying to purchase a Wall Street investment bank for at least three years. Like Hollywood, Wall Street has always been a dangerous place, one where outsiders need to tread carefully, lest they get their heads handed to them. But Jack was intrepid, and especially so after the praise he received for the RCA deal. He plunged ahead. At different times Goldman Sachs, Morgan Stanley, and Lehman Brothers had all told him no. Now he and Bob Wright—who was then still head of what was known as GE Financial Services, the holding company for GE Credit and Employers Reinsurance—decided to buy 80 percent of Kidder Peabody, the privately held Wall Street investment bank, for $600 million. (The other 20 percent of Kidder remained with its partners.)

Initially, even Kidder turned Jack down. The thinking of Kidder's management about a sale began to change slowly in January 1986, with the announcement that Morgan Stanley, the respected investment banking partnership, had decided to go public. According to *The New York Times*, the Morgan Stanley news could only be read one way inside Kidder: it "had to act, and act fast." Max Chapman, the head of Kidder's fixed-income department and a member of the firm's management committee, told Ralph DeNunzio, Kidder's boss, that he felt "like we were falling further and further behind." He said he told DeNunzio, "We just had to address the capital question." Chapman wanted the extra capital to use for his growing, and highly profitable, trading operation. "The more capital we had, the bigger positions we could take," he said in an interview decades later. "Ralph and others accused me of being the real impetus behind 'We needed to do something.'"

On February 1, the "drama immediately intensified," according to the *Times*, when Marty Siegel, Kidder's highest-paid executive and its star M&A adviser, showed up at DeNunzio's Riverside, Connecticut, home to tell him that he was leaving to join a rival, Drexel Burnham Lambert. Whereas Kidder could be paid a million dollars here and there for advising on an M&A deal, Drexel, thanks to junk-bond king Michael Milken, had figured out how to get paid tens of millions of dollars for arranging the financing on a big takeover deal. For instance, while Kidder got paid $1 million for advising GAF Corporation on its unsuccessful bid to buy Union Carbide Corporation, Drexel had earned $43 million for providing the financing. Siegel was voting with his feet. "Marty's leaving shook us up," DeNunzio told the *Times*. "We probably should have moved faster to build our capital. We were looking more like a middle-sized firm in a land of giants."

In any event, the names on DeNunzio's buyers' list were E.F. Hutton

Group, another big brokerage firm; Deutsche Bank; Sanwa Bank; General Motors Acceptance Corp.; and, of course, GE and GE Capital, which had earlier expressed interest in a deal. "They stood out on the list as the obvious choice that really made business sense," DeNunzio said.

On April 9, DeNunzio called Jack. They had dinner in Fairfield on April 14, with Jack, Bossidy, and Wright, who had been designated the "point man" for the Kidder negotiations. Wright crammed all weekend about the securities industry to prepare for the meeting. Over a meal of pasta and shrimp, in a dining room overlooking the Merritt Parkway, the three GE executives grilled DeNunzio. They seemed especially concerned about the growing loss of talent from Kidder to other firms. (It wasn't just Marty Siegel.)

"The whole subject of people was difficult for him," Wright recalled. Still, by the end of the dinner, it was clear a deal might be in the offing. Wright canceled long-standing plans for a GE employee-reward trip to China with his wife. Over the next few days and the weekend, the negotiations intensified, mostly at DeNunzio's Connecticut home. He wanted $600 million in cash and insisted that be the price, whether GE bought 80 percent or 100 percent of Kidder. Wright wanted the top Kidder executives to retain 20 percent of the firm to keep them from bolting and aligned with GE. He also wanted the top thirty-five Kidder executives to sign employment agreements and noncompete agreements. The price was three and a half times Kidder's book value, a historically high valuation for an investment bank.

Furthermore, according to the Kidder partnership agreement, when Kidder's partners bought into the partnership, they bought in at book value; when they left, and sold their shares, they sold at book value. "So this was quite a windfall," Chapman told me. But—windfall aside— Chapman, for one, was increasingly ambivalent about the sale to GE. He was slated to be the CEO of an independent Kidder, whether public or private, and he thought that a public Kidder was better (for him, anyway) than working for GE. But the firm's executive committee voted to move ahead with the deal, against Chapman's preference. He was despondent. He stayed home from work the next day. Bob Wright called him. Wright persuaded Chapman to stay at Kidder and work for him. Chapman persuaded his traders to stay, too. "Bob Wright is a pretty upstanding guy, in my mind," he said. (It's worth noting, though, that Wright failed to mention to Chapman that he was heading to NBC.)

The following Wednesday, Kidder called a board meeting for 7:00 p.m. in the Pierre Hotel in New York City. The firm's sixty-five officers showed up, although many did not know why the meeting had been called. Finally, at nine thirty, with the lawyering done, the GE acquisition was presented to them. They debated the pros and cons into the early-morning hours. Later the next day, the legal papers were ready for signing. That's when Wright told DeNunzio that there was a technical issue. The documents would have to be revised. DeNunzio was infuriated. He turned to Wright to start to complain, only to see that he had reached into his briefcase to bring out the three bottles of champagne that he had snuck into the hotel. He had been joking. "We paid a full price for Kidder," Wright told the *Times*. GE agreed to pay up for Kidder, making millionaires of many partners. "One Kidder partner just giggled when asked if the terms were favorable," the *Times* reported. DeNunzio was of two minds about the deal. "I clearly would have preferred to hand this company over to the next group as a private thing," he said. "I'm still very emotional about it."

According to Jack, the idea behind buying Kidder was "simple," but in reality, it was mind-numbingly misguided. GE Capital, as it now was called, had been financing leveraged buyouts for years by that time. LBOs were "hot" and going to stay that way, Jack decided. From 1983 to 1986, as the LBO craze was gaining momentum, GE Capital financed more than seventy-five buyouts, including one of the most legendary, early deals: the 1982 $80 million leveraged buyout of Gibson Greetings. Wesray Capital Corporation, a partnership created by William Simon, a former Salomon Brothers banker and Gerald Ford's

Treasury secretary, and Raymond Chambers, an investment banker, bought Gibson Greetings from—of course—RCA, which was still trying to shed assets. (Simon and Chambers came across Gibson Greetings when they succeeded in buying yet another company from RCA, TacTec Systems, a mobile telecommunications company.) Wesray put up $1 million of equity to buy Gibson Greetings—or 1.25 percent of the purchase price—with Simon and Chambers each investing another $330,000.

Of the remaining $79 million, an affiliate of Barclays Bank put up $13 million and GE Capital put up $39 million. The balance of the money came from selling off Gibson's real estate. Eighteen months later, Wesray took Gibson Greetings public. Simon and Chambers each reaped $70 million from their $330,000 investment, or 212 times their money. Many Wall Street observers mark the Gibson Greetings deal as the start of the boom in LBOs that continues—bigger and better than ever—to this day. For its part, GE Capital got its money back and a bunch of interest payments. GE Capital had also financed the buyouts of R.H. Macy & Co., Dr. Pepper, and Tiffany.

Jack didn't think that was an equitable distribution of the spoils. He wasn't wrong. GE was taking big risks financing these deals and wasn't getting properly compensated for taking those risks. He wanted to find a way to share in what he decided was going to be a big business. He thought owning Kidder would give him a better shot at getting more of the economics of the LBO boom. "We were getting tired of putting up all the money and taking all of the risk while watching the investment bankers walk away with huge up-front fees," he said. "We thought Kidder would give us first crack at more deals and access to new distribution without paying these big fees to another of Wall Street's brokerage houses."

Jack was right about one thing: the Kidder deal instantly made GE a big player on Wall Street, especially in the businesses of advising on mergers and acquisitions and the creation and trading of mortgage-backed securities. But it also was a huge culture clash with GE Capital. The Kidder bankers acted like prima donnas, expecting to get paid big bonuses each year by making money from their intellects, as opposed to from the firm's small amount of capital. GE Capital employees, who were making boatloads of profit from a combination of their good ideas and GE's seemingly unlimited capital, had much more modest pay expectations. Soon enough one group came to resent the other. There were few instances where the two firms worked well together, which was, of course, the original thinking behind the deal.

Jack had bought Kidder over the objections of the GE board members—several of whom were in a position to know how bad owning an investment bank could get. Walter Wriston, the longtime GE board member and Citibank chairman, told him not to do it. Lewis Preston, the head of J.P. Morgan & Co., told him not to do it. Andy Sigler, the chairman of Champion International, told him not to do it. But at an April 1986 board meeting in Kansas City, Jack convinced his board buying Kidder made sense. He got the board's unanimous approval. A week or so after the deal was announced on April 25, Jack was playing golf at Augusta—GE CEOs are allowed in the famously secretive club if they want to be—and teeing off on the third hole. A foursome came up the seventh fairway, saw Jack, and exclaimed, "For Chrissakes, Jack, what are you going to do next? Buy McDonald's?"

## Chapter Twelve

# THE BANK OF
# GENERAL ELECTRIC

The Kidder deal was all Jack, and his hubris. The fact that Kidder's CEO had not allowed GE to see Kidder's books before the deal closed left several of Jack's key lieutenants wondering what exactly they had just bought. This wouldn't be Bob Wright's problem, though: early September 1986 brought the confirmation that he would run NBC and that a rising star named Gary Wendt would succeed Wright atop GE Capital. Both decisions were fateful.

Whereas Wright had been CEO of GE Financial Services, which was the holding company for GE Capital, ERC, and Kidder Peabody, Jack named Wendt, then forty-four, to be head of just GE Capital alone. He asked his right-hand man, Larry Bossidy, fifty-one, still a GE vice-chairman, to be CEO of GE Financial Services, while Wendt was also named president and chief operating officer of GE Financial Services. Both Ralph DeNunzio, the head of Kidder, and Michael Fitt, the CEO of ERC, would be reporting to Wendt. Wendt would still be reporting to Bossidy, and not directly to Jack. It was as if GE's various financial arms were divisions of a private-equity firm with Jack and Bossidy as the founding partners and co-CEOs. "They really consulted each other," Paul Street, a longtime GE Capital executive, told me. "Jack and Larry. Yin and Yang." He remembered how in one of his first important meetings with Jack, Bossidy, and Wendt after he had been hired at GE Capital from McKinsey, they were all together in Fairfield around seven thirty one evening, rehearsing a big presentation to be given the next day. Bossidy and Wendt were seated in the front row of the auditorium at GE's headquarters, listening intently to Street as he rehearsed. Jack, seated by himself in the back row, had just returned from a vacation. He was going through his mail. He wasn't paying attention to Street. At some point, Bossidy jumped up and said, "Jack, for fuck's sake. This is important. Put your fucking mail down. Come down here and pay attention." Jack moved down to the front row. "Larry had that kind of relationship with him," Street said, "which I think was invaluable because Jack did go on a tear from time to time." Bossidy was the kind of guy who would joke with other GE executives and board members that wine cellars were for people who had too much "extra money" lying around. "Most companies would not chance moving the head of their most profitable and dynamic business out the door on short notice to do something else," Bossidy said, speaking of Wright's move to NBC. "The fact that we did is testimony to the depth of talent at G.E. Financial Services." But Jack figured Wendt still needed adult supervision.

Gary Wendt was a GE success story, albeit a quirky one. His father's family left Germany in the 1920s and settled in the small town of Rio, Wisconsin. They were poor, and his grandfather, whom Wendt never met, was an alcoholic. "I was kept away from him," he said. Wendt's father left school in fourth grade and went to work. He was a hard and accomplished worker. He got himself noticed in Rio. One day, another Rio resident met up with his father; they agreed to make a business partnership. Wendt's father would do the work; his partner would put up the capital. They quarried limestone. The Wisconsin soil was very acidic, and there was good business to be done mining and crushing the limestone and spreading it on the fields to neutralize the soil, making it more amenable to farming. He eventually did his own limestone blasting and crushing and made a good living for himself, his wife, and their only child, Gary, who was born in March 1942. He was raised by his mother "with a yardstick," he told me with a laugh. "She used it amply on me."

Through the osmosis of his father's work, Wendt decided he wanted to be a civil engineer. He read a book about dynamite and blasting that someone gave his father when his father was working on building the interstate highway through Wisconsin. The University of Wisconsin was thirty miles down the road from Rio. "Although at that time, it seemed like London," he said. Anyone who was in the top 75 percent of a graduating high school class in Wisconsin had to be admitted to the university. "I made it," he said. He started pursuing civil engineering. But it was a tough haul for him, since his training in mathematics wasn't the best.

His grades slowly improved while his interest in the subject waned. He started taking classes in the business school and was randomly introduced to a buyer who worked at Sears, back when Sears was a big deal. He encouraged Wendt to think more expansively about his future. His grades were good and kept getting better. The Sears buyer thought Wendt should apply to MBA programs. "Maybe you ought to think about going someplace else rather than Wisconsin," he said. "You've kind of been closed in here in this thirty-mile radius for your whole life."

The fellow got Wendt applications for Harvard, Stanford, Penn, and Berkeley. "To my total amazement, they were looking for somebody who would've been raised on a dairy farm in Wisconsin, I guess," he said. He got into all four schools. He chose Harvard. "There was a war going on and you had your choice," he said. "You could either get to keep your pen and your slide rule, back in those days, or you could go carry a gun. I didn't like guns."

Wendt got married to his high school sweetheart, Lorna, after graduating from Wisconsin, and together they drove to Boston, along with a U-Haul trailer, to enroll at Harvard. He had never flown in a plane. He had $2,500 in the bank. "We relied on each other for everything like young couples," Lorna Wendt remembered. She taught music to pay the rent and typed up his papers. "We also lived on the savings I had before," Gary Wendt recalled. "I worked full time in the summers and part time afterwards."

Harvard freaked him out. "I was totally overwhelmed by the situation at Harvard," he explained. "I was totally overwhelmed by these people who were sons of CEOs of major companies, and I mean, we had the guy at U.S. Steel, his son was there and Reynolds Aluminum, and some of them even actually drove Bentleys around campus." He liked Cambridge well enough and learned a valuable lesson in the process. "It had an awful lot of smart people running around there," he said, "although I learned later on that some of them weren't as smart as they behaved." He graduated in 1967. Next to his MBA, Lorna hung her "PHT," a "Putting Hubby Through" degree given to all the wives by the wife of the dean. The Vietnam War was still raging, and Wendt was wary of being drafted. He was offered a job at Lockheed, the defense contractor, but turned it down. And he wanted nothing to do with Wall Street, unlike many of his peers. He wanted to work for a small business.

One day, he saw a sign on the recruiting board: "Looking for a Harvard Business School graduate to come down and help me build a dome city." Somehow, this appealed to Wendt. It was put there by Harlan Lane, a "wheeler-dealer" from Houston who had made a fortune in the automobile business. One night Lane's brother got in a car accident while driving a company car and killed five people. By morning, the business had been sold. Lane took the proceeds of the sale and bought a bunch of land north of Houston and decided to become a developer. He wanted Wendt's help selling parcels of land.

Wendt flew down to visit Lane. Lane flew Wendt around Texas and Arkansas on his airplane, introduced him to his buddies, and told him about the job opportunity. When Wendt got back to Cambridge and thought about it some more, he decided he wouldn't take the job. But Lane wouldn't take no for an answer and offered Wendt a salary of $15,000 and a new Cadillac. Lane had persuaded him. Gary and Lorna moved thirty miles north of Houston. "I was still not just naïve but fairly stupid at this point," he told me. Lane offered Wendt the opportunity to buy stock in the land development company and took him around to see his banks, several of which agreed to lend Wendt the money to buy the stock. Wendt bought the stock, which, predictably, ended up worthless.

Wendt's job was to sell empty parcels of land. He didn't know how. And Lane refused to tell him what the prices were on the land. "I don't have any prices," he told Wendt. "You go make the best deal you can, bring it in to me, and I'll let you know whether it's enough." Wendt said this was the "best lesson" he ever got in business. He was a clueless Harvard Business School grad. He cried. But then he got good at the job. In the first day, he sold two parcels for $3,200 each. He had no idea if that price was right or if Lane would find $3,200 acceptable. It turned out to be the most anyone had sold a lot for in the subdivision. "Suddenly, I became king of the hill after one day," he said. "I learned to sell lots."

After Lane ran into legal trouble and his company was liquidated, the Wendts moved to Miami so Gary could take a job working for an early version of a real estate investment trust, or REIT. He spent his time getting loans to build subdivisions in Florida. The financing for the REITs was done using commercial paper backed by bank lines of credit. But in 1972, the market soured and the REIT went out of business. He had nothing to do. Wendt again needed a job. A headhunter representing GE Credit called him. "I certainly didn't want to go work for GE," he said. "That was the last thing I wanted." But he had a wife, two kids, and no money. He moved to Stamford, Connecticut. In his

Harvard MBA twenty-fifth reunion yearbook, Wendt joked, "After two false starts, I seem to have found my niche in something too big to capsize, the General Electric Co."

He had been interviewed and then hired by Larry Bossidy. "Bossidy's the reason that I got all my future promotions," he said. "If he hadn't been there, I would probably still be the manager of real estate investments." Wendt benefited from having Bossidy as a rabbi, but he never seemed to be able to click with Jack, who recognized Wendt's enormous talent but never liked him. Maybe they were too much alike: aggressive only children determined to succeed at all costs.

Wendt's first assignment was to clean up a bunch of loans that GE Credit had made to Florida developers. "I could go up and down the Florida Turnpike from Palm Beach to near Miami," he said. "Any exit I could find a GE project." Most were unfinished and few people or companies had the capital to complete them. But GE did. "When nobody else was finishing their units, we finished ours," he said. "We sold a thousand units during that winter, and we got reservations for another thousand. There we were. We'd cleaned up that whole mess."

Another mess Wendt cleaned up early on was the one GE Capital had with Roy Hofheinz, the Texas entrepreneur who built the Houston Astrodome and then lured a major league baseball franchise, the Houston Astros, to town. "He was just a brilliant business guy," Wendt said. GE Credit was part of a consortium of five banks that had made loans to Hofheinz, taking his assets as collateral. Unfortunately, Hofheinz had a terrible stroke that left him incapacitated. He had no other management that could run his business. "The business was bleeding all of the lenders blind," Wendt said. "We were sending money down there all the time to keep it alive." That's when Wendt had an idea: He arranged for a meeting with the other bankers and told them GE Credit would buy their loans for 35 cents on the dollar, take over the managing of Hofheinz's assets, including the Astrodome and the baseball team, and then sell everything when the time was right. Once GE got back its money, plus interest, Wendt told the other lenders, they could split up what was left. The banks took Wendt's offer. Wendt had become the principal owner of the Houston Astros. "It was more fun than I'll ever have the rest of my life," he said. GE Credit owned the Astros for one year but managed to increase attendance dramatically and turned the business into a moneymaker. Wendt sold off Hofheinz's assets, including the team. Wendt wanted GE Credit to keep the Astros. He was overruled. "GE got all its money back, all its interest on the whole damn thing, although it was embarrassing for the GE directors— and the GE CEO was Reg Jones back in that day—because the center fielder made more money than Reg Jones," Wendt said.

Wendt also led the charge, with Bossidy's approval, on a number of "tax credit" deals that helped GE avoid taxes and left GE Capital owning a bunch of railcars, aircraft, and cargo containers, ushering in an era of lucrative equipment leasing at GE Capital. According to *The Wall Street Journal*, "Company lore has him once standing on a table and exhorting the troops: 'I want 'em all!'" Said Wendt about that incident, "It was our objective to be as big as we could." By the time he took over GE Capital, in 1986, Wendt's mandate from Jack was to increase profitability. "Every planning meeting Jack ran was always the same," Wendt recalled. "He told everyone else to cut costs, and he'd tell me to grow the business." GE Capital became the nation's first, and biggest, nonbank bank. There were no GE Capital branches. There were no deposits. There were no depositors. There was little, if any, regulation. GE Capital got the money it needed by borrowing it on the cheap— leveraging GE's AAA credit rating—from the commercial paper market or from other pools of cheap capital around the world. It made money by borrowing short and cheaply and lending long and expensively. It took extra sweeteners in deals—such as warrants that would become very valuable if the company financed succeeded—because Wendt was smart enough and powerful enough to make sure GE got properly compensated for the risks it was taking.

THE BANK OF GENERAL ELECTRIC - 217

The formula worked beautifully for years. He would exhort the troops: "We're going to win the war! You've got to take ground." In the 1970s, GE Credit made around $45 million, or roughly 4 percent of GE's pretax income. By the end of the 1990s, GE Capital was providing nearly 40 percent of GE's pretax income, and by 2003, it was providing more than half of GE's pretax net income. It's fair to say that without the profits that GE Capital was churning out year after year, Jack would never have been named *Fortune*'s "Manager of the Century" and GE would never have become the world's most valuable company.

No one was a bigger champion of growing GE Capital than Gary Wendt. The key to his success, he said, was his ambition. "It wasn't a question of expertise," he said. "It was a question of attitude. It was a question of saying 'We can do more. We can do more. Why don't we do this?' And trying to stay close to the ground to figure out what was going to happen." He shared the example of how he went to Washington to get a firsthand understanding of what Reagan's 1986 tax bill would mean for GE Capital's leasing business. "I knew what was going to happen and nobody else did," he said. Armed with that knowledge, he figured out how to save GE from paying any taxes and to get a $2 billion refund on previously paid taxes. "We were in position to take full advantage," he said. "I was opportunistic as opposed to being a manual guy and rules, rules, rules, committees, committees, committees. Just do it." Wendt once told a professor at Harvard Business School that the key to his success at GE Capital was "the fear that the business could fall off the cliff at any moment. Without that fear, the cliff often appears more suddenly."

As the new head of GE Capital, Wendt found himself stuck between the yin that was Jack and the yang that was Bossidy. Mike Carpenter, the architect of the RCA deal, was an additional—and largely uncontrollable—presence in the leadership team of GE's growing financial supermarket. He went to work for Wendt at GE Capital to oversee commercial financial services and activities in the real estate and asset-based financing markets. The combination of Wendt, Carpenter, and Bossidy turned out to be combustible.

Wendt had been against buying Kidder, although nobody had bothered to ask him for his input. He was off enjoying a GE management trip in China when the deal went down (the same one that Wright begged off of when the Kidder deal revved up). "What they told me [about the Kidder deal]," he told me, "was that they got to a point where DeNunzio would not let them look at the books, and that's a 'Walk away now!' But they didn't. They kept going with it after that.... They never got to do an audit of the books. I don't know that that would've made any difference, frankly. But you just walk away at that point. That's why I didn't like it."

With Wendt away in China at the time of the Kidder deal, Leo Halloran, the chief financial officer of GE Capital, asked Jim Bunt, a deputy, to head to downtown Manhattan to look at Kidder's books. "There was no due diligence," Bunt told me. "And in retrospect, I heard that Gary Wendt was absolutely bullshit that we had bought an investment bank. He was adamantly opposed to it." Halloran asked Bunt to "do a calculation: How much does Kidder have to make to be nondilutive?" Bunt thought that was ridiculous. "That wasn't exactly a sparkly way to put a valuation on a deal," he said. "It should be the other way around: How much could Kidder make, and is it nondilutive? It was, How much do they have to make to be nondilutive?"

The next day, the GE guys were assembled down at Kidder to ask questions. They were trying to figure out who the top fifty producers were at Kidder and then lock them into Kidder and GE with new contracts, as Wright had wanted to do. "And this is where Bob got snookered on the deal," Bunt said. Deciding who the top fifty most valuable employees were at a place like Kidder was not easy, apparently. One of the Kidder partners suggested that the way to go was to just give contracts to the top fifty highest-paid people at the firm, the ones who had been paid the biggest bonuses. "Those are the hitters," GE was told.

Wright agreed. "It turns out that, as with every other place in the world that pays big bonuses, longevity in career was the biggest factor in how big your bonus was," Bunt explained. "Of the fifty, some major proportion were basically over-the-hill, ready-to-call-it-a-day, rich, retired losers. The up-and-comers that were the real hitters weren't on the list." He noted his frustration with some of the Kidder bankers who were tethered by contract to GE and who rarely pitched clients for new business. He would be told, "I've got three people under me who go do that." He observed, "That was one of the mistakes." He recalled that the "the deal was done and done quickly." Then he recalled how "right at that time, the tensions between Gary and Bob Wright rose to DEFCON [1]."

At least at the start, Wendt made a show of trying to find some synergy between GE's financial businesses and Kidder. Jack Myers, at GE Asset Management, remembered a call he got from Wendt soon after GE had bought Kidder. Wendt asked Myers to come by his office in Stamford. "We just bought Kidder Peabody," Wendt told him, "and I want you to put a lot of your investment business through Kidder now." Myers explained why that wouldn't work. "Gary, whatever business I was doing with Kidder I can't do with them anymore," he said, "much less put more business through them, because now Kidder Peabody is a party in interest, and I can't do business with them because of ERISA. You can't self-deal." Myers also remembered how at the beginning Dammerman asked him to go down to Kidder and examine the firm's huge portfolio of mortgage-backed securities, of which Kidder was one of the leading underwriters. Through GE Asset Management, Myers had his own huge book of bonds, making him an expert on the topic. He discovered that Kidder was selling off the least risky part of the mortgage-backed securities and retaining the riskiest part—the part that might not have been paid back in a challenging economic environment—on Kidder's balance sheet. "Kidder was retaining the last level, which involved toxic waste," he said. "They were just out there selling the white meat and retaining the high-risk meat in their mortgage portfolio. The other banks were not doing that. So Kidder was leading because they were taking more risks than anybody." Myers called Larry Fink at BlackRock to help Kidder pare its portfolio of risky mortgage bonds. "We worked out of that portfolio," he said. "It was a disaster that we inherited. Kidder was just out there leading the mortgage-backed-security rankings, booking all types of money for the guys that were doing it, and Kidder was retaining the risk. . . . Of course, that was a disaster, Kidder Peabody."

Just as there was no mention of Kidder Peabody in Bob Wright's memoir, *The Wright Stuff*, there was no mention of Gary Wendt either, even though Jack chose Wright to run GE Financial Services in part because he wanted someone to keep an eye on Wendt. During an interview, Wright explained to me his complicated relationship with Wendt. By the time Jack chose Wright to run GE's financial portfolio, he pointed out that he had run "maybe six businesses" and added, "I was pretty facile with numbers and with accounting because we were independent profit centers. At the cable thing"—when he was running Cox Cable—"it was a big deal. We had big balance sheets. We had all kinds of depreciation issues. I had seen probably a lot more than anybody my age might have seen in larger businesses by that time. I wasn't afraid of that." But, he said, "I was no expert in leveraged leasing. I'll tell you that, though. That's almost like pension funds. It's very tricky. But Gary Wendt was the guy that was doing it. Gary Wendt was a very, very smart guy, but he had such a quirky personality that put off a lot of people, and they"—Jack and Bossidy—"didn't want him to be the guy to run it. They wanted me to come in and do it. They wanted me to control Gary, basically. I'd had enough experience, though, with people, firing people and hiring people, that I felt comfortable in that."

He remembered that Wendt did not like investment bankers. "He had this inferiority complex," Wright said. "He was a very smart guy, went to Harvard Business School, grew up in Wisconsin, had that kind

of Lutheran mentality. He had a chip on his shoulder all the time."
Asked how he got along with Wright, Wendt demurred. "I don't see
him anymore," he told me.

Eight months after closing the Kidder deal, in February 1987, GE
discovered firsthand the dangers of owning an investment bank, especially one—if Wendt was correct—where GE's due diligence was spotty
and Kidder's books opaque. The scandal that exploded onto the front
pages of the nation's newspapers involved allegations of insider trading
against Marty Siegel, who had left Kidder the year before for Drexel
Burnham, as well as against Timothy Tabor and Richard Wigton, two
of Siegel's associates in Kidder's arbitrage department. Also caught up
in the headlines, if not the scandal, was Robert Freeman, the head of
arbitrage at Goldman Sachs.

Back in the mid-1980s, the business of merger arbitrage could be
very profitable indeed. It was a hot corner of finance, and arbitrageurs
were the Steve Schwarzmans of the day. The idea was to buy or sell the
stock or stocks of companies involved in a merger or acquisition after
the deal was announced, and to bet on whether or not the deal would
close and when. Since many shareholders bail out of stocks immediately after a deal is announced—rather than wait around to find out
whether it closes or not—arbitrageurs can make money by making the
bet about whether a deal will actually close. The problem with the
business—at least until the late 1980s, when federal prosecutors started
clamping down on inside information—was that the way arbitrageurs
obtained the latest information about the status of various deals, and
how likely they were to close and when, was to speak with the M&A
bankers, like Marty Siegel, who were involved in putting together the
original deals and mining their information. The problem was compounded when a firm's management—in this case Ralph DeNunzio at
Kidder Peabody—had secretly given Siegel the role of being head of
Kidder's proprietary arbitrage trading desk at the same time he was
the firm's leading M&A adviser. Not only did few people on Wall Street

know that Kidder had an arbitrage department, but fewer still knew that DeNunzio thought it made sense to have Siegel run it. No other firm on Wall Street allowed its M&A bankers to also make principal bets on the outcome of deals.

The consequences of that decision—which, of course, Jack had no knowledge of when he signed the Kidder deal in April 1986—became embarrassingly apparent in mid-February 1987. On February 10, Thomas Doonan, a U.S. marshal, swore out a six-page complaint against Bob Freeman at Goldman Sachs, based on information given to Doonan by "a person who is cooperating in this investigation and to whom I shall refer hereinafter as 'CS-1'" and who was revealed by the news media to be Marty Siegel. According to Doonan's complaint, CS-1—Siegel—had provided Doonan with "very extensive details about an illegal insider trading scheme involving Kidder and Goldman in which CS-1 personally participated with the defendant Robert M. Freeman and other individuals during the period from about June 1984 through about January 1986." Doonan had put his faith in Siegel and the "reliability and trustworthiness" of Siegel's "information" not only "in light of the extensive details that [he] provided" and "in light of [his] admissions of [his] own participation in the above-described scheme" but also because Siegel had "agreed to plead guilty to two felony counts, one pertaining to the" alleged conspiracy with Freeman "and the other . . . related to another scheme involving the misappropriation and stealing of inside information."

In other words, Siegel was a crook and had fingered Freeman, Wigton, and Tabor in exchange for leniency from prosecutors. Specifically, in his complaint, Doonan claimed Freeman and his "co-conspirators" at Kidder had twice used inside information to make illegal profits—one time involving Unocal and the other Storer Communications. In any event, the next day, Siegel pleaded guilty to insider-trading charges.

Marty Siegel was something of a whiz kid at Kidder who made his name on Wall Street defending companies from hostile takeovers. He was highly compensated by Kidder, but apparently that wasn't enough. He had become friends with the wealthy arbitrageur Ivan Boesky, a relationship that involved Boesky paying Siegel enormous sums for inside information—once, $150,000 paid in $100 bills, another time $400,000—at the same time that Kidder was paying him millions more in salary and bonus. Boesky had no qualms about paying Siegel, especially since Siegel's information was worth many millions of dollars in profits to him.

Boesky's profile continued to rise on Wall Street, along with his wealth. He wrote a book, *Merger Mania,* about doing deals. He became a media darling. A few of the articles about Boesky suggested he had an unusually close relationship with Siegel and noted that Boesky was making lots of money on deals where Kidder was an adviser. These innuendos made Siegel extremely concerned, and so, after a $400,000 payment from Boesky for his services in 1984, he decided to stop sharing with him his inside information. Perhaps his transgressions could remain in the past, he hoped. (While miffed at Siegel's decision, Boesky had lined up other M&A bankers and lawyers to feed him a steady stream of illegal tips about deals.)

While Siegel had stopped sharing inside information with Boesky, he could not, apparently, do without the adrenaline rush he derived from illegal behavior. Against his initial instinct, DeNunzio decided, in 1984, to set up a secret arbitrage unit inside Kidder. DeNunzio instructed the group to keep its existence quiet. As if it were not enough of a conflict to have an M&A banker run an arbitrage department that took large ownership stakes in the stocks of companies involved in Kidder's M&A deals, Siegel made DeNunzio's bad decision even worse by speaking to other Wall Street arbitrageurs, including Bob Freeman occasionally, without sharing the relevant fact that he was also arbitraging deals.

The charges against Tabor and Wigton would eventually be dismissed. But that fact was of little concern to Jack in February 1987. As the new owner of 80 percent of Kidder, GE had also bought its liabilities, or potential liabilities. He had a conflagration to snuff out, and fast. Rudolph Giuliani, the U.S. Attorney in the Southern District of New York, had the ability to indict Kidder on criminal insider-trading charges—not just its allegedly bad actors Tabor, Wigton, and Siegel—effectively putting the firm out of business and ending GE's ill-fated Wall Street experiment barely eight months after it began. On February 14, a Saturday, Giuliani and Gary Lynch, head of the SEC's enforcement division, met with Kidder executives and told them an indictment was in the offing. Giuliani urged Kidder to plead guilty. He explained to the Kidder executives that Siegel, one of the firm's former top bankers, had pleaded guilty to two felonies and had told federal prosecutors he had shared inside information with Freeman, regardless of whether his accusations were true or not. Kidder had made "millions" from Wigton's and Tabor's trading, Giuliani asserted in the meeting. He also said he had evidence that Kidder had entered into a "parking" arrangement—the illegal buying and selling of stock with the intent to deceive—with Boesky. Kidder could be held criminally liable for both the parking scheme and what Siegel did, Giuliani said.

Concerned that Kidder's traditional law firm, Sullivan & Cromwell, might be taking too combative an approach with Giuliani, the GE brass decided to hire Gary Naftalis, a former prosecutor at Kramer Levin, to join the Kidder defense team. The day after the meeting with Giuliani, GE unleashed the two law firms to do an investigation into what happened at Kidder. "The results of the investigation were discouraging," *The Wall Street Journal* reported. GE's internal lawyers were concerned about Siegel's admissions and thought Kidder could be "in serious trouble" as a result of them, according to the paper. The GE lawyers were also worried about DeNunzio, the Kidder CEO, and why he had asked Siegel to head up an arbitrage department and keep its existence confidential. GE's investigation also included sending in the famed GE auditors, essentially doing seven months after the fact something that should have been done before agreeing to buy Kidder. "The situation was abysmal," someone involved with the audit told *The Wall Street Journal*. "There were no systems and controls. It was basically run like a family business. DeNunzio ran it as he saw fit, and things had gotten out of control." Just as Wendt had feared.

Jack took control of the negotiations with Giuliani. He cut out the Kidder executives and Sullivan & Cromwell. Jack gave Naftalis, at Kramer Levin, the reins. Jack's idea was not to litigate what had happened but rather to try to convince Giuliani that Kidder going forward under GE's tighter control was not the Kidder of the past, that Kidder had learned its lesson and was contrite. On March 7, Bossidy met with Giuliani at his office in Lower Manhattan. It was the crucial meeting to determine whether GE's argument about a new Kidder could sway the federal prosecutors away from indicting the firm. Bossidy told Giuliani that GE knew nothing about the insider-trading and parking scandals when it bought Kidder. He said GE's investigation had revealed "serious problems" at Kidder but not necessarily "criminal wrongdoing." He said Kidder would get out of the risk arbitrage business and would implement "sweeping" management changes. Bossidy argued it would be unfair to Kidder's seven thousand employees and 300,000 customers to indict the firm—effectively destroying it—for the wrongdoing of a handful. He said GE and Kidder would cooperate with Giuliani. Later in March, the GE negotiators again met with Giuliani. He told them that if Kidder could reach a settlement with the SEC, Kidder would not be indicted. On April 9, Freeman, Wigton, and Tabor were indicted. (The indictment would later be tossed out, and a long-threatened superseding indictment was never filed.) Kidder was not indicted.

In mid-April, Bossidy and Naftalis began negotiating with Gary Lynch at the SEC in Washington. The question before them was how much GE would pay the SEC to settle the Kidder allegations. At one point, the SEC suggested that GE be held liable for Goldman's alleged illegal trading profits. That was an unacceptable demand to Jack (obviously), and the SEC eventually relented. In mid-May, while the settlement negotiations with the SEC were ongoing, GE made good on one of its promises to Giuliani: it replaced DeNunzio as the Kidder CEO, along with two other top executives. In *Jack: Straight from the Gut*, Jack wrote that the Kidder executives "decided to leave." But the reality was more complicated. Implementing sweeping management changes at the firm was a key part of any deal with Giuliani.

Max Chapman said he understood that management changes were a requirement of any deal with the SEC. GE asked him to take over for DeNunzio running Kidder on a day-to-day basis. But Chapman was uncomfortable. "He was always my mentor," he told me, "and I greatly respected Ralph." Instead, Chapman proposed a six-month transition period. But DeNunzio told Chapman that neither GE nor the SEC would accept that idea. That's how DeNunzio left Kidder and Chapman replaced him. "Ralph was really the glue that made Kidder work," Chapman said. "I certainly respected him, which I still do, a great deal. And to make that kind of switch in management, it just didn't happen at the right time."

In its public statement, GE said DeNunzio's departure was a "mutually determined" decision. But by then Jack had already made up his mind that DeNunzio had to go and had already offered his job to former GE board member Silas Cathcart, the longtime chairman of Illinois Tool Works, with Chapman reporting to him. It was an odd choice. Cathcart, then sixty-one years old, had never worked on Wall Street a day in his life. The appointment ruffled the Kidder rank and file. Joked one Kidder executive, "I was thinking what we need around here is a good tool and die man." Remembered Chapman, "Si was a great guy, who knew nothing whatsoever about the securities business, but was certainly a well-connected, and a senior director, at GE. . . . Quite frankly, he was not a lot of help to any of us in terms of what we were trying to do."

But Jack wanted someone he could trust. When Jack first called

Cathcart in Chicago and mentioned to him the idea of heading up Kidder—weeks before the announcement that he would replace DeNunzio—Cathcart thought Jack was crazy. "Are you out of your cotton-pickin' mind?" he asked. A few days later, in March, Jack and Bossidy had dinner with Cathcart at a small Italian restaurant in New York City. He had a list of fifteen reasons why it was a bad idea for him to become chairman of Kidder, reporting to Bossidy. He also suggested the names of other people he thought would be better for the job. Jack looked over Cathcart's notes and then crumpled them up. "We have to stabilize things and get Kidder back on a recovery path," he told Cathcart. "The job won't last much more than a couple of years." Jack asked him to think it over and reminded him that he was too young to retire. He went back to Chicago, talked it over with his wife (who was keen to move to New York City), and called Jack a couple of days later to accept.

On May 14, Bossidy announced to Kidder employees that Cathcart would be the new CEO and president of Kidder. DeNunzio would be chairman of the Kidder board (a meaningless position) for two more weeks and then leave. Chapman would run the firm on a day-to-day basis and report to Cathcart. "They felt a manager that was perhaps not steeped in the tradition of the [investment banking] industry might make some contribution," Cathcart said. GE also took full control of Kidder's board.

The various changes seemed to appease the SEC. In short order, GE agreed to pay a $23.5 million fine to settle the civil charges related to insider trading and stock parking. Kidder also agreed to some new supervisory requirements designed to prevent a recurrence. Everything had been agreed with the SEC by May 31. But GE still had not heard back from Giuliani about the Kidder indictment. Then, finally, in early June, Bossidy got word that Giuliani would not indict Kidder. At a subsequent press conference, Giuliani praised GE as a "responsible corporate citizen." Jack had managed to snuff out—for the moment—the existential threat to Kidder's existence.

*Chapter Thirteen*

# TOP OF THE ROCK

On August 26, 1986, after the formal announcement that Bob Wright would be the new head of NBC, Jack, Wright, and a few dozen NBC executives headed to the Four Seasons Restaurant for a celebratory lunch. "Jack and I had it all orchestrated," Wright remembered. "He played the white hat and later I played the black. He told everybody how happy GE was to have NBC and we were going to have a lot of fun together, and everything was going to be great. Everyone was laughing and cheering, and I was thinking to myself, this is not going to be much fun." After Jack left, Wright met with his new direct reports in his sixth-floor office at 30 Rockefeller Center "and gave them my perspective on what GE was going to do with NBC." He told them to be "aggressive" about growing the business and about "the importance" of selling NBC content to cable television providers and to think about whether NBC should own cable networks. He spoke about needing to resolve conflicts with the unions and the need to figure out NBC's future, if any, in radio. "They didn't like what they heard," he remembered. They liked things the way they were. They couldn't understand why it was important "to embrace cable" when NBC was one of the dominant forces in the media business.

The idea of NBC creating its own cable news channels to compete with CNN was anathema to them. But Wright knew he was correct. He wooed the cable crowd. "I knew they would be important partners in my growth plan for NBC," he said. His strategy was to own cable channels or systems or to create program services from scratch. He had learned about the importance of cable from his years at Cox and was bringing that wisdom to NBC. "To some within NBC, it looked like cavorting with the enemy, but it was the only recourse I had to save the network," he recalled. At the time that Wright took the reins, NBC's only cable asset was a one-third equity stake in the Arts & Entertainment channel. To expand NBC into the cable business, Wright hired Tom Rogers, a longtime legislative aide for Tim Wirth, a member of the House of Representatives from Colorado, who was also the chair of the Telecommunications Subcommittee. Wright and Rogers were a stellar duo, gobbling up cable deals wherever they could in order to fulfill Wright's vision for NBC. The first deal they cut was with cable pioneer Chuck Dolan, who had founded Cablevision Systems on Long Island. Dolan had launched Bravo, a cable channel dedicated to showing artistic films, and was a cofounder of HBO. (He was also, along with his son Jim, the much-detested owner of the New York Knicks.) Wright knew Dolan from his days at Cox. In January 1989, NBC paid $140 million to buy half of Dolan's Rainbow Properties, which was the holding company for Bravo, American Movie Classics, and several sports-focused channels. NBC also bought stakes in Court TV and the History Channel, among others.

"We were hedging," explained David Zaslav, who worked for Wright and Rogers, "and these were the years where it wasn't clear at all that cable was going to be a business, and Jack was convinced that broadcast had to diversify. We had to be in the next generation of media. And no broadcaster was in cable at the time. The cable business was the cable operators, and then many of the cable companies developed their own content and then there were a few entrepreneurs. The broadcasters were the enemy. And Jack was convinced that eventually people were going to spend more time watching cable—even though it was the golden days of broadcasting, that we had to get into that business. He said to me, 'You know, I don't care if we lose money. We have to learn about this business. This is the future.' Which is a theme for Jack: it's about embracing the future."

Part of Wright's vision for NBC getting into the cable business was to launch new cable channels from scratch. He and Jack wanted to build a news channel. But Jack figured out early on that to be successful, or to have the chance to be successful, the new cable channel needed the blessing of John Malone, the powerful head of TCI, the huge cable operator. Wright described Malone as "a Vito Corleone–like figure" who would "give as long as he could get something he wanted of lasting value in return." The new news channel had to be carried on Malone's cable network or it was doomed. Wright and Zaslav flew to Denver to see Malone. "We couldn't do it without John," Zaslav told me. "We all went. I was like the little guy with the suitcases. We went to John and we said, 'We want to launch a news network.' We made a whole presentation about how we have the number one news network in America, and John looked at us and said, 'I already have a news network, it's CNN, and I have another one called Headline News. I don't need another news network.'"

Without missing a beat, Malone suggested that NBC start a business-news channel instead. He said there was one existing business news channel—FNN, Financial News Network—and its content was pathetic, mostly just infomercials. "It's bullshit," Malone told them. If NBC were to start a competitor, he said, TCI would "launch you." Wright and Zaslav went to a conference room to tell Jack what Malone had said. "We got to take whatever real estate we can," Jack told them. "Let's do it." Said Zaslav, "We did this deal with Malone to launch CNBC." Wright observed that, to Malone, NBC was "just another chess opponent" but "he had all the moves." Wright declared himself "satisfied" with the outcome. Malone took an equity stake in CNBC.

In April 1989, NBC joined together with Cablevision, in a fifty-fifty joint venture, to rebrand a struggling cable channel as the Consumer News and Business Channel, or CNBC. "I loved the idea from the start," Jack recalled. When the CNBC set was being built in Fort Lee, New Jersey, Jack would call up and ask to see the design. "They still talk about it," one longtime television executive told me. Jack would get up in the morning, study the ratings, and then call up someone at CNBC and yell, "Why the fuck is CBS doing this?" and "Why the fuck are we doing that?" Jack loved CNBC and he loved cable television. Once, anchors Sue Herera and Joe Kernen were bantering on air about how nice it would be if Jack were to split the GE stock and raise its dividend. It was the end of the day and they figured Jack was not watching. An important CEO would have better things to do, right? "It's late," Kernen said. "He's probably having a nice dinner, doing one of those corporate events." When Herera got off the air, she noticed a fax waiting for her: "Good try—I'm always watching—love, Jack."

Despite Jack's vigilance, CNBC lost $60 million in its first two years of operation. "Business news was not taking off," Jack said. Dolan was getting antsy. But Jack was as jazzed as ever about CNBC and NBC pushing into cable programming. "We were losing a lot of money because there weren't that many cable subscribers in those days," Zaslav explained. "Every cable network in America was losing money except maybe for MTV because they got that content for free. Ted [Turner] was upside down. We were upside down. All the cable networks were upside down. The broadcasters were like attack dogs to us on the cable side because there was a limited amount of money from GE and we were losing money. They set us up in New Jersey. We needed to be lean and mean. We couldn't be union, otherwise it would've busted the whole cost structure."

Early in 1991, FNN filed for bankruptcy. It had access to 32 million homes. CNBC had 20 million subscribers. Dow Jones had already made a bid for FNN, out of bankruptcy. Wright and Rogers had negotiated a deal to buy FNN, but Jack rejected it. "His argument was that it was too risky," Wright recalled, "and that GE as a rule, did not buy broken assets." In May 1991, Rogers and Zaslav went to see Jack—Wright joined by phone—to try to change his mind and persuade him to buy FNN.

"Come on, Jack!" Tom Rogers exhorted. "This is a rare opportunity that will make all the difference. If we didn't think so, we would not have camped out in your office the past two days!" They put together a big presentation: analyzing how many subscribers they could get, where the combined channel would be carried, what the "synergy" would be, and what the internal rate of return would be on the investment. Wright was doing the presentation; Zaslav was feeding him the pages. Nearly two hours were carved out of Jack's schedule for the meeting. About six minutes in, Jack cut Wright off. "Wow, wow, wow, hold on," Jack said. He was clapping his hands together. "Forget the two hundred and fifty pages," he continued. "I read it, forget it. We've got two questions: Number one, is business news a business? Nobody's making money. If we put these two together, is business news on cable a business? And two, more importantly, can we own it?" By that he meant, What happens if Malone suddenly decides to compete with the combined CNBC-FNN? Jack wasn't concerned GE would be outbid by Dow Jones for FNN. ("If we bid, we're going to have to win," he told them. "We're GE.") Rather, he was worried about looking foolish because once owned, the combined channel might get swept asunder by the crafty Malone.

While Wright, Rogers, and Zaslav were getting marching orders from Jack, they heard from Dolan that he wanted out of the CNBC partnership. "I don't believe," Dolan told them. (Dolan wanted GE to pay him something for his money-losing stake in CNBC. But Jack declined to do so, figuring correctly that GE would have to absorb the ongoing losses alone.) But Dolan's bailout just reinforced Jack in his concern. "Okay, Chuck's out," Jack said. "What do we think? Chuck doesn't believe. Do we think it's a business?" Then Jack laid out his "insight" about Malone, "which is what made him extraordinary," Zaslav said.

Jack continued, "What happens if we buy this thing? Now we merge and everything you put in that deck is right. Instead of 20 million subscribers now we put them together, we have 33 million subscribers. We're able to get more money from the cable guys because there's only one business news network, and everything that you guys put in here is right. What happens if the day after we buy this thing Malone announces that he's going to convert Headline News into a business news network? And that's in every home in America and it's in every hotel room in America? What the fuck do we do then? We bought nothing. So, get out of here and come back with the answer to two questions. Take all your numbers and all your internal rate shit. Is Chuck right or are we right? Is this a business and can we own it? Can we own it? Once we get this goddamn thing is it going to be ours to win? Or are we going to wake up and look stupid?" Jack told them he would reverse his decision if they could assure him that Malone wouldn't "screw us."

Zaslav realized Jack was right, that Malone could turn the tables on them if he wanted. Zaslav went back out to Denver to talk to Malone. Malone told him he and Turner had no intention of turning Headline News into a business channel to compete with CNBC. "He likes the news business," Malone said. Zaslav needed some written assurances from Malone. They structured an agreement that if Malone or Turner started competing with CNBC, the payments to carry these channels from the cable companies to the channels would increase dramatically.

Deal in hand, Wright and Zaslav went back to see Jack. "We can't promise anything," they told Jack, "but we think this is a business and here's why." They went through their business logic again. "This could be a real cash business because our costs don't go up," they continued. "And in the meantime, we can't stop someone from competing with us. But we have some economic protection where if there is conversion or another competitor, the distributors would have to pay us a lot more money."

Jack was convinced. "Only Welch would have seen those two issues," Zaslav said. He elaborated about Jack, "Jack was an extraordinary leader in a sense that you both feared him and desperately, desperately wanted his attention and love. And he gave you his attention and love. But it was really purely a meritocracy with him. I never saw a leader like that. Usually, if you don't make your numbers or if you don't have your stuff down, you could be gone quickly. And people disappeared. You went into a meeting and you didn't have your stuff together, you were gone. On the other hand, if he called you in on a Wednesday and said he was talking to the analysts the next day, 'Give me some good stuff,' or he wrote you a note about what you did or how good you were, that somehow you yearned for it." Zaslav said he and others believed they were working for the greatest CEO in the world. They just wanted to please him and stay part of the team. "It was exhilarating," he said. Jack, he said, didn't waste time. "His overall style was basically to get real immediately," Zaslav continued.

And you had better be prepared for your meetings with Jack. "He was an incredible study of everything," Zaslav said. "When we would go up to him to talk about an acquisition, he read everything. He read the exhibits. He knew about the industry. He got it all. When you went up to meet with him, the discussion you had with him was at a completely different level. When you reported to him quarterly about your business, he knew what was good about your business and he also knew what was bad about your business. When you went in there, you had to tell him exactly what was wrong with your business and how you were going to fix it. And if *he* told you what was wrong with your business, you really had a problem."

Since Jack didn't tolerate bullshit, it prevented most people from taking victory laps when meeting with him. The trick was, per Zaslav, to get "him in the boat with you on having this problem with CNBC or having this issue with the regional sports networks or 'We can't figure out this news business.' You could see him getting excited because that's what he was really there for: to try and figure out with you how to get shit fixed, and then he was in the boat with you, and once you had him and he was with you, he was never a diving judge. He was always in there with you."

Jack was with the NBC guys in going after FNN. With Dolan out, GE went after FNN on its own. Jack initially thought he could get the bankrupt cable channel for $50 million. But that was before he found himself in a bidding war with a partnership between Dow Jones and Westinghouse. Their opening bid was $60 million but quickly got to $150 million. Wright and Rogers urged Jack to pony up another $5 million to win. Jack and his top executives "agonized" about bidding $154.3 million, Jack recalled, because it was three times what he thought he would have to pay and $60 million more than he had authorized.

"We weren't breaking champagne bottles open that night," Wright said. Jack fancied himself a disciplined investor, and the FNN process showed that he could get carried away. "Fortunately," he continued, "we badly wanted a financial news network, and the extra $5 million closed the deal." According to Jack, the man who "got CNBC going" was Roger Ailes, the controversial Republican presidential political consultant to both Ronald Reagan and George H. W. Bush and the executive producer of Rush Limbaugh's television show. Wright hired Ailes to be the CEO of CNBC in August 1993. His three-year contract paid him a salary of $550,000 a year, with increases, and a chance to make a bonus of $1.7 million annually, if CNBC hit certain ratings targets. The deal was consummated at Jack's home in Nantucket. "I was an instant fan," Jack said of Ailes. Soon after he took over CNBC, Ailes made it a ratings and profits juggernaut.

After six consecutive years in first place in the network ratings, by the fall of 1992, NBC cratered. Its profits fell to $204 million in 1992, from $603 million three years earlier. Dozens of new shows were introduced that didn't work. "I've never seen anyone predict a sure thing," Jack reflected. "Most of the shows bomb." This was about the same moment Jack and his executive team picked Jay Leno to succeed Johnny Carson on *The Tonight Show*, causing runner-up David Letterman to leave for CBS, which then started getting higher ratings in the time slot with Letterman than NBC was getting with Leno.

"We got beat up over the Leno decision," Jack said. He recalled how people would ask him regularly why GE owned NBC, since Jack knew nothing about what makes a good television show. "That's true," he said, "but I can't build a jet engine or a turbine, either. My job at GE was to deal with resources—people and dollars. I offered as much (or as little) help to our aircraft engine design engineers as I offered to the people picking shows in Hollywood."

Warren Littlefield, then head of NBC's programming, remembered Jack's role in the Leno-Letterman debate: he had an opinion but knew it wasn't his decision to make. "In many ways the decision tore the company apart," Littlefield said. "Jack said to the small group of decision-makers as we sat around the table, 'This is your decision. It's clear that they're both very talented. I think there is something to be said for loyalty and who you're in business with.' That was Jack's way of saying, 'In a close call, if someone were asking me, I think I'd probably lean toward Jay Leno.'"

Another time, Jack suggested to the NBC brass that NBC do the movie *Wall Street*, which had just come out on the silver screen, as a television series. "We pissed all over his idea," Littlefield said, "and he goes, 'I'm going back to businesses where people listen to me.'"

Things were so bad at NBC that rumors abounded about GE considering the sale of the business. Would Martin Davis and Brandon Tartikoff at Paramount buy it? Would Barry Diller, the former chairman of Fox, Inc., be interested? In October 1992, there were newspaper reports that Bill Cosby was putting together a serious bid to buy the network, even though the price tag was thought to be $4 billion— *The New York Times* put Cosby's net worth at the time at around $300 million—but that GE did not want to sell, or at least said it did not want to sell. Jack admitted to trying to sell NBC to Time Warner, Sony, and Viacom. The most serious discussions were with Paramount and Disney.

"We weren't looking for cash," Jack told me. "We were trying to put things together to make NBC a bigger and stronger player." During the summer of 1994, Jack and Dammerman had dinner with Michael Eisner and his team from Disney at GE's headquarters. They reached a tentative agreement whereby GE would sell 49 percent of NBC to Disney, with Disney having operating control of the combined NBC and Disney TV production studios. Jack's one condition was that Bob Wright remain the CEO of the new business. "Michael liked it," Jack remembered, "and Dennis and I were thrilled. By morning, though, Michael had changed his mind and didn't want to do the deal."

Eisner recalled the dinner and how Jack had him mesmerized. The wine had been flowing. "Jack had me wrapped around his little finger," Eisner said. "I was all excited about this." He thought Jack was an incredible salesman. "I was ready to go," he said. But after the dinner, as he was getting into the elevator, he realized that the deal made no economic sense for Disney. "It was a totally one-sided deal," Eisner said.

Even Jack and Dammerman wondered why Eisner was so enthusiastic. "Do you fucking believe it?" Jack said to his team when Eisner left. "That he's going to buy into this deal? That he agreed?" Dammerman thought Eisner would think better of it when he gave it some more sober thought. The next day Eisner called Jack. "I got on the plane last night and my guys jumped all over me," Eisner told him. "They wouldn't let me sleep. I couldn't do anything. For the five hours it took us to get to the West Coast, all they were doing is beating me up. They said, 'How could you be so stupid?'" The deal with Disney was over. Jack and the GE board decided to keep NBC and redouble the efforts to revive it. The next year, Disney instead bought Capital Cities, which owned ABC.

JUST AFTER NEW YEAR'S IN 1995, JACK HAD TO DEAL WITH A POWER play by Roger Ailes, the head of CNBC, for Bob Wright's job atop NBC. Jack should have seen Ailes's move coming—given the man's outsize ambitions—especially since Jack had agreed to participate in a flattering, rejuvenating *New York Times Magazine* profile of Ailes, "Embracing the Enemy," published on January 8. After reviewing how Ailes had taken control of CNBC in August 1993 and then started America's Talking—twelve live shows that were essentially talk radio on TV—which debuted on July 4, 1994, the article revealed that he was directly responsible for thirty-two hours of original programming, "more than anyone else in the country." Said Ailes, "This is the most powerful force in the world. Politics is nothing compared with this." CNBC's profits had tripled in the year since Ailes took over. The turnaround was "utterly astonishing," Jack told the magazine. He also added that Bob Wright's decision to hire Ailes "may well turn out to be the smartest thing Bob Wright has done in his career." The *Times* reported that Ailes was "working his way up the short list" to succeed Wright, atop NBC. Wright didn't extinguish the speculation. "If you had asked me five years ago whether Roger Ailes could run a network, I would have said no," he said. "Back then, most people thought that we needed corporate types to maintain the franchise. Today, it's obvious that what we need are risk takers."

Jack's reaction to the article was hard to gauge. He had participated in it and had long been one of Ailes's biggest fans. But it was Jack and Wright alone—without Ailes—who took the lead in continuing to try to peddle NBC. First up was a deal that Wright and his team thought they had negotiated with Ted Turner, the "free-spirited" founder of CNN, as Wright described him. The idea was for NBC to purchase Turner Broadcasting System, but it was structured as a reverse merger, with TBS being the surviving, publicly traded entity, under the name Turner NBC. GE would control Turner NBC and own at least 51 percent. Turner would remain a board member and an advisory vice-chairman. The combined NBC-TBS would be valued at around $11 billion. "After years of false starts with Turner," Wright recalled, "this was my best shot to bring our companies together." On January 13, Jack and Wright flew in a GE corporate jet to Atlanta to meet Turner at the Ritz-Carlton Hotel in Buckhead. The main purpose of the meeting was to see if the two powerful personalities could find a way to get along.

On the plane ride down, Wright was having his doubts. Warren Jenson, the NBC chief financial officer, had prepared a letter for Jack to sign outlining the deal that ended with the standard phrase "Looking forward to working with you for many years to come." But Jack couldn't imagine Ted Turner, the literal and figurative cowboy, fitting into the GE corporate culture. "I'm not saying that!" Jack proclaimed. Thought Wright, "It was an uneasy precursor for the Ritz-Carlton exchange to come." In the meeting were Jack, Turner, and Wright. It was a disaster. Jack wanted to do the deal but was determined not to move an inch on price or on Turner's role. He also nixed the idea that Turner would be on GE's board.

For his part, Turner wanted more money for TBS and that GE board seat. "I've earned it," Turner told Jack. Turner was afraid he'd just be another GE employee. "We can't do that!" Jack exploded. "We're just not set up for that!" What he meant, Wright translated, was that Jack thought Turner was "a loose cannon" who would "inevitably" say something "outrageous" that would anger investors. Jack could not risk that possibility. As tensions mounted between the two men, "Ted began barking like a dog," Wright reported. "It was his very Turner-like way of demonstrating that he could, when pushed, be subservient to Welch and GE's conservative board." Turner's display horrified Jack. The deal was dead. "An exasperated Welch and a ramped-up Turner gruffly shook hands and bolted for the door," Wright explained. (In October 1996 Turner would eventually sell to Time Warner for $7.5 billion in stock, and NBC would find itself looking on as the other networks expanded exponentially. The new paradigm was Rupert Murdoch's Fox. Murdoch owned a movie studio. He had built a prime-time network and now was starting a news division. He had cable and satellite holdings in Europe and Asia. Soon it would not be enough to be a mere network with some cable channels.)

On June 30, a month after NBC announced a joint venture with Microsoft to create MSNBC, Ailes signed a new four-year contract to stay at NBC. His base salary would increase to $725,000 a year and then again to $900,000. His annual guaranteed bonus was to be no less than $250,000. But despite the support Ailes got from Jack for his work at CNBC, Ailes was again cut out of the loop of the negotiations to create what became MSNBC, which Jack and Wright decided would replace America's Talking. Ailes was not happy that his baby was being taken away from him without his input. "The reality of it was, Bill Gates didn't want [MSNBC] to be under anyone but the president of NBC News," Wright recalled. "That was a crushing blow to Roger. He hated it."

This was the moment Ailes went into "meltdown mode," Wright recalled. In the fall, a long-running feud between him and David Zaslav exploded into the open. At a company dinner in September, according to Gabriel Sherman's *The Loudest Voice in the Room*, he told his "loyalists" about Zaslav, "Let's kill the S.O.B." Then in a later meeting with Zaslav, according to Sherman, during a "tirade," Ailes called Zaslav "a little fucking Jew prick." If true—and Zaslav remained steadfast that it was—Ailes's outburst was disastrous for him personally and professionally and risked being quite the scandal for Jack and GE, had it leaked. "The allegation that a high-profile executive had hurled an anti-Semitic insult against a Jewish employee was just the type of matter that needed to be handled with extreme sensitivity, especially in a media company full of gossipy journalists," Sherman explained. After being briefed on the allegation, Wright hired Howard Ganz, an attorney at Proskauer Rose, to conduct an investigation. Wright admonished Ganz to keep things quiet. Aware of Ganz's inquiry, Ailes tried desperately to disparage Zaslav by sending along to Ganz a "scathing letter" about Zaslav written by CNBC's chief financial officer. But that ploy did not seem to work.

Two weeks later, Ganz made his initial report to the NBC brass: in his opinion and based on his research, Ailes had made the anti-Semitic remark to Zaslav, and it was part of a "history of abusive, offensive and intimidating statements/threats and personal attacks reportedly made to and upon a number of other people." Ganz said that Ailes's comment to Zaslav was grounds for "cause termination." Wright believed Ganz. But NBC asked Ganz to "suspend his investigation," according to Sherman, because had he kept going, the chances of a leak would have risen exponentially. Instead, NBC tried to negotiate a settlement with Ailes.

Option one was for Ailes to resign. A financial settlement of his new contract would be negotiated for less than a full payout. On the other hand, if he wanted to stay at CNBC—if it was even possible, Ganz told Ailes's attorney—Ailes would have to apologize to Zaslav formally by retracting the anti-Semitic comment and would have to commit to never speaking that way again. In the stay scenario, Zaslav would report to both Ailes and Tom Rogers. In the resignation scenario, Ailes would be offered a consulting contract and in return would have to sign a confidentiality agreement, a nondisparagement agreement, a noncompete agreement, and a nonsolicitation agreement. His departure would be announced around Thanksgiving weekend 1995, effective December 31.

On October 20, Ailes hand-delivered a letter to Wright and faxed it to Jack. "We have an opportunity to resolve this matter quickly and effectively," he wrote. He defended his behavior: "the charges are false and despicable," "I have not received a fair hearing," "This is un-American," and "all the lawyering will provoke an untoward outcome." Five days later, Ailes made a point of seeking out Zaslav in his NBC office. He extended a hand in friendship and announced creepily, "I like you and I have always liked you." He added, "There has been nothing personal in this whole matter. At this point we are just hurting each other's careers. . . . We can find peace." The bizarre conversation devolved from there. Zaslav typed up the notes of it and sent them off to NBC HR. "It was a he-said, he-said situation," Wright recalled. "If they could have come together, I would have been satisfied . . . but that didn't happen."

Ailes opted to stay at NBC. But his situation remained precarious and his mood, volatile. At a staff meeting on October 30, Ailes said, "I feel like General Patton. I am afraid they will find out I love war." On November 10, Ailes and NBC formalized a deal that allowed Ailes to keep his job. Ailes agreed to work "constructively and harmoniously" with Zaslav and agreed not to act in any way that could be viewed as "intimidating" or "abusive," or else he would be terminated for cause and his new contract voided. He and Zaslav signed a separate side agreement to bury the hatchet.

Tensions remained high between the two men. At a major cable television industry conference in Anaheim, they could barely speak to each other. Jack and Wright had decided, firmly, against Ailes running MSNBC. That job instead went to Andy Lack, the president of NBC News. America's Talking was kaput. Ailes had fallen from being Wright's heir apparent to being essentially a cable news producer in charge of

CNBC's programming. Over the Christmas holiday, Ailes retreated to Florida to ponder his future and his options. One day, he drove over to North Palm Beach to see Jack at his house in his exclusive CEO enclave—ironically, no Jews allowed—and to discuss his future at the network. Jack defended Wright's decision to let Andy Lack run MSNBC. "We both realized there were people who didn't want him there anymore," Jack said. "We talked a lot. This was a sad thing for both of us."

In January 1996, Jack negotiated Ailes's separation agreement: $1.35 million and a mutual nondisparagement agreement. NBC agreed to say nice things about Roger Ailes and Ailes agreed to say nice things about NBC, Wright, and Jack. He was precluded from taking a new job at CNN, Dow Jones, or Bloomberg. Nothing was said about Ailes going to work for Rupert Murdoch at News Corporation, an oversight that would have historic implications.

## Chapter Fourteen

# WHEELIN' AND DEALIN'

Jack's flurry of dealmaking would continue. In addition to the buying, selling, and swapping of the RCA assets and the Kidder acquisition, GE's cash hoard was still more than $2 billion. Before the 1980s would close, GE would roll up the appliance maker Roper for $507.8 million and Gelco, a Minnesota-based transportation leasing company, for $414 million and add them to GE Capital. GE paid $2.3 billion for Borg-Warner, which made a resin widely used in the manufacture of computers. But GE's stock wouldn't budge. The equity market didn't seem to understand Jack's vision for GE. In a May 1988 Heard on the Street column in *The Wall Street Journal*, Janet Guyon, who covered GE for the paper, reported how badly GE's stock had performed under Jack's leadership. "Poor Jack Welch," she wrote, a phrase rarely used in connection to him. "After seven years of cutting costs, jobs and businesses to make General Electric one of the leanest, most profitable enterprises around, the GE chairman and chief executive finds the stock badly underperforming the market." Jack seemed perplexed. "I cannot understand it for the world," he said. "As far as our earnings outlook and our performance, it couldn't be better." The stock seemed stuck at or near its fifty-two-week low despite Jack's reassuring comments at a recent meeting in Miami for research analysts and after buying 20,000 shares for himself, adding to the 107,000 GE shares he owned, worth around $4.2 million.

To that point in the year, the S&P 500 index was up 3.9 percent but GE's stock was down 10.8 percent and was below where it had closed on the previous October 19, the day the New York Stock Exchange fell 22.6 percent, its biggest one-day percentage decline ever. One possible reason for the stock's performance was growing concern that GE was nothing more than a buyout firm. "Some fear that GE may stub its toe on another multibillion-dollar acquisition," Guyon wrote. "GE's bidding war with Whirlpool for its $507.8 million acquisition of appliance maker Roper makes some fret that it will overpay for its next purchase." She cited, as well, the immediate problems GE faced after buying Kidder. "Kidder doesn't look like the smartest deal that ever occurred," Jack said, "although I think it's doing quite well."

Given his optimism about the company and its prospects, it made sense for Jack to add to his personal holdings of GE stock. But he seemed reluctant to do one of the things analysts and investors wanted him to do: have GE buy back stock in an effort to boost the company's stock price. He told the analysts in Miami that he was leaning "60–40 against" buying back stock because he believed GE had better ways to create shareholder value. He also was careful not to use GE's stock, which he considered undervalued, as the currency for all of his deal-making. Slowly but surely, though, GE's stock began to move up. GE might have looked like a company built on power, jet propulsion, medical equipment, plastics, and appliances. But GE Capital had become an economic juggernaut—not just in the world of big-time banking but as an essential part of GE's own ongoing financial success. GE Capital was outperforming other divisions when it came to earnings gains. Gary Wendt and GE Capital were becoming the driving force behind GE's consistent earnings growth.

What's more, GE Capital had such an extraordinarily large collection of assets, many of which were fairly liquid—a loan here, an equity stake there—that Jack became increasingly confident that when he told Wall Street analysts he was going to hit a certain earnings number for the quarter, he could do it—knowing that, if need be, he could always sell some of the assets in the GE Capital portfolio to make up for any earnings deficiencies on the longer-cycle industrial business for which GE had become justifiably famous. Above all, Jack subscribed to the idea that if he told Wall Street that GE was going to make a certain amount of earnings for the quarter, by God, he was going to deliver those earnings. That kind of dedication to what Wall Street expected and wanted began to earn the company the respect of institutional investors. The stock started to move up.

In 1987, before the market crash, GE's stock market value ranked it third behind IBM and Exxon. Three years later, in July 1990, under Jack's watch, GE was on the verge of surpassing IBM to become the most valuable company in the United States. In its Heard on the Street column on July 23, *The Wall Street Journal* noted the milestone. The market value of GE's stock was $65.8 billion, the paper reported, while the market value of IBM was $67.4 billion, putting GE within spitting distance of the crown. "GE's surging past IBM would be akin to winning the World Series for Mr. Welch," wrote the *Journal*'s reporters David Hilder and Randall Smith. They noted that the changing of the guard was driven by a combination of GE's surging stock price and IBM's sagging one. In the previous eighteen months, GE stock was up 55 percent, nearly double the increase in the S&P 500 stock index. And on July 26, GE's market capitalization—at $65.5 billion—passed IBM's for the first time, for all of three days, before Exxon stormed past it. (GE would not regain the top spot again until May 1993.)

In September 1990, under the headline "Hot Light on GE," James Grant, the founder of the profoundly insightful *Grant's Interest Rate Observer*, explained GE Capital's importance to the GE narrative. GE

Capital had gone from making $115 million in profit at the beginning of the 1980s to making nearly $1 billion at the end of the decade, and its contribution to GE's net income was 20.7 percent in 1989. One fifth of the company's net income was coming from its unregulated bank. Grant went further and suggested that this new reliance on finance would become a potential weakness for GE. Since there was no AAAA rating in the bond market, the debt ratings of companies such as GE and GE Capital—both rated AAA by S&P and Moody's—could go in only one direction, and that was down. He noted that GE was only one of twelve companies rated AAA and that the yield on its debt was virtually the same as that issued by the U.S. government—the "interest-rate spread" between them "is now only thin as a dime."

A month later, in October 1990, David Tice, then a relatively unknown money manager and short seller, openly questioned the wisdom of Jack's expansive foray into financial services. In a more-than-five-thousand-word article in *Barron's*, Tice was struggling to reckon with the fact that while the stocks of many other banks were down as the extent of their poor lending decisions became increasingly known in the burgeoning recession, somehow GE's stock was hitting all-time highs and becoming increasingly valuable. "While most people identify GE with dishwashers, jet engines and the Bill Cosby show, GE, of course, is also one of the largest financial companies in the country," he wrote. "Indeed, through its financial arm, GE Financial Services, GE has a larger portfolio of leveraged-buyout loans and real estate loans than most of the money-center banks."

Actually, Tice had first raised questions about GE Capital earlier in 1990 in his relatively low-circulation *Behind the Numbers* newsletter. He wrote that GE Capital was no different from the biggest money-center banks except that it had no deposits and no foreign loans. Just like big banks, Tice wrote, GE Capital had "a large exposure to commercial real estate loans, LBO loans, credit-card loans and equipment leasing." His analysis in *Behind the Numbers* had little impact. But

*Barron's* was the big time, a national publication, and in a predigital era required reading for most sophisticated portfolio managers. In his *Barron's* essay, Tice explained how important GE Capital had become to GE: its largest division, accounting for 24 percent of net income and 31 percent of its earnings growth.

Tice then zeroed in on the loans GE Capital made to finance leveraged buyouts. He was spot-on, especially for an outsider who wasn't getting much, if any, cooperation from the company. He explained that GE Capital had booked some $8 billion in leveraged loans, growing at a rate of 35 percent per year. He claimed that nearly $2 billion of the loans, or 25 percent of the portfolio, were "potentially troubled" and were at risk of not being paid back. He then examined a few of them. There was the $600 million GE Capital had lent to Patrick Media, a Pennsylvania-based billboard company. After a large acquisition, Tice wrote, Patrick had $850 million in debt and was "suffering severe cash-flow difficulties" and had been forced to sell off some of its assets "to pay current interest expense." But the buyers weren't showing up in the "present soft market." He was right.

There were also problem loans made to Cablevision System, a large cable operator in the Northeast, and a disastrous series of loans made to film production company MCEG, including a $70 million bridge loan Kidder made to the company that was transferred to GE Capital after it was not repaid. Tice also discussed GE Capital's huge $10 billion real estate lending portfolio, its leasing portfolio—the largest part of GE Capital—its credit card portfolio, and its acquisition of a container business from Sam Zell, the crafty Chicago billionaire. "GE has been one of the great success stories among diversified U.S. industrial companies," Tice concluded. "Its ability to make top-quality products in a highly efficient manner remains unquestioned. GE's ability to manage a financial-services giant growing at 26% annually, however, is subject to doubt. GE would not be the first diversified company to encounter severe problems because the one business it understood the least, grew the fastest."

GE stock fell 3 percent on the Monday morning following the publication of Tice's article. Then GE went on the offensive. It held a conference call with investors to refute Tice's analysis. In October 1990, *The Wall Street Journal*'s Heard on the Street column featured the questions Grant and Tice had asked about GE and its finance subsidiary. "Is GE's lending arm its Achilles' heel?" the *Journal* wondered. Suddenly, risks about GE that had been shared narrowly among a small subset of sophisticated investors were aired far more widely.

After refusing to engage with both Grant and Tice, GE decided to respond to the *Journal*. By then, GE's stock had fallen 9 percent in one week as investors worried GE "might stub its toe" in financial services, and was down 27 percent from its July high, when the company was flirting with overtaking IBM as the country's most valuable. "We are confused why these people seem to be concerned," Larry Bossidy responded, stepping out of a quarterly management meeting to chat with the *Journal* reporters. He said GE Capital earnings grew 20 percent in the first half of 1990 and were likely to increase 15 percent to 20 percent in the second half. He did allow that GE Capital was "in a turbulent industry," citing specifically real estate and LBO loans, but asserted, "we are not in the same business as the banks." He defended GE Capital's real estate lending practices and explained that it had warrants in seventy-five different companies, valued at between $750 million and $1 billion, as a result of its LBO lending. Could GE Capital "hobble" GE's earnings "outlook" in the near term? the *Journal* wondered. "As best as I can tell, that's not going to happen," Bossidy said. GE's pushback seemed to have worked, putting the squeeze on Tice and others who had bet GE's stock would fall because of GE Capital. "I can't say this was a great success financially," Tice said of his GE short. "GE went on to do great things for another 10 years. The warts remained hidden because they were able to keep borrowing in the commercial paper market and were able to grow out of their potential problems."

There was a legendary story inside GE about how a young John Flannery—who would go on to become GE's CEO in 2017—was having a "skip-level meeting" with Jack and Bossidy soon after the *Barron's* story appeared. A "skip meeting" was when top executives met with junior employees, cutting out the many layers of management in between them. It was a wonderful, if treacherous, idea. Flannery had just been moved to the LBO workout group to try to restructure—and to salvage—the soured loans GE Capital had made. They had the young leaders come to a conference room in Stamford three hours before Jack and Bossidy were going to show up, to rehearse what they were going to say and what topics to avoid. After three hours of preparation, Flannery was terrified of being in Jack's presence. "If Jack asks you anything about our business," Dennis Carey, Flannery's boss's boss, told him, "it's name, rank and serial number." In other words, the problems in the LBO group were sufficiently significant at that time that Carey didn't want Flannery, at his low level, to inadvertently spill the beans on how bad things *really* were. Carey didn't want Flannery to be the messenger in that meeting.

At the meeting, Jack spoke for five minutes and then told the young executives he was there to answer their questions, not to talk. "Who's got a question here?" Jack said. Silence. Everyone was looking at his shoes. "Oh, shy group, shy group," Jack said. "Okay. All right, who's here from this leveraged buyout group?" Flannery tensed up. "Un-fucking-believable that he lands on that one right out of the gate," he said to himself. He thought about letting his other colleague from the group answer instead. But when he looked over at her, she was literally shaking in fear. Her body was shaking. Her hands were jiggling underneath the table. "She probably can't even breathe at this moment, never mind say something," remembered someone who was there. Flannery raised his hand. Jack started peppering him with questions. "Do we own TV stations?" he asked. "Do we own a billboard company? Do we really own a ceiling fan thing? Do we really own a taco chain in California?" It quickly became obvious to Flannery that Carey's caution had been meritless. Jack already knew the extent of the problem companies at GE Capital. "Now I have a gauntlet," Flannery thought. "I either give you my name, rank and serial number and look like an idiot in front of you or I answer questions." Flannery decided to answer the questions, and he knew the answers to Jack's simple questions. His boss wasn't there. He went for it, and survived. At the end of the session, Jack told the group, "You know, I'm not really even CEO of GE anymore. It's John and his buddies here, John and his real estate buddies, because if we can get through this thing, that's one scenario. If we blow up GE Capital right now, our P/E"—price-to-earnings ratio—"is going to go from eleven to two, and there's nothing in the world I could ever do as CEO that would ever compensate for that. So I'm not even in charge of the company anymore. It's John and his buddies."

As Jack was getting ready to leave, he pointed his finger at Flannery. "Don't fuck up," he said. Message received, Flannery thought.

GE Capital was the fiefdom of Gary Wendt, who, despite the Cassandras, kept pumping out the profits and the acquisitions into the 1990s, opportunistically scooping up the assets that other distressed financial institutions were forced to sell as the excesses of the 1980s became apparent. A few days after the Heard on the Street column, Wendt announced the purchase of the cargo-container-leasing business of Itel Corporation for $825 million. GE Capital also bought Macy's consumer credit card business, another eight million customer accounts to add to its forty-million-customer credit card business. It bought more credit card customers in Europe as well as the mortgage servicing business from Travelers, the big insurance company. GE Capital also bought more than $900 million of auto loans from the Resolution Trust Corporation, the federal agency set up to sell assets from failed thrifts. GE Capital had around $64 billion in assets at that time, more than all but the largest U.S. banks.

GE Capital also had the ability—unlike many other lenders—to take control of a defaulted asset and manage it until markets improved sufficiently to sell it. For instance, when Eastern Airlines defaulted in 1991, GE Capital seized twenty-six jets that it had financed. Instead of taking write-offs, GE Capital paid to maintain the planes and then re-leased them to other airlines. "When we have problems, we like to go in and get the asset and start doing something with it," Wendt said. "We manage the problem." It was the same playbook Wendt had been using at GE Capital since he took over managing the Houston Astros. Wendt also had another huge asset: GE. When GE Capital's nonper-forming assets began creeping up in 1991, causing concern at the rat-ing agencies, GE, with its AAA credit rating, agreed to guarantee some of GE Financial Services' debt and to pledge not to reduce its equity in the subsidiary.

By November 1994, GE Capital had assets of $212.5 billion, mak-ing it the third-largest, by assets, bank company in the country, behind only Citicorp and Bank of America. Wendt and GE Capital were the subjects of a rare two-part front-page article in *The Wall Street Journal* by Steve Lipin and Randall Smith. In the first part, a profile of Wendt, he described himself as "a middle manager in an industrial company." But the paper wasn't having it. "Don't let Gary Wendt's show of mod-esty fool you," Lipin and Smith wrote. They compared him with Jack, actually. Wendt, they wrote, "is just as tough on subordinates and is a brutal negotiator with the same win-at-all-costs attitude."

The article included stories of Wendt tearing up PowerPoint pre-sentations about potential deals if he felt they weren't up to snuff and ordering underlings to "shut up" if they had nothing worth saying. He was also prone to holding up two paddles—a stop sign and a question mark—to save time during presentations by his direct reports. He also never approved deals on first blush. "Gary can be very intense with people—short and abrupt and preoccupied," longtime GE executive Frank Doyle told the *Journal*. When Doyle says hello to him, Wendt "sometimes walks by without responding," the paper reported. The *Journal* portrayed Wendt as a man of the people who liked his expensive toys. "His suits rumpled, his shirttail sometimes out, he carries his own tray through the company cafeteria," Lipin and Smith reported. "Although he does more deals than most investment bankers, he eschews power ties and Armani suits." On the weekends, during which he was happy to work, he wore khaki pants and polo shirts. He drove a Jeep Cherokee and lived in a home appraised at $900,000 in North Stamford on two acres, instead of living in the tonier suburbs New Canaan and Darien, like his GE brethren. But, the paper also reported, he was a member of the "posh" Blind Brook Club, in nearby Purchase, New York, and took family vacations to "exotic" locations such as Mount Kilimanjaro, which he trekked up in 1993. The telling kicker to the article was the story of when he and his friend David Beim, a professor at Columbia Business School, were trying to get back to New York from traveling around Nepal and discovered there were problems with the flight leaving Kathmandu airport. Wendt refused to accept that he could not leave Nepal as he wanted to do. Wendt recalled of the incident, "I was able to get out and he wasn't."

The Wendt profile was about an important and quirky—but relatively unknown—GE executive who was driving GE's profitability year after year. It was timely and essential reporting, and a fun read. The second story, the next day, by the same duo plus the GE beat reporter, was far more devastating. Headlined "Managing Profits," it was a story about how Jack supposedly used one trick after another to smooth out GE's annual earnings to make sure that he never disappointed Wall Street's profit expectations for the company.

That Jack managed GE's earnings on a quarterly basis had been rumored for years inside the company and on Wall Street. The *Journal* seemed to be onto him. It was a charge that infuriated Jack to the bitter end (as he told me repeatedly). Now it was front-page news in *The Wall Street Journal*. The central question: How did GE, under Jack, report quarter after quarter of rising earnings, through both booms and busts? One "undeniable" explanation, the *Journal* wrote, was the impressive performance of GE's eight industrial businesses and its twenty-four finance businesses at GE Capital. "We're the best company in the world," Dennis Dammerman told the paper.

Not so fast, the *Journal* interjected. It suggested another reason: "earnings management," what it described as the "orchestrated timing of gains and losses to smooth out bumps" and to avoid "a decline." Martin Sankey, a respected research analyst at First Boston, agreed that GE was a "relatively aggressive practitioner of earnings management." The idea, according to the *Journal*, was to match both accounting and actual gains with losses so that the overall quarter-to-quarter effect was a slow, steady earnings rise, a pattern that investors could not only love but also count on—and that would result in an ever-increasing stock price and market value. (GE and AT&T were then duking it out for most-valuable-company honors.) Jack didn't want to disappoint investors with an earnings miss or to take a big gain in one quarter that would make a future quarter-to-quarter comparison more difficult. The *Journal* pushed Dammerman to respond to the allegation of "earnings management." He declined. Jack, no surprise, was nowhere near this article.

The *Journal* had plenty of examples to make its case. The paper reported that one way GE futzed with its numbers was by increasing the expected rate of return on its pension-plan portfolio, an increase in profit in one year of $316 million. GE responded that the rate it was using was no different from what others were doing and reflected an increase in returns generally. Another way GE supposedly managed earnings was through the use of "restructuring charges," something it had done to the tune of $4 billion since 1983. GE offset a $1.4 billion gain from the sale of its aerospace business with a $1 billion charge to cover the costs of closing various facilities worldwide. Magically, the after-tax effect of both moves was zero—as both the after-tax gain and the after-tax loss equaled $678 million. GE also assigned a range of different values to the businesses it acquired when it bought RCA.

Some it increased, such as NBC, and some it lowered, such as a home-video joint venture. When the latter was sold, the gain was higher than it otherwise would have been. (On the other hand, by raising the book value of NBC, GE complicated the rumored sale process by needing a higher price from a buyer to be able to book a gain.) Dammerman admitted to the *Journal* that when GE executives perceived that a big one-time gain was in the offing, they got together as a group and asked, "What are some strategic decisions that you would make to make the business better going forward?" But he defended the practice by explaining, "It's not like we're creating some big cookie jar" from which GE could take profits when it needed them.

The *Journal* correctly pointed out, though, that GE Capital *was* like a big cookie jar, filled with all sorts of financial goodies that could be consumed at any time doing so would help the cause. There were the equity warrants that GE Capital received in financing LBOs. There were used commercial jets coming off lease. There were any number of commercial buildings that GE Capital had in its portfolio that it could sell. With more than $200 billion of assets, the list of choices was nearly inexhaustible. Wendt, for one, told the *Journal* about GE's "smoothing" of earnings. "We do a little, not a lot." His CFO, James Parke, was a little more candid. "We have a lot of assets in this business. . . . Obviously, there are timing issues associated with when those assets are sold."

Years later, from the comfort of his Florida home (and his post-GE life), Wendt was more candid about the use of GE Capital's assets to smooth GE's earnings. "We did, we did," he said. "We were able to. I always had a lot of things available for the quarter. I had to because I knew he"—Jack—"was going to call up and say, 'I need another $1 million or another $2 million or whatever,' and so I'd go over to [James]

Parke and I'd say, 'Okay, let's do this one and this one.'" He said under Jack, GE "always" made its numbers. "Making your earnings was just life to us," he continued. "We all knew it. [Jack] was the one that had to get it done, and we knew he had to get it done, and so it was part of the company culture, clearly. Make your numbers. Make your numbers. And stretch. The guys did it."

Over lunch at a Connecticut pub, Denis Nayden, who was one of Wendt's successors at GE Capital, also defended the practice of harvesting GE Capital assets as needed. "What's the job of a public company?" he asked rhetorically. "Produce earnings for shareholders. And, oh, by the way, when you have a portfolio of however many companies—and it was a lot—everything isn't going to go in a straight line. Your job is to realize when A, B, and C have fallen off the cliff this year, somebody else has got to step up. And that was just the attitude. Would you call that managing earnings? No. I'd call it you use the assets that you have in your portfolio to maximize the return for the company. And that's what we did." He saw it as solving problems. "Did we create income because somebody else couldn't deliver in a quarter?" he asked. "Absolutely. There's nothing wrong with that. We didn't do anything illegal ever. We had financial flexibility [at GE Capital] that, if you're making washing machines, you don't have. You can monetize assets. You can sell assets. You can harvest assets. And that's what we did on a regular basis. We securitized assets. We used our financial smarts to maximize the assets, and when called on to deliver income, we did it on a regular basis. And we had a record that's unparalleled year after year. It'd be one thing if you did it a couple of times. That's engineering." GE's philosophy was deliver the earnings you promised you would deliver, every time. No matter what.

Nayden applauded Wendt for pushing people to stretch. "We had at Capital, ultimately, I think there was twenty-six businesses. And every single one of them had a budget. And if you stretched every single one of them to think outside the box and say, 'Stretch yourself. If you think you can only do ninety, develop a plan to deliver a hundred and ten,' recognizing fully that with twenty-six businesses, no way in hell are all twenty-six going to hit the stretch budget. No way. But if a third of them did, then you could offset the guys that were short. And that's what we did all the time."

IN JUNE 1991, JACK LOST HIS WINGMAN WHEN HIS LONGTIME FRIEND and business partner Larry Bossidy decamped somewhat unexpectedly to become CEO of Allied-Signal, Inc., a struggling mini GE based in New Jersey. Bossidy and Jack had been close since 1978, when they met playing Ping-Pong at a GE corporate outing in Hawaii. Bossidy, fifty-six, knew he wasn't ever going to get the chance to run GE given that Jack, then fifty-five, still had a decade left as CEO. Bossidy had been approached in April to take over Allied-Signal from Ed Hennessy, its sixty-three-year-old CEO. But it wasn't until Hennessy agreed to let Bossidy run the company immediately that he decided to take the job.

Hennessy had hoped that his successor would serve an eighteen-month apprenticeship and allow Hennessy to retire at sixty-five. But Bossidy was anxious to run his own show after years serving Jack. He pushed the Allied board for an immediate appointment. The board sided with Bossidy. "We found out that the type of guy we wanted to attract didn't want to come here and sit around for a year and a half at Ed Hennessy school," Hennessy told the *Journal*. "He really wanted to come at a decision-making level."

On a Monday morning in late June, Bossidy told Jack the news. "You know the time has come for me to move on," he said. "I don't want to sit here for the rest of my career." He told Jack he was going to Allied-Signal. He said it appealed to him because it was a "turn-around situation" and because it was in New Jersey, so he wouldn't have to move his family. "It was an emotional meeting," Jack recalled. "A lot of tears fell, and we hugged each other."

When Jack later spoke with Gerry Roche, the headhunter who recruited Bossidy, he told him, "Gerry, half my face is crying because you're taking my best friend and my best guy. The other half is smiling because he can run any company in this country and he deserves to run his own show." Longtime GE Capital bankers recalled Bossidy's impact on the business. For instance, in the late 1980s, GE Capital had the opportunity to buy from the Resolution Trust Corporation a number of failed thrifts in Texas. There were all kinds of tax benefits that GE Capital could realize based on the way the RTC had structured the deal. Wendt loved the deal. "It took Gary about 30 seconds to look at it and say, 'Oh my god. This is printing money,'" recalled one GE Capital banker. "And we rushed up to Fairfield and met with Bossidy. He looked at it and said, 'Okay,' and then went in to talk to Jack about the deal. Then Jack and Bossidy came out and said, 'We can't do this. There are some deals that if we as GE are seen doing, it will just cause an uproar, that we're taking advantage of the taxpayer and the country and all that kind of stuff, and we as a company can't do that.'"

Bossidy's departure meant that the man Jack relied on to manage GE's growing, and increasingly important, financial services business was no longer around. Suddenly, the buck stopped with Wendt at GE Capital. True, there was Dennis Dammerman, the highly regarded CFO, watching out, but it wasn't quite the same, at least for Jack.

BEYOND THE LOSS OF BOSSIDY AND SUGGESTIONS OF GE MASSAGING its earnings, Jack had other headaches. There were cultural problems with the health care business that GE had bought in France by swapping with Thomson. There was the time-sheet cheating scandal in the defense business. There was the ongoing matter of the PCBs in the Hudson—which would go on for decades and reveal that GE had also dumped PCBs on land that homes were built on in places like Pittsfield, Massachusetts, where Jack got his start at GE. There was a scandal about the purloined documents in its valuable "synthetic diamond" business. And then there was the time, in December 1990, that Ben Heineman, GE's general counsel, called Jack at home one Saturday afternoon. "You're not going to believe this," he told Jack, "but we have an employee who has a joint Swiss bank account with an Israeli air force general."

All they knew at that point was what had been reported in the Israeli press and passed along to Heineman by a GE employee in Israel: Herbert Steindler, a GE employee in the aircraft engine business, had "conspired," according to Jack, with Rami Dotan, an Israeli air force general, to "divert money" from contracts to supply GE engines to Israeli F-16 fighter jets. "I nearly choked when I heard Ben's news," Jack remembered. "Imagine having a crook on your payroll." GE suspended Steindler immediately and then fired him three months later when he would not cooperate with an internal GE investigation. He was also under investigation by the Department of Justice. According to Jack, citing GE's nine-month internal investigation conducted by Wilmer, Cutler & Pickering, a Washington law firm, Steindler and Dotan siphoned off for themselves about $11 million into Swiss bank accounts that they controlled.

"By the time this mess was over," Jack recalled, "19 months and many headlines later, we had to discipline 21 GE executives, managers, and employees, pay the U.S. government $69 million in criminal fines and civil penalties, and testify before a congressional committee." Brian Rowe, the head of GE's aircraft engine business, whom Jack described as "a larger-than-life figure," had to plead guilty, on GE's behalf, in federal court, and a GE vice-chairman had to spend a week in Washington "getting our engine business off suspension."

On July 29, 1992, Jack testified before a subcommittee of the House of Representatives Committee on Energy and Commerce, along with Frank Doyle, a senior vice president, to answer questions about the Dotan-Steindler scandal. Jack did not make excuses for what happened. "There are none," he testified. But he did want to put the incident in context. He said that while 12 percent of GE's $60 billion in revenue came from "dealings with the U.S. Government," sales to Israel were less than one tenth of 1 percent. He said there were two "obvious" criminals and two large groups of victims—the American people and GE's 275,000 employees, who were reading in the newspapers "that their company is paying severe penalties for wrongdoing."

Jack gave the committee a full-throated explanation, and defense, of the Jack Welch Way, and why he demanded, and expected, excellence from every one of GE's employees. There had been some suggestion that Jack's aggressive quarterly financial goals, and his insistence that they be met, were driving GE people to misbehave. "This view has largely been supported by interviewing a few employees who have been removed for poor performance, ethical violations or downsizing in businesses whose markets have declined," he said. He asked, rhetorically, what the solution should be when some people cheat. "Tell athletes to run slower, jump lower so they will be above suspicion?" he said. Not a chance. "You must run as fast as you can and jump as high as you can," he said. "But if you break the rules, your medals are gone and you are out of the game for good."

The committee meeting became a soapbox for the Jack Welch Way as he explained that GE's main competitor was not Westinghouse but rather a group of big foreign industrial companies, like Siemens and Philips; that GE had made a positive $6 billion contribution to the trade deficit—something people used to care about—and that GE employees were two to three times as productive as other American workers. He reiterated that GE was "competitive" and "yes, we also keep score in earnings." He said GE had had forty consecutive quarters—ten years—of earnings growth in the 1980s and that earnings growth had continued to date in the 1990s "right through every quarter of this recession."

He said no one at GE got fired for missing a quarterly projection or annual projection. "That is nonsense," he said. People got fired for only one reason: a "clear integrity violation," because "if you commit one of those, you're out." He pointed out that GE had a population equal to Tampa, Florida, and Saint Paul, Minnesota, but without a police force or jails. "We must rely on the integrity of our people as our first defense," he said. There were also hotlines, ombudsmen, and voluntary disclosure policies. He conceded the systems did not work in the Dotan case. "But I take great pride that 99 percent of our 275,000 people get up every morning all over the world and compete with passion and absolute integrity," he said. "They don't need a policeman or a judge. They only need their consciences as they face the mirror each morning."

JACK BARELY HAD TIME TO SAVOR HIS TESTIMONY IN THE DOTAN MATTER before he had a new public relations problem at hand: Debra Chasnoff's taut thirty-minute documentary film about GE and its involvement in the nuclear weapons industry, *Deadly Deception: General Electric, Nuclear Weapons and Our Environment*, openly questioned Jack's integrity. At first, the film seemed like it would have little impact. Chasnoff had never made a film before. She made this one in nine months, starting in September 1990. It was extremely low budget; she spent only $65,000 on it. And according to *The Washington Post*, it was a "frank piece of propaganda" against GE that had been commissioned and produced by Infact, a Boston-based antinuclear organization. Chasnoff's film harshly criticized GE for failing to clean up the Hanford Nuclear Reservation, in Washington State—which it operated under a contract with the federal government from 1946 to 1965—and for the death and decimation of the farms, animals, and people who were Hanford's neighbors. Playing hauntingly in the background were GE's new television spots about "bringing good things to life." The film also focused on the alleged asbestos and radiation poisoning that had occurred at the Knolls Atomic Power Laboratory, in Schenectady, where

GE made triggers for the neutron bomb. *Deadly Deception* and the boycott of GE appliances that Infact engineered along with Chasnoff's film would likely have been ignored, but for the fact that the film won the Academy Award for Best Documentary (Short Subject) in 1992. In her acceptance speech, Chasnoff thanked Infact for making the film possible and for helping her "tell the real story [about GE] that falsely claims it 'brings good things to life.'" Suddenly, Infact was flooded with requests for the film. Jack, who had previously defended GE's role in the nuclear power and arms industries, remained silent. A GE spokesman told *The Wall Street Journal* the film was full of "erroneous allegations," without elaboration. He said, "We haven't any intention of debating them." But Jack did have a desire to leave behind his increasingly controversial aerospace and defense businesses (keeping, of course, the all-important commercial jet engine business).

In October 1992, six months after Chasnoff picked up her Academy Award, Jack went to the Homestead Resort, in Virginia, for a meeting of the Business Council, which one of his predecessors had created. At the meeting, Jack was focused on meeting up with Norm Augustine, the CEO of Martin Marietta. Jack was once again in deal mode. When he found Augustine in the lobby of the hotel, he suggested they get together to talk about what to do with their respective aerospace businesses. Augustine was wary. He feared Jack wanted to buy his company. "We treasure our independence," he told Jack. Jack reassured Augustine that was not what he had in mind. He suggested they have dinner. In Fairfield a few days later, Jack made his pitch while Augustine ate his fish. He wanted Martin Marietta to buy GE's aerospace business. "A deal could give us another graceful exit, this time from a military business I didn't like," he remembered.

The two CEOs decided to proceed in secret. For three nights straight in early November, Augustine attended his company's off-site meeting on Captiva Island, Florida, and then flew up by private jet to Fairfield to negotiate with Jack and Dammerman into the early-morning hours. He then would get back on his jet and fly back to the off-site. In order to reduce the risk of a leak, no investment bankers or lawyers were involved with the deal until the men had reached agreement on the business terms. Three weeks later, the deal was announced. Martin Marietta agreed to buy GE's aerospace and defense businesses for $3.05 billion in Martin Marietta stock. The stock deal made GE a 25 percent owner of Martin Marietta. It was the largest deal in the history of the aerospace industry and had taken twenty-seven days from the dinner Jack hosted for Augustine in Fairfield. The stock markets loved the deal, sending both stocks to their fifty-two-week highs. Two years later, Martin Marietta merged with Lockheed; when GE sold its stake in the company, the stock had doubled in value to $6 billion.

SMACK IN THE MIDDLE OF JACK'S FRENZIED DEALMAKING, HE TOOK time out to focus on his dysfunctional personal life. In 1987, he decided to divorce Carolyn, the mother of their four children. In *Jack: Straight from Gut*, he devoted around fifty words to the subject. "Carolyn and I had been having difficulty in our marriage for many years," he wrote. He credited her with doing a "great job" raising their four children, who were then finishing their schooling at elite universities. He conceded he was the "ultimate workaholic" and that he and Carolyn "simply found ourselves on different paths" and that other than "our friendship and mutual respect, we had little in common." He continued, "She got very, I wouldn't say jealous of my success, but disdainful of it," he continued. "It happens. She was a bigger deal in college than I was, okay? She got mad and she called herself a 'suitcase.'" Of course, he made no mention of his numerous infidelities.

In truth, at least at this stage of his life—the hard-charging leader of one of the most important and valuable companies in the world—Jack was not the most devoted father, as he was the first to admit. Not that he didn't love his four children, two sons and two daughters; he did. But he was more focused on GE than he was on them. In all of our many discussions, the only time he spoke about his children was when he told me that he "loved them to pieces" but that he had made "a mistake" when he gave each of them a bunch of GE stock when he first became CEO. Given how well the stock performed under his tenure, each of the kids ended up with something like $50 million in GE stock, half when they turned thirty-five, the other half when they turned forty-five. Each of them quit their jobs, he told me, as a result, despite two of the four going to Harvard Business School and one going to the Harvard Graduate School of Design. "They turned out differently than I'd hoped," he said. "We're close. But they got too much money. I thought half at thirty-five and half at forty-five was the right thing. If I had to do it all over again, I wouldn't have given it to them. Who would know?" In any event, the Welches divorced "amicably," Jack wrote, after twenty-eight years of marriage in April 1987. "She certainly didn't get any sort of fabulous divorce settlement," Donna Bertaccini, the researcher on *Testosterone, Inc.*, told me.

That was an understatement. After twenty-eight years of marriage, the Welches' net worth was $12 million, according to Bertaccini's research. But the vast majority of it was Jack's. His wife owned only a half interest in their homes in New Canaan, Connecticut, and Sun Valley, Idaho. Nantucket? Jack's. Manhattan apartment? Jack's. Florida property? Jack's. The divorce documents revealed her modest lifestyle and little appetite to fight Jack for a bigger slice of the marital pie. "Reading through the documents," Byron wrote, "one senses in Carolyn Osburn Welch a woman struck dumb by what was happening to her. In the entire divorce file, there is not a single assertion of her entitlement to the lifestyle she had presumably lived as the wife of GE's chairman and CEO." (She noted a weekly expense of $4.42 for an exterminator.)

In the end, Jack gave his ex-wife the house in New Canaan and agreed to buy her a new Toyota. She also got to keep her $90,000 in cash, stocks, and jewelry. He agreed to pay her a lump sum of $2.2 million plus another $1 million in alimony spread out over seven years. For himself, he kept all the other real estate, including Sun Valley, plus the couple's four automobiles, "roughly $11 million worth of assets in all," Byron wrote. In his book, Jack wasn't particularly sympathetic and of course did not share any of the details of the divorce settlement. He noted that Carolyn got her law degree and eventually married her college "sweetheart," who was also a lawyer.

Then it was back to Jack. "Suddenly, I found myself single again," he recalled. "Being single and having money was like standing six feet four with a full head of hair."

He would remain single until 1988 when he was set up with Jane Beasley, then thirty-six, an associate in the mergers and acquisitions group at Shearman & Sterling, the Wall Street law firm. Beasley and Jack had met on a "blind date" arranged by Walter Wriston, the CEO of Citibank, and his wife, Kathy. Beasley, it turned out, worked for Kathy Wriston's brother at the law firm. And Shearman & Sterling had long been Citibank's principal law firm. Wriston, of course, had been on the GE board of directors for three decades. Their meeting certainly seemed like a well-conceived plan. When the idea was first suggested to Beasley, she thought Kathy Wriston was talking about a different fellow named Jack Welch who also worked at Shearman & Sterling. She said she could not date that Jack Welch. When Kathy informed her that she was referring to the CEO of General Electric, Beasley warmed to the idea. Told he was seventeen years her elder, Beasley responded, "That doesn't matter. I'm not going to marry him."

At the time, Beasley was seconded to London on an extended assignment. When she returned to New York, in October 1987, she and Jack and the Wristons went to dinner at Tino's, an Italian restaurant on West Fifty-Sixth Street. "With Walter sitting there, the date was a little stiff," Jack remembered. "I had to be on my best behavior." At 10:00 p.m., Jack and Jane headed up to the small bar at Cafe Luxembourg, on the Upper West Side. They closed the place down. Their second date was at Smith & Wollensky, on Third Avenue. They both arrived in leather jackets and dungarees. And each had a burger. That night "really made the match," Jack said.

It would be inaccurate to say that Jack and Jane couldn't have been more different. They had their similarities. He was the CEO of one of America's most important companies; she was an associate at a Wall Street law firm. He was from a middle-class family in a historic Massachusetts town. She was from rural Clayton, Alabama—actually, along Highway 51 about midway between Clayton and Louisville, Alabama— about seventy miles southeast of Montgomery. Her father, William Durwood Beasley, was a farmer. He raised hogs, grew peanuts and corn, and drove a backhoe for the county. Her mother, Marion, was a teacher who died in May 1987, before Jane and Jack had had their first date. According to Jack, Jane "picked butter beans at 5:30 a.m. on her father's farm until her back ached."

She went to Troy State University, in Troy, Alabama, and then to the University of Kentucky law school, where she was the editor of the law review. From there it was off to New York City and Shearman & Sterling and the setup with Jack. "I wasn't always the ideal date," he confessed. He invited her for the weekend to Nantucket, to his house off Hoicks Hollow Road. "She had to go through a major negotiation with her boss to get the time off," he continued. But exceptions are made for the CEO of GE, who could at any moment become an important client. They went to dinner on Friday night. The next morning, Jack got dressed and headed out the door to play golf at Sankaty Head.

Jane was not happy. "You're kidding," she said. "I had to practically sign away my birthright to get this weekend off and you're going *golfing*?" It was nothing he hadn't done for all the years he was married. But, he said, "I knew this routine was over." As things got more serious between them, they started discussing potential impediments to marriage. He didn't like that she didn't golf or ski. She didn't like that he didn't go to the opera. He agreed to go to the opera; she agreed to try to learn to ski and to golf. He wanted her to give up her job at Shearman & Sterling. He wanted a "full-time partner," he said, "someone who would be willing to put up with my schedule and travel with me on business trips." She agreed, and "luckily for me," he continued, "decided to make" Jack "her full-time occupation."

They were married at Jack's house in Nantucket in April 1989. Jack's four children were in attendance. For the next few years, he did his "husband duties" and went to the opera, until Jane took pity on him. He taught her to golf. She excelled, "even though she had never played golf before meeting me," Jack remembered. She won the club championship at Sankaty Head four years in a row. Jack had only won it twice.

*Chapter Fifteen*

# EARNINGS MANAGEMENT

After briefly surpassing IBM to become the nation's most valuable company in July 1990, GE reclaimed the top spot from Exxon-Mobil in May 1993, with a market value of more than $80 billion. For the next five years, GE owned the top spot. It crossed the $100 billion milestone in July 1995, the $120 billion mark by the end of 1995—and kept going straight up. By the turn of the century, the time had come for Welch, who had taken over in 1981, to start thinking about his legacy in the form of a few final strategic moves, perhaps one last glorious acquisition, and putting in place a succession plan.

It is hard to remember now, but in the midnineties GE and its CEO had attained a near-mythical status for the stock's ability to deliver consistent earnings growth and pay dividends. It wasn't just GE employees whose retirements were vested in the performance of GE's stock; it was often one of the largest holdings in the bread-and-butter mutual funds that made up the retirement savings of middle-class Americans. Delivering these consistent results quarter after quarter, particularly from a conglomerate whose disparate businesses by their nature were destined to have highs and lows, was no mean feat.

Wall Street had long suspected that GE managed earnings, an assertion that Jack bristled at throughout his life, including in our numerous conversations. Even the shocking denouement and liquidation of Kidder Peabody, a longtime thorn in Jack's side and one of his few genuine blunders, was not immune to earnings management. No matter which GE executives Jack installed at Kidder Peabody, there was a cultural divide between the investment bank and GE Capital that could not be bridged. In addition to the insider-trading scandal of the late 1980s, which star-crossed GE's ownership of Kidder from the start, there was also the disastrous scandal perpetrated on Kidder by Joseph Jett, who had started at GE in the plastics division only to return paradoxically years later as a Kidder trader and one of the few Black employees at GE or Kidder. The Jett fiasco was then the largest trading fraud in history. In short, Jett was able to exploit a flaw in Kidder's rickety computer system to make unprofitable trades appear profitable, treating forward-dated transactions as if they were immediately settled. Jett's house of cards came crashing down in the first quarter of 1994, when he was "trading" so frequently that Kidder's computer systems couldn't keep up, and his "profits" had grown to $350 million, large enough that Kidder management feared he was taking unacceptable risks. Computer specialists began noticing that none of Jett's supposed "trades" were ever consummated. Jett kept rolling his trades over two to three days before he was due to settle them up, all the while keeping the "profits" on the books. Eventually, Kidder halted Jett's trading and summoned him to a meeting. When his explanations proved unsatisfactory, he was fired. The losses were large enough that GE had to take a $210 million charge to its first-quarter 1994 earnings. Jack was peeved by the unexpected loss. But it was the lack of contrition and accountability shown by Jett's superiors at Kidder that got Jack to decide that Kidder had to go.

Compounding the massive fallout from the Jett scandal at Kidder was the fact that rising interest rates resulted in mounting losses in

Kidder's massive portfolio of mortgage-backed securities. Denis Nayden, the GE finance veteran who had been tasked with trying to see if anything could be salvaged at Kidder, found himself in an untenable situation—trying to run a business that he barely understood and that was in the process of imploding. He called a series of Kidder management meetings at Crotonville. An early question he wanted to answer: What businesses should Kidder reorganize around? When the two heads of fixed income couldn't answer Nayden's question about how they quantified risk, he went berserk. "What do you mean?" he exploded. "I can't believe that." He irritated more Kidder top executives during another Crotonville session when he decided the energy-futures business, in Houston, should be jettisoned because it wasn't core. When Kidder brass complained that the division had just celebrated its ten-year anniversary, Nayden responded, "So what?" When reports started leaking to the press about what had transpired at Crotonville, Nayden burst into the nineteenth-floor conference room at Kidder's headquarters in Hanover Square, where Kidder's executives were meeting, brandishing the offending article. "Tell me how this is happening!" The annoyed Kidder bankers suggested maybe the leaks did not originate with them. One suggested they take a lie detector test. "I already thought of that," Nayden reportedly said. "It's not legal."

By then, Jack had had quite enough of the storied investment bank. Pretax losses at the firm were at $30 million in both August and September 1994 and were on track for a $400 million loss for the year. Facts on the ground at Hanover Square were forcing Jack into a corner and into changing his mind about a sale of Kidder. Jack and his two senior executives were considering a panoply of options, from okaying a management buyout of Kidder to finding an overseas equity partner to an outright sale. There were conversations with Dresdner Bank, a division of National Westminster Bank in London, as well as Paine-Webber and Donaldson, Lufkin & Jenrette in New York. Without dispatching Kidder, Jack figured GE would have to invest another $500 million into the investment bank.

"The cultures were very different," explained a former GE Capital executive. "Jack more than anything wanted the name on the front of the jersey to be more important than the name on the back of the jersey, and Kidder was a place where the name on the back of the jersey was far more important than the name on the front of it. He just didn't like places where GE stock options weren't the most important compensation you got. It's that simple.... Playing with other people's money is fundamentally different than investing your own. If you really tear apart the culture, that's what it gets down to, the cultural difference."

Time to deal. In October 1994, Jack sold certain Kidder assets—principally the broker-dealer organization—to PaineWebber, the big brokerage firm, for $670 million in PaineWebber stock, equal to about a 24 percent stake in the firm. Jack enlisted the help of Pete Peterson, then one of the founding partners of the Blackstone Group, along with Steve Schwarzman. The deal negotiation was attenuated. Kidder's junk-bond group decided not to join PaineWebber; other top Kidder bankers were jumping ship, too. In early October, the deal fell apart. "In the dead of night," remembered Denis Nayden, "they said, 'We're not going to do it.' It literally almost killed me. I thought I was going to die."

Jack called Don Marron, CEO of PaineWebber, "to see if we could put things back together again," Jack remembered. Marron, a patrician tennis player who had amassed one of the world's great personal collections of modern art, called Peterson and got the deal back on track. "Jack was smart enough to appreciate that this was his chance to surgically get out of a deal he should not have been in in the first place," Peterson said. With a handshake, Jack left for a ten-day business trip to Asia, leaving Dammerman and Nayden to finish the deal. Peterson had to call Jack once at three in the morning in Thailand, but other than that, the revised deal was agreed and announced in ten days. The deal closed in mid-December, a month ahead of schedule. The rest of Kidder, primarily its investment banking and trading operations, was liquidated after the sale of the broker-dealer to Paine-Webber. GE also retained responsibility for the lawsuits that resulted from Jett's $350 million trading loss.

The good news for Jack in the whole Kidder mess, if there was any, was that the initial $210 million after-tax write-down amounted to a 12-cents-a-share hit to GE's earnings. Jack called it "nothing, a peanut." And the market barely noticed. For most of 1994, GE remained the country's most valuable company, with a market value hovering around $90 billion. Jack's executives were exasperated by Kidder. "Kidder turns out, of course, was embarrassing and stupid and tiny," Ben Heineman told me. "It was real money but it was just a fraction of GE's finances. Kidder was an endless mess." Added another senior GE executive, "Kidder was a fucking debacle. By today's Wall Street frauds, Kidder was a pimple on the back of an elephant's ass, okay? But back then, it was a big deal." David Hilder, *The Wall Street Journal* reporter–turned–respected Wall Street research analyst, helped put GE's miserable Kidder chapter into the proper context. "Kidder," he told me, "was a small symptom of a larger problem that Jack Welch and Gary Wendt created: using the GE credit rating to build a massive lending operation that was more short-term funded, including with commercial paper, than any bank would have done after the early 1980s, and certainly not after 1994. A Harvard classmate of mine ran a commercial real estate lending operation at GE and there seems to have been no limit to GE Capital's appetite."

There would be a final coda to the Kidder Peabody saga. Six years later, in 2000, Pete Peterson called Jack to say he wanted "to make your weekend for you" and told him that Marron had agreed to sell PaineWebber to UBS, the big Swiss bank, for $10.8 billion. The deal gave GE more than $2 billion in cash for its stake in PaineWebber. "Make my frigging weekend?" Jack yelled. "You made my whole frigging year." GE's after-tax annualized return on its fourteen-year investment in Kidder was 10 percent, Jack explained. "However," he said, "there's no amount of money that would make us want to go through that again."

But getting Kidder off GE's books was at least one example where Jack's penchant for "smoothing" earnings in one year created a time bomb that went off—with nearly calamitous effect—decades later. According to a former GE director, in and around the time of the sale of Kidder Peabody to PaineWebber, in 1994, GE was facing a $2 billion loss—a December 1994 *Wall Street Journal* article pegged the Kidder loss for 1994 at $1 billion, and the GE SEC filings also show an after-tax loss of $868 million related to the liquidation of Kidder assets, which was substantially completed in mid-1995—which would have put a major dent into Jack's earnings guidance and certainly would have resulted in a dreaded quarter in which the earnings were lower both sequentially and year over year. That would not be acceptable in any ecosystem run by Jack Welch. So, according to the director, an equal of amount of money—$2 billion—was removed from the reserves of GE's insurance subsidiary, Employers Re, to cover the $2 billion Kidder loss and to make negligible the impact on GE's quarterly earnings. "The market thought, 'There's the good old GE machine, so big and so powerful that there are no bumps in the road,'" he said, "and someone buried it for twenty years and said, 'Don't worry, investment returns and discount rates will eventually fix it.' Well, $2 billion grew into ten-plus times that [amount]"—a reference to a major problem with GE's legacy insurance subsidiaries that was revealed to the public in 2018. "Put that on the pile with fifteen other things like that, and there you have it."

Another longtime GE executive told me that "for years" the "word" around the "water cooler" was that GE's insurance companies were used as "shock absorbers" anytime Jack needed a few pennies of earnings to make the numbers he had promised the Wall Street analysts.

He said there were always "billions" in liquid investments—such as government bonds—at the insurance companies with embedded gains in them that could be sold on demand and taken into quarterly earnings. One former senior Wall Street executive told me that Jack was among the last of a dying breed of CEOs. "He did things that today would be considered borderline fraudulent, but in those days it was acceptable practice," he said. "He'd run the company for a long time. And its huge financial arm, where they sold commercial paper without any backup lines [of credit] and lived off the arbitrage [of the AAA credit rating] and got a huge multiple and managed earnings, blatantly managed earnings to smooth earnings. It was beyond good." Added a former senior GE executive, simply, "The rules were different. It was a little looser for sure."

In one of my interviews with him, Jack and I discussed the criticism that he manipulated earnings. He was adamant that he had never done it.

"No," he said. "That's the worst, dumbest—don't you fall for that shit."

I was just asking, I told him.

"No," he said. "I'm not manipulating earnings. Let me tell you how it works. I had stretch targets. You've heard of a stretch target? We never made a stretch target that was an internal target. Everybody missed somewhere, but the stretch target was always five percent more than we promised Wall Street. I'm not going to promise Wall Street my last child, okay? You have to be an idiot to miss a Wall Street estimate. You tell them what the fuck you're going to make, two weeks or four weeks in advance. I had forty businesses. If I can't get a penny here or a penny there for a shortfall, then I'm an idiot. . . . Everybody's got a kitty in every level in every company."

Another time, Jack told me one of the keys to his success at GE was delivering a "consistency of earnings," not "managing earnings." Said Jack, "Consistency of businesses to deliver. You hold back on one, you push another. Always use levers but you always have to focus on consistency. That is a fact. Because the other things didn't count. That is the heart of this bullshit about managing earnings, staying up all night, fooling with the balance sheet. We didn't do that."

———

IF MANAGING EARNINGS WAS A SUB-ROSA PRACTICE OF THE JACK ERA, then Six Sigma was the more public face of how he attempted to maximize efficiency across GE's disparate businesses. In 1996, Jack had asked former wingman Larry Bossidy to come to Crotonville to give a lecture on the wonders of Six Sigma to the assembled executives. By then, Bossidy was the CEO of Allied-Signal and an early convert to the Six Sigma doctrine, which was a way to reduce manufacturing defects first pioneered by Motorola in 1987 after one of its executives traveled to Japan and then adopted the process.

Between 1987 and 1996 Motorola saved $11 billion using Six Sigma principles. "Jack Welch got wind of what Motorola was doing, and he was fascinated," said Thomas Lee, a former GE executive–turned–MIT professor. Manufacturing processes that achieved Six Sigma—a reference to a statistical model, based on standard deviations on a normal curve—had 3.4 defects per one million manufacturing steps. Said another way, at a Six Sigma level of manufacturing prowess, 99.99966 percent of a company's products or services were made or provided without flaws. Most industrial companies are satisfied with around 97 percent success. Very few companies even attempted a Six Sigma level of performance.

But if Bossidy was doing it at Allied-Signal, and it seemed to be working, that was good enough for Jack, especially since it was a way to squeeze more profits from GE's various businesses, assuming Jack could get the troops to buy in to it. GE invested more than $1 billion in training thousands of employees, and the system was adopted by every GE business unit. According to Bill Conaty, Jack's longtime head of HR, Six Sigma was especially useful in reducing defects in complex manufacturing processes, such as building an aircraft engine. "Aircraft guys were miles ahead of the rest of the company [on Six Sigma]," he explained. He said starting around 1993 he and Jack and an outside consultant put together a sophisticated survey of twenty-eight questions that was sent to GE's employees. One of the questions was about satisfaction with the quality of GE's products. "We would debate this topic of quality in the Corporate Executive Council meeting—the thirty, forty guys—every quarter," Conaty explained. But the results from the 1995 employee survey were clear: GE employees were disappointed with the quality of the company's products. "We've got to do something about that," Conaty continued. Jack spoke to Bossidy.

Using Six Sigma, Allied-Signal had improved quality, lowered costs, and increased productivity. Bossidy came to Crotonville; his talk was a big hit. Jack asked two of his senior executives to study what Six Sigma had done for Allied-Signal and for Motorola and what it would mean for GE. "A typical process at GE generates about 35,000 defects per million, which sounds like a lot, and is a lot," Jack told GE shareholders at the 1996 annual meeting, in Charlottesville, Virginia. "But it is consistent with the defect levels of most successful companies. That number of defects per million is referred to in the very precise jargon of statistics as about three and one-half Sigma. For those of you who flew to Charlottesville, you are sitting here in your seats today because the airlines' record in getting passengers safely from one place to another is even better than Six Sigma, with less than one-half failure per million. If you think about airlines, they run two operations. They get you from point A to point B from Seven to Eight Sigma. Your bags get there at Three Sigma." GE's manufacturing skills were ten thousand times below the Six Sigma level. Raising the manufacturing quality up to the Six Sigma level could generate as much as $12 billion in savings—and nearly as much in profits. "With that opportunity, it wasn't rocket science for us to take a big swing at Six Sigma," Jack said.

Jack became Mr. Six Sigma. "When you have an initiative like this around quality, you don't ask for volunteers," he told Conaty. "What you do is you swing the pendulum way the hell out." At the annual GE officers' meeting in Crotonville in October, Jack invited Mikel Harry, a former Motorola manager who was running the Six Sigma Academy in Scottsdale, Arizona. "Harry's the guy," Jack said. Jack canceled the annual golf outing at Crotonville in favor of having his 170 officers listen to Harry proselytize for four hours about Six Sigma. "Most of the crowd, including me, didn't understand much of the statistical language," Jack remembered. "Nonetheless, Harry's presentation succeeded in capturing our imagination." What captivated Jack was the idea the Six Sigma could "turn a company inside out" by "focusing the organization outward on the customer." Jack recognized that some of the Six Sigma dogma was a load of crap. "If I took all the savings that everybody talked about, that they got from Six Sigma, I would have had the GDP of the world, okay?" he told me. He just cared about how focusing on improving manufacturing quality could boost profit margins. "As long as your operating margin goes up, your overall costs go down, down. Give me any fucking projects, I don't care about them," he continued.

At the annual meeting for top managers in Boca Raton in 1996, Jack rolled out Six Sigma. "We can wait no longer," he told them. "Everyone in this room must lead the quality charge. There can be no spectators on this. What took Motorola ten years, we must do in five— not through short-term cuts, but in learning from others." Jack thought the financial rewards from implementing Six Sigma would be profound and called it the "most ambitious undertaking the company" had ever done. "Quality can truly change GE from one of the great companies to absolutely the greatest company in world business."

Yes, Jack conceded, he had gone "over the top" with Six Sigma even as GE was already the most valuable company in America, with a market value in the $125 billion range, far ahead of AT&T, its closest domestic competitor, by $25 billion. Jack was pushing Six Sigma on an already stressed workforce. He wanted more, and he would get it. To drive home the importance to him, and to GE, of Six Sigma, Jack asked his direct reports to find their best people, take them out of their existing positions, and second them for *two years* to be trained in the dark arts of Six Sigma. He wanted them to become "Black Belts" in the subject. These were people who were then deployed throughout the company to solve quality issues—improving call center response times, increasing factory capacity, reducing inventories and billing errors. "A fundamental requirement in our Six Sigma program was that we measured it," Jack said. Six Sigma "Green Belts" were trained in the topic for ten days but stayed in their jobs and then solved problems as they came up in real time.

More important, Jack changed the incentive compensation structure at GE to reward Six Sigma compliance. People are simple: they do what they are rewarded to do. And no one knew that concept better than Jack Welch. With the implementation of Six Sigma, 60 percent of eligible employees' bonuses were based on financial results and 40 percent were based on Six Sigma results. Stock-option grants were given more generously to the Six Sigma Black Belts. "These were supposed to be our best," Jack said. (Not surprisingly, this decision resulted in furor from managers who didn't put their "best people" into the Black Belt program and still wanted stock options for them—and an equal amount of anger from Jack, who wanted the best people in the Black Belt program.) Furthermore, Jack decided that only people who were proficient in Six Sigma—and believers in it—would get promotions into the company's top ranks.

The change in the reward system worked. "The demand was for all executives, if you didn't at least qualify to become Green Belt certified for Six Sigma, you weren't getting a bonus that year," Conaty told me. "If you don't think everybody was all in on this, they were." He told a story about when Mikel Harry was teaching executives at Crotonville about Six Sigma and was droning on. "If you're in a situation where you've got this kind of tolerance and this kind of excess metal, how would you handle this from a statistical process control standpoint?" he asked. One of the manufacturing guys, from Boston, yelled out, "Buy a grinder," with a Boston accent. "Buy a grindah." In other words, before Six Sigma the answer would be purely a pragmatic one: How do I get rid of the excess metal as fast as possible? With Six Sigma, GE reconsidered the whole manufacturing system to prevent in the first place a situation where excess metal would occur. Said Conaty, "Obviously, [buying a grinder] wasn't the answer, but that's how we would have done it [before Six Sigma]. But it really did improve our quality dramatically because we keep asking that question and people would say, 'Yeah, we're seeing it. We're feeling it. We're feeling better.'"

Jack loved Six Sigma and proclaimed its results, albeit after a slow and clunky start. There were five hundred Six Sigma projects in 1995, three thousand Six Sigma projects in 1996, and double that the next year, when, according to Jack, GE achieved $320 million in "productivity gains and profits." That number would continue to grow—$750 million in savings, then $1.5 billion. Operating margins increased to 18.9 percent in 2000, from 14.8 percent in 1996. "Six Sigma was working," Jack proclaimed.

JACK WAS PUSHING PEOPLE AT GE HARD. IT TOOK A TOLL ON THEM AND on him. In the first weekend of May 1995, Jack had a massive heart attack. There were a few signs of trouble along the way. First, of course, was his family history. His mother had her first heart attack—when Jack was a freshman at UMass Amherst—at fifty-nine years old, the same age Jack was when he had his heart attack. She had two more heart attacks, each three years after the previous one, and died at age sixty-six. Second, when he was traveling in India in January 1995 on business, he picked up "a debilitating parasite," the media reported. Jack explained that he "couldn't stop feeling tired all the time" and "felt lousy." Never one to take a nap or to pause, he suddenly found himself curling up on the couch in his office every afternoon. He had doctors "all over New York" testing him, but they could find nothing wrong with him. He complained continually to his wife, Jane. She described Jack's symptoms to another doctor, who prescribed nitroglycerin pills for him, just in case it turned out to be a heart problem.

In late April 1995, soon after the latest hullabaloo about a possible merger partner for NBC had subsided, the Welches went out to dinner with some friends at Spazzi, in Fairfield. "We ate lots of pizza and consumed plenty of wine," Jack remembered. Later that night, when Jack was brushing his teeth, "I felt a bomb hit me in the chest," like nothing before he had ever experienced. "It felt like a massive rock sitting on my chest," he remembered. Jane gave him a nitro pill to relieve some of the pain. He then insisted they hop in the car. Jane drove him to Bridgeport Hospital, where she was on the board of directors. In his frenzied state, though, he insisted she drive to a different Bridgeport hospital instead. When Jane drove through a red light at a high speed— understandably—they got stopped by a cop, who then escorted them to the hospital without giving Jane a ticket. It was 1:00 a.m. Jack jumped out of the car and onto a gurney and announced, "I'm dying! I'm dying!" That got everyone's attention. Tests confirmed he had had a heart attack.

On May 2, Jack had an angioplasty to open his main heart artery. Back in his hospital room soon after the procedure, he again felt the pain of the rock on his chest. "The vein had closed," he explained. He was having another heart attack. Back Jack went to the operating room for another angioplasty. "A priest wanted to give me last rites," he remembered. The same doctor who did Jack's first angioplasty also performed the second one. Jack followed the procedure on the monitor. He could see the doctor was having trouble reopening the vein; a surgeon was standing by, ready to do bypass surgery. "Don't give up!" Jack yelled. "Keep trying." The doctor got the vein open. Jack avoided the open-heart surgery. From his hospital room, Jack sent a memo to his direct reports noting that the GE stock had gone up nearly $1 a share after the news leaked that he been in the hospital. Despite the market's positive reaction, he joked with his guys that "he had no intention of pulling a stunt like that again just to prove them right."

A few days later, back at home, Jack consulted his friends Michael Eisner, the CEO of Disney, and Henry Kissinger, both of whom had had bypass operations. Did they think he might need such an operation? Eisner told him the surgery was "no big deal." Kissinger encouraged him to go to Mass General, in Boston, to get the operation done. The GE corporate doctor flew off to Boston with the films of Jack's angioplasty. On May 10, in the middle of a business meeting, Jack got the news that he needed open-heart surgery, which was scheduled for two days later. "With my family's history and my angina over the last 15 years, I had been dreading this moment, but I didn't have much time to think about it," he remembered.

After Jack had had his heart attack and angioplasty, he called Cary Akins, a cardiac specialist at Mass General, in Boston. "Would you take care of me?" he asked. Akins had also operated on Harold Geneen, the CEO of ITT, one of America's other major industrial and financial conglomerates. Jack wanted—and received—a quick appointment. "Just like Jack," Akins told me. "'Let's solve this problem and get it done.'" He came to the hospital under a pseudonym. And when one of the doctors referred to him by the pseudonym, Jack corrected him. "No, no," he said. "I'm Jack Welch." At one point early the morning of the surgery, Jack supposedly told Jane, "If something goes wrong, don't let them pull the plug. Even if they can't tell, I want you to know I'll be fighting like hell in here."

Akins performed the three-hour surgery on Jack. He had quintuple bypass surgery. Akins was impressed by Jack. "He was incredibly cordial for somebody who was that powerful," he told me. "He was not standoffish. He was not aloof. He was not arrogant. Incredibly personable. Although he wanted a lot of answers, he was also a great listener.

That's probably one of the best things I found out about Jack. He's a great listener."

Jack wanted the surgery to be done on a Friday. "He wanted to make sure that when the stock market found out about it, it would be Monday and it would be three days after the operation," Akins said. "He didn't want the stock to tank." His mother's heart problems motivated Jack to get the operation. "Coronary disease has a lot to do with genetics," Akins continued. "You can be seventy-five years old and smoke and drink, and you may not get a heart attack if you've got the right genes. But you can also live a monastic existence and get a heart attack at forty if you've got the wrong genes. He knew that something in his genetic background might impact his longevity."

Jack was released from the hospital on May 22. "Bypass surgery throws you for a loop at first," he remembered. "Every part of you hurts like hell." Jack recovered soon enough. But not before he and Akins had a conversation about his convalescence. "We had a long talk about things that he should and shouldn't do," Akins said. "Obviously, there was lifestyle changes. I told him that he shouldn't be making any major decisions for a few weeks, just because of general anesthesia. It turns out if you get general anesthesia for anything, I mean even for a hernia repair, your short-term memory is a little bit dinged. So you probably shouldn't be making major decisions for a couple of weeks, and be sure to write everything down."

Jack returned to the office two months later. "I asked him to sort of do it gradually and to be sure to take time for himself, because it wouldn't do any good if he saved GE for a few weeks and then himself was in trouble for years ahead of him," Akins said. By mid-August, Jack was in the finals of the Sankaty Head club championship, which he lost in a thirty-six-hole finale. Jack never had another heart problem in his life. "He'd have a CT scan and his grafts were fine," Akins said. "I felt good about that. I was gratified. I would have been very sad if he had died because he lost a graft."

Akins and Jack became close, longtime friends. "I did about ten thousand heart operations, and there are probably ten patients who made a point of maintaining a really close friendship with me, and Jack was one of them," he told me. They talked not just about Jack and his health but also about Akins and his academic and political problems. "The one thing I found out about Jack was if you were really his friend, you had a friend for life," he said. "But if you were his enemy, you didn't want to be his enemy."

Akins told me that after Jack had recovered from the bypass surgery, he came to see Akins in his office at Mass General.

"You're doing just great," Akins told him.

"Well, go ahead and ask your question," Jack said.

"What?" Akins replied.

"Go ahead and ask your question," he said again.

"What do you mean?" Akins responded, genuinely confused.

"Well, I presume you're gonna want me to give you some money," Jack said.

"You didn't pay your bill?" Akins said.

"Come on, now," Jack said. "You must have thought about this. Do you want me to donate something?"

"Jack, it never crossed my mind," Akins replied.

"Well, I'm going to," Jack said, "so you better figure out what you want to do. Take a look for research and stuff like that."

Akins tried to arrange a meeting between Jack and the hospital administrators to discuss Jack's donation. "They couldn't make time to meet with him," Akins said. Jack then came back to see Akins and asked him to come up with an idea for a research project that Jack could fund. Jack arranged to fund the research through a GE charitable foundation—$100,000 a year for ten years.

Then there was the story about Jack, Akins, and Boston Health Care for the Homeless. Some ten years after Jack's surgery, the two friends went to lunch one day at Topper's, on Nantucket, near Jack's house. Jack told Akins he wanted to donate $1 million to a charity of Akins's choice in Akins's name. "You could have picked me up off the floor," Akins remembered. He introduced Jack to Jim O'Connell, a former medical student of Akins's, who was running Boston Health Care for the Homeless. Akins recalled walking in the pouring rain with O'Connell to see Jack at Jack's beautifully renovated townhouse on Beacon Hill, in Boston. O'Connell told Jack the story of what he was trying to do for Boston's homeless population.

"That guy is the real deal," Jack told Akins. Jack agreed to donate $1 million, the seed contribution to what became a successful $43 million fundraising effort. Jack insisted that Akins's name go on the new medical care facility somewhere. Akins said he objected. Jack wouldn't listen. "If you continue to argue with me," Jack said, "I won't give you the money." Akins relented.

Two days after he left Mass General, word of Jack's heart attack and surgery reached the media. It was big news and immediately raised the specter, at least for the first time publicly, of who would succeed Jack. A year earlier, when Jack had been asked about succession, he had bellowed, "Retire? Retire? A lot of guys are just getting this job at my age." But Jack's emergency open-heart surgery forced GE's hand. The media started asking questions about Jack's successor. GE's PR department shut down the talk of succession. "Absolutely not," Joyce Hergenhan said. "There's no need to." She said Jack would be back at work in three to four weeks.

But a former GE executive told *The Wall Street Journal* that if Jack didn't start the succession process, the GE board of directors would do it for him. "Some of these directors have seen the chaos when a company suddenly loses a CEO to death or he's hired away, and they won't let GE go through that," he said. Then, no surprise, the *Journal* started the speculation about who could succeed Jack. Paolo Fresco, a sixty-one-year-old vice-chairman, was out because he was an Italian citizen, preventing him from being the head of GE because of GE's ties to the Defense Department. And he was too old. Other potential successors, such as Larry Bossidy, Glen Hiner, and Roger Schipke, had already moved on to be the CEOs of other companies.

A "leading contender" for Jack's job, according to the *Journal*, was Gary Rogers, then fifty, who was head of GE's plastics division and who had also successfully run both the lighting and appliances businesses. Another contender was Dammerman, the CFO and a longtime confidant of Jack. The knock against him was that he lacked manufacturing experience. Of course, if either Rogers or Dammerman succeeded Jack, there would be a problem with Gary Wendt, then fifty-three. By 1994, GE Capital was contributing 30 percent of GE's nearly $10 billion of operating income—far more than any other single division of the company.

If Jack chose Rogers, the paper asserted, then Wendt was sure to leave the company, as he had almost done the year before when he tried unsuccessfully to become CEO of Prudential Insurance. But assuming Jack was on the road to recovery, which he was, setting up a horse race among the leading candidates risked losing other promising executives, who would likely think they didn't have a shot. There was also the general concern of how difficult it could be for anyone to succeed Jack. The company continued to make its numbers quarter after quarter. It remained the most valuable corporation in America, with a market value approaching $200 billion.

Jack's "many talents," Harvard Business School professor Samuel Hayes told the *Journal*, would be "extremely difficult" for his successor to replicate. "GE is like the bumblebee that flies—aerodynamically," Hayes said. "It isn't supposed to, but it does." GE's disparate business lines should make it extremely difficult to manage, the professor continued, but Jack had done it, making GE "the most successful conglomerate" in U.S. history.

*Chapter Sixteen*

# THE BAKE-OFF

In June 1995, *Fortune* reported that an "envelope exists" with the names of "two or three" GE executives who would be "capable of immediately taking over" for Jack if he were to become incapacitated. Succession was in the air. But Jack had an intense aversion to the process by which Reg Jones had selected him. He didn't mind creating a competition per se, but what he would not do was bring the finalists to Fairfield headquarters and have them duke it out in front of him for a year or two. He had hated that Jones did that to him and the other finalists, and he wasn't going to repeat Jones's mistake. "It was torture," Jack told me. "I didn't like it."

The possible successors in Jack's mind eventually boiled down to a short list of impressive GE executives: Dave Calhoun, forty-one, the head of lighting; Dave Cote, forty-six, head of appliances; John Rice, forty-two, head of transportation systems; Robert Nardelli, who in 1996 had become head of GE's struggling power business; Jeffrey Immelt, who was then running GE's medical equipment business; and Jim McNerney, who was then running GE's jet engine manufacturing business. Missing from the list were fifty-four-year-old Gary Wendt of

GE Capital, and Bob Wright, who was running NBC and who Jack felt was brilliant but not a bold decision maker. About Wright, Jack told me, "Smart. High IQ. Flexible mind. Can't make a decision if it hit him in the nose."

Even though Wendt's division, GE Capital, by then was contributing about 40 percent of GE's operating profits, Wendt's prospects for succeeding Jack—if there were any—began to unravel once and for all in December 1995. That's when he chose to tell Lorna, his wife of thirty years, he wanted a divorce. He was in love with Rosemarie Adams, whom he had met in Monte Carlo in 1995, when she was in charge of GE Capital's corporate incentive program in Europe. By then, Wendt was very wealthy, especially by the standards of the time, with a net worth of around $100 million. The word around GE was that Jack "loved you in your wallet." He could make his big producers rich if he wanted to. He could also shoot you on a moment's notice. He had made Wendt rich. Wendt offered Lorna a settlement of about $10 million, or "ten cents on the dollar" in dealmaking parlance. She turned it down. She had other ideas. She wanted half of his fortune and she was determined to get it, or as much of it as possible. She filed for divorce from him in Connecticut before he could file for divorce from her.

"He may have earned the bacon," she said during the heat of the battle, "but I shopped for it, I cooked it and I cleaned up." She raised the Wendts' two children. She entertained his GE business associates and traveled with him around the world. In 1986, shortly after he became head of GE Capital, the couple sold their home and moved to a bigger one in Stamford, a four-bedroom center-hall colonial. She set up the house. Two years later, they bought some vacant land, also in Stamford, and for the next two years, Lorna oversaw the construction of a huge five-thousand-square-foot home with four bedrooms, a gym, and many bathrooms. She cooked. She took the kids to camp in the summer. She worked with the children on their homework. She did the family's taxes. If Wendt needed her to go on a GE trip, she dropped everything and joined him. She even maintained the facade of loyal wife, at least at the start of the divorce battle: she kept things together at the Wendts' annual Christmas party in 1995, although she confided in a friend, "I do not know how I can go through with this party because of my marital problems."

What ultimately doomed Gary Wendt at GE was the negative publicity around the divorce battle—and there was a legendary amount of it. There was also the fact that Wendt was so good at his job, and so cocky about his success at GE Capital, that rather than being loved in Fairfield, he was resented. Maybe the divorce was just the excuse Jack needed to defenestrate Wendt, or at least to take him out of the succession race and urge him to move on from GE. "He never liked me," Wendt reminded me about Jack. He said that if it hadn't been for the support (and protection) he received from Larry Bossidy, he would have been gone long before. But, he said, "after a while," Jack had no choice but to at least respect him, given the magnitude of the profits Wendt was sending Jack's way year after year.

The first of the thousand cuts for Wendt came in a December 1996 *Wall Street Journal* article, "Wendt Divorce Dissects Job of 'Corporate Wife,'" which let the world know of the battle that until then had been mostly raging behind the closed doors of a Stamford, Connecticut, state courthouse. Even by the less-woke standards of a generation ago, Wendt's deposition in the case was cringeworthy. "I know what Lorna's needs are, and I want her to be able to live very, very comfortably after we're divorced, and I think that's the kind of number that should allow her to do that," Wendt said in his October 1996 deposition about his first $10 million settlement offer to her. In her own testimony, two months later at the opening of their trial, Lorna said it wasn't a matter of what she "needed" but rather what she "deserved" after "decades" of being "the ultimate hostess" and taking her job "very seriously."

She referenced her PHT degree—"Put Hubby Through"—from Harvard Business School. She testified that Wendt "loved to entertain," including hiring the pianist Marvin Hamlisch to perform once for a party of ninety guests. He "just didn't see that it was a lot of work," she said. At the time, Wendt was making about $2 million a year in salary and bonus. At his deposition, Wendt asserted he was the one who had worked the hardest and deserved the spoils of his hard work, not his wife. "There is no attempt at meanness," he said, "but there is an understanding of how hard I've worked and what I've accomplished and the stress I put myself under and what the rest of my life might in fact be like because of that." That's how he came up with the offer of roughly $10 million, paid over a period of years. "My rewards were financial," he continued, "and I think her rewards were perhaps emotional," including "the satisfaction of being with the children."

*The Wall Street Journal* article was not the kind of publicity Jack appreciated. And it did no favors for Wendt, whose wife had put him on the defensive and in the position of looking like a jerk. What to do? Double down. Wendt agreed to do a national television interview with Lynn Sherr for *20/20*, ABC's newsmagazine. It wasn't the wisest decision. He did not tell Jack, nor did he ask for Jack's permission, which might have been a more politically savvy move. Worse, ABC was NBC's rival. Nor had he done much, if any, television in the past.

Wendt was a brilliant guy but not one the camera was bound to love. Sherr quickly got under Wendt's skin. He talked about how lonely he had been during the last few years of the marriage. "I had achieved some modicum of success in business," he told Sherr. "I didn't have anybody to share it with. And I can remember I had to fly to Budapest and back in forty-eight hours, and I was coming back and I thought, 'Wow, am I ever unhappy.' And I started to cry on the airplane, I literally started to cry. A grown man. And I couldn't stop. I had to go into the lavatory and I just cried for about ten minutes and I said that's it. We have to stop this. And three weeks later I told Lorna."

When Sherr suggested that Lorna had worked very hard over the years, Wendt could have agreed. But he didn't. "Do you think having somebody clean the house for you when you go out and play tennis, do you think having to get dressed up to go to dinner in New York, is hard work?" he asked. "Tell me, please." He wasn't done. "Do feminist causes now want equality without effort?" he continued. "Is that the new battle cry? I don't think so." Sherr also interviewed Lorna. She seemed his opposite—calm, cool, and composed. "I became kind of Mrs. GE Capital," she said, "in that I was expected to go to a dinner in New York, for example, at the drop of a hat. His secretary would call me. 'You do remember you have that dinner tonight, don't you?' And I go, 'No, nobody told me.' And so you'd scramble to rearrange everything, and a driver might pick me up, and you put on your smile and you'd go." Sherr said people listening to the broadcast might find it hard to believe she was complaining. "I wasn't complaining about that," she replied. "I did that willingly. I'm just saying that it is a job. It's as much of a job for me as Gary's job is for him. I have been at dinners with Rupert Murdoch on one side and Jim Robinson on the other and they still talk a little small talk. Their house in California. What are their kids doing? . . . He knew that I could carry on these conversations. I would not embarrass him. I would not embarrass GE."

ABC aired the segment on Friday night, March 14, 1997. Jack happened to be channel surfing. There was Gary Wendt. According to Wendt, Jack called him at home the day after the ill-considered interview. He didn't wait until an already-scheduled Six Sigma meeting at Crotonville. "I want you to leave the company by the end of the year," he said Jack told him. "You're not gonna be the CEO." A few weeks later, Wendt called Jack and suggested that he become a GE vice-chairman, replacing Paolo Fresco, who was soon to retire. He told Jack he could run "around the world, do deals, that kind of stuff." But, again, Jack wouldn't hear it. "Nope," Jack told him, "can't have you around at all, can't have you around at all." There were times when Jack would get furious at Wendt. He remembered how Jack called him once at three in the morning, in Hong Kong, and woke him up. "I don't know what he was hollering about," he said. "But he was swearing, cursing at me. He was not happy. And you know, you tell me. You tell me what the hell the problem was." Despite his falling-out with Jack, Wendt was understandably proud of what he had accomplished at GE Capital. "I'm from this little town of six hundred people," he told me. "As I look back at it now, I can't believe what I did, frankly. I just can't believe it. What I did together with all the other people that were there and GE's AAA [credit rating]."

Wendt's divorce had become national news, and its resolution set a new standard culturally, if not legally, for the way corporate wives were to be treated. Lorna Wendt received about $20 million in settlement with her husband. She had wanted half—or $50 million—of what she believed his net worth to be. Wendt had offered her $10 million; she ended up with $20 million. And Wendt lost his job, despite GE Capital continuing to produce more than 40 percent of GE's profit. In December 1998, word came out that Wendt had "agreed to step down" and leave GE by the end of the year.

In order for Wendt to get his benefits upon leaving GE—his lucrative supplemental pension as well as his unvested stock and options, the total of which his wife argued brought his net worth into the $100 million range—Jack made him sign a "contract," he said, that he wouldn't work "for anybody else" for three years. "He didn't want me in the business," Wendt told me. Wendt left GE in December 1998 and in June 2000 became chairman and CEO of Conseco, Inc., an Indiana-based insurance company in need of a serious turnaround. For Wendt to take the Conseco job, GE had to allow him to get out of his noncompete agreement early, which it did. Jack cut a deal with Conseco in which it would pay Wendt $45 million in cash and Wendt would forfeit another $20 million that GE owed him. GE would also be issued 10.5 million Conseco warrants struck at the market price of Conseco stock. In exchange, GE got relieved of the $65 million it owed Wendt to lie "in a hammock" for another nearly two years, as Wendt told *The Wall Street*

*Journal* would have happened. Jack saw it as a win-win, even though he had rather callously tossed Wendt aside and then figured out a way to have GE benefit from Wendt's misfortune.

A month prior to his defenestration, Wendt left behind a bomb for his successors to try to defuse: the acquisition of Lake Co., an Osaka, Japan, consumer finance company. Wendt had been on a buying tear around the world in an effort to broaden GE Capital's global presence. Lake, then the fifth-largest consumer finance company in Japan, had 564 branch offices, $3.7 billion of loan receivables, and 1.4 million customers. As part of the acquisition, GE Capital agreed to provide Lake with some $3 billion in financing to repay its existing debts and to put it on a sounder financial footing.

Many GE executives, especially in retrospect, saw the Lake acquisition as a mistake that revealed how GE Capital was getting out of hand. "The business where we kind of made drunken Japanese salesmen pay thirty percent and the cost of money was one," was how Jeff Immelt described Lake to me years later. The Lake acquisition, soon to achieve a nearly mythical status inside GE's executive ranks, would balloon into one of many financial disasters that would befall GE and GE Capital. "There's a thing called 'gray-zone liability,'" one senior GE executive explained to me, "and basically what happened in Japan is everybody was charging twenty-nine-percent interest rates. The law specifically permitted eighteen-percent interest rates, and it didn't specifically prohibit higher interest rates; it just only officially sanctioned eighteen percent. So everybody was charging twenty-nine [percent] on big pools of money for long periods of time, and then the litigation started, and in the end, people said, 'Yeah, you're really only entitled to eighteen percent.' We had a massive write-off."

WENDT'S MUCH-PUBLICIZED MARITAL AND EXTRAMARITAL PROBLEMS may have struck a chord with Jack, who probably saw in them what

he feared might be happening to him again. Just as Jack was beginning the process of finding his successor, his second marriage was starting to show early signs of unraveling. As Jack was gearing up for a long
July 4 summer weekend on Nantucket in 1996, came some disturbing
news. According to a July 3 Associated Press story from Clayton, Alabama, William Durwood Beasley, Jane's father, had been brutally murdered, with an "ax wound to the head," at his home. He was eighty-one
years old and lived alone. There was some evidence of a burglary. But
it was not clear that the break-in was related to the murder. Sheriff's
deputies found Beasley at eight o'clock in the morning and then transported him to the local hospital, where he died. Jane Welch was noti-
fied soon thereafter. Mack Houston, the Louisville chief of police, got
to the scene and concluded that Warine Casey, a local woman who
worked for Beasley, had murdered him.

Jack and Jane boarded a GE corporate jet to Alabama. At about
11:00 a.m. on July 4, Jane and Jack were briefed on the murder by
John Hamm, the county sheriff. Forty-five minutes after arriving at
the house, Jack got up and left. Jane followed him. He was her ride,
after all. Without Jack's support, the task of trying to figure out what
happened to her father fell to her alone. She hired private investigators. An investigator from the Alabama Bureau of Investigations said
Beasley worked Warine "like a slave" and that her motivation for the
murder, if she did it, was "out and out anger." The Beasley children
covered the wages that their father owed her. Beasley had promised to
buy her a truck and then backed out of the promise. "Still he expected
her to come to the house every day, and wash and clean, hoe the vegetable garden, and chop the firewood," according to Christopher Byron.
"And sometimes, according to the stories around town, he expected
more than that, and she'd end up spending the night." But she was never
charged with a crime or tried for the murder. The evidence was never
shown to a grand jury.

At the start, Jane made frequent calls to law enforcement in Alabama. But after about a year, she stopped the calls. Then the private investigators were called off, apparently at Jack's insistence, over their cost. Jack never made another call in the matter, or set foot in the county again, by all accounts.

Jane had begun to sour on Jack by this point, and his indifference to her father's murder sealed her disappointment with him. She learned to speak Italian and went on regular shopping "junkets" to Rome and Florence. She started spending more time at an apartment she had purchased overlooking the Arno River in Florence. And then, according to Byron, in her "own version of a score-settling ax through the skull—namely Jack's," she began "a torrid transatlantic love affair" with a man named Giorgio, tall, thin, handsome, and with a full head of hair. Giorgio just happened to be the limo driver for Paolo Fresco, a GE vice-chairman and one of Jack's closest GE confidants. (Jane Beasley did not respond to requests to be interviewed.)

JACK CLAIMED TO HAVE STARTED THE SUCCESSION PROCESS IN NO-vember 1993, soon after appointing Bill Conaty to be his head of human resources. Jack and Conaty had met when Conaty was head of HR for the aircraft engine division. After Conaty's promotion, Jack told him that finding his replacement was their most important project. "The thing you and I will both live with for a long time is getting the right person in this job," Jack told him. Jack said the process "would almost consume us."

In June 1994, Jack and Conaty made their first presentation about succession to a committee of the GE board of directors. There were twenty-three names on a handwritten list, ranging in age from thirty-six to fifty-eight and including the obvious names plus another sixteen "high-potential long shots," Jack recalled. Jack gave "formal" board reviews on succession in June and December and informal, "real-time" reviews in February when he discussed with the board the bonuses he was giving top executives. He had the conversation again in September when he was doling out stock options.

There were several opportunities during the course of the year for getting candidates—both short- and long-listers—in front of GE board members in order for the board to begin to assess the candidates' qualifications. In April, Jack invited the top candidates to play golf at Augusta National with him and selected board members. In July, it was more golf, or tennis, in Fairfield. There was an annual Christmas party with spouses. Jack and his longtime assistant, Rosanne Badowski, would carefully map out the seating arrangements at the parties, dinners, or rounds of golf. He wanted board members to get exposure to the candidates in a variety of settings. Jack kept the seating charts from year to year to make sure to mix up the visits.

He organized all-day meetings for board members with the leading candidates at their offices away from Fairfield. They listened to presentations and visions for the future, had lunches and dinners together, and went to ball games. All without Jack being around. "This is your show," Jack told the candidates. After each visit, Jack got reports back from the board members.

The key to the process, Jack decided, was a combination of rigorous evaluation of the candidates and absolute secrecy. He and Conaty, along with Dammerman, had become a troika who could be trusted to keep the succession discussions private. Unbeknownst to anyone but Jack and his confidants on the board of directors, by the end of 1998 the leading candidates had been whittled down to Immelt, the head of medical systems; McNerney, the head of aircraft engines; and Nardelli, the head of power systems. As planned, Jack kept the three men out in the heartland: Nardelli in Schenectady, Immelt in Milwaukee, and McNerney in Cincinnati. He wanted each of them to stay focused on running their businesses, not trying to suck up to him in Fairfield.

He also made a point of having lunch with the three men, plus Dave Cote, after the quarterly meetings of the GE Capital board, to

which he had added the four men the previous year just to have another excuse to watch them in action. And starting in the spring of
1999, he also had dinner with them and the other leaders of GE's thirteen businesses. The sessions helped Jack think about the team "but
didn't do much to pick a clear successor," he said. He also did his version of the "airplane" question that Jones had used on him, asking each
of them if they liked one another. "No surprises," he said. "They all
liked and respected one another."

ROBERT "BOB" NARDELLI WAS BORN IN OLD FORGE, PENNSYLVANIA, BE-
tween Wilkes-Barre and Scranton. His parents were Italian immigrants.
His father, Raymond, a World War II veteran, worked at GE, first as an
hourly employee and then as a plant manager. He was instrumental in
building GE's dishwasher and garbage disposal businesses in Louisville. Bob's mother, Clelia, was a homemaker and real estate agent. The
family moved from Louisville to Rockford, Illinois, for GE.

Nardelli attended St. Patrick's Grade School in Rockford. He was
a Cub Scout and a Boy Scout and had a paper route. He bagged gro-
ceries at the local Piggly Wiggly. He went to Rockford's Auburn High
School, where he was a member of the ROTC, edited the school yearbook, and played sports. He also joined a rock band. "The Beatles were
big then, and we parlayed their popularity into playing weekend fests
at Lake Louise in Rockford and other places in the area," he said. From
there, he was on to Western Illinois University in Macomb. He was
hoping to go to West Point, but it didn't work out. At Western Illinois,
he was president of his fraternity and played offensive lineman on the
varsity football team. He was voted team cocaptain his senior year,
and his coach said he was the best lineman he had ever coached despite
being the smallest guy on the line. "He was always the first in to study
the films of the games," his coach, Bob McMahan, said. "He was one
of the few kids who really cared that he was carrying his load and do-

ing right." His proudest achievement, though, at Western Illinois was meeting and then marrying his classmate Sue Schmulbach. He sold his motorcycle to get the money to buy her an engagement ring.

They graduated in 1971 into a tough job market. He wanted to be a professional football player, but at five feet ten inches, he realized he didn't have the physical requirements for the job. He also rejected the idea of becoming a football coach: the job was too unpredictable. He was about to take a job at either Ponderosa Steakhouse as a management trainee or selling insurance. Then he got a life-changing job offer at GE to become a manufacturing engineer, for a salary of $9,600 per year. He took the job, in Louisville, and found himself working next to Ivy League graduates. "I was thrilled with getting hired by GE," he said. "For someone from very humble beginnings, without the Ivy League diploma that so many of my co-workers had, I knew I would have to commit myself to continuous improvement just like I had to do on the playing field." He completed a two-year engineering training program and became steeped in the "GE culture." He also enrolled in the MBA program at the University of Louisville, taking classes at night, after work, and on weekends.

"When I first joined GE," he remembered, "I was asked to make a list of all the individuals between the Chair and CEO of GE Reg Jones and me. It took me over two sheets of paper at the time. I remember Sue and I talking about how great it would be if I could reach the position of a unit manager someday, earning $100,000. We'd be in heaven, and it would be a 'Leatherneck makes good' story."

Jack met Nardelli in the late 1970s, a few years before Jack became CEO. He remembered that Nardelli was constantly asking him about ways he could improve his performance. "What am I doing that I need to do better?" he asked Jack. He worked his way up the ladder on the industrial side of GE. When, in 1988, Jack refused to give Nardelli, then a manufacturing vice president, a manager's job, he quit in protest. Jack tried to get him to stay. But Nardelli wouldn't consider it.

"The issue is not between you and I," he told Jack. "It is what is between you and I [sic]."

For the next three years, Nardelli worked at Case, the Wisconsin manufacturer of industrial equipment, then part of Tenneco. He was an executive vice president. First he ran the worldwide parts business and then the construction business. Just as he had once upon a time with Bob Wright, in 1991, Jack lured Nardelli back to GE to run the appliance business in Canada. That same year, Jack promoted Nardelli to run GE Transportation, the locomotive business based in Erie, Pennsylvania. "In three years," according to *Fortune*, "he pacified hostile unions, modernized the product line, expanded into services, took the business global, and more than doubled profits." He flew around the world—Africa, Mexico, China, Eastern Europe—to close deals, while also never missing a family vacation or agreeing to live overseas, which his wife didn't want to do. Jack described Nardelli as "the perfect dad and the perfect husband."

Like Jack, Nardelli was more of a "blue-collar guy," as *The Wall Street Journal* described him, "blunt, earthy, often temperamental and known for driving his managers hard and working long hours." He was also "widely considered to be the best pure industrial operator" at GE, "an executive comfortable schmoozing with CEOs and strutting on the factory floor." Despite his success, he had long felt he wasn't getting the recognition he deserved at GE. "There was always this rap against me about being functionally proficient but not very strategic," he told *Fortune*. But his defenders told *The Wall Street Journal* that criticism was unfair. "[That's] a miserable rap," said Ronald M. DeFeo, chairman and CEO of Terex Corporation, who worked with Nardelli at Case. "He was a manufacturing guy at the start of his career. But that's like saying because you drive a four-wheel bicycle you could never drive a car. He is comfortable in his own suit. He once told me, when I first came to work for him, that 'I am who I am and I'm going to be successful my way and I can't change myself.'"

In 1995, Jack promoted Nardelli to run GE's power systems business and immediately thrust him into the succession race. He excelled in the job. He completed fifty acquisitions, and profits increased sevenfold. Jack said Nardelli never missed targets or made a significant mistake. "Jack and I used to marvel at his ability to execute," Bill Conaty said. "With Bob, it's very, very difficult to have a surprise, because he's into the details down to the level of the shop floor."

Headhunters wanted to recruit Nardelli. And he heard from Lucent, Kodak, and Ford. But he told them no. He had promised Jack he would not talk to other companies until Jack had made his final decision on his successor. One thing was clear: Nardelli wanted to be the CEO of GE, and wanted it badly. He was known as "Little Jack" around GE. "Bob was a grinder," Denis Nayden told me. "A tank. Now at the time did he deliver great results with the power systems business? Absolutely. But talk about a no-prisoners guy." Nardelli used to have staff meetings at 7:00 a.m. on Saturday mornings. That "kills Friday night," remembered one longtime GE executive. Work on Saturday under Nardelli would continue to noon, or 1:00 pm. "Fine," the executive continued. "But not exactly the right balance of work and family."

ANOTHER CONTENDER FOR THE TOP JOB WAS WALTER JAMES MCNERney Jr. He went to New Trier High School, in Winnetka. He told me, "I'm a New Trier boy, Winnetka, Illinois," which he described as "sort of a famous high school back in the old days." He said Winnetka was "an on-the-make" northern suburb of Chicago where people were of the view they were "going to make money, not spend money" and that had an "upwardly mobile culture in every way, shape and form." New Trier High School was big, with about 1,250 students per class. McNerney was a big man on campus. He was a good student. He was captain of the hockey team and head of the Boys Club. Making the varsity baseball team was highly competitive. McNerney did it. "I was the classic well-rounded guy, back in an era where well-roundedness was important," he said. For college, McNerney had his choice of Williams, Princeton, and Yale. There was no pressure from his Yalie father to attend Yale. But he did anyway. He liked it better than the other two. "It was a pretty honest choice," he said. Yale lavished some attention on him, particularly around his athletics, that the other two schools didn't. "It felt good," he said. "It felt like the right thing to do." After Yale, McNerney decided to apply to Harvard Business School, a path taken by few of his classmates during the Vietnam War years, when such a credential seemed absurd to many. He had had "a taste of leadership" at Yale and discovered that he liked it. He wanted the option of being able to go to business school. "I was a pretty good communicator," he said, "[and] I enjoyed working with people. That naturally evolved into sort of an interest in business and big organizations." After he got into Harvard, he deferred his admission for two years.

After two years spent working for first a British insurance company in London and then a big pharmaceutical company in Chicago, he decided the time had come to go to business school. Many of his peers were ahead of him when it came to economics, finance, and statistics. "I had to hustle," he said. "I felt like I was getting tooled up." He was tempted to go into consulting after he graduated in 1975. But instead he took a job in brand management at Procter & Gamble in Cincinnati. "I'm a midwesterner," he explained, "in the sense that I like to be grounded in experiences. I thought I'd get a great grounding experience at P&G. I liked the place." He worked on three brands: Coast soap, Downy fabric softener, and Bounce fabric softener. After three years, he joined McKinsey. "I liked P&G," he said, "but it was a little slow and a little bureaucratic for my tastes. I got a little anxious to be moving faster." McKinsey wanted to promote him. But he began to think that he wanted to run something, as he had in a modest way at P&G. He figured his ability to get along with people, to communicate well, and to lead them was what would "distinguish" him in his business career.

Jack had just become CEO of GE. "He was the hottest thing go-ing," McNerney told me. "I was very flattered when I got a call from the headhunter, who said, 'They've heard about you. They want you to come interview.'" The message from the headhunter was "'Look, Jack is trying to revolutionize the place. We need some people from the out-side. Not all of you are going to make it, but if you have the confidence to take a swing . . .' I know Jack well enough that he wrote the fucking pitch. I said, 'Sure.'" He said he needed "this catalyst" to leave Mc-Kinsey because it was "fun" and "stimulating."

He spoke to a few GE executives. They made him an offer. They gave him the choice between working at GE Capital and GE Infor-mation Services, what was left of GE's computer business—a servicing business of software development and networking—after the bulk of the business had been sold to Honeywell. "Just to show you how smart I am," he said, he chose Information Services. "I chose the wrong one," he continued, "in the sense of which was going to become a big busi-ness. But I didn't choose the wrong one in terms of getting a real tough, tough business opportunity, and tough businesses are the ones you want when you're young. You need to know what tough looks like." He was the business development guy. He moved to Rockville, Maryland. "The business was a good little business, and I ended up running it," he said. "I was the BD guy. Then I ran the software businesses."

But not for long. He was soon off to run GE Mobile Communica-tions, an early manufacturer of cellular phones, in Lynchburg, Vir-ginia, the home to Liberty University and Jerry Falwell. "Me and Jerry Falwell were tight," he said. "I figured that out when I took my first walk around the factory in Lynchburg. I met all the women. There were five thousand who were assembling mobile radios and cellular telephones, and every one of them had a picture of Jerry Falwell in the underlid of their lunch pails, every one of them."

They became friends. He gave Falwell a bunch of radios and a com-munications tower for his church and for Liberty. "I was an enabler," he joked. "But more importantly, I kept my factory folks happy. They would invite me down to their line dances on Friday afternoons. Because it was ninety percent women, it was hysterical. I was this kid who didn't know his ass from a hole in the ground but had fun." (He and his first wife split up in Lynchburg.) GE and Motorola were among the first manufacturers of cellular phones, those brick-like devices that once inspired envy. But Motorola left GE in the dust in sales, and in the end, Jack decided GE couldn't compete. GE sold off the European business and then sold what was left to Ericsson.

Even though these businesses had run aground, McNerney believed he was getting a great experience at GE. He had managed two tough businesses and Jack had taken note. "This was part of Jack's genius," he said. "You know, some guy—me—running businesses that are going nowhere, but Jack developed a relationship with me. Just at the bar . . . informal interactions. This is just the way Jack operated. He made sure I got the message. I felt comfortable. I never panicked." Ericsson made him a "very nice" job offer after it bought GE's mobile phone business. He was tempted. But he decided to stay at GE. "I still felt my future at GE was bright," he said.

After Lynchburg, Jack asked McNerney to return to run GE Information Services, in Washington, DC. "The leadership job was open," he said. "They put me in it." Jack had put the satellite business of RCA into GE Information Services and was still hoping it would all work out. "It was an officer's job," McNerney said. "You know how it goes at GE. It was 'Run your own global business.' More of it was outside the U.S. than inside the U.S. I was a young guy. It was a good job. It was a good job for my peer group. Let's put it that way."

When Mike Carpenter left GE Capital to run Kidder Peabody, Jack and Larry Bossidy asked McNerney to go to GE Capital as a peer to Denis Nayden and to report to Gary Wendt. Like many others, he had a tough time handling Wendt. Wendt wasn't fond of outsiders being plopped into his business, especially if they knew nothing about financial services. He thought McNerney was a spy sent by Jack. McNerney was in charge of GE Capital's plethora of leasing businesses. Anything with an operating lease fell into McNerney's bailiwick: aircraft leasing, consumer leasing, consumer credit cards, shipping containers.

McNerney said he absorbed "a lot" at GE Capital, in large part because the learning curve was so steep. He learned from Nayden and from Jim Fishel, the chief credit officer. But his experience with Wendt frustrated him. He had to find his own way in the job, and he couldn't talk to Jack about his frustrations. It was not in his midwestern chemical makeup to go around his boss to Jack. It was a tough two years. "I still had a fair amount of confidence in myself," he said. "But I also know that getting along with your boss is the most fundamental thing. So yeah, it was a little disquieting. But, again, Jack sort of figured it all out. He plucked me out of there and gave me something else to do. Jack every now and again would say something out of the side of his mouth that told me he understood and, 'Hang in there.' He would always initiate it. That's the genius of him. He understands people and how they're feeling, what they're going through. There are very few people with the kind of empathy and insight that he's got."

From Stamford, Connecticut, Jack moved McNerney to Hartford to run GE Electrical Distribution and Control. McNerney got the job after a rather infamous annual managers' meeting in Boca Raton in January 1991, when Jack fired four division CEOs. "You could have heard a pin drop," McNerney recalled. Someone who was there remembered what Jack said: "I'm taking four guys out of business leadership positions. Here's who they are. Here's why I'm taking them out: Look, you guys don't think I'm serious about you having two jobs. One job is to do your job, and the other job is to integrate across the company to make us stronger together than we are just as a collection of businesses." Nothing focuses the mind at night like an execution in the morning.

A week after the 1991 Boca massacre, Jack called McNerney and told him he wanted him to run the electrical distribution business. It was another tough business for him to run, with small profit margins and lots of competitors. After about a year at the helm of the distribution business, McNerney remembered being at a meeting of Jack's direct reports in Crotonville. It was a Friday afternoon. Jack was berating them for not doing enough to focus on Asian markets and opportunities. "You assholes," Jack admonished. "I've been telling you to get into Asia. You guys are half-assing it. I'm sick and tired of waiting. Asia is going to be the future of two thirds of you guys, and I'm going to do something you're not going to like to make the point to you." The assembled executives dutifully took notes. On Monday, McNerney was back at his desk in Hartford when Jack called him. He told him that a helicopter was on its way to Hartford to pick him up and take him to Fairfield. "Go get on it," Jack told him. "Come back and talk to me." McNerney was clueless about what Jack wanted. "I hadn't connected the dots yet," he said.

"Jim, I want to make you head of Asia," Jack told him.

"Oh, so this is the thing that no one is going to like but you want me to go do," McNerney replied.

"Yeah," Jack said. "I'm going to give you a challenge. Go to Hong Kong and get this company growing in Asia."

"What do you want me to do?" McNerney asked him.

"You figure it out," Jack said.

It was "the absolute best thing that's ever been said to me by a boss," McNerney told me.

"You got it," he told Jack.

The Asia assignment was another difficult one for McNerney. At the time, two thirds of GE's Asia business was in Japan. He realized that two thirds needed to be outside Japan and half should be in China. But he had no operational responsibilities, per se, merely the responsibility to figure out how to generate more revenue in Asia. He had no capital. He also had to contend with the fact that the existing business managers in Asia had their own responsibilities and ways of doing business and didn't want, or think they needed, outside help or inter- ference from "corporate." McNerney trod carefully. He made a team of the heads of GE businesses in Asia. He put together an advisory coun- cil of important CEOs in the region and made a priority of connecting with the governments in the countries in which GE was doing busi- ness. It was McNerney's first job that cut across the entire company.

There were some surprises. For instance, doing business in China required turning over intellectual property in exchange for market ac- cess. But there was no "business leadership" in China in those days. When GE bought China's largest lighting company, in Shanghai, it bought the business from the mayor of Shanghai, since it had been owned by the Chinese people. There were no independent companies, such as Alibaba or Tencent, at that time. Another surprise was that his second wife, who was Dutch but born in Australia, was five months pregnant when the McNerneys moved to Hong Kong. She had the baby in a Hong Kong hospital while McNerney was back in New York on business. No one in the hospital spoke English.

McNerney was in Hong Kong for three and a half years. From his lofty perch in Hong Kong, McNerney returned to . . . Cleveland, to run the lighting business. In 1995, Jack named John Opie, who had been running the lighting division, a GE vice-chairman, along with Paolo Fresco. Jack named McNerney to succeed Opie. Jack had been telling McNerney he wanted him to come back to the United States to run one of GE's bigger businesses. He wasn't exactly thrilled to be moving to Cleveland or to be running the lighting business. He didn't think he needed a spin through Cleveland, at that point, to burnish his experi- ence or his credentials. He had been hoping to run the aircraft engine business. But Gene Murphy was still in charge of it. "Well," he said, "you had to believe in your career, right?" He did. And so did Jack. "Re- member the generation I came from," McNerney continued. "It's all about career progress. It's not about what you have to go through with your family to get it done."

Following Opie at Lighting was not an easy assignment. "He was one tough-ass manager," McNerney said. Opie ran a "tight, tight ship," he continued, and had wrung out of the business the last drops of productivity. McNerney had to figure out a way to run the business better. "I was bound and determined to show Jack and John Opie that I could improve Lighting," he said. His advancement at GE depended upon it, of course. He introduced new products for consumers. He finally got the factory in Hungary working properly.

The wall outside of McNerney's office in Cleveland had ten pictures. "Thomas Edison's was the first one," he said. "Mine was the last one. There was a tremendous amount of pride in this business for all of us at GE at the time. It's hard. You didn't spend enough time at GE to understand that GE Lighting is sort of the core of the culture. It was, but at some point, the technology changed."

Eighteen months later, in 1997, he succeeded Gene Murphy as head of GE's aircraft engine division, the biggest and most admired business in the company. He and his family moved to Cincinnati and he moved into serious contention to be the next CEO. McNerney had gotten the "aviation bug," he said, when he was running the aircraft leasing business years earlier at GE Capital. He was thrilled to move to Cincinnati. "It was a big challenge," he said. "I loved it because it was a tough, tough business."

Jack had called him up one day and said, "Hey, I know you just got" to Cleveland, but there was a problem with the new GE90 jet engine, the largest commercial engine ever designed, which had been slated for the Boeing 777 airplane. This was a huge airplane with two engines. Before the 777 and the GE90, all big jets had four engines. The new engine provided huge productivity gains for the airlines, some 25 percent improvement in "seat-mile cost."

The problem was that although the new engine was a technical marvel, the certification process was rough. GE had to prove that if one engine failed, the plane could keep flying. GE also had to prove that the engine could withstand a lot of heat and lot of power. GE was having problems building the engine. Customers weren't accepting it. And Jack was bad-mouthing it.

Jack called him and told him he had "to fix this thing," and then went public with a threat to kill the GE90 project. "In the aviation business, that's awful because airlines enter twenty-year relationships with you when they buy your engine," McNerney said. "They have to depend on sort of constant commitment from the company that provides them, and when Jack goes public saying . . . you know." His first meeting as head of the jet engine business was with Bob Crandall, the powerful and notoriously cranky CEO of American Airlines. McNerney urged him to buy the new GE engine. Crandall said he wouldn't buy it. American Airlines was buying the Rolls-Royce engine. McNerney told him the GE engine was superior. "Well," Crandall replied, "you've got to go fix that fucking engine before you sell it to me, and fix your boss when you do that."

"Whoa," McNerney thought to himself. "I realized that I had a major challenge."

McNerney spent the next three years fixing the engine and convincing customers it was a superior product worthy of their investment. Part of what he did to make the engine a commercial success was to cut a deal with Boeing's commercial aircraft division, then run by Alan Mulally. He agreed to give Boeing $350 million to finish the development of the 777 jet if Boeing agreed to make the GE90 the exclusive engine on it. That was a tough sell for Boeing and for its customers.

Usually, an airline would decide whether to buy a Boeing jet or an Airbus jet and then have a separate competition among the three engine makers to decide which company would get the lucrative engine contract. Packaging the GE90 with the Boeing 777 exclusively went against convention. McNerney and Mulally made the deal. "It's the thing that saved the engine, because we were taking flak in the marketplace," McNerney recalled. "So then, Boeing and GE together were going to the marketplace, which is what finally convinced customers that we were going to do this and get it done properly."

———

THE THIRD CONTENDER TO SUCCEED JACK WAS JEFFREY IMMELT. Immelt grew up in Cincinnati, Ohio, down the street from Finneytown High School, where he and his older brother, Steve, played sports, did their homework, and absorbed a certain moral rectitude along with an innate sense of optimism that one's goals could be achieved through persistence. "It was a very midwestern upbringing in the sense of we focused on hard work, sincerity, being straight with people," explained Steve Immelt, a partner at Hogan Lovells, the international law firm. (For six years until 2020, he was the firm's CEO.)

Immelt's father, Joseph, spent thirty-eight years at GE Aviation after serving in World War II and attending Ohio State University on the G.I. Bill. He and Immelt's mother, Donna, met at Ohio State. The family's roots go back to Germany and Scotland on both the paternal and maternal sides. Steve Immelt believed his ancestors immigrated to the United States in the years before the Civil War. "The families have been in the country a long time," he said. The Immelt boys were raised Presbyterian. "We went to church every Sunday," Steve said. Politically, the Immelt parents were Republicans, "as was true of most people in that part of Ohio," he continued. But, he emphasized, the Immelts were "very fair-minded" and "traditional" Republicans who believed "in the importance of anti-discrimination" and "of giving people equal opportunity." They were very "practical," he continued, "and focused on getting things done." His mother was uninterested in excuses for why something didn't go the way the Immelt sons wanted. It was more "What do you need to do to make this situation better?" he continued.

Joseph Immelt's career at GE was spent in supply procurement and operations. He worked in building 800 of the jet engine campus, mostly as a supervisor. "He bought buckets and buckets of stuff," Jeff Immelt told me, to build turbine blades. To protect the military and industrial secrets from prying eyes—"in case the Germans were sending bombers," Jeff said—the building where Joseph Immelt worked was buried underground. "He worked for almost forty years," he continued, "and never saw the sun during the day." Jeff Immelt used to go to GE "open houses" with his father in the underground factory. "I was always fascinated with technology, and I very much really dug walking through the factory," he said. His father retired from GE in 1988. "He loved what he did," Jeff continued. "He loved the company, was proud of the company. We were the kind of family that would go down to Lincoln Airport and watch planes land. He could point out different planes. 'This is the 707. It's got Pratt engines, but we should get it someday. It's the same plane that President Kennedy flies,' or something like that. He was that kind of dad that was very engaged and liked what he did." (His parents, both ninety-one years old at the time of our first interview, were living in a nursing home in Mason, Ohio. "They're the Trump base," he said. "He's got a MAGA hat.")

The Immelts gave their two boys free rein while also instilling in them a deep sense of responsibility. Like many, their parents grew up during the Great Depression, an experience that shaped their worldview and one they passed on to their two sons. "For them, it was really important that we had a stable house," Steve said. "And we did. My father had a good job. He worked at General Electric. My mother took care of us and really was always there for us, very supportive. They expected you to behave. My mother was a stickler for that. No prevarication at any level was ever going to be acceptable to her. But we had a lot of fun. My brother is a person who wants to have fun. He can tease people. The only person that I've ever met that is more proficient at that than my brother is my father."

Organized sports were the leitmotif of their existence. "We played every sport, every season," Steve said. Jeff Immelt played football, basketball, and baseball. The consensus seemed to be that his best sports were football—he played offensive line—and basketball, where he played center. He made all-league in both football and basketball. He pitched on the baseball team. "Baseball was probably his worst sport," a teammate explained.

For whatever reason, Jeff Immelt seemed to have little respect for the baseball coach, who wasn't the greatest communicator. One teammate remembered the time when, before the season started outside, the varsity baseball coach asked the team to do some drills in the gym. He split the team into three groups. When he said, "half should do 'x' and another half should do 'y' and another half should do 'z'," Immelt thought it was the funniest thing he had ever heard. "Jeff kind of just rolled around on the ground and started laughing," he remembered. "You know, he didn't take him seriously at all."

The assistant baseball coach tried to get the head coach to cut Immelt from the team. But he wouldn't do it. His teammate said that Immelt "found it amusing" to give the finger to the photographer who was taking the picture of the varsity athlete lettermen to be used in the high school yearbook. "This reflects an enormous amount of disrespect and made the photo unavailable for publication," said the teammate. "His concept for humor and compassion was always off the mark." He added that Immelt was "most assuredly not the smartest, nor kindest, student in our class."

Steve Immelt went to Yale. His high school football coach was from Steubenville, Ohio. One of the football coaches at Yale also grew up in Steubenville. The two men would keep in touch, and the Yale coach often asked the Finneytown coach if he thought any of the football players on his team had the academic chops for a place like Yale. At that time, graduates of Finneytown High went to good colleges in the Big Ten, such as Ohio State and Michigan, or to Northwestern. Rarely did they go to an Ivy League school.

On a trip to Ohio, the Yale football coach met Steve Immelt and then invited him for a visit to Yale's campus in New Haven. "That was kind of an eye-opener," Steve said. "I was used to Ohio State, which is a huge Big Ten campus, but Yale was a very different experience. That's the way it happened." Off to Yale he went. "It was before college tuition took off explosively," he said. "It was before every kid had a BMW on campus. It was easier to come from a pretty middle-class family from Ohio to join up at Yale, because there were a lot of other kids just like that. I'm not sure if that's still the same situation today."

The Finneytown High college counselor also encouraged Jeff Immelt to apply to Yale and the Ivy League. For Jeff, the choice came down to Dartmouth and Vanderbilt. "I wanted to play football, but I didn't want it to dominate what my life was going to be like," he said. Immelt was an applied math major at Dartmouth. He was president of his fraternity, Phi Delta Alpha. He took courses from the Dartmouth president, John Kemeny, a mathematician and the codeveloper of the BASIC computer language. A football player, a frat boy, and a math major. "It was a fantastic academic experience," he said.

One of his best friends at Dartmouth was Tom Garden, a hockey player from Melrose, Massachusetts, who was a member of the next-door fraternity. "We were just great friends," Immelt said. During the summers, Jeff worked on the Ford assembly line near his home. When he graduated from Dartmouth in 1978, he had $15,000 in student debt. "I needed to go earn some money," he said. "It seems like a lot." As McNerney had, he took a job after graduating at Procter & Gamble, back in Cincinnati. He knew someone at the company. They were in touch. "The brand management program looked good," he said. "I was a little bit late to the party. Staying in Cincinnati was not a bad thing to do." He soon found himself on the Duncan Hines brownie mix team. His officemate was Steve Ballmer, who had graduated from Harvard the year before Immelt had graduated from Dartmouth. "We had Formica desks that touched each other, and we would play paper-wad basketball every day," Immelt remembered. "We were complete fuck-wads. That's the best way to say it." He said Ballmer was a blast. "We're like a *Dilbert* cartoon," he said. "Both of us were incorrigible malcontents. I wasn't ready to be a grown-up. I was partying at night. We had great fun wearing Depends to work" at the office.

Ballmer soon had enough of the brownie mix business. He enrolled, briefly, at Stanford Business School before he contacted his former Harvard classmate Bill Gates and told him he wanted to work for Microsoft, since it was then a small company. He said he preferred a start-up to P&G. He didn't even know what software was. When his father asked him why he was joining Microsoft, he reportedly answered, "Right now, Dad, I don't know, and I don't care. All I do know is, if I accept this job at Microsoft, I'll never have to look at another box of brownie mix again!"

Immelt, too, was itching to leave Duncan Hines behind. After a year, he applied to business school: Harvard, Tuck, and Wharton. He got into Harvard, deferred the admission for one year, and, in the fall of 1980, decided to enroll. "Well, it's Harvard," he thought to himself. "So that doesn't suck." During the summer after his first year, he worked at the Boston Consulting Group in Chicago. He liked the people he worked with and enjoyed his assignment—working with the Clark forklift company, which was facing stiff competition from Komatsu, the Japanese company. He worked on a dealer study. "They gave me lots of responsibility," he said. "I went out and did field calls. I loved it. It was great." But he didn't want to be a consultant. He wanted to be a CEO.

He decided to buckle down during his second year at Harvard. He didn't want to get stuck in another Duncan Hines–type job. He took a smattering of courses, including those in finance and marketing. People thought of him as more of a "marketing" guy, rather than a "finance" guy. But his knowledge was broad and catholic. He was part of the famous HBS class of 1982, which included Jamie Dimon, the long-

time CEO of JPMorgan Chase, and Steve Burke, the longtime television executive. Dimon has said of Immelt, "A lot of people crawl over
splintered glass and broken bodies to the top, while others quietly
wish them failure. But everyone I know has always thought Jeff would
be successful, and has always wished him Godspeed."

Burke said he knew Immelt at Harvard and, while they weren't
close there, he liked him. "He was a guy with a football," he remembered. "He was always in sweatpants. Very likable, liked guy. Big kinda
guy. A little fratish." At HBS, Immelt reconnected with Tom Garden,
his Dartmouth buddy. They hung out together in Boston, and Immelt
used to go to dinner at the Garden home in Melrose for Sunday dinner
and for some holidays. During those visits at the Garden "triple-decka,"
Massachusetts slang for a house with three floors for three families—it
was an "extremely blue-collar" neighborhood, Immelt recalled—he became acquainted with Tom's younger brother, Ed, who had gone to Andover and then to Harvard.

Like Immelt, two other members of the HBS class of 1982 ended
up at GE and worked their way up to top executive positions. William
Strittmatter, who spent thirty-four years at GE Capital, was the chief
credit officer for a while. Norman Liu, who spent thirty years at GE,
was the CEO of GECAS, GE Capital Aviation Services. Liu remembered that Immelt hung out with the other jocks at the business school.
But what really distinguished him was that he didn't want to follow
the HBS crowd into investment banking or consulting. "He actually
wanted to be CEO of a Fortune 500 company," Liu told me, "and you're
sort of sitting there, 'Oh, yeah, great, whatever, why? Why would you
do that?' . . . That's when people started going, like, 'What? He actually
wants it?' You don't normally see that at Harvard Business School."

In 1982, Dennis Dammerman, then working at GE Capital for
Larry Bossidy, was in charge of MBA recruiting at Harvard. "I didn't
know him from Adam," Immelt said. Dammerman interviewed Immelt at Harvard. He was impressed. "It was clear he had natural

leadership traits," Dammerman told Harvard Business School in 2006. "I immediately wanted to hire the guy." Dammerman bypassed the usual process of forwarding Immelt's résumé to corporate HR and instead recommended to the head of consumer products that Immelt be hired. Dammerman also alerted Jack to Immelt because Immelt was also being heavily recruited by a Morgan Stanley partner, who told Immelt that if he went to GE he wouldn't have access to Jack for "at least a decade." He also had an offer from Hercules, Inc., an industrial chemicals company.

As usual, the Jack magic worked. Immelt joined GE in "commercial leadership," which included working in several different GE businesses. Within thirty days, he was part of a corporate marketing group that had presented to Jack. He started living in the pool house of a suburban Fairfield County estate. He thought he would stay at GE for a few years and then move on. "It's a good training ground for general management," he thought. "I'll figure out what I want to do. But it'll get you on a general management path. Let's get started."

He moved from Cambridge, Massachusetts, to Bridgeport, Connecticut, where he spent ninety days working for a marketing vice president. Then he was sent to Albany, New York, where he was the product manager for the plastic instrument panels that went into automobiles. GE Plastics, of course, had been Jack's fiefdom for years. Making plastic parts for the automotive industry in the early 1980s— during the so-called Reagan recession—was hard. His first year, the industry in the United States made only 11.5 million cars and light trucks. "It was quite tough," he said. On the other hand, he told me, the auto industry was in the process of substituting plastic for metal wherever it could.

Immelt's job was to try to convince the purchasing manager to make the switch and to buy the plastic from GE. He said that in some respects the sale was like "pushing on an open door," because fuel-efficiency mandates were necessitating a push to lighter cars. But, he said, "I used to get the dry heaves sitting in a Shoney's, waiting to walk across the street to a General Motors purchasing manager's office. You build kind of calluses as you go up in your career in terms of how you deal with the automotive industry."

He did well. He got promoted. He liked his colleagues. He liked GE. "Excuse me while I strut," he signed off in the 1983 Dartmouth class notes. His awareness of Jack was limited to seeing his picture in the annual report. "But it was not like I wanted to be him," he said. "I was not that weird." In 1984, Immelt moved to Dallas as the regional sales manager for the plastics division. His job was to sell plastic beyond the automotive sector. He had fifteen direct reports. "From crusty veterans to young hotshots," he said.

In Dallas, Immelt was under pressure to improve his sales results in a difficult economic environment. His boss scrutinized his revenue and exhorted him to do better. "You had a terrible month," he told Immelt. "I don't care if you have an MBA from Harvard, I'm not keeping you if you have another bad month." He learned that GE was nothing if not results oriented. "I also learned how to sell, and how to manage people who were different from me," he said. One of his subordinates was Andrea Allen, a customer service representative. After eighteen months, Immelt was promoted again to be in charge of selling plastic in the western half of the United States. He moved to Chicago. While there, he reconnected with Allen, who had also, independently, moved to Chicago. They both had booth duty one day at a big plastics show taking place in Chicago. "I was chatting her up and you know," he said. "This was before the #MeToo movement. So I was okay." They were married in 1986. He would have been her boss's boss. She decided to retire from GE. "She never lets me take myself too seriously, and she keeps me balanced," he said of his wife. Soon after, she was pregnant with their only child, Sarah.

The next year, the Immelts moved to Pittsfield, still the headquarters of Plastics. Immelt was tapped to be head of global marketing for

Plastics. He worked for Glen Hiner, who ran the plastics business. But other top executives, such as Dammerman and Bossidy, were keeping an eye on Immelt. He recalled one presentation to Bossidy, who was then overseeing Plastics, that he found particularly bracing: "I proudly told him, 'I've got a thirty percent increase versus last year, which is two percent over my stretch forecast.' Larry only asked, 'To what do you attribute your poor forecasting?'" Immelt and his young family didn't seem to mind the moving around. "Moving is always hard," he said, "but it was kind of like what we did. It wasn't a bad thing." He was making around $90,000 a year, living in Pittsfield. "It wasn't like I was getting rich," he said. "But it was great."

Immelt was gaining recognition across GE. In 1986, he was invited to participate in an executive development course at Crotonville. Jack had to approve his selection to take the course. He was the youngest participant. The Crotonville experience further welded Immelt to GE. "Beyond the knowledge you gain, the best part is just absorbing the culture and the values," he recalled. "And the networking makes you feel that you're part of something great." In the monthlong Crotonville course, Immelt became close friends with John Myers, who would go on to head up GE's pension fund. "We used to go jogging in the mornings," Myers told me. They also spent a week together on a truncated Outward Bound program on Hurricane Island, Maine. "We were doing ropes courses," he recalled. "Anyway, I came back from that month [in Crotonville and Maine] and I said to [his predecessor] Dale Frey, 'I met Jack Welch's successor.' He stood out as a star. And he was 10 years younger than me. He was the youngest guy there."

AFTER CROTONVILLE, THE IMMELTS WERE ON THE MOVE AGAIN, THIS time to the major appliance division, in Louisville, Kentucky. "Out of the blue I got a call from Jack Peiffer"—a senior human resources executive—"who said, 'We're thinking about moving you to the appliances service business. We're just in the beginning phases of a compressor recall, and Jack wants somebody who is not corrupted by the environment.'" Jack decided to parachute Immelt into Louisville in the middle of the crisis involving the new refrigerator compressors that were failing left and right. Around one million compressors would have to be replaced, and the task fell to Immelt to oversee it. "Listen, Jeffrey," Jack told him, "I don't trust any of these guys out here in appliances. I want an outsider to come in. Your boss doesn't even know I'm making this call." At first, Immelt thought a district sales manager was playing a joke on him. But he quickly realized it was Jack. "Then I started maybe one of the most important jobs I had in my career," he said.

He went from managing 150 people to managing an army of 7,000 refrigerator repairmen. Although he said he didn't "know a thing about appliances, or about the service business," he took the job. He quickly realized it was a "sink or swim" opportunity for him. He said it was in Louisville that he "went from being a boy to a man." It was "the middle of a shit storm," he said. GE ended up taking a $450 million charge against earnings, at a time when GE was earning $2 billion a year. "It was a big number," Immelt said.

But the consensus seemed to be that the fiasco could have been much worse and that the repair-and-replace operation was a success. "I learned how to manage in a completely different way," Immelt recalled. "I was in Plastics in the go-go years: I knew how to grow. This taught me how to operate. The degrees of freedom were extremely small, and the difference between success and failure was a penny or two one way or the other." He sometimes drove a forklift or donned the GE repairman uniform and joined the guys as they fixed refrigerators. It was stressful. His weight "ballooned" to 280 pounds, *Time* reported. According to Jim Bunt, GE's treasurer, Jack favored Immelt from the early 1990s but warned him that his career would be truncated if he didn't get his weight under control. "You're never going to be CEO if you don't lose weight," Jack told Immelt at a meeting in Crotonville.

"You've got to get your fucking weight down. Can't have everybody fucking fat."

Immelt recalled attending the annual managers' meeting in Boca Raton when Ed Hood, a GE vice-chairman, told the assembled leaders, "The company had a great year this year, except for appliances. I don't why we invited any of those people to come." Immelt found that humiliating. But he liked Louisville. He was making about $110,000 a year. He bought a $300,000 home. And hobnobbed at the Kentucky Derby. "We lived like kings," he said. When he left Louisville, his colleagues gave him a cartoon showing a harried Immelt sitting at his messy desk surrounded by junk food.

In 1992, he returned to Pittsfield as vice president in the commercial division for the Americas. Jack sent him back to Plastics along with his boss at appliances, Gary Rogers, who had taken over major appliances after Roger Schipke was fired. But initially, Immelt was layered and reported to a manager who reported to Rogers. He thought maybe Jack was trying to send him a message. It worried him. "I know this job is not obviously what you want to do," Jack told him, "but sometimes you win in these moves and sometimes you serve the company. This is a time when you serve the company." After a year, though, his boss was out and Immelt was again reporting to Rogers and running half of the plastics division. Making the situation more challenging was yet another recession hitting the business. "The automotive industry sucked," Immelt said. The good news, Immelt said, was that "it was a quick recession." Early in 1994, Immelt's Plastics operation was floundering. It missed its operating earnings target for the quarter by $30 million, or 30 percent of budget. (Jack said it missed by $50 million.) He found himself on regular weekly calls with Dammerman and with Jack.

Immelt's second stretch in Pittsfield may have been the most challenging in his otherwise charmed ascent at GE. "Nobody wants to be around somebody going through a low period," he said. "I wasn't hitting my numbers. I knew what I had to do. In times like that you've got to be able to draw from within. People can help you, but leadership is one of these great journeys into your own soul. It's not like anybody can give you the answer of how to do it." He found himself in Jack's doghouse as the numbers got worse. The contretemps came to a head at the annual Boca Raton managers' meeting in January 1995. Jack said that Immelt tried hard to avoid him at the meeting. By then, Immelt's numbers had already been improving after he renegotiated some plastics contracts with the car manufacturers. Jack tracked Immelt down on the last night, just as he was scurrying off to the elevator. "Jack did one more bombing run and he was in my face for fifteen minutes," Immelt remembered. "At the end of it, there wasn't one person within a hundred yards of us. Everyone was crowding by the door and in the corner. But I just said, 'Hey, Jack. You're too late. It's already fixed. So have fun.' And we had a good laugh and it was time to move on." Jack's version was that he told Immelt if he didn't fix the problem at Plastics, "I'm going to take you out." Per Jack, Immelt fixed it "and then just nailed every job he held after that."

At the end of 1996, Rosanne Badowski called Immelt and asked him to come down from Pittsfield to 30 Rockefeller Center to see Jack. He had no inkling why. That's the way it was with Jack. He loved to spring surprises on his employees. When Immelt got to Midtown Manhattan, Jack told him that John Trani, then head of GE Medical Systems, was leaving to become CEO of Stanley Black & Decker. Jack wanted Immelt to move to Milwaukee to replace Trani. Did he know anything about medical devices, or medicine? "Zero," he said. "I knew nothing." Jack also asked Immelt to join the board of GE Capital, along with Nardelli, McNerney, and Dave Cote. "I don't know that there was a sense for succession at that time, but McNerney, Nardelli, me, and Cote were all on the Capital board," he said.

Between the promotion to run the medical systems business and the request to join the board of GE Capital along with three other business leaders, Immelt must have thought that Jack was seriously considering him as a successor, right? "To be honest with you, I never really thought about it that much," he replied. He then shared a story about how he had wanted to expand GE's plastics business on the West Coast, with tech companies such as Hewlett-Packard and Sun Microsystems that use a lot of plastic in their products. He invited Jack to join him at a dinner he had arranged with a group of six CEOs in California. He was at the table thinking to himself, "I may not be as good as Welch, but I'm as good as these guys. I can play at this level. But that's as much as I really kind of thought about it. I was too busy working."

Immelt and his family moved to Milwaukee in early 1997. His wife was fine with the move, but Sarah, his only child, found it "hard," he said. Immelt liked Milwaukee. He liked to travel. He liked the medical equipment industry. "It was a good fit," he said. He was managing around 25,000 people. It was a big step up. It was, he said, "a chance to put it all together." He focused on expanding the business globally. He was relentless about winning business and often would not take no for an answer if it appeared that GE lost out to a competitor on the sale of medical equipment. Peter Foss, his longtime friend and Plastics colleague, recalled the time Immelt got down on his knees to beg a purchasing agent to rethink his decision to award a big contract to a competitor. (Jack once did the same thing with the minister of defense of Romania. "Oh, please, Mr. Welch, get up!" he told Jack.) What made Immelt a consensus finalist to succeed Jack was his ability to supercharge the medical systems business during his four years at the helm. Revenues doubled to $8.4 billion and net income increased to $890 million, from $400 million. "When Jeff took over Medical Systems, he let the sunshine in," Jack said.

Jack's annual reviews of Immelt reflected Jack's enthusiasm for his operational and leadership skills. Jack had increased Immelt's 1997 incentive compensation by 50 percent to $600,000. "What a great year!"

he wrote on Jeff's annual review. The previous year he had urged Immelt to focus on the integration of an ultrasound company that GE Medical had bought. "We've never bought a Silicon Valley [company] and made it work," he explained. "It must be nurtured by you personally [and] we can't lose external focus in the process." A year later, Jack told Immelt about the integration, "Appears to be going well."

There were seven other areas where Jack gave Immelt some advice. Overall, Jack gushed: "I loved the year you had and look for another spectacular one in '98. Your concise communication, willingness to learn and grow were very special. I am available to play any role you want—just call on anything." Nineteen ninety-eight was a "sensational year" for Immelt, according to Jack, and his bonus increased to $850,000, up 42 percent, along with his salary of $533,333. "Another great year," Jack concluded. "I like everything you are doing in every way. Call on me for anything." Jack's review of Immelt's 1999 was shorter and more to the point. "Congratulations on a sensational year," he wrote to Immelt in February 2000. His bonus was $1.2 million; his salary was $616,667.

OF THE FOUR OPERATING EXECUTIVES WHOM JACK ADDED TO THE GE Capital board—a bit of a ruse so that he could get a closer and regular look at them while maintaining the farce that the succession process remained a mystery—three were finalists for his job: Nardelli, McNerney, and Immelt. Jack had pared away only one man—David Cote, the head of the major-appliance division in Louisville. To this day, some wonder if that was another of Jack's major mistakes. Others don't. "Cote was not a favorite of anybody's in the top management of GE," explained one executive in that group.

Born in Manchester, New Hampshire, Cote grew up in nearby Suncook, a fact immediately obvious to anyone hearing his mellifluous

New England accent. In that way, and in others, he was very much cut from the Jack Welch mold. He attended Pembroke Academy—a public high school that was once a semiprivate school. He was accepted at the University of New Hampshire. But he hated school. He thought it was a waste of time. He took a year off. He tried being a car mechanic. He tried being a carpenter. He was not good at either. He could conceive of how to do a project well enough, but his hands wouldn't cooperate with his brain sufficiently to actually get it done. He moved to Michigan to work for his uncle. But that failed, too. Out of some desperation, he joined the navy. The day before he was set to report, he called the recruiter and told him he wasn't coming. He had signed up for six years' duty on a nuclear submarine but learned later that he was claustrophobic. When the recruiter told Cote he had made a commitment to the U.S. government and couldn't renege, Cote asked the guy if he would send the cops to arrest him. "No, I can't do that," the recruiter told him. "Well, then I'm not coming," Cote replied. He was eighteen.

Faced with another blank slate, Cote decided that he wanted to go to college after all. He figured UNH would let him in, as it had accepted him the previous year. But he was wrong. The university wanted him to reapply. Cote thought that made no sense. He drove to UNH, about forty minutes away, and was determined to have a conversation with the director of admissions. He didn't have an appointment, which made his chances of seeing the fellow that much more difficult. He said he would catch him on the way to his car, if need be. After waiting about two hours, the secretary gave Cote five minutes to make his case. "I'll tell you what," the guy told him. "If you get an application in to me by Friday and it meets our expectations, I'll let you in." UNH let Cote in.

To make some extra money, Cote applied for a job at a GE aircraft engine factory in Hooksett, New Hampshire. "They paid well," he remembered. He got hired. He worked nights on an hourly basis. He cleaned mechanical tubing. He ran a punch press. He operated roll mills and tumblers. He was paid $3.66 an hour, which included a night-shift bonus of 36 cents per hour and was about double the hourly minimum wage at the time.

After about a year of working at GE at night and going to UNH during the day, he and a friend decided to buy a boat and try to become rich as commercial fishermen in Maine. "An early indication of my business acumen," he joked. The idea was to fish for haddock. But the haddock had disappeared. You could fish for two weeks straight and get one haddock. They started fishing for cod, which were more plentiful. He was working too hard, sleeping too little.

But things were basically good. He was working. His wife was working. Then, about a month later, his wife came home and told him she was pregnant. He didn't understand. She was on the pill. "Well, this is a pill baby," she told him. Three months later, she told Cote that she was quitting her job and they would have to live off his salary. He panicked. "I knew I didn't make enough to support myself," he said, "and now I was going to have to support three of us in this unheated apartment. It ended up being a boy. My son was going to be born in February. I just completely panicked. I looked at it and said, 'My kid's going to die because I'm a screwup. I'm going to the VFW hall to play pool and drink beer and everything seems fine.' That was a complete turning point for me." Cote said he tells his oldest son that he is the reason for his success. "He scared the bejesus out of me," he said. He realized that school was the only thing he was good at and went back to UNH. He sold the fishing boat. "We weren't making any money anyway," he said. He quit smoking—he decided he couldn't afford the cigarettes, and it wasn't healthy to boot—and started exercising. He kept working at GE at night. In 1976, he graduated from UNH with a bachelor's degree in business administration. Liberal arts bored him; engineering seemed too hard.

He wanted to be a stock analyst. He had an interview with John Hancock, the insurance company, to be a bond analyst and an interview with what is now Wellington Asset Management to be a stock analyst. But neither job panned out. "A 3.2 from the University of New Hampshire that took me six years to get, with a 1.8 one semester, was not exactly the most prestigious calling card," he remembered. He didn't know what to do. He was still working the punch press at GE. One day, one of the factory managers told him there was a full-time job that had been posted in accounts payable. He suggested that Cote apply for it. But he resisted. He knew nothing about accounts payable. "I'd just be a mess at it," he thought. But the fellow insisted. He thought Cote was nuts. "You have a college diploma, for Christ's sake," the guy told him. "Why don't you at least give it a try?"

Cote decided he was right. He applied for the accounts payable job. He had three interviews. It came down to Cote and a woman who worked in accounting. Everyone figured the woman would get the job. She did. A manager at the jet engine plant in Lynn, Massachusetts, made the final decision. But he also promised to find another job for Cote. Three weeks later, he was hired as a full-time internal auditor and later entered GE's Financial Management Program. "I was an experiment," he said, "because they'd been wondering what it would be like to take an hourly employee and make [him] an internal auditor." Generally speaking, assuming you were capable, after two years in the financial management program, you were sent out to an operating division as an internal auditor.

Not Dave Cote. At the start of the FMP, his manager told him that his accounting skills were lacking and that he needed to spend another six months studying accounting before he could enter the program. That was not the message Cote wanted to hear. "I still wasn't making enough money, even though I was now an exempt employee, to support my family," he said. It was another one of those moments where Cote had to do some persuading. It was a tough sell. When the manager went on vacation to Florida, Cote asked him if he could borrow his *Principles of Accounting* textbook. Cote's idea was to study the textbook on his own time during the week when his boss was away and to prove to him—somehow—that he had mastered the material. It wasn't the best-conceived plan. But it was the only one he could come up with on short notice. He studied the textbook, made notes in the margins, and answered the accounting problems in the back without cheating. He was sure he understood enough of the concepts to convince his boss he didn't need another six months in the FMP. "I know the stuff," he told his boss. "Ask me anything."

His boss called corporate and asked them what to do. Cote had presented GE with an unusual set of circumstances. The verdict was that if Cote could pass an accounting test, he could avoid the extra six months. Two weeks later, Cote took the accounting test. He scored a 96.5. "You don't have to take the course," he was told. (He kept the marked-up accounting textbook as a memory of his persistence.) He was now a full-fledged member of the FMP. Then his wife left him. He promptly got divorced and returned to living at home with his parents. "This is pathetic," he thought to himself.

The corporate audit function was based in Schenectady. It could be as long as a five-year commitment. The good news, from Cote's perspective, was that he was constantly being sent around the world to take on different assignments. He didn't need a house or an apartment. He didn't need a car. He could live cheaply, with GE paying most of his expenses. When he was in Schenectady, he'd stay in the cheapest hotel he could find and eat in the least expensive restaurant. He spent two of the years outside the U.S. He audited some twenty different GE businesses. He learned how to get along with people. "It ended up being a big career accelerator for me," he said. He had been indoctrinated in the GE way. He was sold. "I was such a GE homeboy," he said. "I loved GE. I loved what it was doing." He loved that it seemed to be at the "forefront of everything that was going on."

When Cote finally got a long-desired posting to an operating company, it was decidedly second tier: he was sent to Syracuse, New York, to work in GE's audio electronics business, which made radios, tape players, and telephones. Eventually the business was combined with consumer housewares, and together they were sold off to Black & Decker, one of Jack's first asset dispositions. As part of his Neutron Jack campaign, Jack cut back on staff everywhere. Cote was suddenly out of a job. He sent out résumés trying to find something new. But he wasn't getting any nibbles. Again he got lucky. A finance executive at headquarters in Fairfield recalled Cote from an audit he had done. He knew that Cote had been hard on him but was also good-natured. People remembered. "We need more of that in this company," he told Cote. He offered him a job at corporate in finance: spreadsheets, discounted-cash-flow analysis, analyzing acquisitions, and, when it made sense, stock buybacks. In that vein. Of course, he didn't want the job. He wanted a job that would give him more operating experience. On the other hand, that was the job offered. "How many alternatives do you have?" the Fairfield man asked. Cote took the job. Along with his new wife, he moved to Fairfield, Connecticut. He worked many sixteen-hour days.

By then, he was a huge Jack fan, despite nearly losing his GE job because of the cuts. "He walked on water," Cote said. "Even though we were all scared to death of him, I just could see what he was doing. I could see how he was thinking about things." He remembered how, during his first weeks in Fairfield, he got invited to a "skip-level" meeting, where Jack would meet with the junior employees to get a feel for what they were thinking. At that time, Jack had eliminated the corporate strategic planning function, preferring operating executives do their own strategic thinking, rather than have it imposed on them from corporate. The gossip was that Jack had decided long-term strategic thinking was not important. Cote decided to ask Jack about that. For some reason, like John Flannery, he was fearless in questioning the imperial CEO in a roomful of corporate types. Jack's answer impressed him. "Strategic planning is critically important to a business,"

Jack replied. "But the business leader has to own it, and what's happened is they've just all delegated responsibility to a group of people who were in the strategic planning function, who work on making prettier and prettier charts to just impress all the other strategic planning bosses, and business leaders no longer own a strategic plan, and they're not doing the strategic thinking the way they need to." Cote said Jack's answer made a lot of sense.

During his time in Fairfield, he worked mostly for Dammerman, analyzing acquisitions. One day, in 1985, his colleague, who worked on board presentations for Jack, was away. Cote was asked to fill in. While he was working on the presentation for Jack, he got an urgent call that Jack wanted to speak with him. "Dave," Jack asked, "is it true we ask the medical systems business for what the ROI"—return on investment—"will be in the ultrasound business [four years later] in 1989?" Trying to collect himself, Cote was wondering, "Where the hell is this coming from?" Thinking fast on his feet, he told Jack he thought it was probably part of the strategic planning forms that were sent out to business development teams in the divisions. The data would come back, as requested, but then never get used. Cote had previously thought the request should be eliminated and had said so to his supervisors. But they shot him down. Now Jack was yelling at him for sending out the forms. "He came right through the phone at me with all the f-bombs and how could I do this?" Cote remembered. "I had to bring that request up right away to his office. He wanted to see it for himself." Cote ran back to his office and retrieved the form. Jack had tried to speak with Dammerman, who wasn't in. He had tried to speak with Jim Costello, the corporate controller, who wasn't in. And he had tried to call Cote's boss, who wasn't in. That left Cote, the man who had sent out a form that neither he nor Jack liked.

He brought the form to Jack's office. Jack then called Cote in to see him. Jack was flipping through the form, getting more and more worked up. "He berates me for at least twenty minutes," Cote said, "and is just yelling at me. How can I do this? With everything he's try-ing to get done, how could I be so stupid? The *F*s are flying all over the place." Finally, Jack said, "Dave, you're a smart guy. Why did you do such a fucking stupid thing?" There was not much Cote could say. "It's just a financial expression, a strategic plan," he said. "We need to have some understanding." He never told Jack he had been against using the form and that he had been overruled by his boss, Tom Hartnett. By the time he was done that day, Jack had also yelled at Dammerman and told him he was "going to have his ass" because of the stupid form. Cote told his wife he probably was going to get fired. "I'm not sure how this works," he told her. "But I think I've been fired."

But two months went by and he wasn't fired. He was thinking he might have somehow avoided disaster. In the meantime, he had spent three weeks in late 1985 secretly working on GE's acquisition of RCA. He was one of two young analysts in headquarters crunching numbers on the deal. The whole thing had been kept so quiet he hadn't been able to tell Hartnett what he was doing. (When he did tell Hartnett, shortly before the deal was announced, Hartnett traded on the insider infor-mation by buying RCA stock options, got caught, and was swiftly fired.)

On the June 1986 day the RCA acquisition closed, Cote and his colleague were invited to the victory party. They went together to the swanky hotel that Jack had built at the Fairfield headquarters, and when he walked in, he heard from across the room, "Dave, Dave, get over here!" It was Jack. He thought for sure he was about to be fired. "I can't believe it," he thought. "At a party?" He "trudged" over to see Jack, with a "fake smile" on his face. His colleague went with him. "I was never so pissed at anybody since I was in Plastics," Jack told him.

"Well, I really appreciated you sharing it with me," Cote said. Jack thought that was funny and started laughing. They started chatting amicably. "Clearly, it was just going to be a convivial kind of occasion," Cote recalled. They chatted some more about the stupid form and how much Jack hated it. That's when Cote's colleague said, "You know, Jack, he never wanted to send out the request in the first place. He had recommended against it, and all of us on the staff and Tom Hartnett, our boss, said no, we were gonna start using this stuff in the future, so we still needed to do it." Jack was taken aback. "So you just took the knife for those guys?" he asked.

"Well, I didn't view it that way," Cote replied. "You don't throw in your friends on something like this. You just don't."

Jack kept shaking his head: "Wow, geez, wow. That's something. Wow. Okay."

When the two young analysts returned to another side of the room, Cote thanked his colleague for sticking up for him. "Well, yeah, I guess," he said, "but just so we're clear, if he hadn't been in a good mood, I wouldn't have told him."

A little bit later, Dammerman pulled him aside for a conversation. "You have no idea how much good you did your career with everything that happened," he said. Cote told him he didn't really see how, given the berating he had received from Jack. "Well, the first thing was the way he yelled at you in his office," Dammerman said. "He has made vice presidents cry, and you never got flustered. You just kept explaining the story and why it was done. He really admired how you stood up. Then when he found out tonight that you didn't throw in your friends . . . Nothing but good is going to happen to you from now on."

At GE, in those days, if you went up one level in your job ranking, it was considered good. If you went up two levels, you were considered a real highflier. A month after the RCA victory party and Cote's conversation with Jack, Cote suddenly found himself interviewing for jobs that were three and four levels above his analyst position in Fairfield. "Which creates its own backlash within the company," Cote said. Before he knew it, Cote was back in Lynn, serving as the finance manager for the production of small jet engines. (The big jet engines were manufactured in Cincinnati.) He had two hundred people working for him. The most he had had before was three people. Once again, his career got a big accelerant. "All of a sudden I was on his"—Jack's—"radar screen and things began to work out," Cote said.

But Cote was still antsy. He wanted out of finance and into operations. He wanted a job managing a business. There was significant resistance from the bureaucracy. Once a finance guy, always a finance guy, the thinking went. He kept pushing and pushing. He told GE he would likely have to leave the company if it didn't try to find him something on the operations side of the business. Of course, if he were to do that, it would have to be in a business he actually could comprehend. And jet engine manufacturing was not a business he was likely to understand—it was too complex, required too much engineering knowhow and the wisdom to manage a workforce that understood how to make the product and make it better.

That's how he ended up in Louisville, Kentucky, running the laundry and dishwasher businesses in the major-appliance division. "I thought, 'Okay, this is going to be great,'" Cote said. "It's my first general-management foray.... This is my shot." It was a $1 billion business. But in GE's matrix management system, it was a billion-dollar business run by three people at the same time. Weird, he thought, but another "good learning experience."

After about a year, the CFO of Major Appliances left. Roger Schipke invited Cote to lunch. He knew something had to be up because Schipke had never before invited him anywhere. They hadn't gone more than a hundred yards in Schipke's car when Cote blurted out, "Roger, just so you know, if this is about me taking the CFO job, I don't want it. I know it pays more, but I don't want it. I'm finally in general management. That's what I want to do and I need to build my credibility." Schipke was caught off guard. "We don't still have to go out to lunch if you don't want to," Cote told him. They had the lunch. It was pleasant enough. But Schipke told Cote, "Just so you know, this is probably not the end of it, so just be prepared." When he got back to the office, Dammerman called him. He was yelling at Cote. "Finance needs you," he said. "This is a bigger job. You need to take this. You're wanted in the finance function." Cote objected. "Dennis, I know," he said. "I get it, but it's not what I want to do. I think I'm finally starting to learn something here." He told Dammerman he wanted to keep climbing the steep learning curve as a general manager. "Okay," Dammerman said, "you can say no to Roger and no to me, but your next discussion is with Jack and we'll see how you do there."

Fully confused, Cote called a recruiter he knew to get some advice. The recruiter told him it was easier to place a CFO of the whole division than it would be to place the manager of the laundry and dishwasher business. He then shared some epic advice. "The second thing you need to understand is the decision has already been made," he told Cote. "They're just waiting for you to catch up." Cote went up to Crotonville to see Jack. Jack was calm. "I don't understand why you wouldn't want to do this," he told Cote. "It's more marbles for your family. You'll do better. It accelerates your career. But it's your choice. If you want to stay in that job, where it's probably not going to get you anywhere, that's your choice, but I just can't understand why you wouldn't want to do this." Cote was convinced. He told Jack he would take the CFO job. "No, no, no," Jack replied. "This is not between you and me. I'm just giving you some career advice. This is between you and Roger. So if you want the job, you just need to tell Roger."

He took the job. When Jack was on a kick to reduce working capital investment in the businesses, Cote helped devise a plan to reduce inventories while improving delivery. Over four years, he cut the inventory in half, to $500 million, and increased product availability to 90 percent from 84 percent. "The thing was a huge success," Cote said, so much so that Jack sent managers from other parts of GE down to Louisville to see what Cote had done. He then ran the consumer service business at Major Appliances—all those GE vans that went to people's homes to fix their appliances. It was a business in decline. But Cote fixed it up. It was the first time he thought he could be a CEO somewhere, if not at GE.

He had earned his chance to run an actual business. Jack offered him the job running GE's silicone business, the stuff that had a variety of uses from caulking to a shampoo ingredient. The business was based in Waterford, New York, near Schenectady. He moved there from Louisville. "Finally, a full, functioning general manager," Cote said. It was yet another difficult assignment. Three of the four previous general managers of the business had been fired for doing a lousy job. He was determined to give it all he had. Over the course of thirty months, what had been a $500 million–a–year revenue business with $50 million of income became a $1 billion business with $105 million of income.

Of course, Jack noticed. When in May 1996 the job running Major Appliances opened up, Cote was a leading contender to get it. But so was Jeff Immelt. There was much scuttlebutt around the company as to which of the two men would win. One late night, Cote got a call from Jack, John Opie, and Bill Conaty telling him he got the job in Louisville. Cote called Immelt. "While I might have got this one," he told Immelt, "I think in the long run you are going to be better off, because they're going to put you in a different job, and we know what's going to happen, and you're actually going to be better off." On the plane ride down to Louisville, where Jack was going to introduce Cote as the division's new CEO, he told Cote that he viewed the major-appliance division as the worst business in the company. Thirteenth out of thirteen. Cote remembered telling his wife about the likelihood he could succeed Jack. "If I'm not put in a different job within the next two years, then it's not me. It's not going to be me."

Three and a half years later, Cote was still in Louisville, running Major Appliances. It had been a tough slog. His predecessor in the job had saddled him with a new washer that was a disaster. There were lawsuits from suppliers to whom GE had made oral promises and had not done what it promised. There was inventory sitting unsold around the globe. "That was my inheritance," Cote said. He invested capital to improve the product offerings, but that depressed the financial performance. Revenue, at around $5.5 billion, and operation profit, at around $750 million, remained flat during Cote's tenure. "I could tell he"— Jack—"wasn't that happy with me, because he wanted the numbers to be better," Cote said. "Hell, I wanted them to be better, but there was just no way around it. This was one where you had to take your lumps and move on."

In June 1999, Cote took his lumps. Conaty called him up one day and told him Jack wanted to have dinner with him. "Well, is this it?" he asked Conaty. "Is this where he tells me I've got to go?" Cote's directness startled Conaty. "Well, you just need to talk to Jack," Conaty replied. Cote had his answer. He flew up to Fairfield and had dinner with Jack in the dining room on the third floor.

"Dave, it's not going to be you," Jack told him. "And I want you out of the company by year-end."

This was harsh. But Cote felt like probing further. "Okay," he told Jack. "But what is it that you saw in me that you didn't like or that you didn't see that you wish you had?"

"You don't understand," Jack said, his voice moving to a higher pitch rapidly. "You need to be out by year-end."

"Jack, really, I'm a big boy," Cote said. "I get it. I think I'm better than you think I am, and if there's something I need to address, I'd like to know what it is."

"Dave, you don't understand," Jack replied. "You need to be out by year-end."

"Okay," Cote said. "I got it."

They spent the next hour eating and talking about anything but GE and succession. Jack told Cote he would help find him a new job. And he did. Jack called both Gerry Roche at Heidrick & Struggles and

Tom Neff at Spencer Stuart. "I have an embarrassment of riches," Jack told them. "Will you help me place this guy?" (It turned out Cote had already been in touch with Roche about a senior job at TRW because he sensed that Jack had soured on him.) Cote kept the news to himself for the next five months. "I didn't want the business not doing well or people thinking I didn't care anymore," he said.

He had job offers from both Waste Management and TRW. He decided to take the job at TRW, which was as chief operating officer. "I thought there was a path to being a CEO of a more multi-industrial kind of company, which is what I wanted," he said. Three weeks after the dinner, Cote saw Jack. He offered him a job at GE Capital. "It's not one of the big jobs," Jack told him. "It's down a few levels. But it's there if you want it." He thought that was a potential safety net but also kind of bullshit. He had been a divisional CEO. To become a general manager of small business at GE Capital was tantamount to a concession that his career was over. At forty-six, he wasn't willing to concede that point, not to Jack or to anyone. He wanted to see if he could become the CEO of a public company. "I want to see what's out there," he told Jack. "I don't want to be sixty years old someday and wonder if I could have. I'd rather know." Cote didn't tell anyone at GE about his sobering conversations with Jack. That's why when Jack came to Louisville to introduce the new leader of the division, there was a look of astonishment on the faces of the Major Appliances employees. In November 1999, Cote left GE for TRW. "It was kind of interesting," he said. "Once you leave GE, you don't hear from anybody anymore, nobody. Even your friends don't call anymore. It was really kind of surprising." When I asked Jack years later about Cote's leadership and management abilities and his capabilities as a CEO, he said, "I underestimated how good he is." We talked some more about Cote and his string of successes, including as the CEO of Honeywell, which eventually had a higher market value than GE. "I just didn't think he was as smart as he turned out to be," Jack said.

AS THE SUCCESSION BATTLE WAS HEATING UP, IN FEBRUARY 1999, JACK
decided to invite Ken Langone, one of the billionaire founders of Home
Depot, onto the GE board of directors. Home Depot was one of GE's
big customers, but that was just one of the many reasons Jack wanted
Langone on the board. Langone also had his own private-equity firm,
Invemed Associates. Like Jack, Langone was a self-made man with a
gargantuan personality. Jack had cornered Langone at a Christmas
party in 1998 at Larry Bossidy's home in Florida, near Jack's in North
Palm Beach. He wanted to talk. "Jack, get off my fucking ass," Langone
told him. "No business tonight. Come on, Jack." GE and Home Depot
had just done a deal for Home Depot to sell private-label GE water
heaters, and he was thinking Jack was "going to beat the shit out of me
about water heaters."

"I need five minutes," Jack insisted. They went to Bossidy's backyard. "The party's inside," Langone said. "He puts me against the fucking wall. He said, 'I want you to go on the GE board.' I said, 'What?!?!'...
So this is big, big, big stuff." But before Langone joined the GE board—
in February 1999—Jack wanted to make sure Bernie Marcus, another
Home Depot cofounder, was okay with the idea. Langone called Marcus. There was no problem. "It's a great honor," Marcus told him. "You
should do it." Langone couldn't believe it. "Let me tell you something,"
he told me, "what breaks my heart about the outcome of the story: to
serve on the GE board at that point was one of the great business accomplishments of my life." He said he was "stunned" to have been asked.
Jack had never mentioned it before, nor alluded to it. "He knows everyone, absolutely everyone," Jack said of Langone.

He wanted his successor to have in Langone what he had had in
Walter Wriston, someone thoroughly plugged in around New York City,
the financial capital of the world, and who could serve as a trusted adviser to the new CEO. GE was at its peak, "the most valuable corporation

ever in the history of corporate history," Langone said. After a few months on the GE board, Jack asked Langone to serve on the powerful Management Development and Compensation Committee, or MDCC, in GE argot. Langone couldn't believe it. The MDCC was the most coveted board committee at GE. Langone figured he was too new to the board to be considered for it.

"I want you on that committee," Jack told him.

"Jack," Langone replied, "they have guys hanging around there for fucking years that aren't on that committee."

"I don't give a fuck," Jack said. "I want you on the committee."

"Put me on the committee, then," Langone replied.

Langone told Jack that other board members would be envious and spiteful. "Jack, you've got a fucking scarlet letter on my head and you've got a bull's-eye on my back that fucking big," he said. "That's okay, Jack. I don't give a shit." Who cares? he thought to himself. "I'm on the board of GE, so I could brag I was on the board of GE, whether it was one month or forty years," he told me. "It's like being president of the United States." It can't be taken away from you once you have it.

But sure enough, shortly after Jack announced Langone's appointment to the MDCC, Sandy Warner, the chairman and CEO of J.P. Morgan & Co, complained. "Jack, Jack," he said, "I was CEO of a major bank. I've been here nine years, and I'm not on that committee?"

"I don't want some fucking banker to decide what my guys should be paid," Jack told Warner. "Langone has no trouble with big numbers." Langone said he knew why Jack put him on the MDCC. "Translation: I'm a serial overpayer," Langone said. "How did I get so rich? Paying people lots of money when they worked their asses off. And I just had a piece of it. That's all I want."

By the time Langone joined the MDCC, the choice to succeed Jack was pretty much down to Immelt, Nardelli, and McNerney. "Loved all three of them," he said. "The aura that it created by this process, and whoever got it, whoever got it was king. But the other two weren't exactly chopped liver." In and around February 2000, according to Langone, Jack heard through his extensive grapevine that McNerney was looking around for other jobs, just in case he wasn't the one Jack selected. "Jack got wind that McNerney was talking," Langone said. "Now, Immelt was talking to nobody and Nardelli was talking to nobody."

"What the fuck are you doing?" Jack said to McNerney.

"I know what I'm doing," McNerney replied.

"Sure," Jack said. "You're out talking about leaving or getting another job."

"Jack," McNerney responded, "you can't promise me I'm going to get your job. I've got to see what's out there."

McNerney shared his perspective on the succession process. "Jack kept it very close to his vest," he told me. "He made it clear to us that—and this was, again, shaped by his experience—he said, 'Look, you guys, you can stay, but I *expect* the two people who don't get the job to leave, because it will screw it up for the person who gets the job.'"

McNerney remembered that he and Immelt and Nardelli would joke about how they were all fielding the same calls from the same top headhunters. All three of them were looking around, although it seemed Immelt and Nardelli did a better job of hiding that information from Jack and the MDCC. "We discovered that all of us were interviewing at Lucent," McNerney told me. "All of us were interviewing at 3M. It was hysterical. We bumped into each other. We were all hedging our bets a little bit."

Jack's brainstorm in the shower one day to install chief operating officers to back up the three finalists—so there would be a continuity of management in the business units after the big decision was made—caused no small degree of further panic inside the company. After he won the board's support for the move, he had to tell the three finalists. "I acknowledged that it might seem unfair," Jack remembered. "But I made the case that it was in the best interests of their employees and our shareowners. Nonetheless, it came as a surprise to them."

Jack was very proud that he had put Dave Calhoun in place to succeed McNerney, John Rice in place to replace Nardelli, and Joe Hogan in place to follow Immelt. "Putting these three stars into these new jobs was a game-changing event," Jack explained. "While 300,000 employees, including me, still didn't know who their chairman was going to be, the people in three of our biggest businesses knew exactly who their new CEOs would be."

He then tried to keep the whole matter of succession quiet. "I continued to turn away reporters who wanted to talk about these moves or anything else involving succession," Jack insisted. "So did Bob, Jeff, and Jim. None of us wanted to help the media make a circus of it." But it was the biggest business story going, by far. The race to succeed Jack was already big news, but after he appointed the three new chief operating officers in Power, Jet Engines, and Medical Systems—thinking he was so clever—the press speculation about his successor ratcheted up exponentially. Within days, there were detailed articles in the business press about the race, filled with plenty of speculation about who was up and who was down, "without any help from us," Jack noted. A late-July 2000 article in *Wired*, of all places, was typical of how focused the press was on who would succeed Jack. "Forget the U.S. presidential race," the magazine reported. "The year's most compelling contest, certainly in the business world, is the race to become the next leader of General Electric."

## Chapter Seventeen

# HABEMUS PAPAM

Things were finally starting to come to a head about succession at the July 2000 board of directors meeting. Jack met with the MDCC—Management Development and Compensation Committee—for nearly four hours, "wrestling with the pros and cons of each candidate," Jack recalled. "It was a wide-open meeting. . . . I was fighting not to make up my mind. I wanted to keep my options open to the very end." As usual, after the conversations, many of the board members played golf with Jack and the business CEOs and then headed off to dinner. The next morning, they spent another ninety minutes talking about succession. In the previous six years of debating Jack's successor, Jack and the board had never before suggested a name. Unable to take it any longer, Frank Rhodes, the president of Cornell University, who had been on the board for sixteen years, suggested to the board that Immelt was the One. "Good," Jack replied. "That's what I think, and that's what Dennis [Dammerman] and Bill [Conaty] think." But despite a growing consensus building around Immelt, Jack and the board still did not want to make it official, not then, anyway. "Even in the summer and fall we said, 'We've still got time,'" Conaty said. "And the other guys were doing fabulous things."

But Langone said the decision had basically been made in July, if not before. He said that's when Jack told the board that McNerney had been "shopping around" and that "effectively we knew we were down to two." He said that after Jack heard about McNerney looking for a job outside GE, McNerney was effectively "out of the loop" and out of consideration. He said Nardelli was never in consideration. "Nardelli was never in the hunt," he said. "He was a stalking horse." Immelt was going to be the next GE CEO. "[Jack] knew who the one was before McNerney was out," Langone said. "But he wanted this aura, this crescendo, this magnificent moment."

FOR HIS PART, IMMELT DIDN'T DARE THINK OF HIMSELF AS A SERIOUS contender until June 2000 when Jack called him to his office and told him, "You're one of three people who are going to get my job. And if you don't get it, you have to leave. That was probably an exclamation point." Immelt said he was friends with McNerney and respected Nardelli, although he didn't know him as well as he knew McNerney. "I really didn't have a personal relationship with Bob," he told me. "We had never really worked together. We had crossed once very briefly in the appliance business. But his numbers were great. It really seemed like he was doing a good job. And with Jim, I would say we were friends. We hung out together at meetings, and I knew him when he was in Asia. I had spent some time there in the businesses I was in, and I'd always stop in and say hi."

At times, Immelt said he thought he was too young to get the job. "If on D-Day, Jack had said, 'Hey, we're going to give Jim or Bob the job, and we'd like you to stay in some big capacity,' I would have said yes. When I sat in Jack's office, and he said, 'You're one of three, and if you don't get it, you have to leave,' my response to him was 'You're kidding.' That was like a punch in the stomach. It wasn't like if he had said, 'Hey, you've been one of three, and you're not going to win,' I would have felt like, 'Okay. Let's go.' But the notion of leaving GE was like a kick in the nuts." He said that although forcing losing candidates to leave GE was "not something I would choose to do," he "never much doubted" Jack's "motives or his reasons why. I had complete trust in him."

D-Day was the October 29 board meeting. On that Sunday evening, Jack convened the GE board in Greenville, South Carolina, the home of GE's power systems business. During the day on Sunday, Jack played golf with a group of current and former directors at Augusta National. Afterward, the group flew to Greenville for dinner in a private room at the Poinsett Club, in what had once been the imposing private mansion of a wealthy textile executive, built in 1904. It was the first word in exclusivity, in Greenville, anyway. The idea for the next day was to have the board visit a turbine power plant in Nardelli's backyard. That was kind of mean. "It probably wasn't the smartest move I ever made," Jack remembered. Jack feared it would put additional pressure on Nardelli at a time when Nardelli's blood pressure was already skyrocketing. But he did it anyway. Plus, the meeting had been scheduled for Greenville a year in advance. "I liked to show the board stuff that was hot," Jack continued. "Thanks to Bob, nothing was hotter than power systems." Nardelli put on "a great show" for the directors, Jack recalled. There was also a Jack sideshow: He was on *60 Minutes* that Sunday night, interviewed by Lesley Stahl. She watched him give a talk to future leaders at Crotonville and interviewed him at his home in Nantucket. She also interviewed Jane, his wife, who said only glowing things about her husband, never once betraying the underlying trouble in their marriage.

In his book, Jack portrayed the dinner at the Poinsett Club as yet another gathering of a happy family. But it wasn't. Jack wanted to address "head-on" the "schism" that had opened up on the board between the McNerney and Immelt camps. (It seemed not all board members had received the same sense as Langone that Jack had already eliminated McNerney.) "Jack seemed to be uncommonly nervous during dinner," Bob Wright remembered. By that time, Wright was not only still the CEO of NBC but, since July 2000, also a GE vice-chairman and board member. He had also spoken with Jack before the Greenville board meeting and asked him who he wanted as his successor. Jack told him he wanted Immelt and asked Wright for his support. Wright told him he would back Jack up. At the Poinsett Club dinner, Jack called in the promise. "Without any chitchat," Wright continued, "he began an informal discussion about succession." Jack turned to Wright and asked him who was his choice for the next chairman and CEO. "I was being thrust in the role of kingmaker with no time to think," he said. He said all three of the candidates had talent, "but in the end it was really a choice between McNerney and Immelt." He thought both men "were extremely qualified" but recommended that Immelt get the job. "Jack gave a sigh of relief," Wright recalled. He remembered that his statement about Immelt "seemed to cut short any debate on the matter" because Immelt was clearly Jack's preference. Jack wanted to choose a younger man who could reign for a long time, as he had done, and according to Wright, Jack believed Immelt "possessed a solid mind" and the "skills" to manage GE's diverse set of businesses.

In an interview with me, Wright elaborated: "It was clearly a choice between McNerney and Immelt. Jack wanted Immelt. I didn't feel that strongly that that was a bad decision. I think if I'd had my own choice, I would have gone with McNerney"—he added that "McNerney turns out to be a superstar and he proves it over and over again"—"but I'd been with Jack for twenty years and he asked me to support him and I said I would. I don't think I was doing any stupid thing. It wasn't like I was supporting a bum. He had leadership characteristics. He's smart. He had a lot of ability to analyze stuff. He had credentials."

Wright added that he had worked with Immelt—a fact that helped him with his decision—and had never worked with McNerney. On the other hand, he said, choosing Nardelli, whom he described as both a friend and a "Little Caesar," would have been a terrible mistake. "He was only a candidate because of the bubble in [the] power [business]," he said. "He got into a bubble and he took advantage of it. He did a hell of a job on the manufacturing side. But the other guys I thought were broader-based."

After *60 Minutes* and the dinner at the Poinsett Club, Jack and the board retreated to a nearby Hilton for a special board meeting. The only non–board member in the room was Bill Conaty. GE security guards were posted outside the conference room to ensure privacy. Sometime after 10:00 p.m., Jack announced, "We've come to a conclusion." For the next fifteen minutes, he extolled the virtues of Jeff Immelt. He had done "great things" at GE Medical. He was the "perfect blend" of "intelligence and edge" and "epitomized" a trait Jack valued highly, being "really comfortable in his own skin." The decision was a tough one, Jack told the board, but Immelt was "the perfect selection." Dammerman and Wright then offered their support for Immelt. Every board member spoke up, "all in unanimous approval," Jack recalled.

He said Frank Rhodes, the Cornell president, "spoke eloquently" about Immelt's capacity to learn and to grow and about his "intellectual bandwidth." Immelt, Rhodes said, was "clearly the right choice." Jack asked Wright to second the nomination of Immelt as his successor. Wright did as Jack asked. "I recognized that Immelt was Jack's candidate, and part of my role was to be supportive," he recalled. There was some discussion about whether choosing Immelt meant losing McNerney and Nardelli. Two board members hated the idea of such talented executives leaving. But Jack stuck to his guns. "I've been there," he said. "I know what it's like. Whoever becomes chairman of this company has to be filled with self-confidence and full of enthusiasm. I want him to feel bigger than life. I don't want him looking over his shoulder." The meeting ended at midnight. Although the decision had been made, Jack wanted the board to sleep on it some more. "Take three weeks to think about it," he told them. "Call me with any concerns you may have." He said he would ask the MDCC for its final endorsement the Wednesday before Thanksgiving.

JACK WANTED TO DELAY THE ANNOUNCEMENT UNTIL AFTER THE Thanksgiving holiday because he figured, correctly, that the holiday break would provide perfect cover, given everyone's preoccupation with their travels and their families. There was also the ongoing vote recount in Florida that would determine the outcome of the 2000 presidential election. Most close observers were expecting Jack to make the announcement after the regular December 15 board meeting. On the Wednesday before Thanksgiving, Jack called the members of the MDCC and secured their approval to recommend Immelt's selection to the full board two days later. Rosanne Badowski called each of Immelt, McNerney, and Nardelli. "Let us know how to reach you," she told Immelt. "It's not just going to be you. We're going to reach all three of you guys over the weekend. Nobody knows what's going to happen. But give me a number." Jack spent Thanksgiving at his home in North Palm Beach. Immelt was at his home on Kiawah Island, South Carolina, his version of what Nantucket was to Jack. After the stock market had closed on Friday, November 24, Jack called the GE board of directors. At 5:00 p.m., he asked them to ratify the selection of Immelt, which they did "unanimously and wholeheartedly," he reported in his memoir. That wasn't precisely true.

According to Ken Langone, Paolo Fresco, a GE vice-chairman who had been on the board since 1990, objected to the selection of Immelt, just as he had objected to Jack's selection twenty years earlier. "Paolo was very emphatic McNerney was the guy," Langone told me. "Very, very emphatic. He said, 'Jack, you're making a big mistake.' He was the only guy to push back."

Although he wasn't asked, Denis Nayden, at GE Capital, agreed with Fresco: McNerney was the best choice. "It was McNerney hands down," he told me. "Not even close. As a leadership matter, as a business matter, as a reform matter [and] results . . . All I heard was that Jack thought that Jeff was viewed as a great salesman—I have no comment on his sales record, that's whatever the record is—but he was viewed as a great sales guy. Jack's view was he was a great people person, which couldn't have been farther from the truth. That was a fatal flaw. He wasn't a good people person as a leader. He didn't listen to people, he didn't establish relationships, he wasn't willing to engage. It was his way or the highway."

Nayden said Jack was as "equally opinionated" as Immelt but was willing to have his mind changed. He remembered a time, at a GE Capital board meeting, when Dave Nissen, the head of the consumer finance business, proposed that GE Capital buy an auto finance company in Thailand. It was based on an idea from Mark Norbon, who headed up GE Capital's office in the country. Jack was looking through the board book the night before the meeting and he spotted Norbon's memo. "This guy, this nut, wants us to invest $1 billion in autos," Jack recalled thinking to himself. "I'm going to blow him out of the water tomorrow. I mean, this guy's got no chance."

Nayden remembered the meeting clearly. "Jack walks into the boardroom," he recalled. "We haven't even said good morning, and he goes, 'Are you guys fu-fu-fu-fu-fucking kidding me?' That's how the board meeting started. Two hours later, after listening to the discussion and after asking a hundred questions, having everybody around the table ask questions and answer them, Jack goes, 'Okay, let's do it.' He always had a point of view, but he also had twenty questions and he always kept an open mind. And he wanted to engage with the people proposing to invest, proposing to take risk. He wanted to test the mettle of your argument, and then he would make up his mind."

Nayden said Jack wanted smart board members and then would

listen to them. "Jack was clearly one of the smartest guys ever," he said. "But he didn't behave like he was the smartest guy ever. He wanted to know what Dammerman thought. He wanted to know what Dave Calhoun thought. He wanted to know what McNerney thought. He listened to them." Jack recalled why he changed his mind about the Thailand idea. "It's almost a perfect example of how it works," he said. "We were convinced. If you asked me five minutes before the meeting, I'd say, 'Get out of here. What are you, nuts?' And yet he made a great case, and you love him for doing it. You love him for knowing he's walking into the jaws of trouble."

But there were other times when Jack could be ruthless. There was a famous story about Jack and Anders Maxwell, who worked (with me) at Lazard before leaving to join GE's equity investment business. Around 1993, as an investment, GE bought around a 30 percent stake in Dreyer's Grand Ice Cream company. Half of GE's equity stake in Dreyer's was owned by GE's pension fund, the other half by GE's equity investment arm, which was part of GE Capital and reported to Wendt. It was the only investment GE Pension and GE Capital did together. Maxwell was running GE Equity at the time. "He's kind of crazy," remembered someone who worked for Maxwell. One day there was a meeting with the Dreyer's CEO in the office of John Myers, the head of GE Pension. Combined, GE had about a $50 million investment in Dreyer's, which was a considerable investment in those days. Dreyer's had gone public and was soon to be bought by Nestlé. The deal was a big success for GE.

An attendee at the Dreyer's meeting remembered that Maxwell came into the meeting with a stack of unopened mail. He then announced his view that the "management of this company are idiots" because Dreyer's EBITDA (earnings before interest, taxes, depreciation, and amortization) had not grown in a few years. That was not a universally held view among the GE executives. Some in the room believed that the Dreyer's management was actually "incredibly strategic and shrewd" because they had invested in a massive electronic distribution network inside the stores, enabling them to tell immediately when their products were out of stock. They could tell how much of a product was selling on any given day. "It was way ahead of its time," remembered one GE executive. "And they used all of that information ultimately to quintuple the business." Maxwell disagreed. "These guys were idiots," he said. "I can't even be bothered with these guys." There was concern that the meeting could be a disaster given Maxwell's view of the Dreyer's management.

When the Dreyer's CEO came in, Maxwell started going through his mail. He used a letter opener, which of course made a distinct, and distinctly annoying, sound, especially in a business meeting. It was blatantly disrespectful. "He'd look at the mail and stuff and then he'd put it down and he'd open up another one," remembered a GE executive. "And I'm looking at him like 'What an asshole. I can't believe he's doing this.' Like I was humiliated to even be with this guy." After the meeting, "as luck would have it," the Dreyer's CEO had a lunch scheduled with Jack in his Fairfield office. The Dreyer's CEO worshipped Jack, especially since GE had come to the rescue when the company was the focus of an unwanted takeover bid. He told Jack about what Maxwell was doing at the meeting. "I get a phone call from Welch," Myers remembered. "He says, 'John,' he says, 'Were you just in a meeting? Tell me what happened.' I tell him what happened. And Welch's quote was 'I've been looking for a reason to fire that son of a bitch. Gary [Wendt] loves him. He's out of here.'" Myers called another member of the deal team, Jon Sprole—no fan of Maxwell's either—and told him, "Your ship has come in, your ship has come in. You're going to buy me a dinner." Sprole asked Myers what he meant. "Just wait," he said. "You'll find out." Jack fired Maxwell the next day.

In other words, Jack was used to getting what he wanted. And Jack wanted Immelt, that's all there was to it, doubters be damned. "Jack was quite enthusiastic about Jeff," Langone recalled. "Wasn't like 'shit.'

And here was the most watched search process, the most watched succession process, ever. This was four or five years focusing on 'What happens when Welch is gone? What happened when the king, the real king, dies? Who's going to become the king?'" Shortly after the board vote, at five thirty, Jack called Immelt in South Carolina. "The board made a decision," he said. "It's great news for you. I'd like you to come down to Palm Beach tomorrow. Bring your family and we'll meet you at noon for lunch."

Immelt said he will always remember the phone call from Jack. "That was when I knew I had the job," he said, without irony. More than anything he was relieved. He was relieved that he got the job. "I was ready for either outcome," he said. "I was happy to have it over, and clearly happy, again, because I didn't want to go do something else. Most of my friends, most of my social life, was GE. That was my whole life. And I wasn't quite ready to start something else then. But I was ready for either outcome. I had no expectation." He was also relieved he would not have to leave the company. He remembered what a difficult moment it had been when Jack told him, and the other two men, that whoever didn't succeed Jack would have to leave. "My wife kind of got teary-eyed, like, 'Gosh, why do you have to get fired if you don't get the job?'" he said. "I was just trying to explain it to her, but I had a hard time explaining."

All along he had figured that McNerney was the leading candidate for the job and that Nardelli never had a chance. At GE, Immelt said, "your peer relationships are really critical, and Bob never had any peer relationships. Jim had good peer relationships. He had really worked on it and was a good guy to hang around with."

It was time for some corporate sleight-of-hand. Immelt, his wife, Andy, and his daughter, Sarah, had to get from Kiawah Island to North Palm Beach without attracting media attention. Instead of using a GE corporate jet—too obvious—GE chartered a jet, which was booked under the name of James Cathcart, Si Cathcart's son. The plane flew to

Stuart, Florida, as opposed to West Palm Beach, where GE jets usually landed, to further confuse any snoopers. Cathcart arranged for a car, in his name, to take the Immelts to Jack's house. "As the car pulled up, I was in the driveway, waiting to greet him and give him the great news," Jack recalled. From there, they had lunch together at Carmine's in North Palm Beach. Afterward, Jack, Immelt, and Conaty, who was also in Florida for the weekend, worked on the announcement process and materials. That night, the Welches, the Immelts, the Dammermans, and the Wrights dined together at Jack's home. "We had a great night together," Jack noted.

Armed with the confidential information that Immelt had won and that both McNerney and Nardelli had lost, Ken Langone called Jack. Langone had been searching for a new CEO at Home Depot, to replace Arthur Blank, one of the Home Depot cofounders. "Jack, okay," Langone said. "Now you've decided. Are you okay with me trying to talk to Nardelli?" Jack was pissed. "Stay the fuck away," he yelled at Langone. "I don't want anybody to know anything. You don't do a fucking thing until it's announced, until we let the world know he's got the job."

"Okay, okay, okay," Langone replied. "But what I want you to know, because we've been out looking for somebody for Home Depot. There was nobody out there like these two guys, McNerney and Nardelli." Langone agreed to wait, for a few days, anyway.

On Sunday, it was time for Jack to face head-on one of the most difficult tasks of his long career at GE: telling both McNerney and Nardelli that they didn't get his job. He waited until 2:00 p.m. that day to make the calls. Both men were at home, Badowski having previously procured their schedules. "The board and I have made a call," Jack told each of them, having rehearsed what he would say to them over and over. "I'd like to come out and review the decision and the rationale behind it." It was a tough call for all three of them. Jack didn't want to give either man false hope or to deliver the fateful news over the phone. He believed he owed each man a conversation in person. But there was no mistaking the message, even if it was not made explicit over the phone. McNerney and Nardelli weren't idiots. "If he was going to make me the guy, it probably would have been a different call," McNerney told me.

At 3:00 p.m. that Sunday, Jack went to the West Palm Beach airport in the middle of a "torrential downpour," he remembered. The weather that afternoon made holiday air travel along the East Coast especially miserable. Jack compounded the difficulty by telling the GE pilots he wasn't going back to Westchester. He was actually off to Cincinnati to see McNerney. "They were shocked," Jack allowed. The whole flight plan had to be refiled at the last second. Between that and the bad weather, Jack had to wait two hours before the jet could take off for Ohio. That gave Jack plenty of time to think about the process he ran and the changes he had wrought in the lives of these three men. "I hated what I had to do," Jack remembered. "It was like having to pick one child over another. It seemed so unfair. They had all busted their butts for the company. They had never played unfair with me or with each other. They had given 1,000 percent." All three men, he continued, "had vastly exceeded our expectations," and "now I had to give two of them the worst news of their careers."

The GE jet took off from West Palm Beach at five thirty and got to Lunken Airport, in Cincinnati, ninety minutes later. "The place was soaked, dreary and dark," Jack recalled. "It was a bone chilling night." Jack felt alone, facing a most unpleasant task. When he entered the general-aviation building, McNerney was already there. Jack could see the disappointment on his face. "Obviously," he told him, "this is going to be the toughest conversation of my life."

He spoke frankly. "I picked Jeff," he said. "If there's anyone to be mad at, be mad at me. Put my picture on the wall and throw darts at it. I can't even tell you why. It's my nose and my gut. We had three Gold Medal winners, and only one Gold Medal to give." Jack did not mention to McNerney that he resented the fact that McNerney had been looking for another job as the process was winding down. Jack took that as a sign of disloyalty and insecurity. But it was just a smart, if impolitic, move on McNerney's part. Jack also did not mention that he didn't think "McNerney was smart enough," as one GE executive told me. The GE executive said that Immelt, McNerney, and Nardelli were each asked to make a presentation to the board "in the final innings" and that McNerney's hadn't gone well. "McNerney just choked," this executive said. "He did a bad job, fumbled it, and almost froze. Immelt's a salesman; McNerney's not a salesman."

As the Florida recount was still going on, McNerney joked about whether he could have a recount, too. "I want you to know I wanted the job," he told Jack. "But I also want to tell you I think the process was fair because you played it straight and gave us every chance." For the next forty minutes, they reminisced about McNerney's upbringing and his successful career at GE, especially his last two years transforming the jet engine business. "It was a lot of reminiscing," he said.

McNerney tried to look on the bright side. "Jack and I, we like each other," he told me. "He made me feel as good as you could make someone feel in these circumstances. He was very open, and I tried to be as gracious as I could." Jack said if McNerney wanted to stay, he could be a vice-chairman, a suggestion that contradicted what the plan had been all along. "Do whatever you want," Jack told him, "but you've got a big future ahead of you if you do leave." McNerney declined to say whether he thought Jack made the right decision or not. They remained friends and spoke frequently during McNerney's successes running both 3M and Boeing. "It turned out to be more than fine," McNerney said. "Losing a competition is one thing, but, on the other hand, I was still young enough that I could go run two great companies and, if I'm being modest, do a pretty good job running both of them and having great experiences. I think if I'd stayed at GE, it would have been a tough job, as has been shown. It wouldn't have been easy. Do I think I could have done it well? Yeah, I think I could have done it well."

Back on the plane, Jack told the pilots he had yet another surprise for them. They were off to Albany to see Nardelli. They arrived earlier than planned, despite the bad weather, at around 9:00 p.m. Nardelli had not arrived yet. Jack was relieved to have some extra time to think. Telling Nardelli he didn't get the job was the tougher of the two conversations. Jack had known him the longest, and Jack could not get over the job Nardelli had done at Power Systems. He had put up the best numbers, "and GE is a numbers company," he said.

Nardelli showed up at the appointed time. "I went with my heart in my throat," Nardelli recalled. Now, late on a stormy Sunday night, the two men sat together in an otherwise empty lounge at the private jet office at the Albany airport. "How do you describe this?" Nardelli continued. "It's something you strive for for 30 years. You're hanging on every word. You're focused on his mouth, and the final words: 'I've elected to give it to Jeff.' And you're like, 'Did I really hear it right?'" It was a bleak scene.

"What more could I have done?" Nardelli asked Jack.

"Bob, you've done more than I ever would have dreamt," he told him. "You've done a great job. Everyone loves you, and you're going to be a great CEO. But I can't answer this question for you. I can't give you satisfaction on it. You did everything and more that was ever asked. I believe Jeff is the right guy for this company going forward. There's only one person to blame here. It's me."

They went round and round some more. Nardelli didn't understand why he was not chosen, given his excellent operating results. The decision had been made. It was over. "Bob, you're going to be an all-star CEO," Jack told him. "There's a big lucky company out there waiting to get you." They shook hands and hugged.

Jack told me that Nardelli almost broke down in tears at the news and that he still had not accepted Jack's decision. Nearly two decades later, Nardelli remains embittered that Jack chose Immelt. Those close to him told me he remains angry and resentful. Ken Langone, for one, said that Nardelli has never gotten over not getting the GE job. "Bob Nardelli woke up this morning, this day, his first thought is, and was, and always will be 'I should have had the GE job,'" Langone said. "Because as of today, no matter how well he could have done, or would have done, or should have done, it would never make up for the fact that he's impaired because he didn't get the job."

But Nardelli's done talking about it, at least publicly. He declined my repeated requests to be interviewed. "Bob does not want to be perceived as a 'Monday morning quarterback,' where it appears that he is second-guessing decisions made by those who were responsible for the company during the past 16–17 years," his spokesman wrote to me in an email. "Bob's career started at GE, where he enjoyed great success. He is very proud of his contributions to GE and of the dedicated people who made the company great. Bob has continued his relationship with many GE executives over the years, and he considers himself a lifetime member of the GE family, comprising thousands of GE associates who have contributed to GE's legacy."

Back on the GE jet and finally heading back to Westchester after a day that had ricocheted between exhilaration and tristesse, Jack "ordered a large vodka on ice" and thought about what he had just unleashed. "I was relieved it was over," he remembered. "I was thrilled for Jeff and totally confident we had picked the best candidate." But, he continued, "I felt really sad to disappoint two friends who had done so much for the company." He "vowed" to help them any way he could to get their next jobs. It wouldn't take long.

Jack's plan was to announce Immelt's appointment at 8:00 a.m. the Monday morning after Thanksgiving at a press conference at 30 Rockefeller Center. At 7:30 a.m., CNBC broke the news that Immelt would succeed Jack. That was not the plan, of course, but a leak to GE-owned CNBC was perhaps inevitable, after years of keeping the information about succession tightly controlled (and preferable to the news being leaked to a CNBC competitor).

The night before, Langone had flown to California for a Monday meeting. He liked to keep himself on East Coast time, especially during short business trips. That meant waking up at 3:00 a.m. Pacific time. "CNBC is live," he said. "They fucking announce Immelt got the job." That was Langone's dog whistle. The race was on. He would have been happy to have either McNerney or Nardelli replace Blank at Home Depot. But by then the buzz on Wall Street was that McNerney was already off the market and heading to be CEO of 3M. That left Nardelli. Langone immediately called Gerry Roche at Heidrick & Struggles. Langone recalled: "I said on the fucking phone, 'Get Nardelli right this minute, and tell him I want to talk to him. Because all bets are off. It's wide open.' That was the deal." Roche called him back and said he had spoken with Nardelli, who was waiting for Langone's call. "Bob," Langone told Nardelli, "we think you're the guy. We have a great opportunity for you. You can make a lot of money. I think you can have a great business. You bring everything to this company that we need. And we'll give you every opportunity you should have."

At the press conference at Rockefeller Center, which CNBC mostly covered live—there was also the ongoing vote count in Florida dueling for attention—Immelt shared the stage with Jack and the two GE vice-chairmen, Dammerman and Wright, who would join Immelt in the executive suite. Each of the four men wore blue blazers and collared shirts without neckties. "In true GE fashion," Beth Comstock remembered, "we had managed the PR and media efforts with the gusto I imagine goes into a papal transition." There were plenty of platitudes and much backslapping, of course. "This is a time of change," Immelt said, "but GE people have always embraced change. The company has never been in better shape than it is today. We have great global businesses and our initiatives have made us faster and put us closer to the customer. And I think this is going to be a time period of unprecedented growth for GE. I'm following a pretty good guy here in Jack Welch, and I look forward to the next year and learning from Jack."

He also said he looked forward to working with Dammerman and Wright.

He praised Jack. "Jack's got a tremendous legacy at GE, but I think the thing that's the best part are the people, the culture, and the focus on the future," Immelt continued. "And that gives me great confidence as I sit here today. We simply have the best team in the world. And I think as we go through this change, this team is going to thrive, and I think all of us believe that the best days for GE are truly ahead of us."

And then Jack praised Jeff. "I've had the good fortune to know Jeff for twenty years or so," Jack said, "and have watched him grow every job, every stretch assignment. He just grew bigger and bigger. But if I really had to say what, to me is the real hallmark of Jeff, it is his focus on people. . . . Every time you go into a meeting with Jeff and his team you can't quite decide whether it's a meeting at the UN or if this really is a business review. It's so global, it's so diverse, and it's so good. And in today's GE, you cannot run today's GE without almost an obsession about people." It was quite a performance.

One questioner in the audience noted that when Jack took over from Reg Jones twenty years earlier, Jones was the most admired CEO and GE the most admired company in America. The day after Thanksgiving 2000, Jack was the most admired CEO of the century and GE was still one of its most admired companies. GE was also the most valuable company in the world, with a market value of $500 billion. What did Jeff think about the legacies he was assuming, and how did he plan to keep the company's streak intact? "I've had about forty-eight hours to think about this in great detail, so it's still a little bit early," Jeff replied. "But I think what you can count on from me is that the standards of performance won't change. I think that's the bedrock of GE, and that will continue with me just as it exists with Jack. I think the values that GE stands for are incredibly important. That's really what binds us together." He said globalization, Six Sigma, and digitization would continue to be among his top priorities. He said he intended to have GE remain the most admired company and to have it be a leader in making change and in adapting to it.

Jack then chimed in about the importance of GE's "values" and how Immelt, McNerney, and Nardelli had displayed them throughout the grueling succession process. "Values sounds soft to a lot of people," Jack said. "It sounds like they're mushy. But we went through a transition that the press made somewhat of a spectacle out of with three candidates. And these three people never once did a political act in the company."

One of the lingering questions was: Why did Jack pick Jeff? Was it his age? one journalist wondered. Jack replied that picking someone who could be in the job a long time was important. "I believe it to my toes," he said. "I believe that you make a mess, you clean it up, and you learn from it." Jack was asked when he realized that Jeff was the choice and also to share the qualities he believed Jeff had that the other two men didn't. There was no way Jack intended to answer that question directly. "We had three gold medalists, okay?" he replied. "We have been working for a long time to work on picking the right person. This isn't about what one had, and one was tall or one was short or one had more hair or was younger. It was a whole bundle of characteristics that make us so proud of Jeff and think he's just the right person to lead us forward. We think that the other two could have led us well, too, and could lead other companies well, too." He said he, Dammerman, Wright, and Conaty had been moving in Jeff's direction for the past sixty days, "and it got serious in the last three weeks." Jack was asked about Jeff's compensation package. "He got a helluva raise, I'll tell you that," he said. And Jack was asked what choosing Jeff Immelt as CEO meant for GE Capital, then about half the company's revenue and earnings. Jeff replied that he had been on GE Capital's board for three years and had a "good working relationship" with the team there.

Before the news conference ended, Jack was asked about how the four men ended up wearing the same uniform of blue shirts and blue blazers. "Jeff didn't have any clothes," Jack said, to laughter, "and his wife went shopping for him this morning. He had a brown jacket . . . and she didn't like it, so she bought him a blazer, so he ended up that way. He couldn't get home. We were all kind of flying around the country last night to do things, the weather was terrible, it was Thanksgiving, we were changing flights on planes and things, and it took us several hours, it took him several hours to get out of where he was. . . . We ended up in the corporate costume without trying."

Within days, both McNerney and Nardelli had been snapped up by other companies. First, on December 5, 3M announced that McNerney would be its new chairman and CEO. He said the only two jobs he was serious about considering were the ones at 3M and Lucent. The next day, in what was more of a surprise, Home Depot announced that Nardelli would be its new chairman and CEO, replacing Blank. "There was nobody out there like these two guys—McNerney and Nardelli," Langone said. "And by the way, to be fair, the first four years Nardelli was at Home Depot, the first four years, everything he touched he made better. Every single thing he touched, he made better."

Jack was thrilled with the way everything had gone down. He had the man he wanted as his successor, and the two men who had lost the race had become the two most sought-after and valuable executives in the marketplace. In a ten-day stretch, Jack had succeeded in naming his successor to general enthusiasm and in finding for the two others new jobs as CEOs of two important American companies. Jack celebrated by inviting them all, as well as his predecessor, Reg Jones, to GE's annual Christmas party at the Rainbow Room, atop 30 Rockefeller Center. "When I mentioned them during my remarks," Jack recalled "our directors and executives gave them standing ovations. No one clapped harder than I did." The next day, the GE board, along with its newest member, Jeff Immelt, met for its quarterly meeting.

The Jeff Immelt Victory Tour continued in early January 2001 at the annual GE manager gathering in Boca Raton. It was Jeff's first as CEO-elect and Jack's last as CEO. Jack gave some heartfelt opening remarks and then flew off to Austin, Texas, to meet with President-elect George W. Bush, who had asked a select group of CEOs to meet with him. It was the first time in thirty-three years that Jack had missed a day of the Boca meeting. "It gave Jeff the chance to do his own thing, without me sitting in the front row," he said. Jeff repaid that confidence, telling the GE faithful, "Only time will tell if Jack Welch is the best business leader ever, but I know he is one of the greatest human beings I have ever met." When Jack returned to Boca that evening, a video copy of what Jeff had said was waiting for him. "Seeing him take command of the company was exhilarating," he remembered. "Jeff was witty, smart, visionary and incredibly powerful. He was the CEO!" He was especially impressed when Jeff told the Boca crowd, "Everybody at GE thinks they work for Jack; every customer thinks they buy from Jack; every political person thinks they deal with Jack. So, the job that Jack and I have for 2001 is to be sure that at the end of the year everybody knows they work for me and everybody knows they buy from me and everybody knows they deal with me."

In his closing comments the next day, Jack shared with the GE managers that he felt the pride of a first-time father watching Jeff onstage. He recalled how he felt walking into his office in the plastics division of GE, in Pittsfield, thirty-nine years earlier with a box of candy to share after the birth of his first child, Katherine. He said he felt like he had been competing with Jeff's father, who had spent thirty-eight years at GE, to "see whose chest could pop out the most." He said he was sure the "new guy" was the "right guy."

THE PLAN WAS FOR JACK TO FORMALLY HAND OVER THE REINS TO JEFF in April 2001. But it turned out not to be all golf matches and accolades for him in his final months on the job. Back in October, just as he was pushing Jeff over the finish line, Jack had felt the need for one last transformative transaction, one last legacy-making deal. It started on

October 19 when he was visiting the floor of the New York Stock Exchange. He was there helping an old friend, Azim Premji, the CEO of Wipro, an Indian internet provider, celebrate the initial public offering of Wipro's stock. As the two men were wandering around the floor—back in the days when trading occurred there—Bob Pisani, the CNBC reporter, shoved a microphone in Jack's face and asked him what he thought of the news that United Technologies, the hated GE rival, might be on the verge of buying Honeywell.

"It's an interesting idea," Jack said.

"What are you going to do about it?" Pisani asked.

"We'll have to go back and think about it," Jack said.

The news shocked Jack. He had no idea that United Technologies wanted to buy Honeywell, which was the successor company to the giant merger between Allied-Signal and Honeywell that had been completed in late 1999. Jack's old friend Larry Bossidy was the chairman of the newly merged Honeywell. GE had considered an acquisition of Honeywell earlier in 2000. The two companies were complementary, especially in avionics, aircraft engines—where Honeywell made engines for small jets—industrial systems, and plastics. Honeywell had $25 billion in revenue—about a quarter of GE's revenues—and 120,000 employees. It would have been a big deal, and an expensive one. With Honeywell's stock then trading between $50 and $60 a share, Jack and his team decided, in February, it was too expensive. But after Bossidy retired in April 2000 and his successor, Michael Bonsignore, had warned Wall Street that Honeywell would miss quarterly estimates, the Honeywell stock had plunged to $36 a share, making it both more affordable and vulnerable to a takeover.

United Technologies' offer for Honeywell, in October, was $40 billion, a little more than $50 a share. After the bid leaked on CNBC at around two o'clock, McNerney and Dave Calhoun, his new deputy, called Jack at the stock exchange and told him GE had to buy Honeywell. "Let's do it," Jack told them. After Pisani had surprised Jack on the floor of the stock exchange with the news of the Honeywell deal, Jack was jacked. He immediately started calling his key board members. He reminded Si Cathcart that GE had analyzed the acquisition earlier in the year and that even at a price of around $50 a share, the deal looked attractive to him. He called Dammerman, then a vice-chairman and head of GE Capital, and ordered him and his team to New York City to begin hashing out the details of an acquisition.

Jack's competitive juices were flowing. He was in deal heat. He called Bill Harrison, the CEO of JPMorgan Chase, and asked him if Geoff Boisi, the former Goldman Sachs M&A partner who was then head of investment banking at JPMorgan Chase, was available to work on a GE deal for Honeywell. He was. He and his team headed to GE's Rockefeller Center office. Then, "since we were in the middle of choosing my successor," Jack allowed, without irony, he called each of Immelt, McNerney, and Nardelli and let them know "what we might do." McNerney and Calhoun were especially enthusiastic about a Honeywell acquisition, given how well Honeywell's avionics and jet engine businesses would fit with GE's jet engine business.

The next morning, Friday, October 20, Jack reported, "teams of GE people piled into a couple of helicopters in Fairfield loaded with data from previous internal reviews and came to New York" to meet with him and Dammerman and Boisi from JPMorgan Chase. McNerney, Calhoun, and Lloyd Trotter, the head of the industrial controls business, joined the discussion via teleconference. They quickly decided to make a bid for Honeywell and believed GE could offer Honeywell more money and more certainty than could United Technologies. And, of course, since GE was GE, it had to win.

Part of the reason Jack was so keen on the deal was because of the fit Honeywell's businesses had with GE's industrial businesses and part of the reason was to try to dilute the impact GE Capital continued to have on GE's profitability. Buying Honeywell would mute the assertions that GE was simply becoming a big bank with some fancy industrial businesses attached to it. "Jack wanted it in the worst way," Conaty told me, "and it would have been a nice legacy play at the end."

One possible impediment to a GE deal for Honeywell emerged nearly immediately: Jack's previously announced decision to retire at the end of April 2001, five months after his sixty-fifth birthday. Jack decided that he was going to have to stay longer at the helm of GE to oversee the integration of Honeywell. "I couldn't throw an acquisition like this on top of a guy in a brand-new job," he recalled. "On the other hand, I couldn't sit on my hands and let the biggest deal in GE history go by." Dammerman agreed with him, as did a number of board members with whom Jack had discussed the topic.

Boisi and Doug Braunstein, the head of M&A at JPMorgan Chase, realized that because of the leak about the deal the day before, there was a high probability that Honeywell and United Technologies were working hard to finalize their merger agreements. Jack had to move fast. "He engineered the wedge into the Honeywell board," Calhoun said. At 10:30 a.m. on October 20, Jack called Bonsignore at Honeywell headquarters, in New Jersey, to make GE's offer. But Bonsignore was already in an executive session with his board discussing the United Technologies offer. He could not be disturbed. But Jack wouldn't stop at that. Jack's longtime assistant, Rosanne Badowski, had known Bonsignore's assistant since she had worked for Bossidy. Badowski convinced her that Jack's message was urgent and that if Bonsignore didn't take his call, Jack would put out a press release announcing his superior offer, which would, of course, make any agreement with United Technologies irrelevant, at least for the moment. Bonsignore came to the phone. He told Jack the board was five minutes away from signing the deal with United Technologies.

"Don't," Jack told him. "I want to make you a better offer."

He said he'd jump in a helicopter and fly to Morristown, New Jersey—thirty-two miles west of Manhattan—and meet with him and his board. Bonsignore said such a dramatic step was not necessary.

Jack could fax his offer. He quickly scrawled it on a piece of paper and faxed it to New Jersey. He offered one GE share for each Honeywell share, or about $44 billion, about 10 percent more than the United Technologies offer that Honeywell was on the verge of accepting. Not only was Jack's offer by far the largest deal in GE's history—and some seven times larger than the 1986 RCA deal—it was also a stunning example of audacious corporate dealmaking.

Calhoun couldn't quite get over it. "In six hours," he told me, "in six hours. Could I have done that when I was sitting there by myself as an aviation company? I doubt it. I doubt it." In fact, few companies could have mobilized for such a large deal so quickly, let alone one that was on the verge of a historic succession process. But that was Jack's GE at the start of the twenty-first century: being the world's most valuable company had its advantages. That, and the fact that Jack and his team had already studied a Honeywell acquisition. Plus, Jack could not let United Technologies win. Another difference, besides sheer size, between the RCA and the Honeywell deals was that to get RCA, GE paid 19 percent of its market value to get 14 percent of its earnings; this time, GE was paying 8 percent of its market value to get 16 percent of earnings. Jack was salivating. "Mike I really want to come to Morristown ASAP to clarify any and all issues on your mind," Jack wrote on the fax.

That wouldn't be necessary. After receiving Jack's fax, the Honeywell board put the United Technologies deal on hold. After the market closed, United Technologies announced it had ended merger talks with Honeywell. "Word began leaking that we were in the deal," Jack recalled. That night he and Jane had dinner with Andy Lack, then the head of NBC News, and his wife, Betsy, at Campagna, an Italian restaurant on Twenty-First Street off Park Avenue. He hadn't been able to reach Jane during the day and began excitedly to tell her the news of the deal. "She didn't take it well," he recalled, "but she understood." (A more diplomatic turn of phrase has rarely been used.) Even though their marriage was pretty much on the skids, they had by then already started working on the designs of a smaller postretirement home in Fairfield and had planned a ten-day vacation in Capri for June 2001. Jack had also signed a lease for office space in Shelton, Connecticut. A deal with Honeywell would change most of those plans.

The next day, a Saturday, after a furtive round of negotiations, Jack agreed to increase GE's offer to $45 billion or 1.055 GE shares per Honeywell share. That did it. He and Bonsignore shook hands. Jack left Skadden just after six o'clock that night and took the D train to Yankee Stadium, in the Bronx, to watch the opening game of the 2000 World Series between the Yankees and the Mets. The legal documents were signed on Sunday night.

The next morning, the press made much of the fact that Jack had agreed to stick around GE past April 2001. In his telling, he made clear that Bonsignore had made Jack's continued involvement at GE a condition of the deal. That request was understandable. After all, if Honeywell shareholders, including Bonsignore and his fellow executives, were going to take GE stock in exchange for their Honeywell stock, they certainly did not want to do anything that would send a shock wave through the GE shareholders. The deal itself was shocking enough; if Jack were to leave before the deal was closed and fully integrated, that might send the GE stock diving, greatly reducing the value of the deal for Honeywell shareholders. To make it clear that he wasn't a CEO who couldn't let go—certainly that argument could be made—Jack made his case for the opposite. "This is the most exciting day in the 188-year history of GE," Jack said. But Beth Comstock, a senior executive at NBC, said there was more to it all. "I always thought he didn't really want to leave the stage," she told me.

In any event, Jack wasn't finished. He was just winding himself up. He said it was not the plan to make the biggest deal by far in GE's history in the bottom of his ninth inning. Aside from RCA, the GE approach to acquisitions had favored smaller, more easily integrated companies—GE had done 125 such deals in 1999 alone. But when Jack and his team looked at the numbers and the strategic fit for Honeywell, he couldn't pass it up. The goalie had been pulled and he had a clear shot.

Jack didn't expect much trouble from the European Union about the Honeywell deal because he was, well, Jack and because he had high-priced lawyers who told him everything would be fine. He also took comfort from the fact that the previous year the EU had approved the merger between Honeywell and Allied-Signal with only requests for minor divestitures in order to appease Thales, a French competitor. Jack was wrong to be so confident. "The first inkling of trouble for us came in January," Jack remembered. Thales was back with more complaints about the GE-Honeywell deal and wanted "all kinds" of divestitures. On January 11, Jack flew to Brussels to meet with Mario Monti, an EU commissioner, and his staff. Jack was accompanied by other GE executives and GE's lawyers. He was hoping to get a quick and positive decision from the EU by March 6. If not, that would likely result in a delay until July. Jack made his case for a March decision. Then he raised the issue that European competitors were salivating at the chance to "extort" a "goodie bag" of assets out of Honeywell for GE to get the EU's approval. Monti gave Jack some comfort. "I assure you that extortionist aspects will remain outside this investigation," he told Jack.

Enrique Gonzalez-Diaz, the head of the EU's merger team, told Jack that while customer and competitor comments were a source of valuable insights and information, he took them with "a pinch of salt." Monti assured Jack that he and GE were doing "everything that is expected." After the public meeting, Jack met alone with Monti for lunch for more than two hours. "I felt good chemistry between us," Jack remembered. Monti insisted on calling Jack "Mr. Welch," despite Jack's request that he call him Jack. "I'll only call you Jack when the deal is over," Monti replied. Jack left Brussels feeling optimistic.

But by mid-February, Jack was getting "bad vibes" out of Brussels about an early regulatory sign-off on the Honeywell deal. He was being told it would take the EU another four months to review the deal. On Sunday, February 25, he flew from Palm Beach to Brussels. He and Ben Heineman met with Monti the next morning. Monti seemed to have already made up his mind to extend the review of the deal until July. Jack made his case to Monti about why the Honeywell deal should be fast-tracked. He offered Monti options to solve complaints, but none of his proposals included divestitures. Jack thought Monti seemed persuaded. He asked Jack and the team to give the commissioners time to digest what he had said.

Around dinnertime in Brussels, Jack got a call to return to the EU headquarters. Monti had not been swayed. The EU was proceeding on the path of a slow evaluation. "More troublesome," Jack remembered, were the questions raised by Monti and his colleagues about what effect the GE-Honeywell combination would have on the aircraft industry. Big European competitors such as Rolls-Royce and Thales were complaining, as were several U.S. competitors such as United Technologies (of course) and Rockwell Collins. Jack said Monti was "pleasant" enough, "but I couldn't move him."

At the end of March, while Jack was still wrestling with Mario Monti in Brussels, Matt Murray at *The Wall Street Journal* wondered what Jack would do after he left GE, either at the end of the year or perhaps as early as September 2001, assuming, as Jack said, the Honeywell deal got EU approval, as he expected it would. Jack said he might want to be a consultant to other CEOs and to help them with big-picture problems that might arise. "I'd like to be an adviser, in the background," he said. "I want to be involved with people, in management training, in broad strategic overviews—stuff I like doing."

On May 2, the United States approved GE's acquisition of Honeywell, after Jack agreed to sell Honeywell's military helicopter engine business and to open up to more competition GE's increasingly profitable engine-servicing business on small jet engines and auxiliary power units. The EU was not so accommodating. On May 8, the EU issued a 155-page report listing its objections to the GE-Honeywell merger. Distilled down, what the EU was worried about was the dominance a combined GE-Honeywell would have in the global market for big commercial jet engines, smaller commercial jet engines, and private-jet engines. Each company alone was a powerful competitor: GE in big jet engines and Honeywell in private-jet engines.

Together, the EU worried, the combination would overwhelm the competition—which was already reduced to a few players, like Rolls-Royce, Thales, and Pratt & Whitney—given the industry's capital and technological intensity. The EU was also worried that GE Capital's role in the airline industry and its financial dominance would give GE-Honeywell even more power to force airlines into buying GE-Honeywell engines.

The EU observed that GE Capital not only could facilitate the purchase of engines—something competitors couldn't do—but also, through GE Capital Aviation Services, was the world's largest purchaser of airplanes. GECAS owned 1,040 jets, twice as many as its nearest competitor. The EU worried that GE Capital would insist that the aircraft it was buying be equipped with GE engines, something it had already been doing. In the previous decade, of the six hundred jets that GE Capital bought, only four did not have GE engines. The EU worried that GE Capital could offer attractive financing to buyers and further enhance GE's dominant position in the jet engine business.

Jack returned to Brussels in late May for two days of hearings, the final stage in the process. "This was where things really began to break down," Jack recalled. There was much to be appalled at from Jack's perspective: the short shrift given to the GE executives and lawyers defending the deal as it was; the time and indulgence given to the GE competitors by the EU officials; and the fact that the EU commissioners had, in Jack's words, "after acting as investigator and prosecutor for several months," become "the judge and the jury." He added, "They ended up making the decision on their own proposal."

Jack made his final trip to Brussels on June 7. He would be in Europe for much of the next week. As he was getting on the plane, Bill Conaty wished him good luck. "We'll get this done," Jack told him. It was a whirlwind of proposals and counterproposals, with GE and Honeywell slowly but surely agreeing to give up more and more of their coveted assets. But the EU seemed unwilling to budge. After he arrived on Thursday evening, he worked until midnight with his team to put together a new divestiture proposal. Jack tripled his divestiture offer to assets with revenue of $1.3 billion. But that offer was deemed so wide of the mark by the EU that neither Jack nor Monti attended the Friday session.

That night, Jack and Jane flew to Capri for the weekend as the guests of Paolo Fresco and his wife. Fresco, the former GE vice-chairman, had become the chairman of Fiat, the big Italian car manufacturer. Jack flew back to Brussels Monday night. Dammerman briefed him on how poorly the EU had reacted to the $1.3 billion divestiture proposal. At the negotiating table on Tuesday morning, June 12, Jack increased GE's divestiture offer to $1.9 billion and included in the package "Honeywell's best avionics." The new proposal still didn't fly. Jack and the team then found another $340 million of assets to divest, bringing GE's total offer to $2.2 billion as the final mid-June deadline approached.

On June 13, Jack showed up alone at Monti's office to hear his "official response." Monti was surprised Jack had come alone but greeted him formally and then escorted him into a big conference room filled with the EU team reviewing the merger. It was Jack versus his jury. Even the greatest CEO of his generation was outmatched. Monti read an opening statement thanking Jack and GE for their "good efforts," which he quickly said were not sufficient. Monti wanted GE to divest "one Honeywell business after another," Jack recalled, totaling somewhere between $5 billion and $6 billion of asset sales, he figured.

"This is very somber," Conaty remembered Ben Heineman telling him. "Monti is reading everything that we're gonna have to give up to do the deal. He goes through this long diatribe, very monotone. Heineman said when he finished, Jack said, 'Are you shitting me?' They're all stunned. Then, Jack says, 'You're asking me to buy an eighteen-hole golf course that's only got fifteen holes.' I don't know whether they got the analogy."

In his book, Jack described the encounter more succinctly. "Mr. Monti, I'm shocked and stunned by these demands," Jack wrote. "There's no way I could consider this. If that's your position, I'll go home tonight. I've got a book to write."

Alexander Schaub, once dubbed Europe's "competition guru," started laughing. "That can be your last chapter, Mr. Welch," he said. "'Go Home, Mr. Welch' is a perfect title." (And Jack did use that phrase as the title of the chapter in his memoir about dealing with the EU on the Honeywell deal.) Schaub's crack broke the tension in the room, briefly. There was still plenty. Later that night, Jack met alone with Monti for twenty minutes. He told the EU commissioner that GE and Honeywell had done all that they could do. The EU's case against the merger "doesn't work on the theory and doesn't work on the fact," Jack told him. GE would make its final offer—the same $2.2 billion—the next day.

Before he left Brussels, Monti called Jack and thanked him for his time and effort.

"Now that the deal is over," Monti told him, "I can say to you, 'Goodbye, Jack.'"

"Well, goodbye, Mario," Jack said.

It was time for the last gasps. On June 25, Jack and his team met in New York with executives from Honeywell. They agreed to offer Monti a new package of divestitures. Jack proposed selling 19.9 percent of GECAS, GE's lucrative aircraft-leasing business, in a private placement to a buyer of GE's choice and to invite one independent director onto the five-member GECAS board. He also agreed to divest Honeywell assets with $1.1 billion in revenue.

The next day, Jack called Monti and said that he and Bonsignore, the Honeywell CEO, would fly back to Brussels to see him. Monti said that it would be inappropriate for that visit to happen after the EU's deadline. He said GE's lawyers should make the new proposal. On June 28, Monti called Jack. The new offer was also "insufficient." Jack was as frustrated as ever. "We tried to be responsive to what we were hearing," he told Monti.

Bonsignore told Jack he wanted to give it one more try, especially since Honeywell without GE would be a wounded company. The conventional wisdom at the time was that Honeywell needed GE far more than GE needed Honeywell. The next morning, Bonsignore sent Jack a new idea. GE would divest $2.2 billion of assets, consistent with the EU's final proposal, as well as sell 19.9 percent of GECAS. In addition, the EU would decide who the investor in GECAS would be as well as the independent board member. To help ease Jack's pain, Honeywell agreed to cut the price of Honeywell by $1.8 billion, agreeing to take 1.01 shares of GE for each share of Honeywell, rather than 1.055 GE shares. "It was unacceptable," Jack recalled. At the time, he wrote to Bonsignore that the revised deal "makes no sense for our share owners" and "What the commission is seeking cuts the heart out of the strategic rationale of our deal."

The GE board concurred with Jack's—and Jeff's—decision to reject the new Honeywell deal and to let the merger die in a Brussels conference room. The July weekend following the deal's collapse, Jack was at a wedding reception at the Country Club of Fairfield, overlooking the Long Island Sound. He was enjoying a cocktail, as he was known to do. People were asking him what had happened in Brussels to force him to scuttle the Honeywell deal. "Just imagine if you bought this beautiful golf course," he told them, "and in order to close the purchase, the city officials demanded that holes two, three, four, five, and eight—the best holes along the water—must be given to another golf course in the area. And then they ask you to give up part of your own house." That metaphor sufficed. It's the same one he used with me years later, only it had been distilled down to Jack telling Monti, "You robbed me." He told me this had been a strategy employed by the EU for years: building up European companies by hiving off assets from more successful American companies.

"They do that all the time," Jack said. "Microsoft faced that for years. IBM faced it. Google faced it. It's nationalistic." I asked him how he would have felt about the Honeywell deal if Monti had given him all eighteen holes. The question arose because several GE executives had suggested to me that Jack was looking for a graceful way out of the deal and Monti's demands had given him the exit strategy he needed. "I would have loved it," Jack said. "Yeah."

The truth was probably more complicated. One longtime GE executive told me that Jack and his team had started to see things about Honeywell they didn't like. There were unaccounted-for liabilities involving asbestos and other environmental matters. Jack was also concerned about Honeywell's "bookkeeping," accounting that didn't add up. For the previous decade, Honeywell had been converting only 69 percent of its net income into free cash. "It gives you a sense of how much bookkeeping was going on," he said. As time went on, Jack's concerns about the deal grew. "He was pretty pleased at the condition that Mario Monti put on to do the deal," the executive continued. "It gave him enough of a reason to say that he wasn't going to proceed with the deal. Everybody says the EU killed the deal, which of course is not true. What the EU did is put conditions on the deal that GE chose not to support or not to agree with. . . . Jack was ecstatic." Ken Langone, the GE board member, agreed that it was a blessing that Monti imposed unwanted conditions on the Honeywell deal. "Thank God they did," he said. "Because we decided on our own, 'We don't want to go ahead. We don't want to do the deal.' We saw things that bothered us. We had a hard look and said, 'Wait a minute. You want to strap the new guy with a big business like this?' There was a lot of reasons."

For his part, Jeff was willing to let Jack make the decision to abandon Honeywell, given that he was not involved in the deal in the first place and the business he was running—Medical Systems—would not have been changed much by it. As the incoming CEO of GE, though, and the man slated to run the company for the next twenty years, shouldn't he have had more of a say in that decision? In short, Jeff said it was complicated. He said at that moment Jack "was king of the world" and was highly motivated to do the Honeywell deal because "he hated David George"—the CEO of United Technologies—and didn't want a rival to get the company. He said he thought the Honeywell deal, if completed, would "have reset the bar" for GE by building up GE's industrial businesses while reducing the spiraling upward impact of GE Capital and GE's other financial businesses. Jeff said he believed the company's industrial businesses needed a boost and Honeywell would have given it to them.

On July 3, the twenty EU commissioners voted unanimously to block GE's acquisition of Honeywell. It was the only merger, to that time, between two U.S. companies that the U.S. had approved and that the European Union blocked. The deal was dead.

GE also got right back in acquisition game, almost as if the Honeywell failure had no impact. In the month after the deal collapsed, GE announced five new acquisitions. "The reason," posited Matt Murray at *The Wall Street Journal*, was that "it needs to. More than many other multinational companies, GE depends on acquisitions, lots of them, to help fuel its growth." On July 30, GE Capital bought Heller Financial for $5.3 billion, the second-largest deal in GE's history after RCA. Denis Nayden, the new CEO of GE Capital, was enthusiastic about the Heller deal, and deals generally.

THE HONEYWELL SETBACK ASIDE, THERE WERE PLENTY OF ENCOMI-
ums and other celebrations for Jack. There was a party for him at Crotonville. GE found a car similar to the first one he had bought out of
college—a convertible—and paraded him around the upstate New York
campus in it. A building at Crotonville was named for him. Despite his
outrageous success, many of those who knew him best thought he managed to keep things in perspective. Matt Lauer recalled a golfing trip
he and Jack, Bob Wright, and Bryant Gumbel took to Augusta National. They all came in from different places on GE's private jets. "It
was a perk because we'd had a particularly good month on the show,"
Lauer said. They played golf and stayed in the cottages, and Jack
treated them to dinners. "We had this fabulous weekend," Lauer said.
They took cars back to the Augusta airport and drove onto the tarmac,
where two GE jets awaited them, nose to nose, engines revving, doors
open, and staircases down. "I was just in awe," Lauer recalled. He
turned to Jack and said, "Do you ever get used to this?" Jack grabbed
Lauer's arm and stopped in the middle of the tarmac and replied, "I
want you to know something. I pinch myself every single day. I never
get used to this. I never lose the thrill of what this job has given me."
Larry Bossidy remembered that Jack always used to carry around an
old tattered brown satchel. He easily could have afforded a new one.
But he carried the sad sack around for forty years. It had become hid-
eous. "But that's what he is," Bossidy said. "He knows where he's from,
and he's always been able to ground himself. It's an eyesore but, you
know, I don't think it's a security blanket. It must be a metaphor for
life. He doesn't feel like he wants to buy a new one. Never has. He says,
'That's the one I came in with. That's the one I'm going out with.'"

## Chapter Eighteen

# ONE GOOD DAY

Jeff Immelt once said the "one good day" he had during his initial year as CEO of GE was September 10, his first business day in the corner office without Jack Welch around. At that moment, GE's stock was trading near $40 a share and GE had a market value of nearly $400 billion. Not only was GE the most valuable company in the world, but it was also named the "most respected" company by the *Financial Times* and the "most admired" company by *Fortune*. But the days, months, and years that followed were not what he had imagined when he won the highly celebrated beauty contest to succeed Jack.

While Jack thought he had handed off to his successor a company perfect in every way—a point he made to me repeatedly, replete with charts and graphs—Jeff would realize that the popular perception of GE at the start of the twenty-first century was not nearly the same as the reality of GE. The new decade would test many of the suppositions upon which Jack's GE had been built, and those legacy problems, many of which once appeared virtuous, and the changing face of American business would challenge Jeff in ways he never imagined, often with unsatisfactory results. If the Jack Welch era had been defined by Six

Sigma, splashy acquisitions in media and financial services, the rise of GE Capital, and an almost religious devotion to delivering steady earn- ings and earnings growth for investors—all of which made GE the most valuable and most admired corporation in the world—Jeff Im- melt would attempt to take GE in a very different direction, one weaned off its growing dependence on financial services and the vicissitudes of Big Media and returned to its roots as a hub of technological in- novation and industrial might. But unlike Jack, Jeff would struggle mightily from the beginning to articulate a clear, simple vision for the sprawling conglomerate—a challenge that was complicated by Jeff's own shortcomings as a leader and the fact that he was put on the defensive from the very moment he stepped into the corner office in Fairfield.

On his first day in the office, Jeff did a video-cast for employees from Crotonville and then flew to Seattle. He had a meeting the next day at Boeing, where he was to speak to its engineering group. The next morning—around 6:00 a.m. on the West Coast—he was working out on a StairMaster in his hotel, watching CNBC. By then, the first plane had hit the North Tower of the World Trade Center, in Lower Manhattan. But like so many others, Jeff thought it was a small Piper Cub–type plane that had mistakenly rammed into the building. While he was working out and watching TV, the second jet hit the South Tower. He went back to his hotel room and called Dave Calhoun, who was with him in Seattle and was running the jet engine business, and told him to cancel the meetings at Boeing. "There's something going on here," he told him. The other top GE executives were "distributed," Jeff told me. Keith Sherin, the CFO, was in Los Angeles. Bob Wright was in Florida. "Triple D was somewhere," he said, referring to Dennis Dammerman. He got his team together by teleconference, a precursor to Zoom, every hour or so, trying to figure out what to do. Andy Lack, the head of NBC News, was the only top GE executive in New York City. "Andy was actually extremely helpful because he was connected to everybody in New York City," Jeff said. Jeff was with Phil Condit, the CEO of Boeing. "The government thought there were six planes that had been hijacked," he said. "I'm tracking all this stuff real time, with Phil. They had more pull than we had in this context." Jeff's laptop computer could not break through GE's firewall, understandably frustrating him. He and Calhoun went to a small GE sales office in Seattle to log on. It was "five cubicles with a small place downstairs to get a doughnut and Diet Coke," Calhoun recalled. But Jeff was in command. The crisis demanded his leadership and he seemed to rise to the occasion. He decided that a previously scheduled investor meeting would be moved by only two days, to September 18. "We knew where we stood," he said. "I felt it was important for us to get out and make our statement."

Jeff's problems were nearly immediate. Two GE employees were killed in the attack, Natalie Lasden, who worked in aircraft engines, and William Steckman, a technician at NBC. Jeff sent GE's mobile generators and medical diagnostic equipment to Lower Manhattan. He decided to cancel all scheduled commercials on NBC for three days, costing GE "a couple hundred million bucks," he said. GE had partially reinsured both towers and all four planes used in the attacks. GE had also reinsured 7 World Trade Center, the small third tower that collapsed as a result of collateral damage from the collapse of the Twin Towers. "It turned out to be a billion-dollar write-off we had to make," Jeff said. And there was the potential ripple effect: GE owned 1,200 aircraft and the world's largest jet engine business. "Nobody had envisioned that a plane could be used as a missile, but in one morning you kind of saw that that's what could happen," he told me. On September 12, GE made a $10 million donation to the families of the rescue workers killed when the Twin Towers collapsed.

Jeff had a chance to fly back to New York City during the first day or two after September 11, when VIP types were getting permission to fly despite the air lanes being shut down by the government. He decided not to do it. (He can't remember why.) He ended up flying out on

Thursday night, arriving back home at around 2:00 a.m. Friday morning. While he was in the Signature terminal at the Seattle airport—where people flying on private jets alight—he got a call from Denis Nayden and thought he had another major problem. Nayden told him that GECAS—GE Capital's aircraft leasing business—owned a few pilot training centers and one of them had trained a guy named Mohamed Atta, one of the leaders of the September 11 attack. "I get on the plane thinking, 'Oh my God, we're fucked. We are so fucked. We were fucked before, but now we're triple fucked,'" Jeff told me. "By the time I landed, they had seen that 'Mohamed Atta' is like 'John Smith.' There's a bunch of Mohamed Attas out there, thank God. There was a whole series of things like that." GE had trained a different Mohamed Atta.

He had been the CEO for a total of four days and found himself facing a near-existential crisis that had never been even remotely contemplated by any GE contingency plan. "If you're in the commercial aviation business at that moment, you really had no idea that there would ever be an industry ever again," he said. "You really felt like this business will never be the same. Who will fly? How will they fly? What's gonna happen?"

Ten days after September 11, President George W. Bush created the Air Transportation Stabilization Board, under the auspices of the U.S. Treasury, to provide up to $10 billion in loan guarantees to U.S. airlines. The committee was comprised of Alan Greenspan, the Federal Reserve chairman, Paul O'Neill, the Treasury secretary, and Norm Mineta, the transportation secretary. The airline industry had already been hurting prior to the September 11 attacks, and a number of airlines looked to be heading toward bankruptcy. The fear of flying that resulted after September 11 exacerbated the financial quagmire. Jeff took the lead in negotiating with the board on behalf of the airlines to allow them to borrow the money they needed to survive. He went to see Greenspan, who was immediately supportive. "Of course we're going to give them money," he told Jeff. "I mean, why wouldn't we?" He went to see O'Neill, who was more dogmatic and iconoclastic. He was against lending the airlines money. O'Neill told Jeff that the airlines were bad actors, were already near bankruptcy, and did a ton of things poorly. He was a no. Jeff then went to see Mineta, who had never expected to be in the position of being a key man at a key moment. Mineta had been thinking the job would be a cushy one. "I can just play hearts at lunch," he told people. "I get to go to the State of the Union." Mineta was the deciding vote, and Jeff needed to convince him to make the loans so that GE's jet engine division—one of its most profitable and important businesses—would still have customers. "Norm, my man, you've got to come through for us right now," he told Mineta.

Mineta agreed with Greenspan. The airlines would get their loans. "The government did its part," Jeff said. GE's customers would have the money to keep buying engines. "That's kind of the way you were living," Jeff told me. "Every day was a new year. You'd go to bed at one a.m. and you didn't want to wake up the next morning, because you just weren't quite sure what was going to happen." He said he was lucky to have a team comprising Dammerman, Nayden, and Henry Hubschman, then the president of GECAS, working with him to solve problems. He remembered times when Nayden would tell him that GE needed to buy $1 billion of American Airlines EETC—enhanced equipment trust certificate—bonds by the next morning. "Or they're going to go bankrupt," Nayden told him. "What do you want to do?" He didn't understand. "Dennis, what's an EETC?" Jeff asked. "Can you just explain it to me for fifteen minutes? Because I don't even know what the fuck it is." He got his explanation. GE bought the bonds. "It was day after day after day of stuff like that," he said. He couldn't quite believe what was happening. "What was surprising to me was not that—I always knew I was gonna have to fight for myself—but very early, like within ninety days, I was fighting for the company," Jeff recalled.

On September 10, GE's market value was close to $400 billion, making it still the most valuable company in America. For the week after the September 11 attacks, the stock exchanges were closed to al-low people to absorb the emotional and psychological impact of the attacks and to allow the New York Stock Exchange to repair the col-lateral damage it had suffered. When the stock exchanges opened again on September 17, GE's stock had lost $40 billion in value. Three days later, it had lost another $50 billion in value. Ten days later, it had been reduced to $300 billion, a 25 percent decline.

That decline was on top of the one-third drop the GE stock had suffered since its all-time high in August 2000, from a combination of recession worries and those about a GE without Jack and Jack's magic. What exacerbated the problem for GE's stock on September 17 was the fact that it was *so* liquid: its ten billion shares outstanding were ubiq-uitous. Investors with wildfires elsewhere in their portfolios as a result of September 11 ended up selling GE stock. It was just easier to cash it out.

Jeff remembered wondering why GE's largest shareholder dumped half its stake in the company after September 11. "The week after the market opens, the biggest shareholder owned four hundred million shares of GE, and probably two days after the market reopens after 9/11, they flush maybe one hundred or two hundred million shares just like that, bounce the stock by like four bucks," Jeff told me. He called the stockholder. "Hey, dude," he said, "slow down a little bit. People are scared. You're pushing around [the stock]." As Jeff recounted the story, without naming the shareholder—it was before such disclosures were required—he was incredulous when the guy told him, "Gosh, I didn't really realize that GE was so big in reinsurance. So I'm trimming my position."

"Look," Jeff told him, "you own four hundred million shares. That's kind of on you, not me, really, in terms of not knowing."

He said the "hardest" realization for him at that moment was dis-covering that "people didn't really know why they owned GE at that moment in time. So in addition to all the externalities, we had to really work hard to reposition the company."

A surprising number of close GE observers, both inside and outside the company, believed that Jeff should have used the time after the September 11 attacks to sit down with the research analysts on Wall Street who were covering the company and also with key GE shareholders and explain to them the world had changed. That GE had changed and the days of continuous double-digit sales and earnings quarterly growth were gone. That earnings growth was likely to become more episodic and less predictable. He could have explained that he was not Jack Welch. He was not the CEO of the Century.

It was time for Jeff to "reset" expectations about GE's earnings and GE's stock. He could have put an end to Jack's practice of managing earnings, taking any write-downs that needed to be taken, and he could have begun to defuse the ticking time bombs tucked away in the recesses of GE's balance sheet. He could have told Wall Street that it was no longer reasonable for GE to have the astronomical price-to-earnings multiple that it had under Jack, often approaching fifty times earnings. He could have said that GE was a bank attached to a huge, slow-growth industrial company that had just suffered a major rupture. "He absolutely should have done that," explained Jeffrey Sprague, a longtime Wall Street research analyst. "I hate to call 9/11 an opportunity. It was a terrible event, but it was a real event that had economic implications that impacted his business and did provide an opening to reset things, and he really chose not to do it. I think not doing it early in his career, then with each passing year it just became harder and harder to even think about doing it."

Another Wall Street analyst, who asked for anonymity, said that early in Jeff's tenure he was shown—either by Wall Street bankers or by big shareholders—a plan that would have reeled GE Capital back in and made it a much smaller and much less significant part of the overall

GE picture. "If he had unwound this business, you would have been sitting there saying, 'Holy bejesus, Jeff Immelt is a fricking genius.... I know he was shown something that told him, 'You're going to take a hit today, but you will be happy in the future if you do this.' He basically said, 'Yeah, thanks for the analysis. We'll talk later.' And didn't take it." He said he remembered that early on in Jeff's tenure as CEO he told Jeff that the investors he spoke with hated GE Capital and didn't understand why GE was in the financial services business. They told him they wouldn't touch GE stock as long as it owned GE Capital. He told Jeff, "Everybody seems scared of it. They just don't want to own your stock." He said Jeff replied, "You don't know what you're talking about. My investors do not want me to sell that asset. It generates too much earnings." The analyst continued, "They used to tell us that. They used to play us against the buy side and say, 'Hey, the buy side thinks this.' It makes you feel stupid, because when you're a young analyst, when a management team tells you that, you're like, 'I guess I just don't know my customers as well as they do, so I'll trust them.' Then you go along. That was a strategy for them. They would divide and conquer, and that was the wrong answer to that question, because I know for a fact that people did not own the stock because of GE Capital."

Looking back, years later, Jeff told me simply, "I wish I had reset the company in 2002." But he decided not to do it. "I can explain to you why I did what I did, but it certainly would have been easier for me [to do the reset] and maybe even better for investors," he said. "[The stock] went from $38 to $25" after September 11.

IN ADDITION TO THE EFFECTS OF SEPTEMBER 11, JEFF WAS ALSO OPERating in a political environment that was very different from Jack's, particularly when it came to corporate transparency. After the dramatic collapses of Enron and other big companies such as WorldCom, Global Crossing, and Adelphia Communications due to accounting scandals and other financial irregularities, the media and investors were increasingly skeptical of nearly everything they had once thought they understood about corporate disclosure and accounting. In that context, it was inevitable that questions would be raised about GE and GE Capital, which was famously opaque, and how GE had managed to hit its earnings numbers quarter after quarter with astonishing precision and consistency. GE, the biggest and most finely tuned earnings machine, became an obvious target of skepticism in a few but influential corners of the financial world.

Even *Fortune*, which had just named GE the country's most admired company for the fifth year in a row, was joining the ranks of the GE skeptics. It was as if once the emperor had exited the stage, his subjects finally figured out he had no clothes. Where once upon a time, nailing quarterly earnings on a consistent basis was seen as a virtue on Wall Street—which was why Jack made sure it happened quarter after quarter—suddenly such financial precision was viewed skeptically. "It's kind of like smoking," one financial analyst told *Fortune*. "Smoking used to be chic and fashionable and cool; now it's not. The companies that reliably deliver 15% earnings growth year after year are [the] new smokers."

The change in the zeitgeist had put Jeff again on the defensive, where he already was anyway after September 11. "Would [an earnings] miss be more honest?" he asked incredulously. He explained to *Fortune* that when he realized the recession following September 11 would hurt GE's "short-cycle" businesses—appliances and lightbulbs—he asked the "long-cycle" businesses—power systems, jet engines, and medical equipment—to "do better."

"We don't manage earnings, we manage businesses," Jeff told the magazine's Justin Fox. But Fox made clear some funky earnings management was still happening. In the fourth quarter of 2001, GE offset a big gain of $642 million from a satellite partnership by taking a $656 million write-down on a variety of failed investments, including a loss of $84 million on Enron bonds after the company filed for bankruptcy. Keith Sherin, the GE CFO, defended the practice, asking Fox why investors would want GE to report inconsistent earnings from quarter to quarter if it could be avoided. "It just doesn't make sense to us in managing a business," he said.

In mid-March 2002, GE Capital's gargantuan $11 billion bond issue—then the second-largest U.S. dollar–denominated bond issue in history—gave critics of GE and wary investors a chance to express themselves publicly. And they did. What started out as a $6 billion bond deal was increased to $11 billion because of investor demand. And the yields on the debt—for a AAA-rated company—looked attractive to investors in the wake of the Fed's decision to lower interest rates dramatically after September 11. But what troubled the savviest bond investors—those who could look beyond a 6.75 percent yield on a thirty-year GE Capital bond—was what the capital raise implied about GE Capital's $150 billion of outstanding commercial-paper borrowings, which were nothing more than unsecured, short-term IOUs issued by companies with the best credit. GE Capital was the largest issuer of commercial paper in the world—a fact Moody's pointed out in a note around the time of GE Capital's bond offering. "Commercial paper is our rocket fuel," was how Mike Neal, who ran GE Capital for many years under Jeff, framed it for Wall Street research analysts. Moody's wrote that GE Capital's AAA credit ratings came as a result of the "explicit and implicit support" of GE, and on a stand-alone basis GE Capital would be rated lower than AAA.

GE Capital was not a bank. It had no bank branches in the United States and no way for anyone to make deposits. Unlike a bank, which relied on deposits to get the money it uses to lend out, GE Capital got the money it lent out primarily from the short-term commercial-paper market. And why not? The borrowings were unsecured, readily available, and cheap. Given that GE and GE Capital had a AAA credit rating, it made abundant sense for GE Capital to borrow in the commercial-paper market. But the market for commercial paper could also be finicky. Like a group of worried depositors lining up to get their money out of a failing bank, the investors in the commercial-paper market might decide that the risks of making those short-term loans—usually repayable in 270 days—outweighed the rewards, closing that as a source of financing for those companies, like GE Capital, that relied upon it for funding. The $11 billion bond offering was a recognition on GE Capital's part that it had become too reliant on the commercial-paper market and needed to diversify its borrowings, in case that market got too rocky at some point and dried up. The cost of the money to GE was higher than in the commercial-paper market, but the big bond offering was a perfectly logical reaction to the mini financial crisis that followed the September 11 attacks.

Only that's not the reason GE Capital gave the market for doing the huge bond deal. Instead, the company said it was merely being opportunistic. "Absolute yield levels are at historic lows, and investor demand for long-term debt from high-quality issuers is strong," Marissa Moretti, a GE Capital spokeswoman, told the Dow Jones Newswires, "so we think now is the right time to be doing an offering."

Moretti's explanation for the GE Capital bond offering—opportunism—irritated Bill Gross, the founder and managing director of Pacific Investment Management Co., or PIMCO, one of the world's largest buyers of bonds. Gross was known on Wall Street as the "Bond King." What he said about bonds and other fixed-income securities mattered. Even though he was considered a bit of an iconoclast by some people—he lived on an estate in Laguna Beach, California, and practiced transcendental meditation—he was the most powerful person in the bond market. In a March 20 note to his clients, Gross expressed his skepticism about Moretti's logic for GE Capital's huge bond deal. Some six months after September 11, Gross no longer bought the explanation of rates being at historic lows—"maybe three months and 100 basis points ago," he wrote. Instead, he argued that Moretti was "trying to mask" the real reason for the bond offering: the fact that the commercial-paper market "is quite sensitive these days and is drying up for some heretofore stellar corporate credits that are then required to fund in the longer-term credit markets."

Gross pointed out that "a lack of market discipline" had allowed GE Capital to have three times more commercial paper outstanding than it had credit lines to back it up. Typically, a company had backup bank lines equal to its outstanding commercial paper, just in case. He wrote that the GE Capital bond offering was proof that GE was "sensing vulnerability" to the "current mercurial opinion" of analysts and fund managers that "any" company "had better clean up their act" or "suffer the consequences."

Like others before him, Gross believed neither Jack nor Jeff had been "totally forthcoming" about how GE had been able to grow its earnings at nearly 15 percent a year, year after year. "The fact is that GE is a conglomerate financed by a money machine," he wrote. The two CEOs kept the machine humming through acquisitions—more than five hundred in the previous five years—using either GE's high-priced stock—he was wrong about this, as most of GE's deal were done with cash—or its low-cost commercial paper. Then, he added, GE Capital was using "near hedge fund leverage" of seven to eight times and paying investors "non–hedge fund risk." In other words, Gross argued, investors were mispricing the risk inherent in GE Capital's portfolio. "But," he wrote, "investor confidence these days is fragile indeed." He suggested that without the "leverage afforded to them by the Street," GE resembled more closely the "failed conglomerates of yesteryear" such as Gulf + Western and LTV. He announced that PIMCO would no longer buy GE's commercial paper. Ouch.

He wanted more disclosure. He wanted companies to "face up to their owners and yes—their creditors," he continued. "I want management to focus not on their options, but on mine and that of other investors." In closing, he borrowed from the 1976 drama about television news, *Network*, and said he was mad as hell and wasn't going to take it anymore. "It's our money," he concluded. "Value and honesty should dominate corporate decision-making."

Gross's unexpected fusillade stunned the markets, and GE. To question the creditworthiness of a AAA-rated company was akin to blasphemy. GE's stock fell 3.5 percent. It issued a press release saying it was in the process of increasing its backup lines of credit to $50 billion, from $33.5 billion, and it said it was reducing its percentage of commercial paper outstanding to total debt. The company reconfirmed its 2002 earnings projection. Gross went on cable TV to complain some more about GE. That's when Gross got a call from Dennis Dammerman. "He was furious" at Gross, the *Journal* reported, primarily because Gross had written that GE often used its high-priced stock for acquisitions, which wasn't true. "We were pissed off that the report was so factually wrong and completely out of the blue," Dammerman said a month later.

Gross admitted to Dammerman he was mistaken on the use of GE stock for acquisitions, and on that point alone. But not on the others. "Soon, the executives were screaming at each other," the *Journal* reported. Gross couldn't figure out why Dammerman was so damn angry. "Why are you giving me such a hard time?" he asked the GE vice-chairman. "I'm an investor in your company, for God's sake." He said the call was like a Fox News segment where one guest tries to shout down the other. "Forget it," he told Dammerman. He hung up the phone. He remained convinced that GE was more of a finance company than an industrial company.

In his interview with the *Journal*, Gross addressed the long-standing belief that GE manipulated earnings every quarter. "What you keep hearing behind the scenes is that they're selling corporate securities to book profits," he said. "Everyone on Wall Street knows GE plays games; it's totally legal but just another example of how companies aren't coming clean with investors." He said PIMCO would not benefit from the dispute—it still owned about $50 million of GE bonds—but rather he wanted GE to "come clean," not "knock them down a notch."

Jim Grant, the publisher of his eponymous quirky and highly respected bond-market publication, couldn't resist commenting on the hornet's nest that Gross had kicked, especially since Grant had been trying for years to focus investors' attention on GE's exploding balance sheet, profitability, and inexplicably consistent earnings growth. In an end of March 2002 report, Grant wrote of GE, "They are an industrial company that has morphed into a financial services company."

In truth, the months after September 11 had roiled the American psyche and the American capital markets, and top GE executives were left with little choice but to begin to rethink the way the company funded itself. The problem was, rightly, that GE Capital was making the classic mistake that banks have made from time immemorial: borrowing short and lending long—borrowing money inexpensively in the short-term markets and lending it out a higher interest rates for the longer term. The difference between what GE paid to borrow the money and what it received to lend the money was a big part of GE Capital's profit. Problems arise when the sources of the short-term funding dry up or get too expensive or are no longer available. That's when the proverbial shit hits the fan in the banking world. It's happened over and over again, decade after decade, generation after generation. Even after it happens, and the consequences are devastating, memories are short, and the cycle repeats itself.

Of course, alternatives are limited. Certainly, the reverse would not work. There is no scenario under which a bank or a financial institution can borrow at a high rate of interest and lend at a low rate of interest and still make money. Without steady and consistent profits, banks cease to exist. Hence the paradox at the center of the so-called fractional banking system. We're willing to put up with the occasional financial crisis—on average once every decade or so—to make it possible for loans to be made, which in turn helps companies grow, hire more employees, and pay them higher and higher wages and wealth be created. In exchange for increasing our standard of living, we have collectively decided we will permit the occasional financial crisis, despite their incredibly high costs, and misery, when they do occur.

Jim Bunt, GE's treasurer, was on the front line of the growing controversy about how GE Capital funded itself. He was the one who had to interact with the rating agencies about GE's credit rating. He was, in effect, the steward of one of GE's most prized assets: its AAA credit rating. While few aside from Bill Gross or Jim Grant—or even Jack Welch or Jeff Immelt—fully grasped the implications of GE Capital being the world's largest issuer of commercial paper, or the fact that by 2001 GE Capital had issued three times as much commercial paper as it had backup lines of credit available, the ratings agencies knew. Jim Bunt knew.

He remembered how in the first few months of Jeff's tenure as CEO, he asked Bunt to set up a series of dinners with the analysts at S&P and Moody's in New York to review with them "what we're doing." Bunt thought that was smart. He hadn't "clashed" with Jeff, he told me, but he often wondered why Jeff wasn't "more panicked" about GE's growing debt load. Bunt said he lived in fear of *The Wall Street Journal* headline that GE had lost its AAA credit rating and for some reason "the treasurer was not available for comment." He said he felt that somehow "this is where they're heading me."

On the twenty-minute helicopter flight down to New York City to meet with the credit ratings agencies, Jeff asked Bunt what he should expect from the meetings. Bunt was incredulous. "God, I'd been telling him fucking about twenty times at least," he recalled. "Liquidity. And cash flow. And the leverage! All this stuff. Because I talked to their representatives that were assigned to the companies like me, and I know what they all felt. And I had saved us from getting hammered by S&P and Moody's twice." Bunt told Jeff again what to expect. "You really think so?" Jeff asked him. "Really, you think so?"

"Yes, Jeff," Bunt told him. "They are going to be concerned."

"Oh, I don't know," Jeff said.

As the dinner with the S&P analysts was winding down, Jeff asked the fellow sitting across from him, "If you were in my shoes, what would you worry about?"

"Well, liquidity," he said. "You don't have your short-term commercial paper covered, and you're sitting here, and if the shit hits the fan and you can't access that market, you might have to do this and that and get short-term loans and call on your bank loans, this and that. And then it spirals down and before you know it, you're bankrupt."

Bunt couldn't believe what he heard. "He actually used the word 'bankrupt,'" he said. Bunt recalled how S&P had embarrassed itself some thirty years earlier when, on a Friday, the ratings agency reconfirmed the AAA rating of the Penn Central Railroad and then, on Monday, the company declared bankruptcy. "It doesn't get much worse than that," he said. In other words, Bunt was trying to convey, as Hemingway wrote in The *Sun Also Rises*, "How did you go bankrupt? Two ways. Gradually, and then suddenly." Bunt said he was concerned about GE's liquidity and tried to warn Jeff that the credit agencies were also concerned about it. John Myers, the head of GE's pension fund, recalled that part of the problem was that GE would exploit its AAA credit rating whenever possible. "It was looked at as free money," he told me. "It was an arbitrage. If we needed to raise $1 billion, we'd raise $1.2 billion and take the extra $200 million and invest in lesser-quality securities, which were still not that bad. You can go from AAA to A and pick up a hundred or a hundred and fifty basis points or whatever as arbitrage. There was quite a bit of discussion." Myers remembered Bunt being particularly worried about the way GE financed itself. "He was probably the most vocal critic internally about what was being done," Myers said. "He was crying wolf for a lot of years, and people said, 'Yeah, Jim, okay, we got it. Blah, blah, blah.'"

After the Gross tempest passed, GE took some steps to reduce its dependence on the commercial-paper market. By April 2002, GE had increased its bank lines, issued long-term debt, and decreased its commercial paper outstanding from $117 billion to $103 billion, or 42 percent of its total debt. The company wanted its commercial paper outstanding to be 25 percent of total debt by the end of 2002.

For his part, Jeff remembered how in 2002 GE almost lost its AAA credit rating—as Bunt had warned—because GE Capital had "for a long time" given the rating agencies "the finger." He added, "We just weren't listening." One of the requirements for keeping the AAA credit rating was that GE's commercial paper could not be above a certain percentage of its overall debt. "It grew and grew," Jeff said. That's when GE decided to issue the long-term debt and to reduce its commercial-paper dependence as a way to get the ratios back in line with what the credit rating agencies wanted for GE to keep the AAA rating. Jeff said PIMCO bought some of GE's newly issued long-term debt. But when GE issued more long-term debt, the value of what Gross had bought fell. "So he crashed us," Jeff remembered.

Jeff's version of the credit-rating-agency conflagration was very different from Bunt's. In his book he wrote about how on the same day—September 28, 2001—that anthrax was mailed to Tom Brokaw's office in 30 Rock, he asked Kathy Cassidy, an executive at GE Capital, to become GE's new treasurer. "GE's Treasury department was suffering from years of neglect," Jeff wrote, digging at Bunt. Her first day in her new job was December 1. After the failures of Enron and WorldCom, Cassidy asked Jeff to do something Jack had never done: meet with the credit-rating agencies. When she went to set up the meetings for Jeff, an analyst at Moody's told her, "I hope you know we're about to downgrade GE."

Understandably, Cassidy was "shocked" and asked the analyst why Moody's was heading in that direction. According to Jeff, Moody's logic was that GE didn't have enough "cash," especially in the context of GE's "huge" insurance business, which GE was levering eight to one—$8 of debt for every $1 of equity—when others in the insurance industry had two-to-one leverage. In Jeff's version of events, there was no mention of having too much commercial paper outstanding that was not sufficiently backed up by bank credit lines. Instead, Jeff blamed Bunt. Moody's had sent Bunt a letter "months earlier" warning him that if GE didn't improve its "liquidity," then GE would lose its AAA credit rating. "Our Treasury department never responded," Jeff wrote. Cassidy had found the letter to Bunt, "but he'd ignored it, thinking it was an idle threat," Jeff continued. "But that had been a mistake and now I had to repair the relationship." (Bunt was not mentioned by name in Jeff's book. He passed away in April 2019, between the time of my interviews with him and the publication of Jeff's book, *Hot Seat*, in February 2021.)

Denis Nayden, the former chairman and CEO of GE Capital, said the idea that GE Capital would have ignored letters from the ratings agencies was "bullshit," as was the idea that either Jim Bunt or Jim Parke, the GE Capital chief financial officer, would have been anything less than highly attentive to the ratings agencies. "We never ignored them," Nayden told me. "And Jim Parke had direct contact with them all the time. I mean, it's bullshit. It's totally not believable." He said GE Capital had a direct connection to the ratings agencies and there was "not a fucking chance" a letter from the ratings agencies to Bunt or to him or to Wendt or to Parke would have been ignored.

The meeting between GE executives and Moody's occurred in March 2002. While he was making the case for not downgrading GE's credit rating, an unnamed senior Moody's executive "lit into me," Jeff wrote, "saying how arrogant and rude GE Capital people were." The Moody's executive showed Jeff a series of emails. "Arrogant" was an "understatement," Jeff continued. "We needed an attitude adjustment. We also needed to shrink GE Capital's $250 billion of debt."

Jeff wrote that he "disagreed" with Gross but acknowledged GE "had a problem" and not "just of perception." He continued that "part of the legacy" he "inherited" was that GE could grow its earnings through acquisitions because "nobody was really watching that closely." He wanted to "operate differently," but in the "short term" GE Capital had become GE's growth strategy. "We had no other engines of growth," he wrote. "We had to keep our heads down and weather the scrutiny." He tasked Cassidy with educating GE Capital executives about how GE Capital funded itself—"even they were insufficiently aware of the specifics," he continued—and to come up with a plan for debt reduction. "For the moment," he wrote, "she was keeping ahead of our critics."

Years later, Gross recalled to me, his note to PIMCO investors about GE's commercial paper exposure still peeved GE executives, as it had Dennis Dammerman in 2002. Gross recalled three separate occasions, when he was playing in golf tournaments along with various GE executives, when each of them made a point of coming up to him and "in clearly angry" and "almost threatening tones" to tell him he "shouldn't have written what I did." It was then, Gross said, that he realized that GE "was not just another company" but rather had become "an institution," one that had to be "protected and sheltered from the foibles and mistakes of other lesser-known or less-revered companies." Gross remembered being shaken up by the criticism from the GE executives. "As the fate of GE became more obvious through time," he concluded, "I believed that 'hubris' was a term well suited for what was once a grand example of American ingenuity and managerial excellence."

Gross's stature at the time was such that the financial press could not ignore what he had written. And some began to take a deeper look at GE's financial statements, rather than just report the dispute. Gretchen Morgenson, then the assistant business and financial editor at *The New York Times*, wrote a technical but devastating piece for the Sunday business section in April 2002, asking the question "What devils lurk in GE's details?" Morgenson had won a Pulitzer Prize for her reporting at the *Times* the week before. Morgenson, a talented investigative reporter, quoted a variety of increasingly skeptical hedge fund managers and research analysts who, in the post–September 11 and post-Enron era, were no longer willing to accept what GE was doling out at face value. In the article Robert Olsen, a hedge fund manager and "accounting expert," according to Morgenson, said that GE's earnings were increasingly derived from "financial engineering," not from its world-class manufacturing businesses. There were earnings gains from asset sales—for instance, a $1.2 billion gain from the sale of GE's satellite business—and gains from pension accounting. There were also gains on the securitization of GE Capital's loans and from lower tax rates. A recent $3 billion cash infusion from GE into GE Capital also caught the eyes of critical observers, along with a $900 million mark-to-market derivatives loss. (GE said the equity infusion was necessary for GE Capital to complete the purchase of Heller Financial.)

There was also a growing weakness in GE's power business, as customers canceled orders in the wake of the economic recession and the Enron blowup. Customers who canceled power orders had to pay fees to GE; these fees resulted in $326 million in profit in the first quarter of 2002. "Who wants to make money from cancellations?" Olsen told Morgenson. Olsen continued, "GE is getting growth from things it can't keep doing. This leads to us believe that the company is going to have problems down the road." She also spoke with Robert Friedman, a financial analyst at S&P, who had been an early critic of GE's financial engineering. He was concerned about the juxtaposition of GE's high price-to-earnings multiple and its "relatively low" return on capital. "Sooner or later the chickens are going to come home to roost," he told Morgenson.

Morgenson's article caused GE's stock price to fall more than 5 percent when the market opened on Monday. Jeff was pissed. GE posted a six-page refutation of her article on its website. It bought a full-page ad in *The New York Times*. Jeff went on CNBC. When he's unhappy, Bill Conaty said, Jeff "puts that football face on—like a player about to make a block, with four-letter words sprinkled in for effect." The genial Jeff Immelt was nowhere to be seen.

For his first annual meeting as CEO, in May 2002, Jeff returned to Waukesha, Wisconsin, the home of GE's medical equipment business. As CEOs often do, he complained about GE's stock price, which was then around $30 a share and 15 percent below where it had been on September 10, the day he became chairman. "I hate where our stock price is today," he told shareholders. "Our stock price today doesn't reflect the performance of this company or its value." He couldn't figure out why the stock had fallen, when GE had grown earnings 17 percent in the first quarter while the earnings of the broader S&P 500 index had fallen 20 percent. "It really is simple and powerful," he said of GE. "You see leadership—number one businesses, a strong and accountable culture, and massive financial strength." Jeff also was on top of the zeitgeist about increased disclosure then roiling the capital markets. "You want to know even more about GE," he said, "and that's fine with me. I welcome it."

Jeff was also leaning heavily into the myths that Jack had nourished during his long tenure. "We're not like everybody else," Jeff said. "We are blessed to have the most talented leadership team in the world. . . . We like to make numbers and we look at the stock price as a way to keep score. But, above all else, we value the reputation of this company. No one, not one leader in this place, would ever put personal gain ahead of the company. We perform and we always do it with integrity."

AGAINST THE BACKDROP OF GROWING INVESTOR AND MEDIA CONCERN, Jeff made his first major strategic move: in July 2002, he split GE Capital into four separate parts. By then, GE Capital was the world's largest nonbank financial institution, with $460 billion of assets. The idea was for Jeff to get a better understanding of how GE Capital worked. Although he disputed the assertion, corporate finance and the business of banking were neither Jeff's inclination nor his particular interest. He was more focused on marketing, manufacturing, and thinking big thoughts. The reorganization of GE Capital was his way of getting a better handle on a business that was growing fast and generating close to 50 percent of GE's earnings. As a result, the heads of the four divisions—insurance, commercial finance, equipment finance, and consumer finance—would report directly to Jeff.

His decision meant the abrupt end of Denis Nayden's nearly twenty-six-year GE Capital career. In announcing the reorganization, Jeff said he had been thinking about it since he took over as CEO in September 2001. In his book, he made the decision sound much more personal; it was clear he and Nayden weren't getting along. He explained how when he met with Nayden in the wake of the Gross commentary and the contentious meetings with the credit-rating agencies, to discuss how "we should respond," Nayden "pushed back, basically telling me to mind my own business." He quoted Nayden telling him, "How about you take care of your stuff and I'll take care of mine." (Nayden told me this meeting never happened and he never said anything remotely like this to Jeff. "Those are absolute lies," he said. "That's bullshit.") Jeff allowed that he knew Nayden was "a pit bull" but decided that "he'd crossed a line" and "seemed determined to keep crossing it." GE Capital had become the "the tail . . . wagging the dog" and "that was the way its leaders liked it." Jeff decided Nayden needed to go. "I felt shitty about that," he wrote. But not so shitty that he would change his mind. "Nayden believed he'd earned the right to run GE Capital his way and I could see why," Jeff continued. "Over the previous decade, nobody had done more for GE than he had. But he didn't think he needed a boss. And times were changing in a way that made his imperious attitude unworkable." To get the job done of executing Nayden, Jeff called Conaty, who was still the HR boss. Jeff referred to him as "Mr. Wolf," after Harvey Keitel's character in *Pulp Fiction*, the man who cleans up messes. "Mr. Wolf, I've got a job for you," Jeff told him. "I just eliminated Nayden's position." It was Conaty's job to figure out the details and to come to terms with Nayden on his unexpected departure. Dammerman put a spin on Nayden's removal and on Nayden's decision to become a paid adviser to GE while he found a new gig. "There is no other comparable job in the company," he told the *Journal.*

Nayden's version of the events that led to his departure—the first major executive change of Jeff's tenure—couldn't have been more different. He said Jeff made the decision unilaterally to split GE Capital into four parts. Nayden was never part of the deliberations. "He was telling me what he decided to do," Nayden told me. It was presented to him as a fait accompli. When Jeff told him his decision, Nayden was so shaken and surprised that he spilled his glass of water onto the conference table. "His explanation was that he thought that they would be better run as individual units, number one, and then number two, if I remember this correctly, he wanted to be closer to the actual business units," Nayden continued. "I'll never forget this. I was shocked when he told me this decision and said, 'Well, what does that mean for me?' He hadn't thought about that." There was a long pause. He told him he wouldn't get to be a GE vice-chairman, which would have been a promotion. "Oh, well, you can run one of the units," Jeff told him. Nayden looked at the new CEO in disbelief. "Really?" he said. "I don't think that's going to work for me." He didn't question the fact that Jeff was the boss and could make whatever decisions he wanted, either unilaterally or consensually. "Smartest guy in the room, right?" Nayden said of Jeff. He just disagreed with the decision. He thought his treatment unfair. "After spending my whole life trying to build this company," he said. There were also rumors floating around GE that Nayden had wanted to be the co-CEO of GE with Jeff, in recognition of the obvious fact that GE Capital was generating half of GE's earnings and that Jeff was not exactly a student of finance. Obviously, there was no way Jeff was going to agree to that request.

After Jeff shared with him the shocking decision, Nayden sat down and wrote Jeff a long, heartfelt five-page handwritten letter. The substance of his message, Nayden said, was that he was a "very dedicated, committed, and successful GE executive" and that "my record, frankly, was unparalleled." He wrote that he had assembled a world-class team and made the company into one of the leading financial institutions. "We delivered all the time," he wrote to Jeff. "So as a business matter, I'm proud to lead that effort and the idea of breaking it up, frankly, I don't think that's the best thing to do." He said, though, that he made sure Jeff understood that Nayden understood, "Hey, you're the boss. I'm a team player, and I think I can contribute materially to the future of this company." Nayden said he told Jeff, "I'm outspoken. Yes, I'm a tough-minded Irishman. Yes, I'm vocal. But two things. One, I always did what I thought was in the best interest of the company, no matter what, and I always voiced the opinion to back that up. That's one. Then two, I completely respect that you're the chairman and the CEO. You're the boss. And as long as I'm treated as your partner, and as long as it's recognized that whatever I do, I always have the best thing for the company in mind, I'm fine." He told me "there was absolutely no demand for power sharing, none. I explicitly recognized that he was the chairman and CEO, that he was the boss, and that I had demonstrated over a very long period of time that I could work with and for very strong, tough managers."

Nayden said Jeff never responded to his letter but did let him know he had no future at GE. Having Nayden stick around GE in another senior role "was not a compelling proposition" for Jeff, Nayden said. Nayden believed Jeff got rid of him because he saw him as a threat. "He considered himself the smartest man in the room," Nayden said. "When he made up his mind, he didn't give a shit what anybody else said. That was his MO. I readily admit I spoke my mind. That was my job. I gave the company, I gave the board, I gave when it was Welch, when it was Wendt, I gave them my best thinking as to what I thought we should do and why, and why I thought it was in the best interest of the company every single time. Did we always agree? Absolutely not. Did we have debates, if not verbal fistfights? Absolutely. . . . They were able to deal with strong-willed, opinionated, and capable people."

Nayden went to see Mr. Wolf as instructed. They were friends. Conaty suggested Nayden go talk to Jack, who "was still watching" what was happening at GE. (He never stopped watching.) Nayden went to Jack's new office in Shelton, Connecticut. Jack said to him, "I'm sorry this is the way this is playing out. . . . Den, when there's a change of leadership and a change in style, sometimes leaders just want to add their guys, and rightly or wrongly, Jeff just believes you would be the second coming of Gary Wendt, who he hated." Nayden explained to Jack that he believed he was a very different leader from Wendt. "You've gotta be kidding," Nayden told Jack. "I worked for Wendt all those years, we were partners, we were very successful, but style-wise, leadership-wise, personal relationship-wise, not even close." But that was Jack's opinion. "You're better off just looking around," Jack told Nayden. He suggested that Nayden look at private equity and, in the end, Nayden decided to join Oak Hill Capital Partners, a private-equity firm started by J. Taylor Crandall and the Bass brothers of Texas. Summing up Immelt's style of leadership, Nayden remarked that Jeff was just "a salesman" who didn't listen to others or take their advice. "He always made up his own mind," Nayden added. "He decided what he wanted to do, regardless of what anybody else said."

———

AS JEFF WAS COMING UP ON THE END OF A BUMPY FIRST YEAR IN OFfice, Jack once again stole the spotlight. And once again in a way that raised serious questions about GE's transparency and ethics. Around Labor Day 2002, Jack's estranged wife, Jane, filed documents with the

Superior Court in Bridgeport, Connecticut, as part of their ongoing divorce, outlining the extent of Jack's retirement package with GE. The absolute size of the retirement package and its quirky details were shocking enough—especially for a man already worth around $900 million. But the fact that GE had failed to make the extent of the agreement public in 1996, the year the GE board and Jack agreed that he would extend his stay atop the company until he was sixty-five, was further distressing. Jack was supposedly thinking of retiring a few years before he turned sixty-five, and these undisclosed perks were part of what the GE board decided was necessary to induce him to stay until 2001 and then to stay on as a paid "consultant" thereafter—at a rate of $86,000 a year for his first thirty days of work and another $17,307 a day for each day spent consulting after the first thirty—although he did no consulting for GE after he retired because, as he told me, Jeff never called him or spoke to him. In any event, for a company that had been preaching transparency and integrity in the wake of a barrage of corporate accounting and deception scandals, the failure to disclose Jack's lucrative retirement package was bound to make headlines and drag GE back in the unwanted spotlight at the very moment Jeff was trying to make amends with the investment and analyst community. To be fair, GE had disclosed in its public filings that for the remainder of his life, Jack would get many of the perks he had received as GE's CEO, "including access to company aircraft, cars, office, apartments and financial-planning services," as well as reimbursement for "reasonable" travel and living expenses. But no cost estimates for these expenses were provided or subsequently disclosed until Jane's filing in the divorce proceedings.

Jane released the details of the agreement between Jack and GE publicly because she wanted to stick it to Jack. Their marriage had come to an end in a public and inelegant fashion. The denouement to Jack's second marriage began as he was promoting his bestselling memoir in the fall of 2001. In fact, At 7:36 a.m. on the morning of September 11, while Jeff was in Seattle, getting ready to hit the StairMaster and his meeting with Boeing engineers, Jack was in the ground floor studio of the *Today* show. It was publication day for *Jack: Straight from the Gut*, and a conversation with his friend and golfing buddy Matt Lauer was his first stop on his planned whirlwind media tour. Lauer asked Jack, looking spiffy in a powder-blue, spread-collar shirt and yellow foulard tie, how it felt to have abdicated his throne four days earlier. He said he felt "terrific" and had had "one helluva party" the previous Thursday night. Lauer asked about many of the obvious touchstones of Jack's career at GE: Neutron Jack, PCBs in the Hudson, the massive increase in the stock price, and the Honeywell miscarriage. Lauer asked if he regretted not leaving as GE's CEO after ten years, as he had once contemplated doing before being talked out of promising a term limit. He didn't.

Lauer also wondered if Jack could have stayed as GE's CEO. He said he could have but that it was time for new blood and that GE was filled with talented executives. He had no regrets. "It's time to repot the plant," he said. "Get some exciting new faces." Ten minutes in, at 7:46 a.m., it was a wrap, and Lauer asked Jack if he could, what, if anything, would he take as a "mulligan"—a do-over, in golfing argot—during his twenty years as GE's CEO? You could see Jack was thinking hard about how to answer. "I don't take mulligans," Jack said. An amused Lauer replied, "I'm going to hold you to that next time we hit the course." It was time for a commercial break. The next guest was Massachusetts senator Ted Kennedy. An hour later, everything would change.

One of the interviews Jack agreed to do as part of his rescheduled October book tour was with Suzy Wetlaufer, the editor of the *Harvard Business Review*. They agreed to meet in New York City. Wetlaufer flew down from Boston. Jack loved hanging around Harvard Business School—even though he pretended to disdain the Ivy League—and Harvard Business School seemed to make a specialty of fawning case studies about Jack and his tenure at GE and how he transformed the place into the world's most valuable and important company. So naturally, Jack couldn't resist giving Wetlaufer and the *Harvard Business Review* an interview, even though its relatively modest circulation—said to be around a quarter of a million influential people—was not likely to result in a big bump in book sales.

What happened next would make headlines for months, sullying what had been Jack's nearly pristine reputation and grabbing attention away from Jeff and what he was trying to accomplish in his early days at GE. There was immediate chemistry between the two of them during the first interview. In the weeks that followed their first meeting, Jack and Suzy spent hours on the phone, working together to edit the interview, an approach between journalist and source that was encouraged at *HBR*.

Born Suzanne Spring in Portland, Oregon, she was the third of four children. Her father was an architect. Her mother had a doctorate in education from Columbia University. She was a painter and a sculptor. The Springs moved to Princeton, New Jersey, and then to Harrison, New York, in Westchester County. Suzy went to Exeter, where she met Eric Wetlaufer, who was three years younger. They were a couple at Exeter. He went to Wesleyan. She went to Harvard, where she was the features editor of *The Harvard Crimson*. After graduating from Wesleyan, Wetlaufer went into the money-management business in Boston; Suzy pursued a career in journalism.

After a short stint at *The Washington Post* and two years at the *Miami Herald*, in 1984 Suzy returned to Boston to work for the Associated Press. The next year, she and Wetlaufer married. In 1986, she enrolled at Harvard Business School. After she graduated as a Baker Scholar—the school's highest honor—in 1988, she took a job at Bain & Company, the management consultancy, in Boston. But the peripatetic Suzy didn't stay long. She left Bain after a year to have her first child. By the time Suzy and Jack met, the Wetlaufers had four children together, aged seven to thirteen. She taught Sunday school at a local

Protestant church. In addition to raising her children, Suzy wrote a novel, *Judgment Call*, in 1992, about a female journalist—a "beautiful, fun sassy-mouthed flirt"—who started having an affair with her source, a Miami coke dealer. According to *New York* magazine, Disney optioned the novel. (A second novel she wrote was never published after a dispute between her and her editor at William Morrow.)

In 1996, Suzy joined *HBR* as a senior editor. She tried to jazz up the publication, founded in 1922 and controlled by Harvard Business School Publishing, to make it less staid and academic. She did interviews with big-time CEOs such as Michael Eisner at Disney, and Jacques Nasser at Ford. In October 2000—a year before her interview with Jack—she was named editor in chief, with a mandate to keep making the publication more relevant and slightly more edgy. Four months after her elevation to the top job at *HBR*, her fifteen-year marriage ended unexpectedly. Citing an "irretrievable breakdown," in June 2000, Eric Wetlaufer up and left their million-dollar home in Lexington, Massachusetts, leaving Suzy to care for their four children.

Suzy, the editor in chief, began a relationship with Joe Maurer, a twenty-four-year-old editorial assistant at *HBR*. It was just a few months after her husband had walked out on her. According to a profile of Suzy in *Vanity Fair* written by Suzanna Andrews, after a "lavish" party at the Harvard Club, in Boston, in January 2001, to celebrate Suzy's decision to increase to ten the number of *HBR*s published each year, the two reportedly engaged in a "lip-lock." They would also make "googly eyes" at each other. When Maurer's apartment flooded, Suzy invited him to stay in the basement of her house in Lexington. Eventually, Maurer left *HBR* to go back to school to become a doctor.

There were other alleged dalliances, too, in the months prior to her October 2001 meeting with Jack. A month after their first visit in New York City, Suzy returned again to see him, under the ruse of needing to have a photo session for the *HBR* article. She had wanted to have her photograph on the page where she wrote her editor's note.

Her boss said okay, as long as Jack was also in the picture. After the photo shoot, the two had a long lunch at the 21 Club and then went dancing that evening. Then it was back to Jack's apartment in Trump Tower for an assignation, or so the consensus seemed to be. "I've thought back on what I could have done differently," a colleague remembered her telling him after the fact, "and I suppose, yes, there was a moment when we were standing on Fifty-Second and Fifth"—down the block from the 21 Club—"when I could have just said, 'No, no, this isn't right.' But I didn't." When she returned to Boston, she told colleagues that Jack had told her he wanted to make her the next "Mrs. Welch." When the gifts of jewelry began arriving for her from Jack, the staff began wondering if a line had been crossed between journalist and source. Confronted about a diamond bracelet from Jack, Suzy replied, "It wasn't *diamonds*."

The burgeoning relationship between Jack and Suzy might have just stayed a matter of local gossip within the confines of the *HBR* offices and not become an issue for GE, had Jack's estranged wife, Jane, not confronted Suzy the day after Christmas 2001. Accounts differ as to what tipped Jane off to Suzy. One rumor had her picking up a phone extension during a December call between Jack and Suzy. Another suggested that Jane had done some reconnaissance work on Jack's computer.

According to *New York* magazine, Jack was so "smitten" with Suzy that he confessed the affair to his wife. Jane reportedly called Suzy and questioned her journalistic integrity. After the call from Jane, "finally aware of how serious the situation was," Suzy called her boss, told him the "nature" of her relationship with Jack "had changed," and "insisted" that the interview be killed. Then, in late December, Jane sent Suzy an email. "You two ought to be more discreet," Jane wrote.

Suzy's boss still wanted to run an interview with Jack, despite the surprising turn of events. On New Year's Eve day, two other senior editors from *HBR* were sent to reinterview Jack for the publication at his home in North Palm Beach. The two editors weren't happy, nor was Jack. He acted like he couldn't figure out why he was being reinterviewed. When he cut the interview off, saying he had to have lunch with Jane, it was all the two editors could do to keep from bursting out laughing. (The interview appeared in February 2002, and later an editor's note was appended sharing the fact that Suzy Wetlaufer had done the original interview, which was later scrapped because "she'd commenced a romantic relationship with Welch after interviewing him for the magazine.")

On January 16, 2002, one of *HBR*'s top editors wrote to Suzy's boss. "Dear Walter," he wrote in an email to Walter Kiechel, the editorial director of Harvard Business Publishing. "Emulating the brave lady at Enron who sent a warning letter to Ken Lay because she thought he should know the facts, I want to make you aware of what life is like here. . . . The *HBR* culture is in shambles." *New York* described the email as "long" and "anguished" and as sharing his concerns about "the reputation of the magazine he had come to love and the editor who loved too much." The editor continued, "The undisguised kick she gets from companioning high-profile, powerful men is one of the reasons some of us on the staff swallowed our qualms about the way she mixes her personal and working lives. I mean, who wants to be the killjoy who spoils her transparent sense of fun?" He signed off "in sorrow and anger both."

In February, the staff insurrection against Suzy Wetlaufer intensified. After polling the staff, two editors went to Suzy and asked her to resign. She refused. Six editors wrote to Kiechel demanding that he remove Suzy from the top job at the publication. By then, word started circulating inside the *Harvard Business Review* that James Bandler, a highly respected reporter in the Boston bureau of *The Wall Street Journal*, was starting to track the growing controversy with the intention of writing about it. Suzy had heard the rumors about Bandler too, and had taken to wondering who the "mole" was.

Jack, this time accompanied by Suzy, again went into "catch and kill" mode, trying to keep Bandler's story about them from being published in the *Journal*. The *New York Daily News* reported—accurately—that Jack called his friend Paul Steiger, the managing editor of the *Journal*, and tried to prevent Bandler's story from being published. Jack tried to convince people that his relationship with Suzy was no different from the ones he had with many reporters, including with Steiger. Few believed him. Suzy was bragging to the *HBR* staff that Jack had gotten the story killed. That further irritated *HBR* staffers. It "made them feel really sort of helpless, because this hapless *Journal* reporter spent weeks and weeks reporting on it and then suddenly, she was saying, 'Jack's killed the story,'" someone familiar with the machinations inside *HBR* said.

But when word got back to the *Journal* that Suzy was going around *HBR* saying that Jack had gotten Bandler's story killed, the *Journal* editors reversed course. "You don't want to have that be true," explained someone familiar with the *Journal*'s thinking. On March 4, Bandler's story appeared in the *Journal* on the front page of the third section. While it was skillfully written and edited to tamp down the more salacious details, the story landed like a bombshell, as one might expect, given that Jack was married and the newly retired CEO of GE, out on the trail plugging his new bestselling book, and that the newly divorced Suzy Wetlaufer was the editor in chief of the *Harvard Business Review* and twenty-four years younger than Jack. It was delicious tabloid fodder appearing in the august pages of *The Wall Street Journal*—years before it was sold to Rupert Murdoch.

Bandler's story described "a mutiny" at the "vaunted" *Harvard Business Review*. And there was no arguing that. But the bigger news was that Jack and Suzy were having a "romantic relationship," which was why the staff was in revolt and why Suzy had urged Kiechel to pull the interview she had done with Jack, which was slated for the February issue and then swiftly replaced by the one done by her colleagues

with Jack on New Year's Eve day. She told Bandler her romantic involvement with Jack began "several weeks" after the interview with
him was written, edited, and sent to production. The situation, she
said, "changed" after the photo session in New York City, followed by
their lunch at the 21 Club. She confirmed she pulled the article soon
after receiving the phone call from Jane Welch in December. She told
Bandler the matter was "private" and added that "personally and professionally, I feel very proud of the way I handled this situation."

For his part, Jack told Bandler—before he tried to kill the story—
that a "friendship" with Suzy developed during the course of the interview and getting it ready for publication. He called her "quick" and
"funny" but declined to comment about whether they had become ro-
mantically involved. "I don't talk about my personal life," he told
Bandler.

The fallout from Bandler's story was nearly immediate. The day
after Bandler's article appeared, Suzy's bosses presented her with what
seemed like a graceful exit: a four-month leave of absence followed
by a new role as a consultant, going into the office one day a week.
That same day, Jack returned from a vacation with Jane in Italy—
that must have been special—and flew up to Boston on the private
jet that GE provided to him. He advised Suzy not to accept Harvard's
offer.

Jack told Suzy to fight. He hired Robert Popeo, a fierce Boston attorney, to represent her. He had represented GE in its ongoing fight
with New York State and the EPA about the PCBs in the Hudson. Jack
also called Kiechel's boss, Jim Cash, the chairman of Harvard Business School Publishing and also a longtime member of the GE board
of directors, to help him try to achieve some sort of soft landing for
Suzy. On March 8, Suzy sent her *HBR* colleagues an email with an at-
tachment outlining the terms of her deal with *HBR*. Under the agreement that Jack had effectively negotiated for Suzy, she would continue
to work full time at *HBR*, keep her office, and forgo her management

responsibilities. She would resume these duties in late April, not af- ter a four-month leave. "Harvard basically cowed to Jack Welch," one *HBR* staffer told *New York*. At a staff meeting with Kiechel that same day, people were furious that Suzy would still be around. "What does someone have to do to get fired from Harvard?" someone yelled. Once again, the terms of Suzy's departure were renegotiated. She would work from home for the following few months.

In any event, she was thrilled. She seemed to be well on her way to rebounding from her divorce to marrying Jack Welch, the former CEO of the Century, with a net worth estimated to be nearly $1 billion. "You've finally met your Aristotle Onassis," one of Suzy's friends told her. Jack was telling people he had met "the love of my life." He met her children in mid-March.

ON MARCH 13, 2002, THE NEWS BROKE THAT JANE WELCH INTENDED to file for divorce from Jack once an amicable division of Jack's assets had been made. Jack, like his old colleague Wendt, made the mistake of lowballing his soon-to-be-ex-wife. Jane found Jack's initial offer of $20 million to be "an insulting amount from a man said to be worth $900 million," according to *The New York Times*. The details that came out in Jane's filings were stunning. There was the Manhattan apart- ment at One Columbus Circle—the former Paramount Communica- tions building that GE financed for Donald Trump's makeover of it into condominiums—and the meals paid for at Jean-Georges, the fancy French restaurant on the building's ground floor. There were courtside seats for the New York Knicks, courtside seats for various tennis tour- naments broadcast on NBC, including the U.S. Open and Wimbledon. There was also the fact that Jack's agreement allowed that GE would pay for all of Jack's expenses at One Columbus Circle, including wine, laundry, food, toiletries, and newspapers, and that GE would also pay the costs of providing Jack with satellite TV at his four homes—

Manhattan, North Palm Beach, Nantucket, and Fairfield County. There was also a subsidy for a car and driver and for security when Jack traveled abroad. There were box seats at the Metropolitan Opera—despite Jack's disdain for it—and for his beloved Red Sox, and for the Sox's rivals, the New York Yankees, plus the membership dues at Jack's four country clubs. There were even discounts on diamonds and jewelry settings. There were other nuggets, too: an aviation expert put the cost to GE of providing Jack with the use of a Boeing 737 at about $3.5 million a year. GE also paid Jack and Jane $7.5 million for the cost to renovate and furnish their homes during their thirteen-year marriage, homes for which they paid $32.5 million, according to Jane's filings. The couple's monthly expenses were more than $126,000.

In a separate affidavit, as part of the divorce proceedings, Jack confirmed that he was raking it in, and that he was fabulously wealthy. His annual pretax pension from GE was $7.4 million, netting him $4.3 million in cash per year. From his various consulting and self-employment schemes, Jack was getting another $8.2 million a year, or $4.5 million a year after taxes. His annual dividend payments were $2.6 million a year. His taxable interest income was $354,000 per year; his nontaxable interest income was $6.7 million per year. His director's fees were $41,000 per year. He was also entitled to Social Security payments of $18,000 per year. Jack helpfully shared that his annual after-tax income was just about $17 million.

Jack also revealed that his annual living expenses were about $4.4 million, the majority of which, $2.4 million, was to reimburse GE for the personal use of the benefits that GE provided for him (after he had decided to give them back in the wake of Jane's filing). His monthly expenses for his homes were more than $50,000. He spent $20,000 a month on accounting services and $5,500 a month on country club memberships. He paid $14,000 a month in life insurance premiums and $2,500 a month in property insurance. He spent nearly $9,000 a month on food and beverages but only around $1,900 a month on clothing. His personal charitable giving was only around $600 per month, but his affidavit explained that the John F. Welch Jr. Foundation donated $3.1 million in 2001.

Things got more interesting when Jack shared the value of his assets. His homes in Fairfield, Connecticut, and in Nantucket were valued at $10 million each. His home in Lost Tree, Florida (just north of Palm Beach), was valued at $3.5 million—Suzy Welch sold it for $21 million in April 2020—and his home in Southport, still being built, was valued at $6.75 million. (It sold in 2007 for $10.75 million.) In sum, Jack valued his real estate at $30.6 million, including two parcels of land in Palm Beach and the one in the Berkshires he had bought when he left Pittsfield for Fairfield. As of September 2002, Jack had nearly $250,000 in cash. He had nearly $250 million invested in bond and equity funds, as well as $14.2 million invested in limited partnerships— essentially private-equity funds. The net present value of Jack's GE options was $73 million. The net present value of his after-tax deferred compensation from GE was about $30 million and of his "executive salary" another $22 million. A grantor trust he set up in 2001 was worth, net of taxes, $36.5 million.

As for his liabilities, Jack said he owed $15.2 million to trusts he had set up for his children and $1.5 million for his taxes. He owed an unspecified amount to GE for benefits the company provided him in 2002 that were personal, not business. Jack also shared that, per an agreement he had made with Carolyn, his first wife, in April 1987, half of the fair market value of his estate at the time of his death, net of various permissible expenses, would go ratably to their four children. In sum, Jack pegged his net worth at around $440 million plus his $12 million in annual after-tax income, less half—or around $220 million— that would go to his four children. What remained unclear from Jack's affidavit was the difference between his net worth of $440 million and the $900 million that the press reported. (His affidavit did not seem to include his 22 million shares of GE, then valued around $330 million.)

The public reaction to Jane's disclosure of Jack's postretirement benefits was one of general outrage. "I would have thought that perks like this had to be disclosed, and they were not," said Nell Minow, a respected governance expert, to *The New York Times*, which broke the story of Jane's court filing. "There is really no justification to pay for any living or traveling expenses at that level, particularly now that he is in retirement." She was just getting revved up. "It is appalling that one of the wealthiest men in America cannot write a check for his own Knicks tickets," she continued. "It is appalling to me that Jack Welch's flowers are being paid for by retired firemen and teachers who are the G.E. shareholders and don't know this is going on." Graef Crystal, an executive compensation expert, put the blame on the GE board of directors. He called the retirement package an "indictment" of the board. "This is the most appalling use of corporate assets," he told the *Times.* "No one had any idea of the magnitude of what the company had been giving him." A spokesman for the International Brotherhood of Teamsters, the union that represented many GE workers and that was also a GE shareholder, said, "The arrangement that Mr. Welch made with GE and with the board is just beyond excess."

Jack fought back. He told the *Journal* that Jane's filing and the subsequent *New York Times* story were taken out of context. "It's divorce posturing," Jack said. He said the lifestyle Jane described in her court filing didn't exist. "It implies that we had GE financing our personal life," he said. "Anything personal I paid for, anything business the company paid for." What annoyed him especially was the implication that somehow someone had done something wrong. "GE made a contract with me, disclosed it, and I and the team delivered on the contract by producing double-digit results, continuing a year after I'm gone, and we had a very smooth succession process," he said. He and Jane often talked about "public divorces and how awful they are, and I've tried everything I know not to have one. I don't want my divorce to be dragged into the papers."

He knew firsthand from the Wendt debacle the debilitating consequences of a public divorce. But it was too late now. "Obviously, one-sided filings don't look that pretty, but they don't reflect reality," he concluded. He told *Boston* magazine that he was used to being criticized. "I've been prince to pig three times probably," he said. "And I think there was a time when business was being treated that way, and I was a great symbol to get some of that. But I know who I am. I know that I'm good to my friends. I really like myself as a human being."

But liking himself was not enough to stem the ongoing press conflagration. *The New York Times* reported that, at a recent Wharton School conference, Jack had been included in a list of "once soaring corporate captains" who "have crashed to earth," along with Ken Lay at Enron and Bernie Ebbers at WorldCom. "Hardly company that Mr. Welch would want to keep," the *Times* wrote. There were the questions raised about GE's accounting and the "eerie consistency" of its earnings increases.

The revelations about Jack's retirement perks also got the attention of the SEC, which announced it was investigating whether GE had properly disclosed the arrangement in 1996, and subsequently, and whether GE was properly accounting for the deal. Jeff was getting irritated, too, by the negative publicity that Jack was again bringing to GE. He wanted to put the matter behind him and the company, and fast. The perks for Jack "didn't play well in the wake of Enron and WorldCom," Jeff explained, "and we unwound them."

With the pressure mounting, Jack took to the pages of *The Wall Street Journal* to explain to the business world that he and the GE board had decided to make material changes to his 1996 agreement. "I want to share a helluva problem that I've been dealing with recently," he wrote, with his usual candor. He wrote that the court filing and the press "grossly misrepresented" the details of his employment agreement. He didn't want to refute every error. But he wanted to set the record straight on a few things. "For the record, I've always paid for my personal meals, don't have a cook, have no personal tickets to cultural and sporting events and rarely use GE or NBC seats for such events," he wrote. "In fact, my favorite team, the Red Sox, has played 162 home games over the past two years, and I've attended just one."

He conceded the world had changed post-Enron. What looked like a fair and equitable arrangement in 1996 didn't look so great in 2002. He didn't know whether to keep his agreement with GE in place and tough out the bad publicity or try to put an end to it by amending his agreement. He talked to GE board members, to Jeff, and to other confidants. "One thing I learned during my years as CEO is that perception matters," he wrote. "And in these times when public confidence and trust have been shaken, I've learned the hard way that perception matters more than ever. In this environment, I don't want a great company with the highest integrity dragged into a public fight because of my divorce proceedings." He agreed to rescind *all* of the perks except for the office space and administrative help that former GE chairmen and top executives traditionally received upon retirement.

Everything that GE was providing to him he would start paying for himself—at a cost he estimated to be between $2 million and $2.5 million per year. He also agreed to consult to GE for free, on an "as-needed basis," which turned out to be nothing, and to teach at Crotonville if asked. He couldn't resist a Jack-style kicker: "This decision may not satisfy everyone, but it sure feels right in my gut."

In the meantime, there was still the Jack, Jane, and Suzy show. No doubt Jack thought, or hoped, Jane would more or less acquiesce to the divorce proceedings and find a way to settle matters quietly, if not amicably. He was mistaken. Jane hired William Zabel, the high-powered attorney who had worked for both Howard Stern and Michael Crichton in their highly publicized divorces. While many of his fellow travelers in CEO world urged him to find a way to settle with Jane, in order to make the increasingly negative publicity go away, Jack, ever feisty, took the opposite tack. He went on the offensive against Jane.

As she continued to rebuff his settlement offers, he began to tell friends about Jane's affair with Paolo Fresco's chauffeur. Fresco, by then the chairman of Fiat, was a former GE vice-chairman and a longtime friend of the Welches.

Word of Jane's affair hit the front page of *The Wall Street Journal* in November 2002. The paper reported that Jack had been telling people that Jane's affair with the "handsome Italian," six-foot-two-inch driver began before his affair with Suzy, "sometime in 2001." They met in Florence and the affair lasted "several months," according to the *Journal.* When Jane was back in the U.S., she kept up with the chauffeur. It was that correspondence that Jack uncovered. Jack's tactic was to embarrass Jane into returning to the negotiating table. The paper reported that Jack told "associates" that he had a "collection of e-mails, phone records and faxes to prove" Jane's infidelity. Some observers found Jack's behavior unsavory under any circumstances but especially so given the rash of corporate scandals that were already sullying the reputations of American CEOs. "The timing of this is troubling, given the current scrutiny of CEOs and the austerity moves of many companies," Jeffrey Sonnenfeld, associate dean of the Yale School of Management, told the *Journal.* "And since it is Jack Welch, the paragon of corporate leadership, who is under the spotlight, executives worry that the public will think, 'What can we possibly expect of everyone else?'"

Part of Jack's problem was that he didn't want to be "a two-time loser," as he told people, and now he was. Walter Wriston, his longtime friend and a GE board member—and the man who had introduced Jack to Jane in the first place—urged Jack to tone it down. His public feud with Jane would "hurt the company," he told Jack. Jack was facing a big financial hit. When Jack and Jane married, she signed a ten-year prenuptial agreement. Had they divorced before the ten years were up, Jack would have had to pay Jane $4.9 million. But they were five years beyond that agreement. In 2002, Jack was facing a settlement in the range of hundreds of millions of dollars, around a quarter of his nearly $900 million fortune. After his decision to renounce the bulk of his GE perks, Jack decided he needed more legal firepower in the fight against Jane. He again turned to Dan Webb at Winston & Strawn as he had in past sticky situations. Webb urged Jack to settle and put the matter behind him as quickly as possible. Webb suggested increasing Jane's monthly payments beyond the existing $35,000—an offer that Zabel accepted—and he suggested increasing her overall settlement to a combination of cash and assets that Webb pegged at a value of $130 million. Zabel rejected the offer.

Webb lashed out, seemingly taking a page from the awful Wendt playbook. "She's a second wife who married a man who had already been CEO for eight years," he told the *Journal*. "Since they married, she's never worked a day, never done a dish, never mowed the yard, never made any substantial contribution to his career at GE—and she had a great life, with no kids." She made no "significant contribution to his extraordinary career and rise to power at GE." Zabel was quick to point out, though, that, in his book, Jack had praised Jane in his acknowledgments. "Jane was the perfect partner," Zabel riffed. "She went on all of his corporate trips. She took up golf so she could play golf with all of his business associates. She saved his life." (Whether contractually obligated to do so or not, Jack decided to excise Jane from the acknowledgments of the 2003 paperback edition of *Jack: Straight from the Gut*.)

On July 3, 2003, four days before Jack and Jane were to appear in a Bridgeport courtroom, the two settled. In a terse press statement, they announced that they were divorced. Jane Welch would be known again as Jane Beasley. According to published reports, Jane walked away from Jack with $180 million. As part of the settlement, Jack revealed in a May 2006 interview with *Boston* magazine that he had contractually agreed never to speak about Jane again. (By then, he and Suzy had married, in April 2004 at the Park Street Church in Boston, and were working on their first book together, *Winning.*) One former Wall Street CEO believed Jack had finally found true love. He said Jack used to tell him, "If it flies, fucks, or floats, rent it, don't buy it." But, he said, "he's very happy with Suzy."

STILL TRYING TO CONTAIN THE DAMAGE FROM BILL GROSS AND THE detailed disclosure of Jack's postretirement-benefits contract and Jack's affair with Suzy, Jeff gave an interview to Ron Insana, a CNBC anchor. "Over the past two years, Wall Street has gone from asking whether Tyco"—a mini-GE-like conglomerate that had run into its own version of an Enron scandal—"is the next GE to whether GE is the next Tyco." Not surprisingly, Jeff did not like the question. "Ron, we're nothing like Tyco," he said. "I'm not here to define Tyco one way or the other, but I can tell you GE has no similarities to what Tyco's doing or has done." Insana also asked about the claim that GE managed earnings, that Jack did it and that he was continuing to do it. He didn't answer the question directly. He said that "one thing we've been very public about" was that GE did offset gains with losses, and vice versa. "As Jack Welch used to say," Jeff said, "'We pay our bills quarterly.' I plan to continue doing that." It wasn't clear what Jeff meant. But it sounded good. Asked about his particularly challenging first year, Jeff deflected here, too. "The way I was brought up in this company is to know that you can never predict the future," he said. "That all you can do is build a culture and have great people and focus on change."

Despite Jeff's efforts at damage control and boosting investor confidence, the skepticism about GE and what it had become continued nearly unabated. At Jim Grant's fall investor conference, scheduled for October 24 at the St. Regis Hotel in Midtown Manhattan, Ravi Suria, a managing director at Duquesne Capital Management, a hedge fund run by famed investor Stanley Druckenmiller, was scheduled to give a presentation about GE, "fact, fiction and book value."

*The Wall Street Journal* described Suria as a hedge fund "star," noting that his claim to fame was a hugely contrarian, and terribly wrong, call that Amazon would file for bankruptcy. (As of this writing, Amazon's market value was around $1.5 trillion.) Suria pored over GE's public filings and scoured the numbers. He was always perplexed why the Wall Street research analysts who covered GE understood industrial businesses but not financial businesses. GE was most definitely a financial business, Suria concluded. He thought GE was benefiting disproportionately by having a finance company attached to an industrial company. GE Capital could create earnings for itself (and for GE) by providing financing to the buyers of GE's big-ticket items, such as jet engines or power plants. Those profits would have gone to a third party if there were no GE Capital. He believed that flexibility gave GE and GE Capital the ability to mark the value of loans however they wanted, based on their own internal assessment of the credit. With a captive industrial business, he believed, you could do wonderful things for the financial unit with creative accounting. Only that could be very misleading to investors. He never believed that the world moved with the smoothness that GE portrayed in its quarterly earnings numbers. That's what he planned to discuss at Grant's conference. It was more of an intellectual exercise than anything else. Duquesne did not have a stake in GE's equity or debt, on either the short or the long side.

But before Suria could give his talk, things took a bizarre turn. That morning, CNBC's Jim Cramer, who had been a client of Suria's at Duquesne, began talking up the fact that Suria was going to give a presentation at the Grant conference questioning GE's accounting and the quality of its earnings. Since Jeff took over as CEO, betting against GE stock had become a favorite topic among Wall Street short sellers. A presentation by a hedge fund manager from Druckenmiller's firm questioning GE's accounting and earnings was bound to get people talking, and GE's stock falling. Several Wall Street equity analysts picked up on Cramer's comments and inserted them into their morning "blasts" for the traders to think about, just as the markets were opening for the day.

Suddenly, on trading desks across Wall Street, at firms such as Goldman Sachs and First Boston, traders were buzzing about what Suria might say about GE. By the time Suria got to the office, still half asleep, he more or less knew what he wanted to talk about. He had the slides. But there wasn't a script or anything. Still, there was no question the content of Suria's presentation had the potential to be explosive.

The phone calls were flying around Wall Street. Grant was flooded with calls from people who wanted to hear what Suria was going to say. Calls also went to the GE board about Suria's talk. Ken Langone, the outspoken GE director, got calls about Suria's presentation. Langone and Druckenmiller were longtime friends. Langone phoned him. "What the fuck is your analyst doing?" Langone screamed. Druckenmiller called Suria into his office. "What the hell are you doing?" Druckenmiller asked him. "I didn't know this thing was going to happen." (This wasn't exactly accurate; Suria's presentation had been much discussed around Duquesne before the Grant conference and wasn't created solely for Grant. It had been used with a variety of Duquesne's clients.) They decided it probably wasn't a good idea for Suria to give his presentation at the Grant conference, stoking the flames of concern about GE. Suria called Grant and told him he was not going to be able to give his talk after all.

Grant was pissed. Suria apologized. His job was on the line and his boss had decided the presentation could not be made. In dramatic fashion, Grant announced that Suria would not be making his presentation. CNBC picked it up and for the next thirty minutes ran the chyron "Where in the world is Ravi Suria?" People were calling him, wondering, "What the fuck is going on?" As a compromise of sorts, Suria agreed to give his GE presentation to a select group of Grant's biggest supporters at a dinner the night of the conference at a private club. By

then, GE's stock had fallen more than 3 percent for the day. At the dinner, as he gave his presentation, Suria was seated next to Seth Klarman,
the renowned Boston-based hedge fund manager. Instead of giving his
talk before seventy people at the Grant conference, he gave his speech
to fifteen people at the private dinner.

But thanks to CNBC, the ramifications of Druckenmiller barring
Suria from giving the talk were more significant than if he had given
the talk. When Donald Sussman, the hedge fund manager who spoke
in the slot after Suria's, said, "It's too bad we live in a society where
people are afraid to say what they're thinking," the crowd erupted in
applause. After the dinner, Grant asked Suria if Grant could publish
his slides. Suria gave his permission. The main point of Suria's presentation was that, by 2002, GE was little more than a huge unregulated
bank—in fact, the fourth-largest in the U.S., with $473 billion of assets—
and that some 86 percent of GE's assets and 90 percent of its liabilities
were related to GE Financial Services, the holding company for GE
Capital and GE's insurance businesses. As had others, Suria questioned
the quality of GE's earnings.

He had slides on GE's gains from pension accounting, on GE's
gains from asset sales, on GE's gains from asset securitizations, and on
the impact of accounting depreciation and amortization. He had a slide
showing how little GE paid in taxes (an effective tax rate of 20.5 percent in the third quarter of 2002). He referred to GE's financial services business as "a rapidly growing black box." He accused GE of
shrinking its equity and "turbo charging" its "trading book." He had a
slide on the growing risks in GE Capital's portfolio of leveraged leases.
His deck had an amusing tagline: he quoted GE's slogan "We bring good
things to life" and added, "We bury the bad ones."

One of Suria's main concerns about GE was the way Jack, and then
Jeff, had mesmerized people about the company. They had convinced
Wall Street research analysts—who by and large were industrial ana-
lysts, not financial analysts—that GE Capital was only 40 percent of

GE's earnings. Or 45 percent of GE's earnings. And then 50 percent of GE's earnings. But that was the wrong way to think about it, Suria argued. Given that GE Capital was nearly 90 percent of GE's assets and that those assets were leveraged eight to one, there was a gargantuan unappreciated risk on GE Capital's balance sheet.

In effect, a small change in the valuation of the assets on GE Capital's balance sheet could wipe out GE Capital's equity, which in turn would wipe out a big chunk of GE's equity. Suria's insightful point was that if Jack or Jeff told people that GE was 40 percent financial and 60 percent industrial, they were implicitly passing on the message that only 40 percent of GE's risk was financial. But in reality, it was much higher because of the leverage on GE Capital. Jack's greatest trick, Suria conveyed in his presentation, was to find the right messengers and to speak to them in their language. Jack spoke to the industrial research analysts; if GE's Wall Street analysts had had a financial focus instead, they theoretically would have been less likely to fall so consistently head over heels with what Jack and Jeff were feeding them. That would change.

## Chapter Nineteen

# DILUTING THE BLOB

By the time Jeff addressed the meeting of GE's top six hundred executives at Boca Raton in January 2003, GE's stock was down 62 percent from its August 2000 high. "There wasn't one normal day in the last twelve months," he told the group in Boca. What used to be like a "Roman victory celebration" under Jack, wrote Matt Murray at the *Journal*, had turned "subdued" and "slightly anxious" under his successor. Jeff told his managers that 2003 was "going to be brutal" and that "we've got to change the company and we will change the company."

Jeff's vision for GE was becoming apparent. His GE would be "mining the growth in the old, core industrial business," he said, that "bore the brunt" of Jack's cutting; selling more GE products globally; and investing in GE's research and development centers. Focusing on manufacturing and innovation, the argument went, would help to dispel investors' concerns that GE was a black box. In any event, Jeff was much more of a salesman and marketing expert than a financial engineer; the change in direction would better suit his talents. Jack was on board. "The reason for management succession and change is always the opportunity for fresh eyes to look at things and take the company in the direction that the times require," he told Murray. "Whatever he's doing, it'll be right for the times. That's what change is about." Jeff's lieutenants—or at least the ones he allowed to speak up—were supportive. Bob Wright, still the head of NBC and a GE vice-chairman, said that Jeff was "undoing what Jack did" because the "markets are different" and his "situation" called for a "different response." Dave Calhoun, the head of the aircraft engine business, told Murray that Jeff was growing up right before our eyes. "He looks like a seasoned veteran now," he said. He added that it "was trying for sure," but the challenging economic and political environment would make GE and GE's management better. "When things are booming," Calhoun said, "nobody learns a damn thing. It's now that we'll find out if we were really serious about change being our best friend."

Jeff's determination to remake GE began to be seen in 2003. Whether he was driven by a desire to undo what Jack had done or by the slow return of investor confidence in the company almost didn't matter. After a tough eighteen months, he decided the time had come to do some big, transformative deals in an effort to remake GE with his bright ideas. He put priority on doing deals in the medical-equipment and alternative-energy industries, while also taking opportunities to expand what NBC had to offer. After his rough education in the credit markets, he also remained determined, if he could, to reduce GE Capital's outsize weight on the conglomerate's balance sheet. He needed to dilute what Jack had called "the Blob."

Longtime GE executives were beginning to notice that Jeff's approach to dealmaking was different from what they had witnessed with Jack. Quite different, in fact. One longtime GE executive told me that Jeff was the "antithesis" of Jack when it came to doing deals. "He doesn't want to talk substance," he continued. "He hasn't really read the material." Like many other GE executives who worked for both Jack and Jeff, he couldn't stop himself from drawing a comparison between the two men. "You had a team that built credibility by working for maybe the greatest leader of our generation, who respected your work," he said of Jack. Jack read everything sent to him. He knew more than you. "When you went up there, you had a discussion that exhilarated you, and you learned something, or he told you something you didn't know," he continued. "And you had to go back and figure out the answer to his question, and it lifted all of us." But it was different working with Jeff. "It was like walking into a cold shower when you walked into your first meeting with Immelt, because he didn't read the deck," he said. That seeming indifference was noticed. "He lost us," he said. "And this is one thing that I know as a leader: if you want your team's respect, you've got to read their work. I think Jeff lost most of us in the first meeting. We all turned to each other and said, 'He didn't read the deck.'" There would be consequences.

Jeff's dealmaking began in earnest in February 2002, with the $1.8 billion all-cash acquisition of the water-treatment business of BetzDearborn, which was a division of Hercules, Inc. Jeff believed the water-treatment business could be a new growth area for GE. "There was a growing awareness that water scarcity was the next catastrophe that was going to befall the earth," he explained. But what he really wanted were the two thousand engineers in the business who could help customers develop water-purification systems. GE quickly added Osmonics, a fluid filtration equipment manufacturer; Ionics Ultrapure, a desalination company; and Zenon, a company that made membranes that filtered impurities out of water. The combination of them made GE the second-biggest water-treatment company in the world. (But, Jeff later conceded, the effort was "a mistake." GE overpaid for each business and found itself unable to create a differentiated product. The business was sold to Suez in 2017 for a gain.)

And in April 2002, GE bought Enron's wind turbine–manufacturing assets for $358 million in a bankruptcy auction. The Enron business was one of the world's largest wind turbine businesses, with sales of around $800 million in 2001. At first, Jeff resisted the idea of buying the business, which had bubbled up to him from Mark Little, who would soon become GE's chief technology officer. Little had been recommending the acquisition to Jeff, and Jeff kept hanging up the phone on him. "This was a Hula-Hoop, a fad!" he kept telling Little. He thought it was a mom-and-pop industry, heavily reliant on government subsidies. What could GE do for a wind turbine business? he wondered. After the constant nagging from the troops, Jeff decided GE could bring its expertise to bear: it knew something about gas turbines—both windmills and turbines made electricity from spinning parts—and about propeller blades that were strong and could withstand all sorts of pressures. "The numbers made it an easier bet," Jeff decided. "The beleaguered Enron was selling its wind power assets at a fire sale price."

At first, Enron's wind business was a flop inside GE. It lost money "at a steady clip," Jeff explained. "We used to say it was as if we bought the business in 2002, then bought it again in 2003, and then again in 2004." At first, the GE engineers and business leaders could not solve the problems. Jeff put Little in charge of it. He and his team fixed the technical problems. They scrapped Enron's 70-meter windmill and built one that was more than 150 meters. They improved the electrical control systems and figured out how to turn the blades so that they would always be moving—and generating electricity—regardless of the wind direction. They figured out how to monitor and reset the windmills remotely. What had been a bad bet turned into a big winner and one that Jeff liked to point to as a singular success. After five years, he liked to say, "GE Wind went from a money loser to a tremendous winner. By improving both the technology and the business model, we created a $12 billion renewable wind business" that brought together the many components of GE's unique expertise to make it possible, from the power and locomotive businesses to aircraft engines and GE Capital.

There were dispositions, too. In addition to the oddly structured sale of Global eXchange Services to Francisco Partners for $800 million—in which GE Capital financed the acquisition for the buyout firm—Jeff spoke openly about selling GE's lighting and appliance divisions, two of the company's best-known and legacy businesses. He wanted to sell off GE's $3 billion portfolio of private-equity investments. He was anxious to get rid of the money-losing Employers Reinsurance. "Insurance is the easiest business in the world to grow earnings and the toughest business to fix when you've made the wrong moves," explained Bob Wright, who bought ERC when he ran GE Capital. "It's just impossible to repair." The policies get written, the premiums roll in, "and then somebody has to pay for it three years from now, two years from now, whatever," he continued. "Reinsurance, the tail is so long. It's enticing to people." Wright continued that the "day of reckoning" for ERC was September 11. "When 9/11 happened, it seemed like each plane and each building was reinsured through ERC." He laughed. "I'm laughing because it was no laughing matter." (It would take years, but in December 2005, Jeff sold most of ERC to Swiss Re for $6.8 billion, resulting in a $3 billion loss. "I couldn't have been happier," Jeff admitted.)

There were a bunch of other financial services–related deals during the summer of 2003. For nearly $5 billion, Jeff sold three other insurance businesses—Financial Guaranty Insurance Co., known as FGIC; GE Edison Life Insurance Co.; and its domestic auto and homeowner's insurance business. At the same time Jeff was selling, he was also buying commercial and consumer finance businesses. He really liked these businesses. He paid $2.9 billion for Deutsche Bank's inventory-finance business and $2.3 billion for ABB Ltd.'s structured-finance group. In August, Jeff paid $5.4 billion for the Dutch insurer Aegon's commercial-finance business in the U.S. and in Europe. Aegon's business added around $8.5 billion of assets to GE Capital's commercial finance unit, which had around $200 billion in assets. It was clear that Jeff was doing all he could to clear GE's decks of the legacy insurance businesses that Jack had assembled—and that Jeff hated—while at the same time continuing to beef up the highest-growth legs of the four-legged GE Capital stool.

During the late summer of 2003, Jeff got his first taste of deal heat, negotiating, at the same time, two of the largest multibillion-dollar deals in GE's history. It turned out to be a coincidence of timing more than anything else. But the deals—some $25 billion worth announced in a few weeks—also confirmed Jeff's determination to try to lessen the growing percentage of GE Capital's earnings in GE's bottom line. Not being a finance guy, Jeff remained wary of GE Capital's increasing impact on GE's financial statements and of the post-Enron questions being asked about GE's black box. The acquisitions were Jeff's first clear steps to change that calculus.

There was the $9.8 billion stock deal for Amersham, a British company that was one of the world's largest manufacturers of contrast agents used in medical diagnostics. Jeff's idea was to combine Amersham with GE's medical equipment business. His interest in Amersham began soon after he became CEO. He asked Joe Hogan, his successor at GE Medical, if he could buy any medical company in the world, what would it be? His business development manager was at the meeting, too. "They answered in unison, 'Amersham,'" Jeff recalled. Jeff "knew all about" the company, he explained. "We'd had our eye on it for years." Amersham was a leader in what Jeff referred to as "precision medicine," injecting into the human body chemical agents that, say, bind with molecules on a tumor, giving doctors a better sense of what bad things might be going on. The idea behind Amersham's technology was to replace invasive biopsies and catch cancerous tumors earlier. But the part of the business that got Jeff really excited was the division that helped pharmaceutical companies manufacture new drugs. "As drug development exploded," Jeff observed, "so would our business." Together that comprised about 80 percent of Amersham's business. There was also a much smaller so-called biopharma business that would, in time, prove very valuable.

Jeff heard that Sir William Castell, Amersham's chairman and CEO, might be willing to sell. He hired Goldman Sachs to advise GE. Goldman contacted Castell, a former accountant, who relayed that he was open to a conversation, as were most CEOs when GE came calling, in the same way that they were when Warren Buffett expressed interest. Six weeks after September 11, on October 23, Jeff and Castell had their first meeting at 30 Rock, in Jeff's New York City office. "I liked him right away," Jeff recalled. "But this wouldn't be a quick negotiation." The wooing went on for most of the next two years. The two men would talk occasionally on the phone. Other times, Jeff would hop on one of the GE jets and fly to Biggin Hill Airport, outside London. Castell would climb aboard the jet. They would talk for a few hours and then Jeff would fly back home. Jeff was hooked. The price talk for the deal was originally around $4 billion to $5 billion. But slowly but surely the acquisition got pricier and pricier. Castell was a tough negotiator. "I wanted to close the Amersham deal, though not at any price," Jeff allowed.

In September 2003, Jeff decided to make one last attempt to cut a deal with Castell. One Thursday around 5:00 p.m., Jeff and five of his colleagues boarded one of GE's two Boeing Business Jets, which Jack had ordered up during the end of his reign. The planes were huge, essentially a reconfigured Boeing 737 that would normally seat 175 people. This one was basically a flying office, with a living room, a conference room section, two spacious bedrooms, and U-shaped banquettes for 20 people to sit. (Jeff sold the two jets in 2005 "to save money," he wrote.) On Friday morning, after a six-hour flight, the GE team landed at Luton Airport, another small private-jet airport outside London. Castell and his team boarded the GE plane to see if they could make a deal. There were group discussions and then the two sides would retreat: the GE team headed to the back of the plane; the Amersham team was in the front. "Castell's daughter was getting married the next day," Jeff remembered, "but that seemed to motivate him, not distract him."

Five hours later, without lunch having been served, Jeff announced: "That's as far as I can go." Castell looked down "for what seemed like an eternity," Jeff remembered. He didn't move. When he looked up, he announced, "Let's do it." They shook on the price and the terms and turned the agreement over to the lawyers to document. Minutes later, the GE jet was in the air, heading back to White Plains. They landed before 5:00 p.m. on Friday, barely twenty-four hours after leaving. "Our Amersham acquisition surprised many people," Jeff recalled. "The company was high-tech, global and expensive." At $9.8 billion, it was one of the largest stock deals ever in the UK and a 45 percent premium to where Amersham had been trading before GE hammered out the deal. Jeff named Castell a GE vice-chairman and a GE board member while he continued to run what became known as GE Healthcare Technologies. Jeff had agreed to pay around four times Amersham's 2002 revenue and around twenty times its earnings, pricey but not inconceivable. He said he was nervous the night before GE did the deal. He thought to himself, "I'm going to get fucking creamed on this one. People aren't going to get it. And I did, I did take a little shit." He advertised that the deal would be "non-dilutive" to GE's earnings in 2004 and slightly accretive in 2005. "My team and I took a lot of external criticism for the price tag," he admitted. He got criticized for using GE stock to pay for the acquisition, something Jack rarely did. Jeff also got grief internally about his lack of discipline on price.

Jim Bunt, the longtime GE treasurer who retired five days after first hearing about Jeff's desire to buy Amersham, said he was "apoplectic." He couldn't believe how little the GE deal team seemed to know about the company. He said they didn't have a list of its products or of its patents or how long its patents would last. "What they sell is contrast material," Bunt told me. "You're not going to put water in [our bodies] and say, 'Oh my, water is special. It's worth $1,000 an ounce.' If you're going to spend billions to buy a company, don't you think you might want to know a list of their products; where they are on patents; how much time is remaining on the patent; what have they got in the pipeline with patent applications. . . . Jeff didn't do that. Jeff was more interested in buying a business in euros and paying in euros and not hedging it because the accounting rules would not let you."

Bunt said that to quiet him, the M&A deal team sent him a two-hundred-page Wall Street analyst's report on the industry, which he said barely mentioned Amersham. He wanted to see the hard numbers, especially because at that time the price tag that Jeff was considering for Amersham was around $6 billion, and rising. Bunt objected to the deal. "So," he told me, "I retired." Other former GE executives told me the Amersham deal was "a stupid deal." Even Jack, in one of our interviews, pissed on the deal. He called it "shit" and said, "We don't even know what Amersham is. That's gone. $10 billion. $10 billion." (At around the same time, Jeff also spent another $2.3 billion to buy Instrumentarium Corp., a Finnish maker of medical devices.)

ASIDE FROM ATTEMPTING TO BEEF UP GE'S MEDICAL AND PHARMA businesses, Jeff also wanted to bolster its position in the entertainment business by adding Universal Studios to NBC and creating a company that could compete in a rapidly changing industry, where merely being a television and cable network was no longer enough. The year before, he had rejected the idea of growing NBC but then changed his mind. Jeff's deal with Vivendi, for Universal's entertainment assets, had a much shorter gestation period than did the Amersham deal. More like four months, rather than two years. Due to its own internal machinations—including a debt-driven, ill-fated buying spree—Vivendi had decided to auction Universal, the entertainment conglomerate that Vivendi had bought three years earlier from the Bronfman family of Canada, who controlled Seagram Co. On the block were Universal's Hollywood movie studio, its theme parks, a television production operation, and two cable channels, USA Network and the Sci Fi Channel.

Bob Wright had been urging Jeff to grow its media business through acquisition for some time. The previous year, he had asked Jeff to consider buying AOL Time Warner. "This fits in with us," he told Jeff, while also conceding that MSNBC would have to be sold to accommodate CNN. Wright thought he had negotiated a deal with AOL Time Warner executives to buy the company for $15 a share. But they decided they could do better, and then Jeff rejected the idea of going after the ailing company in a hostile deal. In December 2002, Jeff did green-light NBC's $1.25 billion acquisition of Bravo, taking full control of the cable channel that GE had owned in partnership with Chuck Dolan at Cablevision. And in the month after September 11, NBC also scooped up Telemundo for $2.7 billion.

Universal was Bob Wright's second choice after AOL Time Warner. And Jeff's view about doing a deal to increase the heft of NBC changed as the importance of cable channels in the media landscape increased dramatically and with Vivendi's decision to put most of Universal up for sale. Vivendi courted GE as a buyer, along with a few others, including a group put together by Edgar Bronfman Jr., the Seagram's heir who had sold the assets to Vivendi in the first place. But once again, with GE front and center in a sale process, other bidders seemed like long shots. And like Jack, Jeff was not the type to lose an M&A auction once he put his mind to winning it. Suddenly, "breadth" was valuable. "If we could make a deal, a merger would give NBC the breadth it needed," he wrote in *Hot Seat*.

In one of our first interviews, Jeff was blunter. The Universal deal and the Amersham deal were part of his drive to revitalize GE's industrial businesses. Jack, he said, "had let the industrial company go to seed." He said the alchemy that led to GE having at one point a price-to-earnings ratio of more than fifty was "hocus-pocus" and that it was "pretty obvious to anybody" that it made little sense. He continued about Jack and his stewardship of NBC: "He says he loves NBC. NBC was a fucking mess. We hadn't invested in cable. You kind of sit there and say, he's just like"—and here he started imitating Jack's distinctive New England twang—"'You're fucking up NBC,' [but] if we hadn't done Universal with NBC, we would have been [in deep shit]. Our one cable asset was CNBC. We had a network that was flailing." He explained to me that at most of his meetings early on his tenure with Wall Street M&A bankers, their advice was to sell NBC. "You're not doing anything with it," he was told. "It's a wasted asset. You're not investing in it." There were still suitors floating around, wondering if Jeff would sell it instead of investing in it.

Wright urged Jeff to go after the Universal assets. Jeff left Wright to lead the negotiations, while "offering to help where I could." He knew Jean-René Fourtou, the Vivendi CEO, from the days when Jeff worked in Plastics and Fourtou ran Rhône-Poulenc, a French-owned chemical company. Jeff figured their relationship would give GE "an advantage over other bidders." Yes, that and GE's big checkbook. The structure of GE's deal for Universal would have allowed Vivendi to keep a 20 percent stake in the assets, giving it a second bite of the apple if the business did well. GE had an option to buy the remaining 20 percent in five years. In the meantime, GE would pay $3.8 billion in cash, assume $1.9 billion in Vivendi debt, and pay Vivendi $800 million a year from the business's $3.5 billion of annual earnings. GE's complex offer valued the Universal assets at $14.5 billion, some 50 percent more than what GE had agreed to pay for Amersham. The structure of the Universal deal plus GE's stature made GE Fourtou's preferred bidder, according to what "people familiar with his thinking" told the *Journal* in August. Fourtou believed the Universal assets were a good fit with NBC and he "also likes the patina of respectability that GE would lend such a deal."

It turned out that Wright and his team had first met with Vivendi executives at 30 Rock, in New York, on May 28. According to Wright, Vivendi saw the logic for the combination with NBC but was a bit wary of GE's deal structure—why did Vivendi have to wait so long for its cash?—and its price. Vivendi wanted $16 billion, not $14.5 billion. They kept talking. There was another meeting on July 8, in Wright's executive dining room at 30 Rock. It was another tough negotiation.

A week later, the teams expanded to twenty-five on each side, including principals, advisers, and specialists. Credit Suisse and Alan Mnuchin, the brother of Steve Mnuchin, who would later become Donald Trump's Treasury secretary, advised GE. Citigroup and Goldman Sachs advised Vivendi. The deal became known as Project Vineyard. "I struggled to keep control of the process," Wright confided. He and Fourtou had "many involved private meetings over lunch in my private executive dining room."

As July rolled on, the sessions became all-day marathons. GE was working hard to turn an auction process into an exclusive buying opportunity, even though at one point, the other bidders included John Malone, Marvin Davis, Comcast, Viacom, and Chuck Dolan at Cablevision, who had teamed up with Edgar Bronfman and was suddenly interested in cable assets again. Wright and Fourtou had to swat away Barry Diller, the billionaire entertainment mogul who had sold USA Network and Sci Fi to Vivendi and retained a 5 percent stake in Universal, along with the ability to renew licensing agreements for the cable networks.

They opted to buy out Diller for nearly $1 billion; it hurt, but at least Diller was gone. Jack and Wright had danced with Diller once before, in 1999, when they were exploring with Bronfman a merger between NBC and the Bronfmans' media assets, including Universal,

USA, and Sci Fi. Had that deal gone though, GE would have owned 60 percent of the combination and Bronfman 40 percent. Diller would have been chairman and CEO. Wright would have been his COO.

The GE board approved GE's offer for the bulk of Universal on July 24. But more deal twists remained. First, the GE board did not want to purchase Vivendi's $2.4 billion electronic games division. Brandon Burgess, who was Wright's business development guy, argued to keep the gaming business in the deal. He stayed up late one night beta-testing *World of Warcraft*, which would go on to become a global gaming phenomenon. (In the end, GE did not buy Vivendi's gaming unit, which was merged four years later with Activision, Inc., to create Activision Blizzard. In January 2022, Microsoft bought Activision Blizzard for nearly $70 billion in cash. Oops.) The other twist occurred on August 14, the day before Vivendi wanted second-round bids in the ongoing auction. That's when the lights went out in New York City in what *The New York Times* described as the biggest blackout in American history. Burgess decided he had little choice but to keep negotiating with his Vivendi counterpart, Jean-Bernard Lévy. They climbed down the fifty-two flights of stairs at 30 Rock, walked to Burgess's apartment building, and climbed up the nine flights of stairs there. They kept negotiating by candlelight, eating crackers and cheese.

"At that moment, I knew the deal was sealed," Burgess recalled. Why else would the French be sitting in Burgess's Midtown Manhattan apartment with the lights out? During the last week of August, lubricated by foie gras that Wright had had flown in from France, the two sides continued negotiating, slowly but surely checking off the open issues between them. On August 29, Jeff and Fourtou signed off on a "preliminary" deal, Burgess reported, "meaning there was still a lot to do." Everyone kept working through the Labor Day weekend, except for Burgess. "I took time out Friday evening to get married," he remembered.

On the Sunday of Labor Day weekend, Vivendi went through what certainly must have been a bit of annoying M&A theater by bringing the various bidders, including GE and Comcast, to the law offices of Weil Gotshal, on Fifth Avenue, to make their final bids. Surely the Universal assets were GE's to lose, right? At one point, Jeff was walking down the hall to get a cup of coffee when he ran into Fourtou, his old acquaintance. "I could tell by his body language that he liked GE's bid," Jeff recalled. "Nothing overt was said." When he returned to his corner of the office floor, he told the GE team, "I think we've got a shot here." Fourtou urged Jeff to keep driving hard. "Let's figure out a way to do this," he told the GE CEO.

Steve Burke, a senior Comcast executive, remembered that after Comcast completed its $72 billion acquisition of AT&T Broadband—buying the cable-distribution assets of AT&T in one of the largest deals of all time—the company wanted to get into the content business. "We always believed content and distribution worked together, even though a lot of people say it doesn't," Burke told me. "But we looked at what Murdoch did and we looked at what Malone did and we were like, 'The smartest guys have done distribution and content.'" Comcast's original hope was to get ten million cable subscribers and then maybe buy Scripps or Discovery for their content. Comcast never expected to pull off the Universal deal. "Comcast was the cover bid on Universal," Burke said. "They actually wanted to sell to GE because of the name, and we pulled out. We said, 'You're gonna give this to GE so that you look great in Paris, and we're just here to bid everything up. We're leaving,' and they said, 'Oh, no, please don't leave! Please don't leave!'" After losing Universal to GE, Comcast made an unsuccessful hostile bid for Disney.

But Brian Roberts, the Comcast CEO, never lost his hope that he would end up with NBC Universal. Burke was in Jeff's class at Harvard Business School. Roberts and Burke made sure to stay in touch with Jeff. They played golf together regularly and had dinners. Roberts called Jeff up for dinner before NBC bought Universal. On the agenda?

"We want to buy NBC," Burke reported Roberts told Jeff. "We get five minutes into the dinner and this is kind of what Brian does. He's like a serial acquirer. And Jeff says, 'I'm never gonna sell. I love our team. We're number one in every single aspect of the business. We're never selling.' This was on the fifty-third or fifty-fourth floor [of 30 Rock]. There was a little dining room and everything. We leave, and Brian's devastated. He's like, 'Oh, my God.' I said, 'Brian, don't worry about it.' He said, 'What do you mean?' I said, 'You can't stay number one at NBC forever. There will be a day when they're number three or number four, and then he's going to call us, right? If your reason for keeping it is "I love the television business" or "I love the movie business" or "I believe it's important; it has synergy with lightbulbs" or whatever, that's one thing. But if your reason for staying in is "We're better than everybody else and we're number one; I don't want to sell," that'll change.'" Roberts never lost hope. He played golf with Jeff ten times and had six dinners with him in the ensuing years, biding his time.

On September 3, GE and Vivendi announced a preliminary agreement on the complicated deal, valued at roughly $14 billion. GE would own 80 percent of what was called NBCUniversal; Vivendi would own the remaining 20 percent. GE structured the deal this way, Jeff said, because he was sure the business would end up being spun out of GE into a stand-alone, publicly traded company. It would be another four months before the definitive merger agreement was signed. Wright remembered many points where the pact could have fallen apart as the finish line approached. But it didn't. He chalked up the determination to get the deal done both to a belief that without the Universal deal NBC would be in trouble and to his growing friendship with Fourtou. On September 28, Wright flew off to Paris to meet with Fourtou alone for a few days of final negotiations. They bonded and developed a close friendship. "It turned out Jean-René was not some artificial lump of bricks any more than I was," Wright recalled. Fourtou recalled that there were many times when his team wanted to abandon the negotiations with GE, given how "very, very hard" they were. But, he said, his belief in Bob Wright saved the day. "He has the personal qualities that make you want to do a deal with him," Fourtou remembered. "He is not one of the California guys."

The deal finally closed on May 12, 2004. Wright said it was one of the happiest days of his professional life. As the principals from GE and Vivendi assembled in Wright's New York office to sign the final agreements, they celebrated with champagne. "Universal was a home-run acquisition," Wright recalled. "It was about as good as you could get in the media business." He said he thought the deal was his "crowning jewel" because it gave NBC "all of the tools to be successful over a long period of time as a very large media company," regardless of who owned it. In a stellar career that included stints working under Jack's watchful and encouraging eye at Plastics, Small Appliances, and GE Capital, and then at NBC for seventeen years, the Universal deal was Wright's crowning achievement—a point he made repeatedly in the media coverage at the time, in his book, and in our conversations. He helped create both CNBC and MSNBC but was derided by most television and movie executives as a bland corporate suit. According to a September 2003 profile in *Fortune*, Wright "has never gotten the acclaim or attention that naturally accrues to other media moguls." But with GE's pending acquisition of the Universal entertainment assets, "that's no longer the case under Jeff Immelt . . . who is giving Wright much more running room" than Jack did. The article also included revelations about Wright: that his wife, Suzanne, was a regular presence around NBC and liked to participate in decision making; that the Wrights were close friends with Johnny and Joanne Carson; that Wright was diagnosed with skin cancer in 2000—it was treated successfully—causing him to think about what was really important to him. He decided he loved NBC too much to leave. "He wanted to finish what he started at NBC," the magazine reported.

There was also the news that, on his way out the door, Jack had selected his friend Andy Lack, head of NBC News, to become Wright's second-in-command, setting Lack up to succeed Wright. "Wright was stunned," *Fortune* reported. When, according to the magazine, Lack "quickly began to assert himself," upsetting other executives, Wright appealed to Jeff, who quickly sided with Wright and sidelined Lack, who soon thereafter left to run Sony Music Entertainment. For his part, Lack said that everything changed between him and Wright when Jack retired. Before then, they had worked well enough together, with Wright as CEO and Lack as COO. Jack's plan for NBC was that Lack would take it over and Wright would go to Fairfield as a vice-chairman, to help Jeff run GE. When Jack retired, Wright told Lack, "I don't give a shit about being vice-chair and I don't want to be up in Fairfield. I mean, I care about doing the right thing for the company, but this is what I love. I love NBC." As Jack said later, he had left two guys in the same seat. He thought Wright would be happy to go to Fairfield. He wasn't. "Pretty quickly after Jack was gone, Bob was a new Bob," Lack said. "Everybody can tell their version of that." Lack felt that his relationship with Wright changed instantly. "Suddenly I was in a political kind of cross fire," he recalled. "That was rough."

Lack told Jeff and Bill Conaty that it became clear that Wright did not want him around. "I don't think Bob wants me here for another day," he told them, "and it's embarrassing. He's kind of polarizing NBC: those who are for him and those who don't want to see me get whacked." That was Lack's first exposure to Jeff, who pledged to make it work for Lack to succeed Wright. Jeff told him, "I'll work this through with Bob." But, Lack continued, "it never happened." He added that Jack had been watching the drama play out and began to realize that for various reasons—some driven by Jeff, others by Wright—the GE that Jack had put together and left for Jeff to operate and to build upon was "going to fall apart." Lack got quiet after he told me this. "I was one of the early cards that got flipped," he said.

It was during this time that Jack and Wright also had a falling-out.

*Fortune* suggested the reason was Jack's divorce from Jane, to whom the Wrights were also close. They were neighbors in both Connecticut and Nantucket. Wright remembered getting a call from Jack in February 2002. "We are in the shit," he said, speaking of NBC and GE. He then recounted the call. "He's saying, 'Bob, I'm just calling you to tell you that I'm divorcing Jane,'" Wright told me. "Jane was so happy that he was going to retire. She would tell everybody she'd made all these plans, where they were going to go. . . . Then here it is, only a few months later, not many months later, and she's getting a divorce." When the divorce became contested, and Jane filed the details of Jack's postretirement package, that was "horrible, just horrible," Wright continued. "Many of us knew about it, but not all of it, just that it existed. . . . Ben Heineman, who was our general counsel, may have been, honestly, the only one in the company that had seen the whole thing. He did it in pieces. When it came out, it was just disgraceful." Then came a Wells notice from the SEC about whether GE had properly disclosed the details of Jack's retirement perks. The divorce was bad for GE. Remembered Wright, "Now everybody knows what's in the contract and everybody is horrified. And instead of settling himself, I mean, for his own purposes, he drags this thing on for like a year and it never gets out of the papers. That really steamed me. It really steamed me. I just thought that that was unfair for all of us. That's a guy that's gone over the top. So I said some things like that and that became very—you know, you can't say anything about Jack. So that's unfortunate. I got over it. It took a long time to heal."

FORTUNE PORTRAYED THE STRUCTURE OF THE DEAL WITH VIVENDI—that the French company would keep a 20 percent stake in NBC Universal and there might be an IPO of the business three years down the road—as "the most radical departure from the Welch era," when there was only one issue of GE stock, not a publicly traded stock of a majority-owned subsidiary. Jack wanted "everybody's life raft . . . tied to the same boat," as he wrote in his autobiography. But Jeff and Wright countered that the possibility of a NBC Universal IPO wasn't so unusual and was just a way to accommodate Vivendi in the short term. Jeff told *Fortune* that GE would likely buy Vivendi's 20 percent stake for $3 billion—valuing NBC Universal at $15 billion—in three years.

The night the NBC Universal combination was announced, Jeff went on *Charlie Rose*. He conceded that he had changed his mind about bulking up NBC after Bob Wright made a convincing argument for why NBC needed to get deeper into the business of cable programming. He said that thanks to Jack's incredible 1986 deal for RCA, GE ended up with NBC for free, and that made a material difference in his thinking about going after Universal once those assets became available. "NBC was the house's money," he said. "That's the way to think about it." He said he got the logic of the combination. GE knew how to run a TV studio and became convinced that Ron Meyer was running Universal's film studio well. "These are our kinds of people," he said. "I didn't see it as outside the bandwidth of what we're already managing." In his book, Jeff wrote that he "loved" Meyer because he "blended business savvy with pure creativity." Recounting Meyer's background—he was the son of Holocaust survivors and dropped out of high school at fifteen, joined the marines, and then cofounded Creative Artists Agency (with Mike Ovitz)—Jeff explained that "in a town that seemed to run on air-kisses and subterfuge, Meyer was always surprisingly candid with me." Jeff got a taste of Meyer's frankness early on when he reported that the movie *King Kong* would be a $200 million flop.

"Can't we do something to help?" Jeff asked, noting that for that kind of money GE could build a plastics plant.

"Nope," Meyer replied. "We are toast. That's the movie business!"

At first, Jeff was not convinced it made sense for GE to own Universal's theme parks business. He didn't like the idea of employing minimum-wage workers or having a business that depended on tourists.

He also didn't like the idea that a major mishap could happen. "To me," he relayed, "that meant red flashing lights. . . . I could see only liability and risk." But after meeting and spending time with Tom Williams, who ran the theme park business, he was won over. Williams convinced him that the theme parks were an extension of the movie studio business. Jeff was a convert. "Today, the theme parks remain one of the company's most durable profit centers," he allowed.

IN THE FIRST FEW MONTHS OF 2004, FRESH FROM HIS VICTORY LAPS IN buying Amersham (too pricey, the market concluded) and most of Vivendi's Universal assets (a steal, the consensus seemed to be), Jeff made two fateful decisions that would plague GE for years to come. Jeff considered GE's slew of insurance businesses to be its "biggest mess," and he vowed to clean them up. Whereas Jack saw arbitrage opportunities with GE's insurance business—essentially ways to make money from money—Jeff again saw only liabilities. He had sold a bunch of these businesses, and rumors were swirling about his desire to get rid of the money-losing Employers Re. But after Moody's warned him about the potential credit downgrade, Jeff decided that insurance was to blame. It accounted for some 40 percent of GE Capital's earnings and therefore about 20 percent of GE's earnings. But it had to go. "It was an over-levered mess," he complained. He seemed to dislike everything about the insurance businesses. There were too many of them—from pet insurance to property reinsurance—and the boys at GE Capital had overpaid to get them and then run them poorly. From his perch on the GE Capital board, Jeff knew the insurance acquisitions were mistakes, he allowed, but he never spoke up. No doubt he figured the GE Capital management wouldn't listen to him, since they poohpoohed his knowledge of finance. He also believed that both McNerney and Nardelli also thought the insurance deals were bad. They didn't speak up either. "It was clear to me that we needed to get out," he wrote.

Out of Jeff's passion to be rid of Jack's insurance "mess" came something called Genworth Financial, the first time GE had ever spun out a new company as a stand-alone public entity. The planning for the creation of Genworth began in the fall of 2003 with the hiring of underwriters at Morgan Stanley and Goldman Sachs, the bluest of the Wall Street blue-chip firms. And not just anyone from those firms, either, as Jeff was eager to point out. At Morgan Stanley, both Ruth Porat (now the CFO of Google) and copresident Stephen Crawford (now a senior executive at Capital One) worked on the Genworth IPO. At Goldman, it was David Solomon, now the Goldman CEO, and John Weinberg, grandson of Sidney Weinberg, the longtime Goldman senior partner, and the CEO of Evercore, a boutique investment bank. "Here is a good rule," Jeff recalled. "The quality of the asset is inversely proportional to the amount of talent it takes to sell it."

In late January 2004, GE filed an S-1 with the SEC for Genworth Financial, essentially a prospectus for an initial public offering of a brand-new, independent company comprising a large swath of GE's insurance businesses, including those providing mortgage insurance, life insurance, long-term-care insurance, and group health and life insurance for companies with fewer than one thousand employees. It was a hodgepodge of insurance businesses, and pretty much everything that GE Capital had left in the insurance aside from Employers Re, its big reinsurance business, which Jeff still hoped to sell. Genworth was a big business, with $11.2 billion in revenue and net income of $1.4 billion.

Along the way to the IPO, Jeff made two fateful decisions, one a sin of commission (fateful for Genworth), the other a sin of omission (fateful for GE). The first was to include in Genworth GE Capital's highly profitable mortgage insurance business. "I would always be grateful that we did," he wrote later. "This business would be hard hit during the financial crisis due to its sensitivity to housing prices." (That decision pretty much sank Genworth.) The other decision he made was to exclude from Genworth—and to keep at GE Capital—something called Union Fidelity Life Insurance Company, based in Overland Park, Kansas. Union Fidelity was in the business of providing individuals with insurance benefits to cover the costs associated with nursing homes, assisted living facilities, and home health care. GE Capital's financial assumptions for the business related to how long people will pay required premiums, how long people will live, and how well they live.

Union Fidelity's prospects were tied to the likelihood that policyholders would need care as they aged and then for how long they would make claims. United Fidelity also agreed to reinsure some of Genworth's insurance risks. Although it would be another thirteen years before the problems in this long-term-care-insurance portfolio would show up in a dramatic way, there are many people who—with the passage of time—believe Jeff should have made the decision to include Union Fidelity in the Genworth IPO, at whatever the cost, and get rid of it.

Jeff had no doubt he made the right decision to keep Union Fidelity at the time, despite the second-guessers. On this topic, he remained defensive. "We tried to get it out the door with the Genworth IPO," he told me. Originally, GE and the underwriters were shooting for an IPO price of $22 a share. But the demand was weaker than expected at that price. The underwriters reduced the IPO price to $18.50 a share. Jeff's argument to me was that had Union Life been included in the Genworth IPO, the underwriters would have priced the stock further below a price that was already lower than he hoped it would be. But, he said, "I would have done anything Triple D"—his name for Dennis Dammerman—"had recommended, anything."

After the May IPO, which raised $3.53 billion for GE and was the largest IPO of 2004, GE owned 70 percent of the company, with the express intention of selling the rest of its stake as soon as was practicable. (Genworth received none of the IPO proceeds.) GE sold three batches of Genworth stock in 2005 and a final chunk, some $2.3 billion worth, in February 2006, bringing its total proceeds from the sale of Genworth to about $9 billion.

Jeff was focused on the fact that GE sold its stake in Genworth for a price of around $35 per share, after the stock increased in price following the IPO. Now "it trades at fucking 2," he said to me. "Do you think there's a lot of people sitting out there saying, 'Give me a piece of that action. Give me some of that action'?" He remained defensive. "I was in the meeting and we could have done the whole thing," he continued. "We could have done all this stuff. Look, if Dennis had told me to bark like a dog, I would have barked like a dog. I would say buying these insurance businesses in the late 1990s was fucking stupid. By the way, there were lots of people on the GE Capital board that agreed with that, but Jack didn't want to hear any of it. We bought them. I recognized it was a problem. I got rid of as much as I possibly could. I had no experience with doing any of this, so I had Weil Gotshal. I had Morgan Stanley. I had Goldman Sachs. I took a business that was earning $2 billion, 20 cents a share, and shoved it out in May of 2004. It was supposed to trade at $22.50. We finally got it at $18."

In his book, Jeff wrote that the decision to retain Union Fidelity, and its book of long-term health insurance, was approved by GE's board and disclosed to investors and was "necessary to get the Genworth deal done." He wrote that "as was common" at GE when retaining stakes in insurance businesses sold off, "we would wind these businesses down without writing any new policies." And then, writing with a surgeon's exacting precision, he continued to deflect responsibility, and blame, for the decision. "In retrospect," he continued, "through a 2020 prism, it would have been better to attempt to dump LTC"—long-term care—"in 2004 even at what everyone would have regarded as a stiff and unfavorable price. Despite extensive oversight, none of us saw then, or over the intervening years, how toxic the investments GE had made in the late 1990s"—take that, Jack!—"would eventually become."

Then Jeff made another mistake. When GE sold Employers Re to Swiss Re for nearly $7 billion in December 2005, Jeff agreed to keep a part of Employers Re called ERAC, Employers Reassurance Corporation, a reinsurer of more than 200,000 premium term-life insurance policies. ERAC had $23 billion of statutory reserves and underwrote lifetime benefits on 70 percent of its policies. GE also agreed to add $3.4 billion to its reserves for the cost of future claims. Regardless, Jeff was ecstatic. He thought he was finally out of the insurance business. "If you see the smile on my face," he told Wall Street analysts about the Swiss Re deal, "you will know I how I feel about this deal." After all, what trouble could two little stub businesses—Union Fidelity and ERAC—cause the mighty GE? "Perhaps Immelt's insurance troubles aren't behind him quite yet, after all," Carol Loomis wrote presciently in *Fortune*.

It would take years to play out fully, but Jack, for one, counted the failure to include the long-term health care liabilities into Genworth as one of Jeff's major blunders. "They never talked about it," he told me about Jeff and his team. "And their liability was building." He said Jeff had an opportunity to "defease" the liability—essentially alleviate the future risks by selling the liability—at a cost to GE of around $1 billion. According to one GE executive, Swiss Re offered to insure the growing risk in GE's legacy health care liabilities in exchange for a premium of $1.2 billion. But Jeff didn't do it. "They left behind a conscious decision," Jack said. "We can make more money on this. . . . That's a sin there. That's a decision. That's a judgment. . . . The judgment of thinking you can make more money by holding it and playing with it. He kept the long-term liabilities. He wanted gains. . . . They could have defeased it for whatever, a billion dollars or whatever it is. They didn't do it. No, that's the big one. He hid that to try and make more money on it than they offered."

Other former GE executives were mystified as well by Jeff's decisions on the insurance front. "There's a lot of that kicking-the-can-down-the-road shit," one said, "and not recognizing that some things come home to roost—which long-term care is the biggest example—that could kill the company." The decision to retain Union Fidelity in GE Capital "and not [tell] anybody" was one that Ken Langone, then still on the GE board, could not resist criticizing. "[Jeff] thought he was going to make money," he said. "I tell you what he hasn't got. You understand Yiddish? He hasn't got a Yiddish *kop*. He's got a goyish *kop*. It's that simple. He went to the right schools. He did the right things. He checked off all the boxes. He ended up getting the job, and now he doesn't know what to do with it. The biggest thing he didn't have that Jack had in spades was people skills." Added another GE executive about Jeff's decision, "The real sad story is it sits out with some actuaries in Kansas City and we knew it was a problem, but anytime you would even ask about it you would get smacked in the face by senior management. It was 'Don't lift up that rock.'" Tony James, a legendary Wall Street executive who became the executive vice-chairman of the Blackstone Group, used to tell people, "Great leaders run toward problems, not away from them." GE let the problems at Union Fidelity and ERAC fester in an office park in Kansas City. Others were more sympathetic to Jeff's plight with Genworth. A former GE board member told me, "I think that the decision was a big and open decision. The board was involved. The finance groups were involved. The underwriters were involved. And Jeff owns the decision since he was CEO, but nobody was saying, 'Get rid of this thing because in ten years it'll be toxic.'"

AND THEN THERE WAS JEFF'S DECISION TO BUY WMC, A SUBPRIME mortgage broker that GE Capital bought for $500 million in April 2004. Jeff's logic for the purchase, although a bit convoluted, went something like this: Having made the decision to jettison Genworth, he decided to replace those missing earnings by increasing GE Capital's investment in consumer finance businesses, which he believed the market rewarded at a higher valuation than it did commercial finance businesses. Buying WMC was part of his bet on consumer finance. He was thinking that WMC would underwrite mortgages and then package them into securities and sell them to investors, as was all the rage on Wall Street at the time. "It didn't turn out to be that way," Jeff told me.

He said there was "a big debate" inside GE about whether to buy WMC. Some executives, he said, such as Jim Colica, the chief risk officer at GE Capital, didn't want to do the deal. But, Jeff said, as he had done with Genworth, "anytime there was close to a jump ball, I leaned hard on Dennis, Triple D. Triple D did a deep dive on it, and we talked about it and decided to do it." The idea was to start small, see how things would go, and then ramp it up. "We were at the scene," Jeff said. "There's no sugarcoating it." Others had a different recollection of how the WMC deal came about. According to Denis Nayden, who although gone by then from GE Capital retained his sources inside at the highest level, "The only two guys who wanted to do it were Mark Begor, the president and CEO of GE Capital's retail finance business, and Jeff Immelt. Everybody else said, 'No. Don't do it.'" . . . How stupid can you be when you have your chorus of 'no, no, no'?" Bill Conaty, the longtime head of HR, remembered that Begor was "in deal heat" about WMC "and appealed to Jeff in a big way. They did the deal. There you go." (Begor, who spent more than thirty-five years at GE and is now the CEO of Equifax, did not respond to a request for comment. Dennis Dammerman passed away in 2013 and could not be interviewed, making him a convenient whipping boy.)

For reasons that they would all regret, nearly every bank on Wall Street made the fateful decision to vertically integrate their mortgage-backed-securities factories by buying mortgage brokers, such as WMC, that were down in the trenches making mortgages to people in order for them to buy homes. Most of the time, the mortgages were made to people who were able to pay them back. That was fine. But other times, mortgages were made to people who said they would pay them back but for one reason or another found themselves unable to do so. That would become a problem.

In any event, the Wall Street investment banks had taken to buying these mortgages from the mortgage brokers and packaging them up into securities, which were then sold to investors all over the world, once the SEC and the ratings agencies had signed off on them. The idea was to give investors a way to get a higher return on their money by investing in the mortgage-backed securities, which often were rated AAA. It seemed too good to be true: investors got the higher yields they were looking for while buying what seemed like AAA-rated securities. And it was.

At first, the Wall Street investment banks bought the mortgages from the mortgage brokers and packaged them up into securities and then sold them off for big fees. The manufacture and sale of mortgage-backed securities was very, very big business on Wall Street in the years leading up to 2008. As demand rose for the securities and the supply of new mortgages began to dwindle, the quality of the mortgages packaged into securities began to slip, meaning that mortgages made to people who might not be able to pay them back were finding their way into mortgage-backed securities. To try to guarantee a steady supply of mortgages—the raw material of the mortgage-backed security—Wall Street banks began buying up mortgage brokers.

In its effort to become vertically integrated, Bear Stearns, for instance, bought Encore Credit Corporation and EMC Mortgage Corporation. Merrill Lynch bought First Franklin Financial Corp. for $1.3 billion. GE Capital bought WMC. Even though GE Capital was not a bank in the traditional sense—it didn't take deposits and wasn't regulated by the Federal Reserve or the SEC—by the mid-2000s it was doing pretty much everything banks do for corporations and for small and medium-sized businesses, making loans in nearly every imaginable configuration to nearly every imaginable industry. It also decided it could make money by selling the mortgages WMC originated to the Wall Street investment banks that were busy packaging them up into securities.

The problem for GE was that, as with the other mortgage originators, the quality of WMC's mortgages was deteriorating rapidly in the years after GE Capital bought it. WMC was violating its own underwriting standards, making mortgages nearly willy-nilly, seemingly to anyone who walked in the door. GE executives described WMC as an "alt-A" lender, targeting borrowers with less-than-stellar credit ratings.

A GE internal auditor observed, in April 2007, that WMC, the fifth-largest originator of subprime mortgages, "jacked up volume without controls." What's worse, when the investment banks inspected the WMC mortgages—to decide whether to include them in the manufactured mortgage-backed securities—and rejected them, or "kicked them out" in Wall Street parlance, WMC tried to resell them to other banks without informing them the mortgages had already been rejected.

The mortgages that WMC originated started defaulting at "alarming rates," Jeff recalled. The securities that contained the WMC mortgages began defaulting, too, opening WMC—and GE—to "claims that the quality of the mortgages was deficient," he continued. By the first quarter of 2007, GE had decided that WMC would stop writing mortgages and that GE would focus on selling at whatever price it could the $3.5 billion of mortgages that WMC still had on its books. (According to one GE executive, GE Capital sold its mortgage portfolio to Wachovia and Bank of America in May 2007 for 85 cents on the dollar; a month later the same kinds of mortgages were trading at 45 cents on the dollar.) "I'm not declaring any kind of victory here," Jeff told me, "But we actually shut the place down in, like, the end of 2006. We basically went cold."

In the end, WMC was a disaster. GE bought it for $500 million, paid a $1.5 billion fine, and took a substantial loss on the mortgages it had to dump into the market to get rid of them. "We bought into a business that many were entering at the time," Jeff continued, "and we learned the hard way that we'd made a mistake. . . . I wish we'd never gotten into the business."

There was no question, by April 2004, that Jeff was focused on transforming GE and putting his own imprimatur on the company. He desperately wanted to reduce GE's reliance on GE Capital, but he never could sever the Gordian knot. Despite his efforts, GE Capital's percentage of GE's overall earnings kept growing and growing, eventually reaching 51 percent of GE's $15 billion 2003 profit. "Partly that was because our Power business"—previously the business that was biggest contributor of profit to GE—"had cratered," Jeff remembered.

But in truth, Jeff was beefing up GE Capital in new and different ways. He started rebranding GE's more than three thousand banking and consumer lending offices around the world—in forty-one countries, though not the U.S. or the UK, which he believed to be too competitive—as "GE Money," a kinder and gentler way to bank. It turned out GE owned banks in Germany, in the Czech Republic, in Hungary, and in Switzerland. Who knew? In Poland, GE Capital bought Solidarity Bank in 1995 and rebranded it as GE Capital Bank, with more than $2 billion in assets and fifty-eight branches by 2004. Over the summer of 2004, GE Capital bought the credit card receivables of Mervyn's for $475 million; 38 percent of Hyundai Capital Services, the financing arm of the South Korean carmaker, for $375 million; the Russian credit card lender DeltaBank for $150 million; and the credit card business of Dillard's, the department store, for $850 million. There was much more, very little of which got much attention. "We're an acquisition machine," boasted David Nissen, who ran GE Capital's consumer finance business.

AT AROUND THE SAME TIME THAT JEFF WAS BUSY UNWINDING JACK'S GE, he showed that he could play political hardball with the best. He "made another difficult decision," as he put it, to fire Ken Langone, "Jack Welch's best friend," from the GE board after five years and before his term was up. Jeff had his reasons. First was that Home Depot, still being run by ex–GE superstar Bob Nardelli, had stopped selling most GE products in its stores. "Imagine doing a town hall," Jeff explained, "and having an employee ask, 'We have a board member who founded the Home Depot and we still are losing business there? How is that possible?' That actually happened." Jeff also blamed Langone for approving "all of the disastrous" insurance acquisitions that Jack had made, although since Langone had joined the GE board only two years before Jack's retirement, it's not exactly clear to which insurance acquisitions Jeff was referring (and he didn't elaborate). The third strike, in Jeff's mind, was the tumultuous and very public feud that Langone was having with Eliot Spitzer, then the New York State attorney general, about the exorbitant $140 million compensation that the New York Stock Exchange had paid Richard Grasso, its CEO.

Langone, an outspoken member of the NYSE board of directors, had been part of the group that approved Grasso's contract. When Grasso resigned in September 2003, a huge public conflagration erupted, pitting Langone, Grasso's most outspoken defender, against Spitzer and other Wall Street titans such as Hank Paulson, CEO of Goldman Sachs, and John Reed, CEO of Citigroup. Jeff was unhappy. "It didn't help GE to have one of our directors arguing so publicly with Spitzer, who was riding high at the time as the new 'Sheriff of Wall Street,'" Jeff explained. Jeff also was fearful of alienating the heads of the major banks who had lined up against Richard Grasso, and with whom Langone was publicly feuding. He believed he needed Goldman Sachs's Hank

Paulson, Citi's John Reed, and Bill Harrison, the CEO of JPMorgan Chase, to help GE "execute" its "portfolio strategy. . . . We needed to build trust with the banks at this time and our board knew it."

In June 2004, Langone wrote an opinion piece in *The Wall Street Journal* titled "Let's Bring on the Jury, Mr. Spitzer." (A framed copy of the article lives on Langone's conference room wall.) It got worse. In *Fortune* in October 2004, Langone said that Paulson was "the guy that's going to have not only egg, but shit all over his face" and Harrison was "the fucking genius that gave Enron all the money." Then Langone took on the whole lot. "They got the wrong fucking guy," he continued. "I'm nuts, I'm rich, and boy, do I love a fight. I'm going to make them shit in their pants. When I get through with these fucking captains of industry, they're going to wish they were in a Cuisinart—at high speed." But Langone could also see the handwriting on the wall. He was nothing if not street-smart. In March 2004, Jeff removed Langone from the powerful Management Development and Compensation Committee of the board—the very committee that had recommended Jeff's promotion a few years earlier. Jeff sent out a memo that suggested that he had spoken to Langone about removing him from the MDCC. "I called Jeff up," Langone said, "and I said, 'Jeff, I got the memo. I got no problem with going off the MDCC, but what did you and I ever talk about?'"

"What do you mean?" Jeff said.

"I don't remember you and me talking about it," Langone replied.

"Oh, I thought we did," Jeff said.

"That's fine," Langone continued. "Hey, look. All you had to do was pick up the phone and say, 'Ken, I'm uncomfortable. Are you okay?' Yeah. If I would have thought I wasn't okay, you know what, Jeff? I'm going to go off the [MDCC]. . . . Okay. But understand something. I'm easy. I early on began to get very uncomfortable about the way you treated the board."

Langone told me, "He talked down to the board. He had this attitude of superiority. And the board let him have it, including me. . . . This fucking guy lied."

Langone remained on GE's audit committee. Various labor leaders and big shareholders were anxious for GE to get rid of Langone, too. Nearly 6 percent of shareholders withheld their support from Langone at the April 2004 shareholders' meeting. The AFL-CIO wanted him gone.

The October 2004 *Fortune* article was the last straw for Langone. He was heading to his home in the North Carolina mountains for a few days over the Columbus Day weekend. He told his assistant, Pam, that he had a "hunch" Jeff would be calling. "Sure enough," Langone recalled, "Friday morning Pam says to me, 'Jeff Immelt needs to talk to you.' I said, 'Put him through.' I said to my wife before I took the call, 'I'm going to put my head under the fucking blade of the guillotine right now.'"

Jeff told him he couldn't sleep. He hadn't slept in three nights. "Jeff, stop it. Stop it. Stop it. The *Fortune* article?"

"Yes," Jeff replied. "Ken, these guys are my friends."

"That's right," Langone said. "They're your friends. And I'm your boss. So what? I didn't intend for this to come out the way it did. In fact, I made provisions that it wouldn't. Be that as it may, I said it and I'm standing by it. I wish I had used different language, but I didn't. But I can guarantee you one thing, Jeff. At the end of the day, everything I said is going to happen is going to happen. And that's all that matters to me. What are the facts, ma'am? Just the facts."

"I know, Ken," Jeff said. "But I got to—"

"Jeff, hold it," Langone said. "Make it easy."

"Yeah," he said.

"What do you want to do?" Langone asked.

"Well," Jeff said, "I don't know how to put it to you."

"I'll resign," Langone said. "I'll resign. No problem. You know, Jeff, on two or three occasions before this one—over the Jack Welch contract, over the lightbulbs—I offered to resign. The first thing was to do right by the company. Fine, Jeff. It's done."

"Well, somebody will be in touch with you," Jeff said.

"What?" Langone said. "I'm here. I'll be around."

Langone got a call from Claudio Gonzalez, chairman of the nominating committee. He wanted to meet with Langone. They were on the board of Home Depot together. "Claudio came to put the bullet through my head," Langone said. Then Langone heard from Ralph Larsen, the CEO of Johnson & Johnson, a GE board member since August 2002, and the lead director. Larsen wanted to see him, too.

Like a scene out of *The Godfather*, Jeff sent Larsen to put the squeeze on Langone. The *Fortune* article was too much. Langone had to go. Langone described Larsen as "a lovely man" and "a decent man." He came by Langone's New York City apartment one day and asked him when he was going to leave the GE board. "I'm leaving at the end of my term," Langone told Larsen.

Larsen replied, "Why not now?"

Langone was pissed. "I didn't do nothing wrong," he told Larsen. "I hear some bad language that some guy printed. That doesn't make me a bad director. I didn't do anything where I committed a crime or [damaged] my reputation. I'm not leaving. Shareholders elected me for a year. I'm staying. You're stuck, Ralph."

Larsen got up to leave. He was "a little agitated," Langone told me.

"You know, the board may not see it this way," Larsen told him.

"Wait a minute," Langone responded. "You're telling me the board might try to evict me. Tell them to fucking bring it on."

"No, no, I'm not saying that," Larsen replied.

"Well, Ralph, what do you mean when you say the board might not see it that way?" Langone wondered. "Ralph, the last thing you want to do is corner me, because when you do, I'll fight you. You tell me."

"No, no, no," Larsen said. "I didn't mean it that way."

"Okay, no problem," Langone replied.

Langone told Larsen he would attend the upcoming committee meetings and the upcoming board meetings. But he wouldn't go to the upcoming annual GE board Christmas party. "I can't stand the fucking noise," he told Larsen. "My wife would be very uncomfortable. I wouldn't want to do that to my wife. My wife's an elegant lady, and she would be very uncomfortable knowing that I'm a bastard son of the family reunion." He said that in December he would write Larsen a letter informing him that he wouldn't be standing for reelection to the board. "It's going to be a nice letter," he said. "That's how I'm leaving. I'm not going to be put up for reelection."

In his December 2004 letter to GE, Langone wrote that he "would be energetically defending his work at the New York Stock Exchange and that he didn't want his issues to affect GE's relationships or be a distraction to the board and the shareholders." Jeff told *The Wall Street Journal*—disingenuously—that Langone "has been a good friend to me and I value our relationship."

Langone said he was shocked when he got a call telling him that GE wanted to have a retirement dinner for him at 30 Rock. He recalled, "I'm like, 'Holy shit. This is fucking bizarre. I'm going to go away. That's it. Turn my picture to the wall and get on with your life.'" A few other recently retired directors were also invited to the dinner, people whom Langone had become friendly with during his five years on the board.

One after another, the distinguished guests lavished praise on Langone and what he had done for GE. "They all get up," he recalled. "They all go around. I can't fucking believe it. They're talking about me? I'm this great guy. What the hell am I leaving for if I'm so good?" GE gave him an expensive Steuben glass sculpture of Excalibur. "You know that big chunk of glass with the sword in it?" he said. "GE logo, the meatball on the blade, plus years of service. Nice touch."

Langone was sitting next to Jeff. "Geez, if this is the way you get executed, it ain't bad," Langone said. Larsen spoke. "He went on and on about my integrity, about my involvement, about my commitment, my passion for business and doing it the right way," Langone remembered. "Shit, this is someone who is about to get the Nobel Prize." Larsen continued, "Damn it, you're stubborn." Langone said Larsen was referring to Langone's unwillingness to resign the previous fall when he had come by Langone's apartment. "He clearly was holding it in," Langone said.

Larsen asked him if he wanted to say something. "Yeah, I want to say something," Langone said. He stood up. "I'll always remember my years of service on the GE board, one of the highlights of my business career. It's an honor that I will carry with me. It means a lot to me. It's certainly a distinction that very few people have, and I'm blessed that I had it. I wish all of you well. I'll do whatever I can to help the company. If there's ever any way at all I can be of help to the company, I'm here and I'll do whatever I can."

Then he turned to Larsen. "Ralph, let me tell you about stubborn," he said. He shared the story of how in 1979 two of Ross Perot's Electronic Data Systems employees were held hostage in Iran and the U.S. State Department "decided not to cooperate with Perot to get them out." He said, "We had a need for documents to get them out, forged documents. I got them for them. I'm not going to tell you where I got them. As a result of that, Ross Perot had made a big—it's ironic—big Excalibur sword, jewels in the handle and everything in the case. Up on the left-hand corner of this encased sword, in the background of the case, was [the] motto 'On matters of principle, never give in to expediency.'

"Ralph," Langone concluded, "that's the only way I can describe my stubbornness. Thank you all very much. It's been a great honor to be among you and with you. I wish you all Godspeed." Jeff sent Langone's wife flowers and to him a note. "Thank you for your great service on the GE board," Jeff wrote. "Your judgment and support has [sic] been invaluable. I admire your courage. I look forward to a sustained friendship. I have learned a lot from you. Thank you for all your support." Bill Conaty wrote, "Dear Ken, thanks for your support. I would love to remain friends. I have such great thoughts about you and respect. That will never change." Recalled Jeff of the incident, "I knew I was turning Langone into a lifelong enemy, but it was the right thing to do."

Indeed, he had made of Langone an enemy for life. "This guy was a combination of hubris, pampered life, I think," Langone told me about Jeff. "He was always president of his fraternity, captain of the football team. This guy plus the sheer white-hotness of the selection process, he bought in to the notion that it was really a coronation, not a replacement, Jack's replacement. He was being coronated. He acted accordingly from day one. Nobody ever challenged him, including me."

He then shared the story of the time when he was still on the GE board and how NYU Langone Health, the vast medical center in Manhattan that Langone had endowed, wanted to buy state-of-the-art medical equipment for "a complete makeover" of the medical facilities at the school to make them world-class. "Everything," he said. "Big, big buy." One day the dean of the medical school and the new head of radiology told Langone that they were a little "uncomfortable." They wanted the process of deciding which equipment to buy to be "an open process," and they were nervous because Langone was on the GE board.

He immediately recused himself. He deputized Tom Murphy, the head of the ABC television network, and Andy Pearson, the president of PepsiCo, to work with the two NYU administrators to decide from which company to buy the new medical equipment. "The only thing is, when you're done, if GE didn't get the business," Langone told them, "I want to know why GE didn't get the business." He also called Jeff and told him that he had left the decision to Murphy and Pearson. "I have no other way to handle this," he told the GE CEO. "I've got to be purer than Caesar's wife." It was a $100 million order. "Huge, huge, but more than the dollars, it was the imprimatur that comes with this huge institution throwing everything out and going with a brand," Langone explained. One day he gets a call from the NYU Medical administrators. They want to come by and speak with Langone, along with Murphy and Pearson. "That tells you one thing," Langone said. "GE ain't getting the fucking business."

It turned out that Siemens, the main GE competitor, had offered to loan NYU $500 million to build a new tower on the downtown Manhattan campus, along with a favorable deal on the new medical equipment. "Jesus Christ," Langone said. "That's how bad they wanted it." Siemens also wanted to make the NYU facility a "beta site," meaning that once the new medical equipment was installed, other potential Siemens customers could have access to the site to evaluate the equipment in situ, as part of a potential purchase decision.

At the time, Klaus Kleinfeld was head of Siemens in the United States. He made the deal with NYU. The university bought the state-of-the-art medical imaging equipment from Siemens, agreed to make the facility a beta site, but decided against the $500 million loan. Siemens threw into the deal a seven-tesla magnet—one of the most powerful magnets in the world—which makes MRI images much better. There were only five in the world at that time. Siemens also paid for installation of the equipment. "It's fabulous stuff," Langone said. "It's world-class. It's state-of-the-art. What I learned was how far behind GE was in technology, because it wasn't just the dollars involved. It was the features."

By the time the deal closed, Kleinfeld had become the CEO of Siemens Global. Langone had a dinner party at his Manhattan home to celebrate the deal. He seated Kleinfeld next to him at the big dinner table. "Okay, how much the fuck did you lose?" he asked Kleinfeld.

"What do you mean lose?" Kleinfeld replied. "Made a fortune."

"How do you make a fortune?" Langone asked, incredulous.

"You agreed to be a beta site," Kleinfeld said. "We sold over $2 billion worth of equipment at list [price] as a result of people coming. Our best salesmen weren't as good as your docs that were using the stuff and were passionate about how it was doing and what good stuff they were doing."

Langone called Jeff and told him GE had lost the business to Siemens. "Anytime my competitor wants to lose that kind of money, I have no trouble with that," Jeff replied.

Langone was incredulous. "You know, Jeff," he replied, "if Jack were in your seat, he'd be driving me out of my fucking mind every day. Even though I told him I had to recuse myself, he'd be screaming and bitching, and he'd be down there himself trying to make that deal because of the image of a guy on the General Electric board that is chairman of an institution that gave all the business to GE's competitor."

*Chapter Twenty*

# IMAGINATION AT WORK

Jeff's early years atop the company were happening against a paradigm shift in how people worked and played. The iPod had debuted in 2001 and the iPhone in 2007. *Google* was added to *The Oxford English Dictionary* as a verb in 2006. The sort of inventions that had been GE's bread and butter for the first fifty years of its existence now seemed to be happening elsewhere.

One way that Jeff differed from Jack—aside from wanting GE to have a different portfolio of companies and businesses—was that he liked to think BIG. He liked to imagine ways in which GE could be in faster-growing businesses, in parts of the world that were growing faster than the United States, and pursuing initiatives—BIG initiatives—that a nuts-and-bolts guy like Jack wouldn't have imagined. Whereas Jack was a chemical engineer, focused on ways to manufacture more efficiently or to lend more money at higher margins—all in the service of continuing to make the numbers quarter in and quarter out—Jeff was more of a supersalesman who believed he could sell anything to anyone, with the right combination of charm, bravado, and leverage.

Whereas Jack had three or four initiatives in twenty years—Six

Sigma, going global, and developing more service revenue—he was never "an idea-a-minute guy," as one former GE executive explained to me. Jeff was always coming up with new ideas, new initiatives. He wanted to out-initiative Jack.

To help him pull this off, Jeff leaned heavily on Beth Comstock, a former NBC senior vice president of corporate communications whom Jack had promoted to be head of GE's communications effort. Born in Morgantown, West Virginia, she moved to the Shenandoah Valley, in Virginia, with her family when she was young. She graduated from William & Mary in 1982 with a degree in biology. She had stints in public radio, local television, CBS, and Turner Broadcasting before settling into the job at NBC, until one day Rosanne Badowski called her and said Jack wanted to see her. Comstock replaced Joyce Hergenhan, who had pretty much been with Jack from the beginning, as head of communications at GE.

At the end of his first year as CEO, and in the wake of September 11 and the Enron scandal, Jeff realized the role of the CEO was changing and that greater transparency was needed. As had been written many times, investors believed GE was akin to a black box. In the face of demands for more transparency, Jack had told people that if he had to give people more transparency, "then it's over," with the implication, according to someone who heard him say it, that GE is "a black box and we like it that way." But Jeff realized that—at least superficially—GE had to be more transparent. He bulked up the size of GE's annual report "to a phone book," as one executive recalled, and tried to answer more questions from analysts and journalists. He also realized he needed "someone to figure out how to explain the role GE played in a way people could understand," he said. He asked Comstock to take the job. He named her GE's chief marketing officer, a position than had been unfilled for two decades. Comstock was a controversial choice.

Her first assignment was to "rethink" GE's motto, "We bring good things to life," which had been around since 1979. "We didn't need rebranding exactly," Jeff explained. "We needed to express ourselves in a new way for a new century." To many inside GE, the change seemed impetuous and felt like a self-inflicted wound. As the story goes, Jeff asked BBDO, GE's longtime advertising agency, to do a study of brands and whether GE should change the slogan. BBDO came back and said that GE and Coca-Cola were the two most recognized brands. "Your slogan is fantastic," the admen told Jeff. "Don't change it." ("They probably sent us a large bill," Bill Conaty recalled.) Jeff didn't like BBDO's answer. He told the firm he would bring in another advertising agency to give the slogan another look-see. BBDO quickly shifted gears and said it would rethink "We bring good things to life." BBDO came back with "fucking 'Imagination at Work,'" recalled one former top GE executive. The GE Capital executives blew a gasket. In the realm of Wall Street finance, pairing *imagination* with *work* was not the message to be conveyed, especially in the post–Enron and WorldCom era, when it was clear that these companies had had way too much imagination at work.

"This is crazy," the former top executive said he told Jeff. But Jeff swatted away the naysayers. "You're old school," he told them. Recalled the senior executive, "Whenever you'd resist something like that, it was 'Old school. You're old school. You're not nuanced.' If you brought a hundred people in this room today and asked them what GE's slogan is, they'd say, 'We bring good things to life.' Yeah. In fact, you still see that with some of these analysts' reports, using it as a dink. That stuff's not productive." No one asked, but Jack hated the new slogan, with a passion. "It's wrong," he told me. "Imagination with financial statements? No. And he put it, 'Imagination at Work.' Unbelievable." In his book, Jeff makes no mention of the controversy surrounding the new motto and BBDO. Instead, he wrote about the debut of the new $100 million "GE: Imagination at Work" campaign in a television commercial during the January 2003 Golden Globe Awards. (It was a bit of convoluted reimagining of the first Wright brothers

flight with a powerful new GE jet engine.) Jeff loved it. He widened
Comstock's "mandate" to include brainstorming in teams of GE executives to figure out how GE could fill the gaps in a broad array of product
categories that otherwise would go unfilled.

Jeff hatched "Imagination Breakthroughs," or "IBs," as he called
them. He gave GE executives, led by Comstock, six weeks to come up
with two or three new products or services or application ideas that
could generate meaningful revenue for the mother ship, either in new
geographies or with new customers. The idea was to push GE's army
to come up with new ideas, to nurture them with a little money, and
to protect them from a bureaucracy that otherwise would have crushed
them. A year in, eighty "IBs" had been given the "green light," and by
2005, twenty-five of them were generating revenue. "I was proud of
everyone," Jeff remembered.

Like many of these initiatives, one Jeff pushed for, "Ecomagination," was plenty controversial inside GE. "Smoke and mirrors" was how
one former GE communications executive described the initiative to
me. "Imagine it works," was the way people inside GE's medical equipment division thought about the initiative. Jeff loved Ecomagination.
The inspiration for the initiative, it seemed, was GE's long-standing
PCBs-in-the-Hudson debacle and Jeff's determination to once and for
all get the matter behind the company. By 2004, the political and environmental mess, which Jack thought he had settled decades earlier,
had turned into the "biggest Superfund site in history," Jeff recalled.
GE still had to clean it up, even though the debate continued to rage
on about whether the PCBs were actually harmful and whether it
would be worse for the Hudson environmentally to stir up the riverbed
mud—as part of the required dredging—where the PCBs had long ago
settled, or the whether the better solution, politically and environmentally, would be for GE to donate the initial $460 million required for
the cleanup to build a bunch of public schools. (Jeff said that Eliot
Spitzer, then the New York governor, rejected this proposal as being

too "complicated" politically.) In the end, GE agreed to spend $3 billion over a decade to complete the cleanup of the river.

Somehow, the ongoing PCB conundrum got Jeff "thinking." GE was in the wind business. GE was in the water-purification business. GE had improved the "energy-saving" standards on its major appliances. GE would have to clean up the Hudson. "Was there something we could do to tie all these pieces together?" he wondered. He presented Comstock with "half an idea" about how GE could implement Jeff's inchoate thoughts about an environmental strategy across many, if not all, of the company's product lines. "Comstock knew I always had my eye out for internal innovations that could be implemented horizontally," he continued. Jeff wanted to grow GE's revenues, which was not an easy thing for a company with revenue of around $150 billion already, in 2004. He also wanted GE, then 112 years old, to "look and act younger." After speaking with Sarah, his seventeen-year-old daughter, and her friends, he realized that her generation cared "more about the planet than their fuddy-duddy parents ever did."

Vic Abate, who ran the wind business, told Jeff that when GE employees were working on something they hoped would improve the world, they had a sense of "renewitude." Jeff wanted to lead by example. He wanted to make a difference and to change GE's image from that of being a polluter of the Hudson River to that of helping to clean up the planet. "And yet, I made clear to Comstock that mere greenwashing wouldn't pass muster with me," he explained. "Unless we could do something with measurable outcomes that could be scrutinized by the public, I wasn't interested." Jeff's slogan for the initiative was "Green is green."

In late 2004, under Jeff's stewardship, Comstock presented "Ecomagination" to the Corporate Executive Council, the top thirty executives, in Crotonville. "I think we're on to something," Jeff told his direct reports. But, he added quickly, "this is somewhat controversial. We're taking a stand on something." Comstock remembered that the "meeting started ugly and got worse." She tried to anticipate the criticism she would get from the old guard about trying to "go green" and then ended with a new television commercial featuring an elephant dancing in a rain forest, while "Singin' in the Rain" played in the background. It was charming and optimistic. It was hard to feel anything but whimsical and uplifted after the forty-five-second ad finished. But this audience hated it. "There was near pandemonium," Jeff recalled. One guy yelled from the back of the auditorium, "You're going to make us look like idiots."

The consensus in the room was that the name was stupid—"Oh, God, that's ridiculous"—that GE would be humiliated for trying to pull this kind of thing off at the same time the company was dredging the Hudson, that GE was getting ahead of its customers, or "worse," Comstock remembered, "make our customers *not* want to buy from us" and would ratchet up the regulation of GE. "There were only two people in the room who wanted to give Ecomagination a chance: Comstock and me," Jeff said. Said one person who would have been in the room, had Jeff not fired him, about Ecomagination, "Give me a fucking break. How about, 'Do you remember how to make money?' One of the biggest crises that the company had was they forgot how to make money. He was spending all this time on bullshit."

Comstock worried that at this moment, Jeff might simply throw her and Ecomagination under the bus and move on. But he didn't. He was a believer. When his top executives continued to object, he shut them down. "Hey, I've got a really good idea right now," he'd tell them. "I've listened to all of you, but here's where we're going. Get in line. We're doing it my way." There was fallout, of course. John Wilder, the CEO of TXU, a big energy company, worried that Ecomagination would mean unnecessary new government regulation. Wilder called Jeff to tell him he was withdrawing $100 million of orders from GE. "And you know what?" Comstock told me. "Jeff bravely asked us to push Ecomagination forward. . . . That's what good leaders do; they absorb the shock waves and anxiety in moments of radical change." The way Jeff liked to frame such debates was "In the beginning, change has a constituency of one."

By August 2005, Ecomagination was in full swing. Comstock and her team had launched seventeen Ecomagination products that were, she said, generating $10 billion in sales. The goal was $20 billion in revenue by 2010. "Ecomagination offered a big lift both operationally and in terms of our brand and reputation," she wrote. And she was happy to take on the skeptics, of whom there were many, who believed Ecomagination was merely sophisticated public relations.

In addition to the dancing elephant and "Singin' in the Rain," there was an ad featuring attractive men and women sweating it out in a coal mine while Merle Travis's "Sixteen Tons" played in the background. "Harnessing the power of coal is looking more beautiful every day," the advertisement boasted. Incredibly, Jeff and GE were waxing poetic about the promise of clean coal and of building a coal-fired electricity plant that coughed out water vapor and nothing else.

In the end, during its twelve-year life span, Jeff reported that Ecomagination generated $270 billion in revenues for GE, without saying how much of that revenue would have been generated without lumping a bunch of products under the so-called eco-umbrella. Ecomagination was "absolutely" a "marketing ploy," he allowed. "We wanted to change our image, but we wanted to earn that shift." Comstock, whom Jeff would make the first female GE vice-chairman (earning $10 million a year), said eventually the skepticism was too much for even Jeff Immelt to overcome. "This is the change-maker's dilemma," she wrote in her book. In the end, Ecomagination couldn't crash "through the layers and layers of complexity and denials of change—across GE, across our customers, across the ecosystem, across financial markets." Yes, Ecomagination was "marketing," she said, but "marketing" done right, by creating a "new business strategy," by meeting "change early," and by building a "coalition of those willing to work for a new future."

But inertia can be a very powerful force, especially at the world's biggest and most valuable company. "Jeff, you seem like you were just miles ahead of everybody's seeing how digital is going to affect industrial companies, jumping in with both feet," said one former GE senior executive about Ecomagination. "Kudos to you for being a visionary. But then it all imploded basically. Tons of money spent and sort of a pile of dust at the end of it all. . . . That's kind of the story of Jeff's life. He's very good at saying, 'We should go here' or 'My thesis is X.' And he's kind of unwavering in pursuing that. And those are all pretty laudable, all things being equal. But then he becomes very blunt force, very unthinking, very not sensitive to any economic details, and it just ends up as a pile of shit."

## Chapter Twenty-One

# THE BLIND SPOT

For all his efforts to make GE a visionary company, Jeff was developing a blind spot as a CEO, and he was losing (or driving out) the executives who dared to tell him the truth as well as the seasoned veterans who could help him navigate a crisis. The departures of David Calhoun and Bob Wright were typical of the brain drain under Jeff.

Dave Calhoun worked for both Jack and Jeff. After graduating from Virginia Tech in 1979 with a degree in accounting, he had a choice between working for GE in Philadelphia and working for Westinghouse in Pittsburgh. Back then, he told me, the two companies "were identical, identical." He wanted a job in industrial America, not in accounting. Since he had grown up in Allentown, he decided to take the GE job to be closer to home. He joined the elite Financial Management Program, which in two years trained recent college graduates in the fundamentals and the nuances of finance and put their GE careers on a higher trajectory. After Calhoun excelled in the FMP program, Dammerman asked him to go to GE Capital, in Stamford, to work for the group that was financing leveraged buyouts. He then worked for Larry Bossidy as his staff executive. "That was a training job," he said.

"That was me working with the CEOs of all of the businesses Larry looked over and trying to help them do projects and trying to keep a finger on GE Capital." Then Jack started taking control of Calhoun's career: after Calhoun worked for Bossidy, Jack put him into the job at Plastics, in marketing, that Jeff had had. "That was a coveted job," Calhoun said. He found himself in that group of executives whom Jack moved from job to job to test their mettle. Calhoun asked Jack to consider sending him to Asia when it came time for his next move. Jack obliged. "I could see that that was exciting and the future, and there was a resourcefulness required that I don't think GE folks were practiced at," Calhoun said. "I figured I'd better put myself in that situation if I could."

From there, Calhoun had big jobs in locomotives, and then as the number two to McNerney at jet engines when McNerney was under serious consideration to succeed Jack. When Jeff got the job and McNerney left, Calhoun became head of the jet engine division, the job he had when Jeff took over right before September 11. Calhoun, GE's best golfer, played a lot of golf with Jack. He and Jack often spoke about the three candidates while on the golf course. "The selection of Jeff at that time was not a bad call," he told me. "If you'd have pinned me in a corner and said which of the three, I probably would have said Jeff." For a while after Jeff became CEO, Calhoun's name was one of three in the envelope that the board kept handy in case Jeff got hit by a bus. The other two names in the envelope were John Rice, who had been head of both GE's power business and its transportation business, and Keith Sherin, the CFO. Jeff had to reflect on the names on the envelope every year—in case he died or got "whacked," as he said—and so they would change every now and then.

In June 2005, as part of another corporate reorganization that clustered GE's businesses into six divisions, from eleven, Jeff named Calhoun a GE vice-chairman and head of GE Infrastructure, which included aviation, power, transportation, oil and gas, wind, and water. In other words, a big job. Perhaps the biggest at GE, aside from Jeff's. "I bent myself into a pretzel to keep Dave," Jeff told me. "And he knows it." At the time, Calhoun was one of the most sought-after potential CEOs in the country: "the No. 1 draft pick in the game of grabbing top executive talent," was how *Fortune* described him.

It was Calhoun's enviable combination of intense competitiveness, global business experience, and knowledge that he would never succeed Jeff as GE CEO that made him a headhunter's dream. In August 2006, he left to become the chairman and CEO of VNU, a media company that controlled A.C. Nielsen and *Billboard* magazine, among others. VNU had been purchased earlier in the year by a consortium of private-equity firms, including Blackstone, KKR, Hellman & Friedman, and the Carlyle Group. Calhoun's pay package was estimated to be around $100 million. (After successfully turning around VNU and leading its IPO in 2010 as Nielsen Holdings, Calhoun left the company in 2013 and became an operating partner at Blackstone. In 2020, he became chairman and CEO of Boeing, after serving on Boeing's board for many years.)

Calhoun said he saw "early signs" of trouble at GE under Jeff's stewardship. "I've stayed close enough to everybody to sort of watch it unravel," he said. It didn't happen overnight. "It took all of fifteen years to do it," he said, adding that if anyone had raised his or her hand during the first eight years, "a lot of it was fixable." But no one did, in part because Jeff didn't like to hear dissent or bad news and praised loyalty, perhaps to a fault. Plus, as ever, it was very difficult to push back on a CEO—unless that CEO, like Jack, mostly encouraged and rewarded outspoken executives. With Jeff, "it was impossible," Calhoun said.

IN EARLY AUGUST 2006, BOB WRIGHT, A GE VICE-CHAIRMAN, MET DAvid Rubenstein, the billionaire founder of the Carlyle Group, for lunch in the outside garden at the Chanticleer, the pricey American restaurant in the village of Sconset, on the eastern edge of Nantucket. Wright's house was nearby, on the same Sconset street as Jack's. Rubenstein's massive home was farther away, off Polpis Road, overlooking the Nantucket harbor. On their agenda, in addition to the restaurant's fabulous Caesar salad, was a discussion about Carlyle, a hugely successful private-equity firm based in Washington, buying NBCUniversal, with Wright remaining as CEO. "It was a window of opportunity I had imagined for much of my NBC tenure," Wright recalled. Wright, of course, conveyed GE's official position that NBCU was not for sale but then proceeded to discuss with Rubenstein a possible deal for the company anyway. He said GE valued NBCU at $45 billion. "He didn't even flinch," Wright remembered about Rubenstein's reaction.

By the end of the ninety-minute lunch, the two men had sketched out a deal. Both men figured that financing for a $45 billion deal would be plentiful—interest rates were low, and banks had plenty of money to lend to leveraged companies in order to be able to charge higher interest rates. That the deal would have been the largest buyout ever didn't seem to matter. "Every bank wanted to be private-equity's partner," Wright allowed. It was an audacious move on his part. Unauthorized conversations with a serious buyer about the sale of an important business like NBCU without permission of the CEO or the board of directors would surely qualify as grounds for dismissal, with cause. There was little doubt that Jeff was pissed about Wright's behavior, lighting the fuse for what would be his unceremonious departure from GE in short order, after forty years. It would take a few more months to play out. As a first step, Jeff simply nixed the proposed deal, despite the attractive $45 billion price tag and the fact that, at least in Wright's opinion, Jeff seemed to be ignoring NBCU. At the same time as Wright was trying to gin up the NBCU buyout with Rubenstein, he was also trying to get Jeff to green-light NBCU's acquisition of DreamWorks, the movie studio started in 1994 by the incomparable trio of Steven

Spielberg, Jeffrey Katzenberg, and David Geffen. Wright thought DreamWorks was a "natural extension" of NBC's merger with Universal, since Spielberg and Universal Studios had a long-standing relationship.

Earlier in 2006, Wright had negotiated a deal with Geffen for DreamWorks at a price of around $1.5 billion, "what we both thought was a very fair agreement," Wright explained. But that resulted in more frustration for Wright. He claimed Jeff "procrastinated" on the deal for the better part of a year, "stringing all of us along," and that the GE board "was uneasy with the ebb and flow of Hollywood economics." When finally, in December 2006, the GE board signed off on the deal, allowing Wright to make a bid, "it was too late." Viacom, the Hollywood fiefdom of Sumner Redstone, and its Paramount division had swiped DreamWorks away for around $1.6 billion, a bit more than GE had authorized NBC to offer. Wright was embarrassed and hurt. "We felt like we had lost out to GE's notorious bean counters," he remembered. It was one of the few times GE had lost a deal, over either price or indifference.

By then, Wright's fate at NBC (and at GE) had been all but sealed. He was sixty-three years old, two years shy of what many considered the GE retirement age. In truth, Wright had run afoul of the boss, Jeff Immelt. In the third quarter of 2006, NBCU's operating earnings had declined 10 percent, to $542 million, the fourth consecutive quarter-over-quarter decline at the business, while the stock of rivals News Corporation, which owned Fox, and Disney, which owned ABC and a variety of other Hollywood studios, had soared. Jeff had wanted to replace Wright years before with Andy Lack, but Wright had seen off that threat. Now Jeff wanted to replace Wright with Jeff Zucker, the television wunderkind who was then all of forty-one years old.

"I had to fight to make Jeff Zucker the CEO of NBCU," Jeff recalled. He believed that Wright looked around and saw that Rupert Murdoch, then seventy-five, still ran Fox, and that Sumner Redstone, then eighty-three, still ran Viacom and CBS, and thought he should continue to run NBCU. Perhaps Wright forgot that NBCU was part of GE, which was run by Jeff Immelt. Jeff wanted Wright gone. In November, he told Wright that Zucker was replacing him in February 2007. "It's over," he told Wright. Wright tried to convince Jeff that was "a big mistake," that Zucker could never run NBCU and that he should spin out NBCU. "I've already worked up a deal with John Malone," Jeff remembered Wright telling him. But Jeff said he didn't want to sell NBCU because it would make GE even more reliant on GE Capital for its earnings. Jeff worried about what being more dependent on GE Capital would mean for GE's stock price. "Wright was desperate to keep his powerful position," Jeff recalled, "but I was done with him."

Jeff dispatched Wright from NBCU in January 2007. He remained a vice-chairman of GE, with minimal responsibilities, for another year. He had been at GE for forty years, the last twenty-one of which were as head of NBC.

As the departures of some of GE's top executives were mounting—among them Nayden, Calhoun, and Wright—Jeff seemed determined to play down their importance. This was a point he made repeatedly, to me and to others, despite ample evidence to the contrary. "Look, love me or hate me," he told me, "I didn't lose good people. I kept the reputation and the talent in the company. The prism I put it through was you want to keep the spirit, the pride, everything about the place. If you're in 2002 or 2003, it wasn't clear there was going to be a commercial aviation industry in the future. You went from 9/11 to SARS and stuff like that. Believe it or not, I put the company first in everything I did, and by 2007, we had a $42 stock, before the financial crisis. We were building our way back up."

Instead of the team of rivals that had surrounded Jack, Jeff had created a team of sycophants, aided and abetted by the unintended consequences of a highly lucrative supplemental pension program, which had the ability to silence even the most outspoken GE executives, unless, like Calhoun, they left. The GE Supplementary Pension Plan, known around GE colloquially as the "SUP," was awarded solely by the GE CEO, in this case, Jeff, to the high-ranking GE employees he selected, once they reached fifty-one years old. As long as GE remained solvent, the SUP guaranteed these lucky employees an additional pension from GE equal to 80 percent of their salary and bonus *for life*. Given that the SUP could mean millions in additional annual postretirement compensation from GE, regardless of whether you took another job or not, most people coveted the SUP if they were in a position to get it and, once awarded it, chose to keep quiet rather than risk losing the SUP by getting on the wrong side of the CEO. "I don't want to kid you," Calhoun told me. "The pension issue is a huge factor in the unwillingness of people to speak up, a huge factor." Calhoun described the SUP as akin to "handcuffs" on senior GE executives once they reached fifty-one and became eligible for it. And to get the SUP, you had to stay at GE until you reached the age of fifty-five. "You get canned when you're fifty-three, you got nothing," Calhoun said. "If you say you're going to leave because you don't agree with where the company's going or whatever, you got nothing. . . . You've got to make it to fifty-five." He said that's why he left GE for VNU. "They hold the keys to the kingdom," he said. "But if you left in your fifties with a SUP hanging out there, you were nuts. I left when I was forty-nine years old. A huge factor for me leaving was I did not want to get caught in the SUP-counting days. I did not want to be hostage to my pension. That's a fact." Explained one longtime GE executive, "Jeff did not surround himself with a level of fiercely intelligent people who could—and would—on a regular basis be combative with him. People were combative with Jack. Jack loved it. Jack wanted it. He wanted to be in a fight. He wanted to have an intellectual throwdown. And Jeff didn't. He wanted to participate. He wanted to read out his conclusions and he did. If you look at the who's who of the GE leadership team and compare them, they were people who understood and were willing to

debate, and debate is what went away, I think. Debate no longer occurred." Another GE executive derided the unintended consequences
of the "incredibly rich" supplemental pension plan, even though he was
a beneficiary of it for the rest of his life, assuming GE did not file for
bankruptcy. "It's the worst thing in the world, because once you get to
the gravitational pull of the SUP, what you have to do is avoid being
fired," he said. "Therefore, you take no risks."

THE ULTIMATE STRESS TEST FOR JEFF'S LEADERSHIP WOULD COME
with the housing crisis of 2007, which ultimately led to the financial
crisis of 2008. In April 2007, two Bear Stearns hedge funds were among
the very first big investors in the various flavors of mortgage-backed
securities to begin to understand on a granular level that the party in
commercial and residential real estate might be ending. By then, sadly,
it was too late for the Bear Stearns hedge fund managers to do anything about their predicament. Within months, the two hedge funds
would be gated, then closed, and then liquidated, leading to a cascading series of events.

If the two Bear Stearns hedge funds were among the first on the
*buy side* to grapple with the collapse of the housing market, GE, from
a different perspective—the *sell side*—early on also had a front-row seat
on the unfolding crisis. Inside GE, Michael Pralle, the man who ran
the real estate business and built it into a multibillion-dollar colossus,
with Jeff's prodding, knew that the real estate business was getting
very toppy. He was nervous. His gut told him it was time to sell a big
chunk of the portfolio. On paper, Pralle seemed like someone to whom
Jeff would be sure to listen.

In 1978, he graduated from Harvard magna cum laude with a degree in economics. At Harvard, he played squash, participated in student government, and wrote a single article in *The Harvard Crimson*
(about Daniel Patrick Moynihan and whether he could retain his posi-

tion at Harvard as well as serve as U.S. ambassador to the United Nations). After Harvard, Pralle spent two years in the decision analysis and strategic methodologies group at the Stanford Research Institute. In 1980, he enrolled at Stanford's Graduate School of Business, where he was an Arthur D. Little Fellow. After getting his MBA, Pralle was off to McKinsey, where he spent seven years, focusing on the financial services industry. He worked with a Japanese investment bank, a commercial bank in Hong Kong, and major insurance companies in Europe. He worked in London and in Hong Kong, where he and two McKinsey partners opened a new office.

He joined GE in August 1989. Over the next nearly eighteen years, he had a variety of senior executive positions. He was president and CEO of GE Asia Pacific, where he was in charge of financial-services activities in eighteen countries, including Korea and China. In his three years there, he built up the business to $100 million in net income, from zero. Pralle had a complicated personal life. His first wife died in childbirth. Cancer claimed his second wife. He was a fitness fanatic, worked out like a demon, and was a great skier. He was a fierce bike rider. He owned a home in the South of France.

In May 1996, he took over GE Equity—the business that made investments in other companies—when it was a break-even business. In his four years at GE Equity, it grew into a portfolio of three hundred companies and $525 million in net income. GE asked him to take over the commercial real estate business. "I knew virtually nothing about commercial real estate, because I had been running the private-equity business," he told me. "But you do what you're told to do." When he started as the president and CEO of GE Real Estate, it generated $360 million of net income. One of his main initiatives was to invest in more of the equity of real estate deals, not just the senior secured debt. Investing in senior debt, at the top of the capital structure, was safe but boring, generating a contractually fixed rate of return. Investing in the equity, at the bottom of the capital structure, was very risky but more exciting. You could make a lot of money, yes, but you could also lose a lot of money. "That's where the money is," Pralle said of taking equity in these deals. In his seven years as head of GE Real Estate, he built it into a global business with some $70 billion of assets and $2.2 billion of net income. There were thirty-nine offices around the world and five thousand employees. GE Real Estate owned two operating business—self-storage and parking garages—and then owned a variety of commercial real estate properties and lent to a bunch of others, including in Sweden, Germany, and Mexico. Pralle's group also did a $3 billion deal buying all of France Télécom's offices around France and leasing them back to the company.

By the spring of 2007, Pralle and his team believed the time had come to pare the portfolio, which would result in large capital gains for GE but likely also reduce the ongoing profitability of the business. "We tried to do it, but the problem was every time we tried, the board and Jeff in particular wanted more net income," Pralle recalled. "That was a struggle. We said, 'Isn't a billion and a half enough this year?' [And Jeff would reply], 'No, I want more.' Or he would agree to the billion and a half and he'd say, 'We've got a problem this quarter in China, and I want you to sell something or do something to get another $50 million in net income.'"

He remembered the challenges he had with Jeff when it came to selling GE's nearly half of 30 Rockefeller Center to Jerry Speyer, who owned the other half and was therefore the logical buyer of GE's equity stake in the building. "I knew I had to very quickly drum up some money for Jeff," he said. "I went to Jerry. I said, 'Jerry, can we talk about selling our half to you?' He and I had agreed to a deal basically, right then. I went to Jeff with the deal, and Jeff says, 'Well, try and get some more money out of him.'

"I said, 'No. You don't know who this is. This guy is the chairman of the board of the Museum of Modern Art. He runs one of the most respected real estate companies in the country. I can't just go and recut the deal with Jerry Speyer.' He said, 'No, no, I want you to go do that.' I go back to Jerry and I start talking to him. It was a very short conversation. It was not going to happen. But that's what we ended up doing. We ended up selling our share of 30 Rock to him and made a nice profit. In the grand scheme of GE, it wasn't really important. It didn't even require a footnote [in the annual report]. But it allowed Jeff to cover a gap in earnings from health care in China." Some version of the same scenario would play out every quarter and every year. Pralle did the same thing he had done with 30 Rock with 125 Park Avenue, across from Grand Central Station. He knew a guy in California who coveted the building, which GE had refurbished, and cut a profitable deal with him. "In the case of real estate, he just used it as a piggy bank," Pralle said of Jeff. "He knew there was a place where he could go and get his income gaps filled. It's true."

Few at GE seemed to question his professional judgment. "Even on his worst day, I'd take him on my team," said one of Pralle's former colleagues. In the months leading up to May 2007, Pralle decided GE needed to sell half of its real estate portfolio. Pralle could have been reacting to the cracks that were occurring, starting in February 2007, in the ABX Index, which measured the trading of mortgage-related securities, or what he saw happening at the Bear Stearns hedge funds or what he witnessed firsthand with a $4.2 billion GE commercial-mortgage-backed-securities deal where investors rejected a bunch of mortgages. Or he could simply have relied on his smarts and market experience. "The crux of the recommendation that we made was to basically take some chips off the table," he said.

He explained his thinking to me: "I really don't know what's going to happen to the real estate market. I don't know what's going to happen to pricing. I don't think anybody does. I think forecasting that is very difficult. Based on my experiences, cycles tend to be longer than most people think. The market stays at a high longer than most people expect it to. The market stays at a bottom longer than most people expect it to. And there can be serendipitous events that affect the market that are almost impossible to forecast."

He told Jeff that buyers were willing to pay a historically high multiple for real estate cash flow. "In some cases, thirty times cash flow on nice properties," he said. "Like a nice office building in London could be thirty or thirty-five times cash flow. Is that too much money? I don't know, but it's the most I'd ever seen." His idea was to sell 51 percent of the business, deconsolidate it from GE, and book a gain of around $12 billion. "The Street will not give you credit for that," he told Jeff, "but that's real money that you can use to restructure some other industrial businesses." GE's real estate group was working informally with Goldman Sachs to help Pralle and his team think through what should be done to reduce the exposure, what the assets might be worth, and how the assets could be sold. (In December 2006, Goldman had its own concerns about the mortgage market and had begun to implement a strategy of shorting the mortgage market that would pay off hugely in April 2007.)

In May, convinced the market was at a peak, Pralle had his team put together a presentation for Jeff recommending that GE begin aggressively selling its portfolio of real estate assets. Others had been telling Jeff the same thing. Jayne Day, the chief risk officer of the real estate group, among others, met with Jeff and told him about the growing risk in the real estate portfolio. Their message was clear: trouble was imminent. Said someone at the meeting, "His response was always 'I just need to ride this pony one more year.' And by that he meant he needs the power business to start producing income" in order to stop relying so heavily on harvesting the gains in Pralle's real estate business every quarter. "Every time we had a financial meeting, Jeff wanted to continue to grow the real estate book," recalled one GE executive. "The leadership team of the real estate business would say, 'You're taking a much higher risk doing that now versus a few years ago.'" Jeff didn't listen. "He needed the income and he felt it was like magic dust," the executive continued. "Every time we bought a portfolio, we'd sell a few pieces of the portfolio and generate the income."

Before a group of about eighty GE executives in Crotonville, at something called LIG—for Leadership, Innovation, Growth—Pralle presented his recommendation to sell half the portfolio. "There are lots of people who are very interested," he told Jeff. The purpose of the LIG was to have GE executives from across the company join together to swap ideas about how to think about their businesses in a new and different way. "We'll book a $5 billion to $10 billion gain," he said. "And we'll have much less exposure to real estate, particularly to those things that we think will be a risk during the next downturn." Jim Colica, the GE Capital chief credit officer, agreed with Pralle: GE had too much exposure to real estate. Goldman Sachs, which was informally advising Pralle, agreed, too, and was prepared to sell a big chunk of the portfolio. But Jeff said no. "He had a very, very negative response," Pralle told me. "I still remember exactly what he said, which was 'That's the dumbest idea I've ever heard. Why would I sell half of the best real estate business in the world?'"

Pralle asked Jeff to hear him out. "Why not be conservative here?" he asked. "No one is going to criticize you for taking profit off the table, for reducing the size of a business that everybody is nervous about." But Jeff kept saying to him, "What are you doing? Why are you saying this?" His team tried to signal Pralle to cut short the presentation, given Jeff's opposition to the idea. They were waving their hands. Giving him the time-out sign. Eventually, he stopped. "We really didn't get very far with that," he remembered. "That was probably the beginning of the end for me at GE."

In meeting after meeting—and in particular at the LIG meeting in May 2007—Jeff held up a McKinsey presentation he had commissioned that predicted the real estate market would continue to boom. "Jeff kept telling everyone that we need to load up because McKinsey said it was okay," said the GE executive. "And everywhere he went, he carried around this copy of the McKinsey presentation where the tree's growing to the sky, and he would hold it up and say, 'McKinsey says . . . and they are smarter than you are.'"

Jeff rejected Pralle's recommendation. A month later, at the end of June, Jeff abruptly fired Pralle. That decision left the rank and file befuddled. "Everybody would say, 'Well, why are they firing this guy?'" recalled the GE executive. "A lot of people who were rising up through the ranks who could have potentially taken over as CEO seemed to be disappearing from the ranks." For his part, Pralle said by then he had already started looking around at real estate investing opportunities with private-equity firms and with investment banks. He was talking to Evercore, TPG, Credit Suisse, and the Quadrangle Group. He went to work for Joe Robert, at JER Partners.

"I didn't leave GE because of an issue with Jeff or anything like that or because he wouldn't do what I said," Pralle said. "That happens in business. You don't always get your way. I understood how an organization chart works, and he was the boss. I *was* disappointed he didn't take my advice."

Pralle said he didn't remember Jeff waving the McKinsey report around but said it wouldn't have surprised him if he had, and it revealed to him the essence of the difference between Jack and Jeff. When Pralle presented a business opportunity to Jack—say, for $2 billion of capital to be used for some real estate opportunities—"Jack would say, 'Okay, Michael, you've given me the presentation. I understand how you think it's gonna go. But if everything goes wrong, how much can we lose?' That was his most important question. What is the real downside here? And I'd have to describe it and have a good idea of why I got to that description." Other times he could be tougher. In the late 1980s, GE Capital was in the process of merging its private-label credit card business with Citibank's much-bigger business. Pralle wanted to do the deal and also to make a $500 million preferred stock investment in the bank. Citibank was in financial trouble. "Some people were nervous about its survivability," Pralle said.

It was one of his first presentations to Jack. He was working for Jim McNerney at the time. "I'm up in Fairfield," he recalled. "I go into the boardroom. The board table must be thirty-five feet long. Jack is in the middle on the other side. He's kind of hunched over, the way he is, a little sort of gnome-like guy. There's three or four guys on either side of him, the usual senior staff people. I walk in the door and Jack says, 'Don't even fucking sit down. This is the dumbest fucking idea I've ever seen. I don't know what you're thinking here, putting $500 million into Citibank. I would never do that in a million years.' He tears off the cover page of the pitch. He crumbles it up into a little ball and he throws it at me. Jim says, 'Jack, come on. Michael has done a lot of work on this. Let him sit down and go through it, and we'll see what you think after that.'" Pralle was nervous. "I sit down and I go through the pitch as best I can, being sure that I'm going to get fired," he continued. "At the end, Jack says, 'You did a nice job, but I still think it's a really dumb idea.' We walk out and we close the door behind us. Jim turns to me and he says, 'I think he liked it, Michael.'"

Jeff, Pralle said, cared about other things. "For Jeff, his most important question was 'How is Wall Street going to view this deal? Will we look smart? Should we bring in Blackstone as a partner, because they're smart? How will the analysts see this deal?' When I would present to Jeff, I would start with the press release and work backwards from the press release to the presentation on the deal and the argument in favor of the deal. It was just a completely different kind of approach to persuasion than I had with Jack, because that's what Jeff was worried about: what other people thought about it. Therefore, McKinsey would validate in the way that Blackstone would validate: by being a part of what he was thinking about or what other people were recommending he should do."

Jeff did not mention Pralle in his book. He wrote that several of his "deputies" had recommended selling the commercial real estate business in the fall of 2006 but he said no, believing the team "could guarantee exceptional returns." GE stopped "making any new deals" in 2007, he continued, but again he nixed the idea of spinning the business out of GE. ("We were in the process of replacing the leader of the business," he wrote.) He conceded he had made a mistake by not following Pralle's recommendation. He considered commercial real estate a "core platform" for GE Capital, "but we let it get too big." He claimed the "coming storm" was "not yet visible," even though something should have clicked together in his mind after what had happened at Bear Stearns and what had happened with GE's own mortgage-backed securities.

When I asked Jeff about the May 2007 LIG meeting and Pralle's recommendation, he said, "We had one or two meetings where we did talk about selling it, and we just missed the window. We really missed the window. This was in the second quarter of 2007 or something like that, but it was after the mortgage stuff had started hitting the fan. I remember in the growth playbook in 2007 we talked about spitting out commercial real estate, which had grown to be extremely big. We were having leadership problems there at the time. We had to fire the CEO and put a new CEO in, and we just missed the window." He recalled the conversations. But then turned defensive. "Let's be clear," he said. "Commercial real estate through the crisis made a shitload of money. Believe me. That's not our problem. In other words, we could have maybe top-tipped the market if we had sold it in 2006 or 2007." He said compared with the mistakes of buying Lake—the Japanese credit business—or of not including the long-term-care liabilities in the Genworth IPO, not selling the commercial real estate assets in the spring of 2007 was a mere blip on the screen. "Commercial real estate was a great team who ran the business really well through the crisis," he said.

But others in and around GE and GE Capital thought Jeff made a big mistake, especially since he didn't listen to his well-paid professionals, including Michael Pralle and Mike Neal, who ran GE Capital after Jeff fired Denis Nayden, and what they recommended to him should be done. "If you're not willing to listen to the experts on something that big, I just don't get it," said one former GE Capital executive. "Then what are they there for? Just hire some really cheap people. Pay somebody $50,000. Save the company a lot of money, if you are always right."

WHEN THE CREDIT-DEFAULT SWAPS LINKED TO YOUR DEBT START blowing out, there's bound to be trouble brewing. Credit-default swaps, or CDS in Wall Street argot, are a form of insurance allowing creditors to protect themselves when they start worrying about whether they will get paid back. The price of the CDS on a particular piece of debt fluctuates based on the market's perceived risk of a payment default. If the probability of getting repaid seems high, CDS prices are low; as the perceived risk of not getting repaid increases, the price of CDS increases correspondingly. In other words, rising CDS prices for a company's debt could be an early warning signal that creditors are getting nervous about whether they will get repaid or, more generally, if a company's credit is not all that it's cracked to be. Unlike regular insurance, though, anyone can buy a CDS on a particular piece of debt. It is not a requirement that the buyer of a CDS own the underlying liability. Unlike other forms of insurance, a CDS buyer does not have to have an insurable risk. It's as if someone could buy fire insurance on someone else's home, which obviously cannot happen. But you can buy CDS on a bond even if you don't own the underlying bond. Hence, buying or selling CDSs is another way for investors to express their optimism or pessimism about a company's prospects. It is a form of speculation, or gambling, on a company's financial prospects. Even the AAA-rated GE was not immune to such speculation. After all, as Jim Grant had been writing for years, when a company's debt is rated AAA, its ratings can only go in one direction: down. There are no AAAA-rated companies.

Through the course of 2007, as the housing market became increasingly tenuous, along with the securities tied to the housing market, the financial markets became increasingly jittery. In addition to the eventual liquidation of the two Bear Stearns hedge funds after some forty-five months of positive performance, the contagion was becoming global. On August 9, evidence of the international spread of America's subprime crisis showed up in Paris when BNP Paribas, France's largest bank, ceased withdrawals from three investment funds, which had about $2 billion in assets on August 7, because the bank could no longer "fairly" value them due to a "complete evaporation of liquidity in certain market segments of the U.S. securitization market." BNP's action followed an August 3 announcement by Union Investment Management, Germany's third-largest mutual fund manager, that it had stopped permitting withdrawals from one of its funds after investors pulled out 10 percent of the fund's assets. On August 9, the European Central Bank injected £95 billion into the overnight lending market "in an unprecedented response to a sudden demand for cash from banks roiled by the subprime crisis," *Bloomberg* reported.

On August 17, the Federal Reserve began to take its first steps to try to stanch the bleeding. The central bank cut interest rates by fifty basis points in recognition that "financial market conditions have deteriorated, and tighter credit conditions and increased uncertainty have the potential to restrain economic growth going forward." The Fed pledged to "act as needed to mitigate the adverse effects on the economy arising from the disruptions in financial markets." The Fed also announced that banks could borrow from the discount window "for as long as 30 days, renewable by the borrower," in order for banks to have "greater assurance about the cost and availability of funding." The new plan would remain in effect "until the Federal Reserve determines that market liquidity has improved materially." The two-pronged approach of lowering interest rates and effectively substituting the Fed's balance sheet for the balance sheets of the country's financial institutions, whether troubled or not, was set at the annual Fed off-site in Jackson Hole, Wyoming, during the third week of August 2007. New York Fed president Tim Geithner dubbed this new approach to the growing crisis "the Bernanke doctrine."

AAA-rated GE was not immune to the growing credit conflagration, even if Jeff and Keith Sherin, the GE chief financial officer, talked about GE's prospects as if it were. Earlier in 2007, in springtime, the cost of buying insurance—CDSs—on GE's debt for five years of protection was cheap: around twelve basis points per year—one basis point is equal to one one-hundredth of 1 percent, or 0.01 percent, or 0.0001—or 0.0012 times the principal amount of the debt being insured. By Labor Day, the cost of insuring GE debt had risen to forty basis points. By November, it had risen to sixty-four basis points, a price increase of more than five times. A similar ratcheting up of CDS cost had occurred at GE Capital. Investors were getting increasingly nervous about GE's credit, even though it was still rated AAA. Jim Grant understood. "As the world learned to its cost," he wrote on November 30, 2007, "triple-A is a judgment handed up by fallible human beings who draw their pay from the very borrowers they rate. GE is a hugely successful globe-girdling corporation that—contrary to the popular perception— is not so much an industrial giant as a financial one." Based on the skyrocketing price of GE's CDSs, Grant figured that Moody's should be rating GE's debt at Baa1, or seven notches below its AAA rating. "You wouldn't suppose that a company with a $384 billion market cap, a $762 billion balance sheet and a storied 115-year history could hide much from the scores of analysts who swarm around it like the Lilliputians poking at Gulliver," Grant wrote. "But GE has done it. It hides its secrets in plain sight."

IN THE MIDDLE OF 2007, JEFF HIRED MCKINSEY, HE WROTE IN HIS book, because he was "determined to monitor GE Capital's risks." (He made no mention of asking McKinsey to look at the commercial real estate business specifically.) Two months later, McKinsey reported back to Jeff that GE Capital would be fine, that between robust sovereign wealth funds and countries, like China, with huge trade surpluses with the United States, there would be plenty of capital to "fuel GE Capital's lending and leverage for the foreseeable future." Concluded Jeff, "According to McKinsey, we were okay." What sovereign wealth funds or trade surpluses had to do with the commercial-paper market was not exactly clear, other than perhaps they would need a place to park their money on a short-term basis, and GE Capital could then use that money to continue to fuel its growth. But when confidence in markets erodes, that's it; it doesn't matter how big the trade deficit is or how much money is in a sovereign wealth fund.

In any event, Jeff declared himself satisfied. "For a little while, it looked as if McKinsey was right," Jeff continued. How he could have missed the gathering storm is hard to imagine, given what he had already seen at WMC, in the pleadings of his commercial real estate executives, and in the blowout of GE's CDSs. To say nothing of the warning that Bill Gross and the ratings agencies had issued to him five years earlier. Or what Jim Grant had been writing about GE *for years*. Part of the problem for Jeff was that GE's ongoing financial performance, at least through 2007, blinded him to the massive explosive device sitting on GE's balance sheet. That and the fact that—despite his claims to the contrary—he had a blind spot for the nuances of finance. When you are, as Jim Grant articulated, an industrial company inside a bank, and you don't understand in your DNA how a bank actually makes money and the ever-present, gargantuan risks that exist when you are in the business of borrowing short and lending long, then one cannot be shocked by the devastating consequences.

On December 11, 2007, at the annual "investor outlook" conference, Jeff said that GE's 2008 earnings growth would be at least 10 percent higher than 2007. "What I really want to give investors is a sense that 10 percent is in the bag," he said. "You can do this." Nearly a month later, on January 18, GE announced its 2007 earnings and Jeff reiterated both that 2008 earnings would be up 10 percent compared with 2007 and that earnings for the first quarter of 2008 would also be up 10 percent above the first quarter of 2007. "I think our 2008 guidance is solid," he said on the call accompanying the earnings release.

Jeff's optimism was both understandable—he was human, after all—but also a shocking example of cognitive dissonance. The fourth quarter of 2007 was a disaster for most Wall Street banks (except, most notably, for Goldman Sachs). Bear Stearns, for one, suffered its first losing quarter in the eighty-five-year history of the firm. Huge losses had already been realized across the mortgage market and in the mortgage securities that had larded Wall Street balance sheets. GE Capital wasn't on Wall Street, or in Manhattan. Nor was it regulated like a Wall Street bank or like a commercial bank. It was largely unregulated. At best, the SEC was the one to regulate GE in its entirety, including GE Capital.

Jeff knew that trouble was afoot for big financial institutions. In Jeff's letter to shareholders, written on February 20, he shared as much: "Bubbles burst and excess ends in an ugly fashion," he wrote. "The easy credit cycles that defined the recent past have given way to a tidal wave of financial crises. As I am writing, banks have written off almost $150 billion, entire classes of securities have disappeared, and rating agencies have been criticized. This transition—from easy credit to no liquidity—seemed to occur in the blink of an eye." How in the world Jeff believed GE—one of the largest banks in the United States—could survive unscathed is mind-boggling. But he did.

On March 13, 2008, Jeff held a live webcast for GE shareholders to offer "answers and reassurance," he recalled. He again reaffirmed GE's prior guidance for 2008. "Delivering financial results, hitting 10 percent EPS growth, getting $2.42 a share or greater . . . is what we believe and what we're committed to do this year, and what we are on track to do this year," he said. As for GE Capital, he said what "I knew to be true: it was strong and profitable." He said that while half of GE was financial services and—understatement alert—"there is a lot of volatility in the market right now for financial services," GE didn't have the "same exposures" as Wall Street firms. Incredibly, he also said that the "turmoil in the financial markets" wasn't "cause for concern at GE." Instead, he said, the misery of other financial institutions was just another buying opportunity for GE's formidable economic machine. He told me there were many different data points coming at him at once, "but none of them led to 'Shut everything down and let's start burning furniture.'" He said relying on the commercial paper market to fund GE Capital was "a model that worked until it didn't, and then it became definitely the wrong model. It didn't go from white to gray to chartreuse. It went like white to black in like seven seconds." (If Jeff had understood the risks of borrowing short and lending long—the inherent risks of finance—he might have anticipated how something like this could happen, even at the mighty GE. But he didn't.)

Two days after Jeff's webcast, on March 15, Bear Stearns narrowly avoided a bankruptcy filing by agreeing to sell itself to JPMorgan Chase for $2 a share, 99 percent below where it had been trading fifteen months earlier. The collapse of one of the most profitable and respected Wall Street securities firms in what seemed like a week's time sent a massive shock wave not only across Wall Street but also across global financial markets.

At the time, the powerful troika of Ben Bernanke, the Fed chairman, Hank Paulson, the Treasury secretary, and Tim Geithner, president of the Federal Reserve Bank of New York, thought that by keeping Bear Stearns out of bankruptcy they had staved off a massive financial crisis. They couldn't know it on the ides of March 2008, but they would be wrong, at least about preventing the inevitable, and painful, correction. Observed Jeff, with more understatement, "Just thinking about what happened next makes my stomach clench. Three days after my webcast, on March 16, the global investment bank Bear Stearns collapsed, overwhelmed by its exposure to mortgage-backed assets. The credit market froze, and that would hurt GE."

What really hurt GE was that some two weeks after Jeff repeatedly told investors that the company's first-quarter earnings would be up 10 percent year over year, he lost a large chunk of his credibility when GE missed those earnings by a country mile. He blamed the market turmoil that followed the shocking near collapse of Bear Stearns. On April 11, Jeff announced that GE had missed its earnings projections by $700 million. What was supposed to be roughly $5 billion of net income in the first quarter of 2008 came in at $4.3 billion. While those earnings kept GE one of the most profitable companies in the world, it was not the formula that GE had perfected under Jack. If you are the CEO of GE and you tell Wall Street analysts and investors two weeks before the end of the quarter that you are going to make a profit of $5 billion, you'd better deliver $5 billion in profit, not $4.3 billion. GE's stock fell 13 percent, the largest one-day percentage drop in the stock in twenty years. Investors were stunned.

Jeff said that, while it would have been worse if not for the scrambling—going to the cookie jar and cashing in a variety of assets, lightning fast—that GE Capital undertook in the last two weeks of March, "it was [still] an unthinkable miss for GE." In explaining the miss, he finally conceded, in writing, what investors had long suspected: that GE Capital had been used for decades as an earnings candy store for GE. "GE Capital had been our most reliable performer for decades," he confirmed. "Since financial assets are, under normal conditions, far more liquid than tangible assets, GE Capital could opportunistically sell them at a profit." But after Bear failed and the capital markets started seizing up, GE Capital began to the face the same reality that many other large financial institutions face in times of trouble. "Losses were growing, and it was harder to sell assets at a gain, which had been a normal practice for years," he continued.

"In the wake of Bear Stearns's failure, finding willing buyers for those assets was nearly impossible." That's about as close as Jeff would get to admitting that the likes of Bill Gross and Jim Grant had been right, after all.

Explained one GE executive, "The first quarter of 2008 was a shocker. That was just a big number and a late miss, and a magnitude that said to me, 'Okay, we're not able to forecast this thing. Something's going on.'" He said normally GE could see how the earnings were developing in a quarter, and if there was a hiccup, the guidance could be reset. But this rupture came so late in the quarter, nothing could be done but to accept the "torpedo in the engine room," as Jeff referred to it, and move on. "In the last two weeks of the quarter, GE Capital missed by like $500 million," the GE executive continued. "It wasn't like you had any chance to communicate it. It happened late and it was big. So that just turned into a surprise, not only for the Street. It turned into a surprise for us. That's when you're like, 'Okay. You're running the place. You've got your hands around it. What's going on?'"

Jack, meanwhile, was apoplectic about the earnings miss. Perhaps that's what losing 13 percent of your net worth in a day will do to your mood. He went on CNBC, still owned by GE, of course, and lit into Jeff. It was an utterly inappropriate move by a former GE CEO. But Jack was Jack, and therefore unsuggestible. He said that if Jeff missed his earnings guidance again, "I'd be shocked beyond belief, and I'd get a gun out and shoot him if he doesn't make what he promised now. Just deliver the earnings. Tell them you're going to grow twelve percent, and deliver twelve percent." Jack said on national television that Jeff had hurt his own and the company's credibility. "Here's the screwup: you made a promise that you'd deliver this, and you missed three weeks later," Jack also said. "Jeff has a credibility issue. He's getting his ass kicked. He apologized." He continued, "I'm not here defending what happened. This was a bad miss. This is a credibility crash. He's got to earn it back. He will earn it back. But to take GE apart, [and to ask, as many had,] 'What's wrong with the company? It's falling apart, blow up the model, sell NBC, sell real off estate, let's go out of business because we had twenty-four hours of a mess?'"—that Jack could not condone.

Jack told me he "flipped his lid" when Jeff missed the first quarter 2008 earnings guidance. He said "to shoot" someone if they missed their numbers was "a common phrase around GE at that time, as in 'We'll shoot him if he doesn't do it.'" He said he "blew my top" but that Jeff had "overpromised," adding "and you can put that on his gravestone: *overpromised and underdelivered.*" He then quoted from the letter he had written Reg Jones in 1979 as he was campaigning to become CEO. He wrote that GE needed earnings consistency: "Make the numbers. Make the cash flow if we're ever going to be recognized as a valuable stock." Jack was unforgiving. "He missed," he said. "You couldn't make a couple of cents in the pile of money that's around there? I spent forty-seven years on a theory that you make your commitments. You promised me a month earlier." In the midst of the widespread backlash, Jack went back on CNBC a day or so later and apologized. He called Jeff "a helluva CEO." But according to Jeff, "the damage had been done."

Jeff called Jack after Jack's outrageous CNBC appearance and told him, "Look, you're dead to me. You're dead to me." He said Jack "chose not to be a true friend." For the first time in the seven years since Jeff became CEO, he privately addressed with Jack his view of the legacy Jack had left him. "Following you has been no fun," he told him. "I've kept my mouth shut about the problems you left me. I bolstered your legacy, when I could easily have shot it full of holes. And because I've done so, you are still 'Jack Welch, CEO of the Century.' But now, when I need your help, you stab me in the back? I just don't get it." Jack tried to apologize to him again. "Look, I'm sorry," he said. "You fucked up, but I shouldn't have said what I said on air." Jeff agreed with what Jack had said about his having "a credibility issue," because he had, in fact, three weeks earlier, made promises about earnings that weren't kept three weeks later. Jeff's point to Jack was that he had never criticized Jack publicly when he could have, or should have, and one earnings miss—albeit one that caused the stock to crater 13 percent—got Jack gunning him down on national TV. "That essentially ended our relationship," Jeff recalled, "because I'd finally realized that he was using his criticisms of me to promote his own brand."

What also seemed to peeve Jeff was how often Jack went on CNBC—some seventy-five times, he told me—after leaving GE and why people such as Ken Langone and Larry Bossidy did, too, often to criticize Jeff and what was happening at GE. "If you had a child the day he left GE," Jeff said of Bossidy, "he would be retiring from the NFL right now. In other words, it was thirty years ago! But he goes on CNBC like the expert, and he says, 'GE Capital was always so great, and Jeff fucked up.' Dude, we were giving Japanese salesmen loans at thirty percent! That was GE Capital. But if your narrative is Bob Wright, Bob Nardelli, and Larry on one side, and on the other side is silence from the company, you know, I'm in the middle. Look, I'm not blameless on this whole thing."

On April 11, Jeff announced GE's first-quarter earnings punt, and then he revised downward—to between no growth and 5 percent growth—the earnings projections for the rest of the year. Asked about the first-quarter miss, Jeff said he "had a chance to review in some detail what the businesses were doing" and that at "that point in time, we still felt like we were on track for the quarter and for the year." On April 17, *The New York Times* added to Jeff's woes when it called into question his credibility after the $700 million earnings miss. After seven years of giving him the benefit of the doubt, investor patience had evaporated, driving the stock to a four-year low.

When the news broke, at 6:00 a.m., that GE had missed its first-quarter earnings, Steve Tusa, a JPMorgan Chase analyst, was on a commuter train heading to the office. "I almost fell out of my seat," he said. "It was a shock. People thought it was a misprint." He added there was "no doubt" the earnings miss was "an historic event" and that "the company has to convince investors that something is going to change." Nicole Parent at Credit Suisse had been covering the company since 1996. "I've never seen a miss this big," she said. "You have to ask what is the driving force behind the miss? Is the company too big to manage?" There were renewed calls to start to break up the company by spinning off NBCUniversal, or the major appliance business, or consumer finance, or GE Money.

In July, the markets appeared to have stabilized—it was more of a mirage than anything else—and Jeff was in Abu Dhabi to announce an $8 billion joint venture between GE and Mubadala, which is often described as a sovereign wealth fund, although Jeff referred to it as a "commercial company," with investments in industrial and financial companies. In any event, both GE and Mubadala agreed to invest $4 billion into the new venture, which would make investments and provide financing for commercial finance opportunities as well as invest in clean-energy and water projects. Such was the enthusiasm around the joint venture that Jeff predicted it would reach $40 billion in assets in about eighteen months. As part of the agreement, Mubadala announced that it intended to become one of GE's top ten shareholders by buying hundreds of millions of dollars' worth of shares in the open market. (At a dinner the night before the announcement, Jeff told his hosts that Mubadala was a "GE in the making.") "We like the Company," said Khaldoon Al Mubarak, the CEO of Mubadala. "We think under Jeff's management, this is an institution that has done extremely well. We think it's valued today attractively from our perspective."

At the press conference, if Jeff had any concerns about GE Capital he did not convey them. He said GE Capital would contribute about 45 percent of GE's earnings and said the Mubadala deal gave the company "good, high-margin origination opportunities right now." He also said

GE's earnings guidance for 2008 "remains unchanged." Asked whether there might be any political ramifications to GE for partnering with a Middle East sovereign wealth fund, Jeff said he had spoken repeatedly with Hank Paulson, the U.S. Treasury secretary, about the idea. "This is viewed to be really extremely important to globalize our companies and other companies around the world, and a very strong desire to have great partners as we do that," he said. "I really don't think this has a political angle to it. I think this is really all about business, all about growth, all about making money for our investors, and just like we've done for decades. We've got French partners, Japanese partners, Chinese partners, Brazilian partners, and now we have Mubadala as well. That is the nature of being an American company that is trying to grow every place in the world, and that is what we want to be."

By August, concern had returned to the financial markets. The Federal Reserve Bank of New York was increasingly worried about two trillion-dollar holding companies—GE Capital and AIG—and how they financed themselves. At that time GE Capital had borrowed $90 billion in the commercial-paper market. In August, the New York Fed set up a team to study the "funding and liquidity risk" of both GE Capital and AIG. In late summer, Mike Neal, the head of GE Capital, remembered, "we actually benefitted I think from a flight to quality in some cases in our [commercial paper] program. Now we're not naive to what was going on in the market, particularly as you moved more into September, but we were able to sell our quota every day, what we were trying to raise. . . . The markets were choppy. We were concerned about the markets and the direction of the markets and where they might ultimately end up. But having said that, we were doing okay."

Each successive weekend brought a crisis at a different powerful pillar of American society. Whatever solution the Fed and the Treasury thought they had crafted by orchestrating the sale of Bear Stearns to JPMorgan Chase was unwinding rapidly. The financial system is nothing more than an elaborate confidence game; once the confidence in the banks at the center of the system is lost, all hell breaks loose, every time. In early September, Paulson decided to put the Federal Home Loan Mortgage Corporation, known as Freddie Mac, and the Federal National Mortgage Association, known as Fannie Mae, into conservatorship. Then the crisis hit the Big Three automobile companies. Then there was Merrill Lynch, the nation's biggest brokerage and investment bank, which would have filed for bankruptcy if it hadn't been scooped up by Bank of America.

As the financial tsunami came closer to shore, Jim Grant, for one (and maybe for only), saw what it would mean for GE, after years of articulating the risks that were building up at the company, even as it was the most valuable and most admired in the world. To his critics, like Jeff Immelt, Grant was a broken clock, which is to say he was right twice a day. But the clock was still broken. In fact, Grant had been right about GE for years. On September 5, he wrote a new piece, "Not Your Father's GE." Grant intoned, "General Electric under Jack Welch, like the Fed under Alan Greenspan, was an enterprise seemingly touched by the gods. It launched a thousand case studies—for growth, profitability, reinvention, globalization and all-around, Six Sigma excellence. And a wonderful business only became more lucrative when a finance subsidiary was grafted on to a manufacturing superstructure."

His beef in early September was that GE Capital had failed to put up sufficient reserves for the coming credit losses and that the inevitable economic recession would hurt GE's industrial businesses. He reflected on his observation, from the previous November, that GE's AAA credit rating was at least two notches too high. "Some scoffed," he wrote, "but we've been proven overly sanguine, not hypercritical." He looked again at GE's and GE Capital's credit-default swaps and saw that they had blown further out, about doubling in cost from where they had been the previous November, further bolstering his view that GE was no longer a AAA-rated company. "If, as we judge, the CDS market is on to something, the stock market is missing something," he continued. "They can't both be right." He believed that, by September 2008, GE was only marginally investment grade. "You'd suppose it walks on water the way it defies the worldwide credit storm," he wrote. "It is as if there were no such blow—or, rather, that the seas crashed and the winds shrieked only for others." Even in "this year of salty credit tears," he continued, GE Capital picked over the carcasses of the other victims, adding $10 billion of assets from Merrill Lynch's middle-market lending business and another $13.2 billion of equipment and commercial finance assets from a struggling Citigroup.

WHAT HAPPENED NEXT REMAINS A MATTER OF SOME CONSIDERABLE dispute between Jeff Immelt, the CEO of General Electric, and Hank Paulson, the Treasury secretary of the United States. In *On the Brink*, his memoir about his time as Treasury secretary and the 2008 financial crisis, Paulson recalled how late in the morning of September 8, after he had received the "troubling news" that Dick Fuld's efforts to sell Lehman Brothers to the Korea Development Bank weren't going well, Jeff called to tell him that GE was having trouble selling its commercial paper in the market. "This stunned me," Paulson wrote. "If GE"—one of the few AAA-rated companies in the country—"couldn't sell its paper, what did that mean for other U.S. companies?"

This, no surprise, was the kind of thing that Treasury secretaries worry about. A week later, at 6:00 p.m. on September 15—the day Lehman Brothers filed for bankruptcy and all hell *did* break loose in the world's financial markets—Jeff "happened" to be in Paulson's office at the Treasury to talk about—of all things—GE's desire to be able to repatriate cash held in other countries back to the United States without incurring the higher corporate tax rate of 35 percent. The discrepancy in corporate tax rates at that time gave companies such as GE and Apple incentives to keep the cash offshore. Jeff wanted Paulson to give GE an incentive to bring the money home. Afterward, Paulson recalled, Jeff asked to talk privately to him in his office. According to Paulson's account, Jeff reiterated the same message about GE having difficulty selling its commercial paper—essentially the company's life-blood. GE had been issuing commercial paper, without a hiccup, since 1952. "I'd known Jeff for years," Paulson wrote, "and admired the cool, unflappable demeanor he had displayed as CEO of the biggest, most prestigious company in America."

Paulson knew it was yet another portentous moment, just as the failures of Bear Stearns, Lehman, and AIG, the big insurer, had been. If GE could not get access to the short-term credit markets, then Paulson figured the company would have to "curtail normal" business operations. "Now here was Jeff telling me that GE was finding it very difficult to sell its commercial paper for any term longer than overnight," he continued. "The fact that the single-biggest issuer in this $1.8 trillion market was having trouble with its funding was startling." He worried that if "mighty" GE was having problems with its commercial paper, then other companies without its AAA credit rating, such as McDonald's or Coca-Cola, would also start having problems. "Jeff," Paulson said, "we have got to put out this fire." Who knows what kind of further market turmoil could have been unleashed by a GE bankruptcy filing?

Jeff's recollection of the meeting with Paulson was very different. He said he knew Paulson "pretty well"—they had played the same position on the Dartmouth football team, although a decade apart—and found him "understandably distracted" late in the day on September 15 when he showed up to talk about cash repatriation. In *Hot Seat*, Jeff wrote that Paulson was "mistaken" about his raising the matter of GE's inability to sell its commercial paper, first on the September 8 phone call and then again during the September 15 meeting. Paulson "had cautioned in his book that he was relying on date books that listed the participants in calls and meetings but not their subject matter," he wrote. He recalled that GE was "not having any such problems at the

time" and that the September 15 conversation was solely about repatriating cash from overseas, which was why Jeff had his "tax director"
with him. He wrote that Paulson's "mistake" caused the SEC to investigate a September 14 "investor blast" that GE had sent out saying its
$90 billion commercial-paper program was "robust." (That letter has
since disappeared from GE's website.)

Obviously, if GE told investors on September 14 that it had "robust" access to the commercial-paper market and then the next day
GE's CEO told the Treasury secretary of the United States that its access to the commercial paper was the opposite of robust, that could be
a big problem. To support his recollection, Jeff recalled how GE's lawyers had produced a "huge, multicolor exhibit" laying out what happened in September 2008. Jeff reproduced a black-and-white version
of the exhibit in his book. But it does not include the specifics of what
Jeff said to Paulson on September 15. Instead, Jeff wrote that "the SEC
subsequently closed the investigation without taking any action against
GE." Other senior GE executives take Jeff's side on this dispute. "I don't
think on September 15, we were having an issue with CP," said one.

But Paulson's recollection is just as clear that Jeff *did* talk to him
about his commercial-paper problems during the late-day meeting on
September 15, 2008. In an interview, the former Treasury secretary
said the impact of a GE failure on the world's economy would have
been "just colossal," given its size and its large footprint as both a major lender and borrower, especially in the commercial-paper market.
And "secondly," he said, "the symbolism." He continued, "Talk about
panic and worldwide panic. If the great GE would have failed, it would
have been amazing." He remembered the September 15 meeting with
Jeff well. He said that, yes, Jeff came to see him about repatriating cash
in a tax-efficient way and then, after the others in the meeting had left,
Jeff asked Paulson if he could stay behind for a minute.

"By the way, I can't sell my commercial paper," Jeff told him.

"Gee, that's pretty bad, isn't it?" Paulson asked.

"It's terrible," Jeff said.

Paulson said Jeff made a good case for why the Treasury should help GE: "that they were making loans to industrial companies and small companies, and a lot of people that banks weren't giving loans to," he said. "There was a good argument to support them from that standpoint, and I use that argument. But it wasn't just GE." There was McDonald's, Coca-Cola, and P&G, which all were having trouble selling commercial paper. "If the [commercial] paper market had gone, and big companies couldn't borrow on a short-term basis, it's just amazing how quickly that would have gone," he said. "If General Motors had gone—Ford was sitting there begging me on General Motors, because there's no debtor-in-possession funding, and if General Motors had gone, all the suppliers would have gone, because they were hanging on and there was the whole ecosystem."

Paulson didn't ask Jeff any more questions. "I didn't say, 'Well, if you can't sell it, how many days can you last? Will you have to declare bankruptcy or what will happen?' I didn't ask him any of that. I just couldn't get him out of my office fast enough." He said "for a long time" Jeff "begged" him not to say in his book what happened in the September 15 conversation between the two of them. "I had to write the truth in the book," he said. "So I wrote it."

As hugely destabilizing as was the Lehman bankruptcy—the largest bankruptcy in history—the next day, when AIG failed and the money-market fund the Reserve Primary Fund "broke the buck" (meaning that the money people had in the fund, which was essentially supposed to be as safe as a bank savings account, was no longer worth what they had invested). The idea of the Reserve Primary Fund was that if you put $1 in it, you would always have that $1 when you wanted to take it out, just as you think when you put your money in a checking account or a savings account at a bank. In September 2008, suddenly, at the Reserve Primary Fund $1 was worth only 97 cents. It turned out that the Reserve Primary Fund had invested money in, among other things, the bonds of Lehman Brothers, which had become essentially worthless overnight. The combination of the Lehman and AIG failures and the Reserve Primary Fund fiasco shattered investor confidence in the financial markets. Even a AAA-rated company such as GE, which was wholly dependent for its continued existence on its ongoing access to the commercial-paper market, found itself in dire financial straits. "Dominoes were starting to topple," Jeff recalled.

Jeff tried his best to exude confidence, both because that was his natural inclination—he was a perpetual optimist and the CEO, after all—and because half of GE's earnings came from a huge bank that was wholly dependent on the confidence of increasingly jittery short-term investors. Money-market funds such as the Reserve Primary Fund were big buyers of commercial paper. But if the value of that paper could incinerate in twenty-four hours, investors were inevitably going to shy away from continuing to buy it. That meant potentially big financial trouble for GE. Bear Stearns evaporated in a week's time in March 2008 after big institutional investors, such as Fidelity and Federated, stopped providing short-term, overnight financing secured by the financial assets on Bear's balance sheet. The investors had lost confidence in the security that Bear was offering them and refused to provide the firm the money it needed, even overnight.

If a version of the same thing had happened to GE, that could certainly have led to the bankruptcy of GE Capital, which would have likely meant the bankruptcy of GE, too, given the implicit guarantee that investors believed existed between GE and GE Capital when it came to repaying GE Capital's billions of dollars in debt. What was previously inconceivable—that a AAA-rated company could file for bankruptcy, let alone a company named General Electric—was quickly becoming a possibility. Jeff was facing in 2008 his own version of what his predecessor, Charles Coffin, had faced 115 years earlier, in 1893. Paulson remained incredulous. He didn't think people fully comprehended what GE "going down" would have meant. "Just the sheer size of it," Paulson told me, "and the symbolism. This was American capitalism. GE was America." He said that when he visited with Xi Jinping, the president of China, and he wanted to show off his English, "It was 'GE, Boeing, IBM.'"

Jeff tried to keep cool. His first test came on September 16, the same day AIG and the Reserve Primary Fund ruptured. He had previously committed to be in Silicon Valley that day for Google's annual Zeitgeist conference. Eric Schmidt, Google's CEO, was scheduled to interview Jeff as part of the conference. The two companies were set to announce a collaboration to create a twenty-first-century electricity system, a "new, smarter grid," Schmidt said, yet another of Jeff's BIG initiatives. But between AIG and the Reserve Primary Fund, Jeff was freaking out. He had flown three thousand miles from Washington to Silicon Valley, along with Gary Sheffer, GE's new top communications executive. Suddenly, Jeff was "overcome by the need to get back to GE's headquarters." He told Sheffer to cancel his speech. They were going back to Fairfield. But back on the jet, right before takeoff, Sheffer urged caution. "This is Google," he told Jeff. "Lots of media. If you leave, people will know that you're terrified." He was "adamant" that Jeff return to Google. Could his presentation be moved up a few hours? Google was amenable. Jeff called Sherin, his CFO, who was moving his son into his Northwestern University dormitory room. Should Jeff stay at Google or return to Fairfield? "Stay put," Sherin told him. "Exude calm." Still on the plane, Jeff looked at Sheffer, "who looked so tense he could've bit a pencil in half," and decided to return to Google. He and Schmidt spoke and made the announcement about their partnership. From the front row, Al Gore, the former vice president, applauded the deal.

Finally back on the jet heading back to Fairfield, Jeff was worried about whether the commercial-paper market would still be there for GE. He called Paulson. He called Tim Geithner, president of the Federal

Reserve Bank of New York. "Both said they were worried about the commercial paper market," he recalled. "I told them GE wasn't seeing any cracks in its ability to sell its commercial paper." He had the pilots turn off CNBC. "I couldn't bear to listen," he remembered. When he got back to Westchester County Airport, it was nine o'clock at night. He headed to his office in Fairfield to meet with Sherin and with Kathy Cassidy, the GE treasurer. "Okay," he asked them, "what CP do we have to roll tomorrow?" In *Hot Seat*, Jeff conceded what many others had figured out long before: "Rolling commercial paper was a key engine that enabled GE Capital's profit machine." Without it, the jig was up. Fortunately for GE and other big corporations, on September 19, the general counsel at Treasury approved the use of $50 billion from the Exchange Stabilization Fund, a discretionary emergency Treasury fund, to provide a temporary guarantee to a variety of money-market funds to restore confidence in that market, easing the crisis for Jeff and GE, at least briefly. "It's hard to explain how terrified I was," he told me. "It's hard to explain because there was no place to dock the boat." He said that during this period he would speak to Kathy Cassidy every day, awaiting the news about whether GE was able to "roll" its commercial paper. "Each day," Jeff recalled, "that call felt like being granted a twenty-four-hour stay of execution." To "calm himself," he would tell his assistant to hold all his calls so that he could take a long, hot shower in the bathroom off his office, "trying to relax." If only he had listened to Bill Gross six years earlier.

He likened the experience to the story depicted in the 1964 British film *Zulu*, about an epic battle, in January 1879, during the Anglo-Zulu War, between a small group of 150 British soldiers defending a fort against a force of 4,000 Zulu warriors. It features the actor Michael Caine in his first major role, as well as Richard Burton. "It's wave after wave of Zulu warriors," he told me. "It's a two-hour movie with eight Zulu attacks—that's what the financial crisis was."

On September 25, against a backdrop of cascading dominoes, Jeff and Keith Sherin gave Wall Street security analysts an update on GE's seemingly precarious financial situation. By this point in the crisis, Bear Stearns and Merrill Lynch had been forced to merge with JPMorgan Chase and Bank of America, respectively, while AIG and Lehman Brothers collapsed. AIG would be rescued by the federal government, while Lehman Brothers filed for bankruptcy and was liquidated, its parts disposed of around the world. Wachovia, too, had failed and had been bought by Wells Fargo, and Washington Mutual failed and was bought by JPMorgan Chase. Even mighty Citigroup, once the biggest bank in the world, nearly went down the tubes but for a massive bailout courtesy of the federal government.

The survivors needed financial lifelines, and fast. Crucially, Warren Buffet bought $5 billion of equity in Goldman Sachs, and Mitsubishi UFJ Financial Group, a large Japanese commercial bank, paid $9 billion for a 21 percent stake in Morgan Stanley. The biggest lifeline of all, the $700 billion Troubled Asset Relief Program, or TARP, was used to invest between $10 billion and $25 billion (or more) into a group of Wall Street investment and commercial banks to try to restore confidence in Wall Street, the left ventricle of capitalism. All eyes, understandably, were on Wall Street. But fifty-seven miles away, another existential crisis was unfolding in Fairfield, Connecticut.

What Jeff and Sherin's Wall Street audience did not know was that GE, the world's most valuable and respected company, had flirted with bankruptcy, not once but twice during the six months or so after the most acute phase of the 2008 financial crisis. Paulson told me he couldn't understand why Jeff had allowed GE to come so close to the brink at that moment. "He had a lot of time before the financial crisis to clear it up," he said. "I, frankly, don't know because I came out of finance, how any smart CEO could even be overweighted in commercial paper or have commercial paper without backup lines. Do you?"

Jeff remembered that for him the most shocking day of the crisis was September 25, when JPMorgan Chase bought the assets of Washington

Mutual, leaving behind a carcass for debt holders to pick over. "The bondholders got crushed," he said. "So if you're in the unsecured debt market like we are, that was brutal, that was the toughest day."

By the end of the month the last two major independent investment banks—Goldman Sachs and Morgan Stanley—had agreed to become bank holding companies. The implications of the move were huge: overnight, they would have access to Federal Reserve funding, meaning they could borrow inexpensively from the Fed and have steady access to capital and liquidity. But they would also forevermore be regulated by the Fed, which would mean over time they would look more and more like big commercial banks and less and less like investment banks. They traded off security for more regulation, a swap that was easy to make under the dire circumstances. Jeff had tried to get the Fed to allow GE Capital to become a bank holding company as well. But the Fed wouldn't go for that, just as it had previously turned down similar requests from Bear Stearns and Lehman Brothers that might have saved those two firms from obliteration. "We had every indication that we couldn't become a bank holding company," Jeff told me. "That basically the government wouldn't know what to do with us, it would be hard to do."

At the September 25 analyst call, the top two GE executives tried to project GE as a pillar of financial fortitude amid the financial chaos. But what was also clear was that—six years after Bill Gross, six years after Ravi Suria, and after year upon year of pounding from Jim Grant—Jeff Immelt finally had gotten the message that relying on the short-term, unsecured commercial-paper market for GE Capital's funding needs embedded huge risks into both GE Capital and GE.

It was a precarious financial high-wire act: Jeff had to act as if GE were basically unaffected by the disaster that had befallen the other big Wall Street banks while also signaling that GE Capital had gotten a funding scare and would pare back its reliance on commercial paper. To reassure investors and with more than one quarter left in the year,

Jeff said GE would earn $20 billion in profit in 2009, more than $9 billion of which would come from GE Capital "dramatically outperforming," its peers. He also said the GE board had agreed to keep the dividend "secure" at $1.24 a share through 2009. Other measures meant to reassure investors included the decision to reduce the annual 40 percent dividend that GE Capital sent to GE down to a 10 percent dividend, allowing GE Capital to retain 90 percent of its earnings to shore up its balance sheet. He also said GE Capital would reduce its leverage from 7.2–1 to 6–1 by the end of 2008. GE said it was suspending its stock buyback program, another way the company was husbanding its capital during the dark days of the crisis.

But the main event was the discussion of GE's reliance on the commercial-paper market and whether that was something about which investors should be worried. And whether GE itself was worried about it. But Sherin did his best to make sure the analysts knew he wasn't the least bit concerned. His job, and not for the last time, was to project GE's financial strength, a calm port in the raging storm. "We have great CP programs," he said. He added that GE and GE Capital had no funding issues. "Even in the last 10 days where you've had some significant disruptive days, we continue to see a flight to quality," he said, implying that GE's AAA credit rating made it part of the "flight to quality." The subtle message being that in times of financial crisis, investors were fleeing what they believed to be risky financial instruments in favor of what they perceived to be safer harbors, such as gold and Treasuries, and even, yes, GE.

He also said GE intended to reduce its $90 billion of commercial paper down to $80 billion, so that it was only 10 percent to 15 percent of GE Capital's $550 billion of debt. (This is not a typo—like many big financial institutions, GE Capital *did* have this much debt, much of it matched to liquidating assets, such as credit-card receivables and other monthly payments. The risk at GE Capital wasn't too much debt necessarily, despite the huge number, but rather liquidity, caused by borrowing money in the short-term commercial-paper markets and lending it out to borrowers for years. Hiccups on either side could be fatal.) To further try to assuage any concerns, Sherin said GE Capital *had* bank deposits—who knew?—some $43 billion, in fact, up $20 billion from the start of the year. "And we can take that higher," he said.

He discussed analysts' concerns about GE Capital's commercial real estate business—the sale of which Jeff had nixed the previous year. Now, Sherin said, although it was performing well—delinquencies were a mere 0.27 percent of the $90 billion portfolio (some $243 million)—the intention was to reduce the commercial real estate portfolio down to $80 billion. He said the goal, as ever—and as ever elusive—was to "de-emphasize the size of GE Capital," which "makes GE even safer."

Jeff, too, spoke (using gobbledygook) about how GE Capital's contribution to GE's earnings would somehow be reduced to just 40 percent "with global origination, diversified risk and deep domain." He said he believed GE Capital would "outperform" in the difficult business cycle and would continue to do well when the economy improved. "We've never seen really a time of volatility like we've seen in the last month or so," he told the analysts. "But I think the fact that we're so committed to the dividend and have made that strong commitment, I think shows you the immense confidence we have in the company."

But behind closed doors—away from the highly attuned ears of Wall Street's research analysts—Jeff was exploring a different scenario for GE Capital. Only it wasn't how to go about spinning it off, as the research community was suggesting he do. Rather, it was to contemplate what a *bankruptcy filing* for GE Capital would look like. This was not something anyone at GE dared mention publicly, for obvious reasons. In fact, the weekend following the Thursday call with analysts, there wasn't anything that wasn't on the table for Jeff and his management team to consider.

By then, Jeff knew that GE Capital wouldn't be saved by becoming a bank holding company. GE and GE Capital wouldn't be beneficiaries of the proposed TARP—that $700 billion being contemplated would go to the Wall Street banks and several car companies—once it was passed. GE was looking increasingly isolated amid the financial wreckage: it was wholly dependent on a clogged short-term financing market but was not regulated by the Fed, which could have helped give the company access to capital, and for some reason it was not going to be part of TARP—and yet it was one of the country's largest financial institutions, with more than $600 billion of assets, some of which were indeed troubled. GE was on its own island.

Jeff called in Rodge Cohen at Sullivan & Cromwell for help. By then, Cohen had become one of the leitmotifs of the crisis, in nearly every room during nearly every aspect of the rolling financial tsunami. His calm, wizened demeanor made him one of the deans of the Wall Street bar, especially during a crisis. He was close with bank executives and close with Hank Paulson, Ben Bernanke, and Tim Geithner. "We saw him as our resident wise man—at once wonky and creative," Jeff recalled. Cohen came to Jeff's rescue. Jeff said Rodge Cohen was the one who had to raise the prospect of putting GE Capital into bankruptcy on that late September weekend. "I couldn't even form the word," Jeff said.

The conversation occurred on the last Sunday in September in Fairfield. Brackett Denniston, GE's general counsel—he had replaced Ben Heineman—told Jeff that on account of "good governance," they had no choice but to consider the option. Jeff told me, "My first reaction was 'What the fuck are you talking about?'" Cohen came up to Fairfield to discuss the possibility with Jeff. "I was just apoplectic," he said. One GE executive, living overseas, remembered getting a call from Mike Neal, the head of GE Capital, at some ungodly early morning hour. "Hey, can you go to the office right now?" Neal asked.

"Sure, what's going on?" he replied.

"We are really having trouble rolling our commercial paper," Neal said. "I think we can, but I'm not sure. You know, and if we can't, we're going to have to draw down on our bank lines and all hell will break loose." The backup lines of credit were a damned-if-you-do, damned-if-you-don't scenario. If you don't have them, investors freak out because there is no backup liquidity; if you do have them, and use them, then the market freaks out because you've used them.

"Is there anything I can do?" the executive wondered.

"Nope," Neal told him. "I'd just feel better if I knew you were at your desk."

His wife, now awakened, wondered if everything was okay.

"Um, not really," he told her, "but I gotta go to work."

"What?" she asked.

"Right then and there," he said, "in my own mind, I was like, 'Wow, we might be going under.'" He said he knew that if GE Capital couldn't roll its commercial paper, that would lead to a cross-default at GE. "If this goes down," he thought, "we're bringing the whole machine down. Part of me was like, 'Never in a million years could I have conceived that GE could go bankrupt.'"

Cohen recalled that the discussion of the possibility of putting GE Capital into bankruptcy arose because Jeff was concerned that GE Capital might not be able to continue to fund itself in the commercial-paper market. "GE, at this point, was like the gold standard," he told me. "And the concerns were massive that they couldn't roll over their commercial paper. . . . The implications for corporate America were astonishing. This was, in a sense, a bridge between the financial sector and the industrial sector. Jeff's obligation, of course, was to protect GE. But I don't think anybody, including the government, lost sight of the broader implications."

Cohen said putting GE Capital into bankruptcy had to be a consideration because it was simply not "tenable" that financial difficulties at GE Capital might also lead to a bankruptcy at GE. "So it had to be on the table but only as an absolute last resort," he continued. There was also a great concern that GE Capital could not survive a Chapter 11 bankruptcy filing—in which a company reorganizes its debts and then emerges again as an independent, viable company—and that GE Capital might end up in Chapter 7 liquidation, which would be a massive feeding frenzy, with creditors likely getting pennies on the dollar. At the time, GE had not explicitly guaranteed the obligations of GE Capital (that would come later). There was only an implicit guarantee of sorts—or what creditors took as an implicit guarantee—through something called an "income maintenance agreement," which was how GE would make up for shortfalls at GE Capital.

According to Cohen, the federal government was made aware of the possible bankruptcy filing of GE Capital. "Again, this was a last resort," he said. "The first meeting that I remember was with the Federal Reserve, and we explained the situation and we were able to say [that it was a last resort] and, if I remember correctly, that at that point, the paper was still rolling. But you couldn't predict what would happen the next day or the next week at all with the financial markets in turmoil." In retrospect, Jeff said that "we'd made a mistake" in how GE executives thought about the reliance on commercial paper. While it was "just 15 percent" of GE's debt, "15 percent of a very large number is a very large number." And then, "At times, we'd failed to contextualize our own aggregate size."

One GE Capital executive recalled for me how harrowing it was to live through this part of the financial crisis. "Interestingly, it was scary and invigorating at the same time," he said. "I remember getting all the GE Capital people on the phone and saying, 'We're going to fucking do this. We're going to get through this. Here's what we're doing. A, B, C, D, and E. Don't look anywhere else. If you're not working on A, B, C, D, and E, you're wasting yours and our time. Don't do it. We're absolutely going to pull through, we're going to be better on the other side of this, we're going to be more differentiated. Don't forget, anybody who doesn't have our financial strength and the quality of franchise, they ain't going to exist in four months. Play to win.'"

While both Jeff and Sherin told the Wall Street analysts that GE was not going to spin off GE Capital as an independent, stand-alone company, there were no structural impediments to its happening, other than the fact that a spin-off would take months to become effective—requiring a multitude of SEC filings and SEC approvals. In late September 2008, there simply was no time for a spin-off to occur, or at least not enough time to make that a viable possibility. Making the spin-off option even less tenable was the fact that, according to Jeff anyway, for a stand-alone GE Capital to receive a single-B credit rating—the lowest investment-grade credit rating—would have required GE Capital to have $30 billion of new equity, which GE didn't have and seemed unlikely to obtain during those tense times. Without that capital infusion, Jeff believed, a stand-alone GE Capital might have to file for bankruptcy anyway.

Instead, Jeff chose the opposite path from what he had told Wall Street days earlier: he decided to raise new equity after all—some $15 billion worth of new equity, to be precise—something GE had not done in decades. The idea had come from David Solomon and John Weinberg at Goldman Sachs. The two Goldman executives wanted GE to do what Jeff and Sherin had just told analysts they would not do: raise equity. "I pushed back," Jeff recalled, "but Sherin was insistent." Jeff listened to his CFO. "If he was concerned, I knew I had to be, too." Within hours, Solomon and Weinberg were mapping out a plan for a $15 billion secondary offering, then the largest secondary in history. (A few months earlier, in March 2008, Visa raised $18 billion in equity, the largest IPO ever to that time.) "I'm extremely fond of David because really, it's kind of like you use bankers all the time and mainly just for friendship," Jeff told me. "But there's, like, twice in your career that you really need a banker, and this was one of them, and I'm happy that I had David because [Goldman] helped immensely." On Saturday, September 27, Jeff headed to the office and arranged for a conference call with the GE board. He told the board "how treacherous and unknowable the environment suddenly had become for GE," he recalled.

Jeff told his board the company needed to raise at least $15 billion in new equity and the process to do that needed to start immediately. "There was dead silence on the other end of the phone that felt like it would never end," he remembered. Roger Penske, the automobile mogul, broke the ice. "Let's go get the money," he said. Jeff had to make the decision to launch the deal Sunday night at 7:00 p.m.—morning in Asia, where GE hoped to raise some of the money. "I've never felt as much pressure before or after as that moment," Jeff told me.

There was a fire hose of so much disturbing financial news that he didn't know whether the deal would fly. "Failing to raise the equity would have just put you right in the middle of unsolvable stuff," he said. At six thirty that evening, he told the team he needed to think about what to do. He retreated to his corner office in GE's Fairfield headquarters to give the matter more thought. He was worried what the failure to pass the TARP bill would mean for the stock market on Monday. He headed back to the first-floor conference room, passing the men's room along the way. "All I wanted to do was slip inside, lock myself in a stall, and never come out," he remembered.

He called off the offering. The next day the market fell some one thousand points at one point after the TARP bill failed to pass, closing down more than seven hundred points, its largest one-day points decline ever. The specialists couldn't close GE's stock, which had fallen to around $25 per share, for the day.

"Had we launched" on Sunday night, "we would have gotten crushed," Jeff observed. One idea that the Goldman bankers were kicking around with Jeff was to get an anchor investor for the equity offering, along the lines of what Goldman had done the week before when Warren Buffett invested $5 billion in Goldman at the same time Goldman also did a $5 billion offering to the public. Jeff loved the idea. He knew Buffett a little, but rather than call the famous investor himself, he relied on Byron Trott, Buffett's banker at Goldman, to suggest the idea to Buffett. On Tuesday, Trott worked to convince Buffett to make the GE investment. "It was unclear whether he would agree," Jeff remembered. The next day, October 1, Trott and Buffett had scheduled a call for 8:00 a.m. "All we could do was hope," Jeff continued.

Increasingly, the success of the massive equity offering—and GE's fate—rested on Warren Buffett. "If Buffett says no, we're fucked," Sherin told Jeff. He wasn't wrong. Said one senior Goldman banker who was involved in the stock offering, "They were freaking out, because had they not got the stock issued, they were probably toast." Thirty minutes later, John Weinberg called Jeff to tell him the good news that Buffett had agreed to buy $3 billion of a new GE preferred stock with a 10 percent dividend. Buffett also received five-year warrants to buy an additional $3 billion of GE stock at a strike price of $22.50 per share. It was an expensive deal—Buffett's money was always expensive, especially in a crisis—but it also carried the Warren Buffett seal of approval. "GE is the symbol of American business to the world," Buffett said after making the investment. "I have been a friend and admirer of GE and its leaders for decades. They have strong global brands and businesses with which I am quite familiar. I am confident that GE will continue to be successful in the years to come." Jeff then called Buffett. "Thanks, Warren," he said. "We won't let you down." Buffett later told the Financial Crisis Inquiry Commission that he invested in both Goldman and GE because he figured "they were going to be fine over time" because he was betting that the federal government "would not shirk its responsibility at the time like that" in the same way that it had during the 1930s. He was essentially betting that Bernanke's Fed would do what it needed to do to save the financial system. He bet right.

Jeff was focused not only on how Buffett's endorsement would help GE raise another $12 billion from equity investors but also on how the new equity infusion would calm the company's creditors. "We remain committed to the Triple A rating and in the recent market volatility, we continue to successfully meet our commercial paper needs," he told investors. He also added, in GE's filings with the SEC, that "a large portion of GE Capital's borrowings have been issued in the commercial paper markets, and, although GE Capital has continued to issue commercial paper, there can be no assurance that such markets will continue to be a reliable source of short-term financing for GE Capital."

Jeff told me that Buffett's investment in GE was "like having an underwriter in a sea of shit." But as helpful as it was, the cost of the Buffett imprimatur was shockingly high, especially for a AAA-rated company such as GE, which made its low cost of capital a matter of pride and competitive advantage. "Equity is not low-cost—it's more expensive than debt—and the special preferred shares issued to Buffett carry a 10% coupon, which is paid out of after-tax profits," the ever-astute Geoff Colvin and Katie Benner pointed out in *Fortune*. "So that $3 billion was about the most expensive capital it's possible to get." As part of the deal with Buffett, both GE's CEO and CFO agreed to retain 90 percent of their stock in GE until Buffett's preferred had been redeemed. (Buffett had cut a similar deal with top Goldman executives.)

Another indication of poor judgment—and one that Jeff would continue to repeat—was his decision to have GE buy back its own stock, at prices that turned out to be high, not low. In 2007, GE bought back about $14 billion worth of its own stock; in 2008, it bought back about $1 billion of its stock—all at prices above $30 per share. Then, six days after suspending the 2008 buyback program, Jeff made the decision to *sell* stock to the public at $22.50 per share. "That is," deadpanned Colvin and Benner, "buying high and selling low."

At around 1:45 p.m. on October 1, GE announced the Buffett

investment and that it wanted to raise an additional $12 billion in equity. Jeff called Mubadala. He called mutual funds. "It was like an extended episode of *Dialing for Dollars*," he recalled. The $12 billion offering—547,825,000 shares at $22.50 a share—was completed in the early-morning hours of October 2. "Hey, guys, we made it," David Solomon told Jeff. (The underwriters, led by Goldman, split some $182 million in fees.) GE had raised $15 billion of new equity in about twenty-four hours, a few days after telling investors it would do no such thing. The next day, President George W. Bush signed into law the $700 billion TARP bill, which Congress had passed on its second try. "Maybe," Jeff thought, "we'd get a little break." The respite, such as it was, didn't last long. The Zulu warriors were merely regrouping for another coordinated attack.

Within days, Jeff would face another stern test from the roiling markets. "This is like, you could never anticipate any of this stuff," Jeff told me. On October 7, the Federal Reserve created the Commercial Paper Funding Facility, enabling the Federal Reserve to buy the commercial paper issued by creditworthy companies, such as GE, which was especially important after the Reserve Primary Fund "broke the buck" and the usual buyers of commercial paper wigged out. The program began on October 27. GE paid the Fed a *$100 million* fee to join the program and then, of course, paid interest to the Fed on the commercial paper that the Fed bought. The program created liquidity in the commercial-paper market after it had evaporated. The cost was high for GE to join the CPFF, but it was better than the alternative.

October 14 turned out to be a momentous day for Wall Street—and for GE. That was when the federal government—the Fed, the Treasury, the New York Federal Reserve Bank, and the FDIC, the Federal Deposit Insurance Corporation—launched a coordinated multibillion-dollar rescue for the country's biggest banks. There was a 3:00 p.m. meeting at the Treasury; Hank Paulson forced the banks—but not GE—to attend and then to take huge equity infusions from the recently passed TARP, whether they wanted to or not. Paulson believed it was important not to pick winners and losers among the banks. Rather, he, Bernanke, and Geithner wanted to restore confidence across the financial markets by ensuring that each big bank had a new infusion of capital and that arbitragers couldn't separate weak banks from the pack. A rising tide lifted all boats was the idea.

At the same time, the FDIC, run by Sheila Bair, implemented the Temporary Liquidity Guarantee Program, or TLGP. The program had two parts: One guaranteed $500 billion of bank deposits of businesses to ensure that employee payroll would still be met and, in effect, encouraged local businesses to continue to make deposits at banks, which could then be loaned out, keeping intact the miracle of fractional banking. In the other part—the Debt Guarantee Program, or DGP—the FDIC guaranteed through maturity, or until June 30, 2012 (whichever came first), $1.5 trillion of senior unsecured debt issued between October 14 and the following June 30 by a bank participating in the program. Both the TARP equity infusions and the DGP were lifelines the big banks needed, whether they chose to acknowledge that or not. At the start of the DGP, the FDIC later reported, banks relied heavily on it "to roll over short-term liabilities because of the fragility of the credit markets and investors' continued aversion to risk. By providing the ability to issue debt guaranteed by the FDIC, the DGP allowed institutions to extend maturities and obtain more stable unsecured funding."

GE, one of the country's largest financial institutions, was not included in either the TARP or the DGP. To federal regulators it was not a bank. Jeff cared less about not being included in the TARP, especially as two weeks earlier GE had raised $15 billion in new equity from Warren Buffett and investors around the globe. Being excluded from the DGP, however, posed an existential risk to GE, a fact Jeff and his most senior executives realized immediately after the FDIC announced the program. "What it did was it created a circle," explained Jeff

Bornstein, then the CFO of GE Capital. "You were either in or out of the circle. It was picking winners or losers." GE was out of the circle. Bornstein remembered being at Crotonville for a GE officers' meeting. The TVs around the room were tuned to CNBC. "On CNBC, they announced the TLGP program and the raw construct of it, including who it pertained to," Bornstein said. "I remember walking over to Kathy [Cassidy] and saying, 'This is a fucking disaster.' Not just for us. For anybody not in the circle."

Since GE was not a bank—and wasn't regulated by the Fed or the FDIC (it was regulated by the weak Office of Thrift Regulation, which would soon be put out of its misery)—there was never any expectation that it would be a beneficiary of the TARP or the DGP. The problem became for GE that when the senior debt of its competitors suddenly *had* a government guarantee, there was no way investors would buy GE's paper if it *didn't* have a guarantee, too. Bornstein had intuited this problem instantly. Jeff quickly realized the devastating implications of the DGP, too. "By leaving us out of the program, the government effectively made our long-term debt worthless," Jeff recalled. "It's not that anyone thought GE couldn't make good on its debts. It's that no one in their right mind would buy GE's debt when the debt being sold by banks had a government guarantee" and GE's debt didn't. He was right about that. The Zulu warriors were breaching the fort. As predicted, investors were suddenly reluctant to lend to GE, and those that were willing to do it demanded higher rates of interest. "While GE didn't have a liquidity crisis, it looked as if these new programs were about to cause one for us," Jeff continued. "The ramifications for us were frightening."

For his part, Paulson said he knew putting billions of dollars of capital into the Wall Street banks and creating the DGP was going to be "distortional" and potentially hurt companies such as GE. David Nason, one of Paulson's top deputies, warned him that several companies, including GE Capital, Prudential Insurance, and MetLife, that had been excluded from TARP and from DGP could possibly suffer from that exclusion. "That's great," Paulson told Nason. "We don't have time to deal with that now, so we'll table that for another day." But he did call Jeff and asked him how he thought the TARP and the DGP would affect GE. "Clearly," Jeff told him, "[you] saved the economy. GE is not going to be left unless you put money in the banks." Paulson thought that Jeff "couldn't have handled himself better" during that conversation, "but a day later he was back begging for help."

Jeff remembered that as the October 14 announcements and meetings took place, the Dow was up a thousand points. He told me, "The only stock in the S&P that was down was GE, because basically people could say: 'Why would anybody buy GE Capital debt when all the bank debt is guaranteed?'" He said he told Paulson during their call, "Do what's best for the country, really. But you've now put me on an island of one, you know? I'm, like, where I don't want to be, which is kind of like the only person in financial services—in some cases bigger than the other people—without the ability to raise funds." Jeff was basically freaking out. "I didn't want to go to bed at night because I didn't want to wake up the next morning and turn on the TV and see what the futures [market was doing]," he told me. "It's kind of endless. The thing that was different for me was I'm dealing with this and then going to say, 'Okay, what's our aviation pipeline look like? Do we fund the Leap engine? How do we deal with XYZ that's going on?' There's time when running a conglomerate is great fun. This time, when it's like 'Oh my God, I owned a network.'" He said unlike September 11, where by noon the basic outlines of what had happened were known—for instance, who did it and who died—the financial crisis was "opaque."

During the fall of 2008, it was unclear which companies would survive. It was unclear who in the government was going to play a key role, if anyone. There was no predicting anything like the TLGP or that

GE Capital would be excluded from it. Suddenly, Jeff recalled, GE was back to where it was before the company raised $15 billion of fresh equity. "We sucked in a little bit of oxygen and now our head's below water again," he said. "Slowly it began to sink in how long lasting this would be and how many waves you'll have."

It was time for Jeff to go back to Washington to see Paulson. "TALP, TARP, anything with a *T* they would have taken, and they were taking as much of it as you were giving them," remembered someone who knew Paulson well, about GE's predicament. "They had as many troubled assets as anyone." Jeff was armed with a one-page "Killer Chart" that he had asked Bornstein to prepare that showed how crucial GE Capital's lending was for small- and medium-sized businesses around the country, as well as that GE was a leader in aircraft financing, equipment leasing, private-label credit card finance, commercial real estate, and on and on. "We were on an island of one," Jeff reiterated, "and our business model took a direct hit." The idea was to show the Treasury secretary that by inadvertently excluding GE Capital from the DGP, the Treasury would be, as feared, hurting not only GE Capital but also the broader U.S. economy. GE Capital, Bornstein's chart declared, was "continuing to provide liquidity to critical areas of the economy." It was clever spin, but GE was in dire straits. Jeff told himself, "I just have to protect us. I just have to protect the company."

Jeff didn't want to divulge just how bad he thought things might get. He said a more "authentic"—a word he hated—leader would have told the troops, "Okay, folks, we're fucked, okay? I can't even tell you how fucked we are. Really, I have no idea what's happening. GE Capital is going to sink us all." He preferred to work with the GE Capital executives to come up with a viable plan that wouldn't mean the end of the business. "They did nothing wrong," he said. "We're going to work on this together." He quickly realized that not having deposits—a cheap source of capital for the Wall Street commercial banks—had hurt GE

Capital. "Deposits were the dock," he said. "If you had to put your boat in the dock, that was it."

Jeff took Brackett Denniston, GE's general counsel, with him to see Paulson in mid-October. They had ten minutes with the harried Treasury secretary to make their case. "I know you've got a thousand shit burgers on your hands," Jeff told Paulson, "but you've got to think about the customers of the products we finance. We keep everyone from airlines to small business owners in business." Jeff said he was "heartened" when Paulson quickly absorbed the implications of the "Killer Chart" and offered to call Sheila Bair, the head of the FDIC, on Jeff's behalf.

After the meeting with Paulson, Jeff and his general counsel headed over to the FDIC, on Seventeenth Street, to see Bair. It was already well past office hours—eight o'clock at night—but Bair was still there. An assistant came out and told the two men that the head of the FDIC had no time for them. No time to see the CEO of GE? Jeff insisted on waiting to see Bair. The threat to GE was existential. They bided their time in the FDIC lobby while the cleaning staff polished the marble floor. "At 240 pounds, I'd be difficult to force out the door," Jeff allowed.

After more than an hour of waiting, Bair agreed to see Jeff. He made the case, as outlined in the "Killer Chart," about how important GE Capital was to the proper functioning of the economy, by making loans that few others would, especially in the middle of a financial crisis. "Our argument wasn't that Bair should feel sorry for us," Jeff remembered. "It was that punishing us would prove disastrous for the economy." He told Bair that the big Wall Street banks would "love to see us go away" and wouldn't tell her that GE "keeps the economy going." He was making the argument that GE was too interconnected to fail, the same argument that Wall Street had made so successfully to get its bailout. It turned out that GE Capital (and thus GE) was every bit as essential to the proper functioning of the capital markets, but because it was an unregulated nonbank bank, essentially it had been overlooked in the madness of the unfolding crisis. "People are coming to us," Jeff told Bair, "because there's nobody else out there who can help. If you don't help us continue to do it, how are these guys going to survive?"

The meeting was brief. Jeff couldn't tell whether he'd made his case to Bair. On the way back to the airport, Paulson called Jeff on his cell phone. "We're going to get this done," he told Jeff. Over the next three weeks, Denniston and Rodge Cohen, on behalf of GE, worked with the Treasury and the FDIC to dial GE into the TLGP. Treasury officials worked closely with the FDIC to make it possible for GE to partake of the DGC.

As part of the initial negotiations about the TLGP with Paulson, Bernanke, and Geithner, Bair said she had agreed to consider nonbank financial institutions, such as GE Capital, on a case-by-case basis. "GE Capital was overreliant on short-term funding, and they did that to maximize their spreads," Bair told me. "You borrow short, and lenders invest long. You widen your spreads. At least under normal times, you don't have an inverted yield curve. They did that, and shame on them. So that was a problem. Every time we have a crisis, that turns out to be central to what the issues were." After hearing from Jeff, Cohen, and Denniston, government officials were more than a little concerned about what would happen if GE Capital went down the tubes. "They didn't go so far as to say GE could file bankruptcy, because I think they were afraid to even utter those words, but they were certainly con- cerned that GE Capital would be in grave peril and could go into bank- ruptcy, and at this moment in time, there wasn't a guarantee of GE Capital's liabilities from GE," recalled one person familiar with the discussions between GE officials and the government. "They didn't say GE could go bankrupt, but they certainly made a very impassioned case as to why they needed the assistance and why they should get it, without going so far as say, 'This could hurt the largest multinational company in the United States.'"

Bair was sympathetic to GE's plight. "This debt-guarantee program was distorting the market," she said. "All the investors are saying, 'Gee, we're going to buy commercial paper that is guaranteed by the FDIC.' It was creating even more problems for them, because you had trouble in the commercial-paper market to begin with, but now you had a guarantee for part of the market, and they weren't part of the market that had the guarantee. That was on us. I mean, that was on the government. That was creating a distortion that was hurting them."

She remembered that Paulson called her. "He was all upset about this," she said, "and [that] we needed to let GE in the program." She told the Treasury secretary she would have to think about the request. GE "had not been completely transparent with me about the scope of their problems," Bair recalled. "I don't recall them ever telling me they considered bankruptcy filing [for GE Capital]. . . . I did not know that that had ever been on the plate, and probably would have changed my calculus." But she appreciated that Jeff had called her directly and then had come in to see her. She had grown tired of representatives of the big banks going around her to the Fed and the Treasury. "It was my decision and my board's decision," she said.

Bair wanted GE, the AAA-rated parent of GE Capital, to guarantee the FDIC against any losses that the FDIC might suffer by allowing GE Capital into the TLGP. "He readily agreed to do it," she said. "He didn't even argue about that, probably a sign that he wasn't in a position to haggle." Bair and the FDIC figured that the industrial side of GE was "quite healthy, quite profitable," and had a AAA credit rating. They joked, "Well, if GE goes down, we've got bigger problems than losses on our Debt Guarantee Program." Bair approved Jeff's request. Rodge Cohen said her decision was a wise and underappreciated one at a crucial moment in the financial crisis tsunami. Bair clearly grasped that a GE bankruptcy would have taken the crisis to a new nadir.

Thanks to Bair, on November 12, GE Capital had access to the DGC—after paying $2 billion in fees to get it—and turned out to be its second-biggest user, after Citigroup, issuing some $131 billion of debt with an FDIC guarantee, none of which required the FDIC to make good on the guarantee. "One of the reasons that I was a little more sympathetic was that the problems were not completely of their own making," Bair said. "We had distorted the market by guaranteeing the debt of all these bank holding companies, and they competed in the commercial-paper market with those bank holding companies. So that did influence me in favor of helping them and letting them into the program."

"And here's a little-reported fact," Jeff wrote in *Hot Seat*. "GE was also the first to exit these programs." He gave the federal government "a lot of credit" for engineering the "genius" programs that restored confidence in the capital markets. Still, few had any idea how close GE Capital and GE had come to the edge of the abyss. "Living through the global financial crisis was like having the stomach flu for eighteen straight months," he recalled. Paulson, for one, thought Jeff displayed leadership during that very tense period. "He handled himself about as well as anybody could in those circumstances," Paulson told me, "without ever telegraphing to me the extent of the problems that I subsequently found out existed. I clearly knew there were problems, and he was concerned. But you never really sensed the fear in his voice."

During that hectic time, Jeff was one of five Harvard Business School alumni awarded the school's highest honor as part of its centennial celebration. In the roundtable discussion that followed, moderated by Charlie Rose, Jeff was his usual optimistic self, despite the adversity. He said he used to tell his daughter, who had just been home from college, that Friday was his favorite day of the week. But that was no longer true. "Now Friday is just when the fun begins," he said. He then became more introspective, without betraying any of his growing concerns. "The last thing I'd say about leadership, Charlie, is that in some ways it's an intense journey into yourself," he said. "It's about how much you want to learn. It's about how much you want to give. It's about personal change, and just being willing to kind of renew yourself almost every day. I take every criticism personally, and I go to bed at night and [say to myself], 'Gosh, I'm such failure.' And I get up the next morning and say, 'Hello, handsome. Let's go get them.' You've got to be able to renew yourself. You've got to be able to renew yourself and go fight the battle again, you know. If I stand up in front of 320,000 people and say, 'Oh, god, I just don't know what we're going to do today,' [that's a disaster]. [Instead] you stand up and say, 'We've got it nailed. This is our best day. Let me tell you why.'"

If the financial crisis of 2008 wasn't enough of a reminder of the perils of GE's reliance on GE Capital, another came the following year, in August 2009, when the SEC finally settled the question of whether GE had manipulated its earnings. As part of the settlement, GE agreed to pay a $50 million fine for allegedly having "misled investors by reporting materially false and misleading results in its financial statements," including using "improper accounting methods to increase its reported earnings and revenues and [to] avoid reporting negative financial results." In other words, the SEC was alleging—and GE was agreeing to pay $50 million to settle, after incurring some $200 million in professional fees along the way—the longtime concern that GE was manipulating its earnings to ensure a steadily increasing stream of profits, quarter after quarter (the very assertion that Jack told me not to fall for). In its complaint, the SEC noted that GE had met or exceeded earnings expectations every quarter between 1995 and 2004. "However, on four separate occasions in 2002 and 2003"—during Jeff's tenure—"high-level GE accounting executives or other finance personnel approved accounting" not in compliance with GAAP, generally accepted accounting principles. In January 2003, some accounting shenanigans related to GE's commercial-paper program had allowed the company to avoid a $200 million pretax charge. In 2003, more problems occurred with GE's interest-rate swaps. In both 2002 and 2003, GE accelerated some $370 million in the sales of locomotives that had not yet occurred. In 2002, an improper accounting of the sales of aircraft engine parts increased GE's 2002 net income by $585 million.

# BURNING FURNITURE

The flipping of the calendar year did nothing to ease Jeff's growing anguish. As the slings and arrows found their mark, he was looking more and more like St. Sebastian. The concern now was one Jim Grant had been highlighting for years: Was GE really a AAA credit, and when would the ratings agency make de jure what was already de facto? And would GE still be able to maintain its annual $1.24-a-share dividend for shareholders, costing the company more than $13 billion a year? That money could perhaps better be spent elsewhere. Of course, cutting the GE dividend—if he did it—would be another black mark for Jeff, since the company had been paying dividends since 1899 and had cut it only once, in 1938. But the drumbeat on Wall Street and in the financial press was getting louder about when the dividend would inevitably be cut, and by how much. If in public Jeff was portraying the facade of the confident CEO of one of the world's most valuable companies, in private he was a mess. Even though he looked svelte in October during his appearances at the business schools, he soon started packing on the pounds, falling back into his bad habit of nervously eating junk food. "I remember opening my closet one morning to find just one suit I could squeeze into," he said. He was losing it. "I was completely fried," he recalled, adding that while the scrutiny on GE was no more than usual, his tolerance for the ongoing attacks had been greatly diminished by relentlessness of the crisis. "I was exhausted," he continued, "so my tolerance for it was waning." But he maintained his optimism publicly. On January 24, he announced that GE would keep its dividend at $1.24 a share for 2009 and would perhaps cut it in 2010, but only if necessary.

Once again his optimism was misplaced. The GE stock was in free fall, hitting around $9 per share by the time of the February board meeting. The board was antsy. The pressure was mounting on everyone to cut the dividend. A few weeks earlier, Moody's had said it was considering cutting GE's AAA rating, and S&P had said it was putting GE on negative credit watch, signaling that it, too, might lower the rating. Jeff wasn't sure what the ratings agencies were going to do about GE's AAA rating; they had themselves come under much criticism for how they had rated AAA so many mortgage-backed securities that later failed.

GE's credit-default swaps—blowing out again past $1,000—also continued to point to GE having a much-lower-than-AAA credit rating, hovering at a level that implied a junk credit rating. Hedge funds were having fun playing GE's stock. "You point out to people, 'We can borrow money with the TLGP as much as we can; we're safe in that regard,'" Jeff told me. "But it just didn't matter, because people just could manipulate almost anything at that point in the middle of a crisis." After forty-five minutes of debate at the board meeting, Ralph Larsen, the lead director and former CEO of Johnson & Johnson, announced the dividend would have to be cut. "That's just the way it is," Larsen told Jeff. The board had overruled the CEO and forced him to fall on his sword for the second time in six months (the first being when he had to backtrack on not raising new equity). Jeff recalled feeling an "odd combination" of "disappointment and gratitude."

He didn't want to cut the dividend. Not only had he said publicly he wouldn't just weeks before, but he also knew how well the industrial side of the business was doing. It turned out that GE's nonfinancial businesses had record cash flow in 2009. It didn't seem to matter. The sentiment among investors had changed completely in six months. Jeff said that GE had gone from trading at a premium to other industrial companies to a business where "you have a first lien on all your cash, GE Capital is on oxygen support, and you run the risk of not being able to grow your industrial company because of financial services."

After Larsen ruled on the dividend, Jeff said he felt like "a cub getting my ears boxed by an elder lion," a far cry indeed from the quiet confidence he conveyed to Paulson, Bair, and others. "No matter how dire the extenuating circumstances, no matter how much people kept telling me it wasn't my fault, I didn't want to be the CEO who cut GE's storied dividend for the first time since the Great Depression," he recalled. He felt awful. He had been overruled by his board and was again made to look like he didn't have a handle on the company. He called Jack for advice, even though less than a year earlier Jack had said on national TV he'd shoot Jeff if he missed earnings again. Cutting the dividend after saying just weeks before that you wouldn't was another version of missing earnings.

Jeff was still pissed at Jack after that crack on CNBC. But, he said, he "didn't have the luxury of holding a grudge." He called Jack and they talked about the wisdom, or lack thereof, of cutting the dividend and reversing himself yet again. Jack told him, "Jeff, you can go back on your word and be a smart guy or be a consistent dumb guy, okay?" (Given that Jack told me he and Jeff never spoke after Jeff became CEO, am not quite sure what to make of this conversation.) Jeff agreed to cut the dividend—not that he had much choice. He decided not to take his more than $12 million bonus.

He also thought about resigning. Before announcing the dividend cut, he was in a cab in New York City, riding with Gary Sheffer, his communications guru. "I want you to get ready," he told Sheffer. "Once we cut the dividend, I'm stepping down." He claimed he was "so shaken by the feeling of failure" that he believed he should resign. (If so, this might have been a small detail worth sharing with shareholders.) Sheffer was horrified. "This isn't your fault," he told Jeff. "You're not resigning. And don't ever say that again to anybody."

On February 27, GE announced it was cutting the dividend by more than two thirds, to 10 cents a share. The move saved the company some $9 billion annually. The stock traded down to $8.50 a share. Ironically, about a week before Jeff's momentous decision to cut the GE dividend for the first time since the Great Depression, he had been at a swanky party in Georgetown thrown by the *Financial Times* U.S. managing editor, Chrystia Freeland, and shared with the assembled journalists, on the record, his frustration about running GE in such a difficult period. In the 1990s, he said, digging at Jack, "anyone could have run GE and done well." He wasn't done. "Not only could anyone have run GE in the 1990s, his dog could have run GE, a German Shepherd could have run GE," he said. By contrast, he continued, GE's future "will be really, really, really hard," and it and the country should focus on "clean energy, health care, and education."

But cutting the dividend did little more than freak out investors further. If GE, once the world's most valuable, most respected, and most powerful company, was hemorrhaging, what hope was there? On Wednesday, March 4, GE's shares hit an eighteen-year low, at one point trading down to $5.73 a share, an 18 percent decline from the day before. It finished the day at $6.69, down 4.5 percent. Egan-Jones, an independent credit-rating agency, said GE needed to raise another $20 billion to $40 billion to "support its supposed 'AAA' rating." GE issued a statement denying that it was raising fresh capital or needed to: "This is pure speculation, is inaccurate and is not based on any input from our company."

Jeff saw a conspiracy, brought on by "speculators" who shorted GE's stock, bought credit-default swaps on GE's debt, and then, he asserted, called CNBC and Fox Business to suggest that GE was in financial trouble, a gambit that would drive down the value of GE's stock—making their short positions more valuable—while also pushing up the value of their credit-default swaps. "We couldn't counter inaccuracies—neither factual errors nor contextual ones—with silence," he argued. "We had to seize our own narrative."

Raising more cash *was* one potential way to silence the barbarians. There were others. After one of the October 2008 weekend sessions with Rodge Cohen, when everything was on the table about how Jeff could navigate GE out of its predicament—including putting GE Capital into bankruptcy—Jeff called Jimmy Lee at JPMorgan Chase. By then, Lee had proved himself to be a preeminent Wall Street banker and one of the architects, along with Jamie Dimon, of the rise of JPMorgan Chase to worldwide prominence. "I caught him on a weekend," Jeff told me, "and said, 'Hey, look, I'm thinking I may need to burn some furniture here.'"

He asked Jimmy to put "a soft lob" out to Comcast about buying NBCUniversal to see if Brian Roberts might be interested, more of a feeler than anything else. It was a call Roberts had been waiting years to receive. "I needed something that was sellable," Jeff told me. "And I knew people coveted NBC, and so I kind of always felt like that was my parachute, right? If I needed $10 to $15 billion," which, of course, in October 2008 he did (and got from Warren Buffett and the market). Around this time, one Comcast executive remembered being in the JPMorgan Chase offices with Dimon and Lee and Brian Roberts. Lee asked Roberts about Comcast's strategic imperatives. "Jeez, we really want to develop more content," Roberts said. Lee responded that he was meeting with Jeff in a few days. "They're under some pressure because of the financial crisis and things that were going on," Lee said. "Would you be interested in meeting with them?"

The Comcast executives welcomed the chance to meet with Jeff and Keith Sherin about buying NBCU. It seemed that Brian Roberts's years of dinners and golf with Jeff were finally going to pay off. "This is what we'd like to do," Jeff told them. "Listen, I'm not sure this is core to us anymore, but it's a valuable asset. I don't really want to sell it at the trough of the market. We know where we are in the economic cycle. Would you buy a minority stake?" The Comcast executives said they weren't interested in buying a minority stake. But they were interested in buying a 51 percent stake. "We'd pay for fifty-one percent now," they told Jeff, "and you could be at forty-nine percent. If the business performs really well and recovers, you're getting a bunch of cash now, plus you'll benefit at a later date as the business performs." The discussions continued off and on without a resolution. "There was all sorts of shit going on," one of the participants said.

By March 2009, Jeff's predicament was even worse: the GE stock was spiraling down, investors were fleeing, and credit-rating agencies were suggesting that GE again needed to raise more equity. Once more, Jeff called Jimmy Lee. It was the night before Jeff was scheduled to be interviewed by Charlie Rose at Jimmy's CEO conference at JPMorgan Chase. Jeff had agreed to speak months earlier. Again he wanted to back out, given what was happening to GE's stock. But he couldn't; that would be even worse.

"I'm ready," Jeff told Jimmy about wanting to sell NBCU. "Is there a possible deal to sell NBC to Brian? Can you get me in a room with him?" Lee called his boss, Jamie Dimon. Together they called Brian Roberts. Roberts, who lived in Philadelphia, agreed to meet Jeff the next morning at seven o'clock, before his scheduled appearance at the 8:00 a.m. Jimmy Lee conference. They agreed to meet in a conference room on a different floor from where the CEO conference would be held.

Meanwhile, Jack and David Zaslav, now CEO of Warner Bros. Discovery, were also at the CEO conference, sitting outside of where Jeff was scheduled to speak, watching Keith Sherin on CNBC trying to explain to viewers why GE's predicament wasn't as bad as it seemed. Jack was not talking to Jeff. He was worried generally about whether the world was going to end and whether GE—and all that he had accomplished during his twenty-year reign—was going to go down the tubes with it. Word had already leaked to him and Zaslav that Jeff was thinking of selling NBCU. "It was the lowest point in GE history," recalled one former GE executive.

In the meeting with Roberts, Dimon, and Lee, Jeff told Roberts, "Let's go. I'll sell you NBC. I'm ready. I want to sell this thing to you." Brian Roberts had been waiting nearly a decade for this moment—one he had considered during his golf games and lunches with Jeff since GE had defeated Comcast in the battle to acquire Universal. "Finally, after 2008, 2009, he called," Steve Burke said. "The GE people would never admit it, but we think they had to sell some stuff, and here's a sellable thing. And he called." Burke said it was a bit of "a funny meeting" where the assembled Comcast brass was trying to convince Jeff that "we were totally serious, and we would not screw around." He said he thought Jeff was more worried about the "headline price" and "what the public reaction would be if he sold NBC than he was about doing the right thing."

Jack, meanwhile, was so appalled that he refused to go into the JPMorgan Chase meeting to listen to Jeff speak to Charlie Rose. He couldn't even look at Jeff. He loved NBC and was infuriated that Jeff had decided to sell it. "Talk about deer in the headlights," said someone who heard the interview. "Jeff came out and said, 'The world will never be the same. Everything has changed.' His voice is quivering. His eyes are wide open. He's scared to fucking death. He's up there and literally the buzz was 'Oh my God, this guy is completely shaken.' If you wanted to embody fear, you would show a video of Jeff Immelt that morning. . . . It was [done] out of fear. So we've got fear and beyond fear. It was jaw-shaking that morning. That's how scared he was. That was the way he sold the company." (The CEO conference was a private meet-ing; JPMorgan Chase declined to share a recording of the event or a transcript.)

For his part, Jeff told me he did call Jimmy Lee the night before the CEO conference to urge him to put another feeler about NBCU out to Comcast. But he has no recollection of a meeting with Brian Rob-erts early the next morning. He said the interview with Charlie Rose was "grueling." Rose had marked up a copy of that day's *Wall Street Journal* with a yellow highlighter. "On that particular day I was *The Wall Street Journal*," Jeff said. "I was the news." He added, "People were in a bad mood anyhow, so I come in and I'm like Darth Vader or some-thing like that. They're all like, 'That poor fuck.'" Years later he told Rose, in a separate interview, that that "was the worst day of my life."

While Jeff was contemplating the sale of NBCUniversal and then being grilled by Charlie Rose, Keith Sherin was on CNBC defending GE. The day before, Sherin was in his office at 30 Rock when Russell Wilk-erson, the GE communications executive who handled the financial press, burst into his office and told him he had to go on CNBC—"conveniently" owned by GE, as one *New York Times* columnist put it—and try to stanch the bleeding. After the unexpected dividend cut, the stock was in free fall. The consensus was that someone had to try to stop it. Sherin was the man chosen.

He had been living the nightmare every day for nearly seven months, since the previous September. "It was the worst seven months of my life," he told me. "It was twenty-four seven for seven months by the time it got to March." Sherin has a congenital tremor, which gets worse when he's especially nervous. On the morning he was going on CNBC, he was nervous, and the tremor kicked in. He was shaking. He thought to himself, "I'm going to be on TV wearing little pieces of tis-sue paper with blood spots all over my face. This is really bad." But he was fine after he exercised. He drove to CNBC headquarters in New Jersey, where no one was there to let him into the building until he saw David Faber arrive. He didn't realize he would be on with Faber, who was known for asking difficult questions.

"David, what are you doing here?" Sherin said.

"I'm here because you're here," he told Sherin.

"I am so fucked," Sherin thought to himself. "I am just so fucked here."

For forty-five minutes, Joe Kernen and David Faber peppered Sherin with questions about GE's viability. Over and over, Sherin reiterated what he and Jeff had been saying for much of the previous year, despite the increasing investor doubts: GE was fine, well capitalized, and not at risk. The GE of Jack's twenty-year reign had not changed. It was still dependably safe and secure. That wasn't particularly true, of course, but it had nonetheless become the mantra of the senior GE executives.

Kernen wasted little time lunging for Sherin's jugular. The market was worried about something when it came to GE, he said. What was it? "We're getting a lot of speculation about the risk in GE Capital, obviously," Sherin said, "and I think it's overdone." He pointed to the company's $45 billion cash hoard and another $60 billion of cash availability, thanks to the TLGP program. He added that "the credit volatility" was also "overdone" and was driven by a smaller-than-usual volume in the trading of GE's credit-default swaps. "Is that really market fundamentals?" he wondered. "Or is it just some sort of disruption based on very narrow trades in a volatile time?"

He said GE Capital would be profitable in the first quarter and that he did not envision that the company would need to raise any additional capital, because the $15 billion in equity that GE had raised the previous October had been transferred into GE Capital. "We do not have a time bomb at GE Capital," he said. He also said he was not particularly worried about the ratings agencies and how many notches below AAA GE's credit rating would fall. GE was already—finally—proactively shrinking GE Capital as a percentage of the overall GE picture. He said by cutting the dividend, GE would save $14 billion over the coming eighteen months. "We have a capital cushion building that we can use in the event that losses are higher in GE Capital," he continued.

He referred opaquely—no one picked it up, not that they should have—to the meeting that Jeff was having concurrently with his CNBC appearance with Brian Roberts and Mike Angelakis, the Comcast CFO, in a JPMorgan Chase conference room. He was talking about the relative robustness of GE's industrial businesses. The infrastructure business had a "big backlog," he said. The service business was both global and "huge." The "long-cycle" energy business was poised to have "a very strong year," as was the aviation business. "Probably the toughest place really, honestly, is CNBC and NBC because of the advertising market," he said. "The local advertising market is very pressured based on the economy, and in the national advertising market we're also seeing some pressure."

Michelle Caruso-Cabrera asked whether Jeff's credibility had been hurt by having had to reverse himself on first-quarter-2008 earnings, the statement he made in September 2008 about GE not needing more equity, and the early-2009 statement about the dividend being safe. Sherin didn't flinch. "It's a real issue that we're facing," he answered. "I think if you make a declarative, confident statement today you get second-guessed. If you make a wobbly statement, you get speculated. We recognize that we've made statements about both not raising equity and not cutting the dividend and we've had to backtrack on those. I think the only thing I can say is, in our defense, we're trying to run the company for the long term. We've got a real franchise here. We're proud of it. It's going to be around another hundred years." He added that GE did have a "credibility" issue. "We've got to earn that trust back," he said.

Jeff characterized the purpose of Sherin's CNBC interview as "dispel[ling] false rumors that some of the more unscrupulous hedge funds were spreading about GE Capital." He didn't specify which hedge funds or what they were doing. Jeff said that Sherin had "nailed it."

—————

BEFORE MARCH WAS OUT, BOTH S&P AND MOODY'S HAD STRIPPED GE of its AAA credit rating, forever altering GE's business model, or at least the one that relied on GE Capital's ability to arbitrage GE's access to cheap money and then to lend it out expensively to a hodgepodge of corporate borrowers. S&P now rated GE AA+, a one-notch decrease. In its note about the downgrade, Moody's said the "primary credit risks" at GE were at GE Capital, where it believed "the risk profile" had "increased" due to "the long-term risks associated with [GE Capital's] wholesale funding model" and the "earnings volatility resulting from deteriorating asset-quality trends." The firm added that the financial crisis had "exposed the vulnerability of even the best risk-managed financial institutions to confidence sensitivity and capital market disruption."

It could have been much worse. GE stock traded up more than 10 percent on the news of the downgrade—but only because investors were relieved the downgrade was only one notch. Jeff, of course, put the best spin possible on the situation. "As we have previously said, we are prepared to fund the company as a double-A, but we will continue to run GE with the disciplines of a triple-A company, which means low leverage, high liquidity and strong risk disciplines," he said. "While no one likes a downgrade, this review and rating reaffirm the relative strength of the company."

—————

ON JUNE 17, TIM GEITHNER, OBAMA'S TREASURY SECRETARY, RELEASED the opening salvo in what soon enough become the so-called Dodd-Frank law, reregulating Wall Street in the aftermath of the financial crisis. Geithner's eighty-nine-page Financial Regulatory Reform proposal left Jeff, and his team of other top GE executives, dumb-founded. While GE was not mentioned by name in the document, Jeff and his team quickly concluded that the draft "contained directives" that seemed "aimed straight at us." It was déjà vu all over again. He believed that, once the bureaucratese was pared away, Geithner's draft "strongly implied" that big financial institutions would not be allowed to have an industrial arm. "Well," Jeff continued, "there were no big financial-services companies, other than GE, that had industrial arms. He was describing an island of one."

After the draft was made public, GE's stock dropped 7 percent. Jeff was fearful that the government might force GE to spin off GE Capital into a separate company. He had already done enough analysis on that possibility—the previous year—to know that for GE Capital to survive as a viable, creditworthy, publicly traded, independent entity would require a capital infusion of around $30 billion, in cash, from GE—money that GE had but could not afford to pump into GE Capital in June 2009. "That's when I had a second conversation with Rodge Cohen about taking GE Capital into bankruptcy," Jeff told me. He happened to be with GE board member Sandy Warner when Geithner's draft appeared. "We just can't keep living this way, kind of ruling to ruling," he told Warner. At that moment, Jeff made two big decisions. One was to use GE's influence in Washington to get Geithner to revise the language in his draft law to eliminate the risk that GE Capital would have to be separated from GE. He may have wanted to reduce GE Capital's importance to GE, but the last thing he wanted was to be forced by the federal government to spin off GE Capital. The second decision was to jettison NBCUniversal. Enough with the feelers and the testing of the market. It was time for NBCU to go. It was GE's most expendable, most salable asset—he knew that Comcast wanted it badly—and the best way for GE to raise more capital without further rattling the market by raising more equity.

Besides, NBC was Jack's thing, not Jeff's. Jack was the one who delighted in going on CNBC and who studied the ratings and hung out with the personalities. RCA had been Jack's deal, not Jeff's. It had been one of the best acquisitions of all time; there was no doubt about that. But for Jeff and his team, it was expendable, rightly or wrongly. "In the dead of winter, even a favorite armchair begins to look like kindling," he said. Incredibly, he believed NBCU "had always been an outlier" within GE and had brought to mind the old *Sesame Street* song—"one of these things is not like the other." He said the financial crisis "made NBCU a luxury we could no longer afford."

This was one of the most momentous decisions in Jeff's first eight years as CEO. Whether it was a wise one, or the correct one, remained to be seen. Was he acting too impulsively in the wake of a series of decisions that had embarrassed him and the company? Was the threat existential or did it just seem that way to him? At least some of Jeff's top executives believed GE had run out of answers for NBCU. "What business did we have running NBC?" one of them said to me. "For Welch, it was 'must-see TV.' It was number one. We were number four for how many years in a row? We were terrible." He then reeled off the myriad of problems that NBC, the network, faced: a slew of different directors of programming, none of whom seemed to be able to pick winning content; stuck in last place in the ratings; losing money year after year. "Without Universal," he continued, "we couldn't have sold it for anything." USA Network alone was making $1 billion in annual profit, covering up a lot of sins at NBC. "Thank God," he said. (In other words, the very scenario that Steve Burke had years before predicted to Brian Roberts would happen, did happen.)

That is not to say he didn't enjoy the fact that GE owned it. "I loved NBC," the executive continued. "It was fun. It was great, entertaining. People were interesting. Everything about it was great. But financially, we were getting squeezed by the cable companies. For whatever it's worth, Brian Roberts and the cable guys knew more about what they could do with our content and our pricing than we'll ever know, and they did a great job of that. But the thing for me was the capital required." He then recalled how the deal to buy Universal was "stitched together with all the creativity" GE could muster "without actually paying all the capital" for it.

As another example of how capital-constrained GE had made NBC, he cited the July 2008 acquisition of the Weather Channel for $3.5 billion. To do that deal, NBCU partnered with two private-equity firms, Blackstone and Bain Capital. That was not the typical way GE did deals. He took that as a sign that Jeff was reluctant to invest more money into NBCU. "I was like, 'Okay, if we can't have enough capital to even give them [the money] to buy a stupid weather channel to make the business bigger, we don't have enough capital to grow [NBCU] in the world they're competing in.' We're also not distinguishing ourselves with any of our performance in terms of picking shows or movies or whatever it was. The movies, they did okay. Some years were terrible, other years were okay. But there was no synergy with the rest of GE. The capital requirements were very large for uncertain returns. And the guys who we were selling to"—Comcast—"knew more about how to leverage it than we did."

This executive seemed to become a little wistful. After all, RCA was a business that GE had started after World War I, then been forced to divest in the 1930s, and that Jack had then bought back in 1986 and made into one of the most successful M&A deals of all time. He didn't think GE could "fix" NBC. He said there was no viable "path" to get it to first place in the ratings, from fourth place. "Look," he said, "we just survived a near-death experience. Capital allocation and capital constraints are in everything I'm thinking about. I didn't think it was a great business in the GE portfolio. Going forward, I didn't think it was something that we were going to be able to grow the way it needed to grow and compete the way it needed to compete. I thought we were going to constrain it and probably have it continue to wither." Quite an admission from a top executive of supposedly the world's greatest talent factory.

The deal GE cut with Comcast was incredibly complex. First, as Comcast had said when the subject was first broached at JPMorgan Chase, in October 2008, it was structured as a joint venture, with Comcast owning 51 percent of the business and managing it day to day and GE owning 49 percent. There were tough negotiations on price and value, since Comcast was contributing to the deal its cable assets, including the Golf Channel, E! Entertainment, and its regional sports networks. There were complications with Universal's theme-park business, which was partially owned by Blackstone, and with Universal Studios, which was a small but important part of what Comcast wanted in order to have content for a burgeoning streaming business. There was a tough negotiation of how and when and for how much Comcast could buy the other 49 percent of the business from GE. And because it was not an outright sale, the two partners had to set up a board of directors. "You have to hope that you like these guys," explained one of the Comcast executives, "because you're going to deal with them the day after, too. It's not like they're selling the business."

And then there was the brutal negotiation with Vivendi, which still owned 20 percent of NBCU and which received around $5.8 billion for its stake, in line with the deal's overall $30 billion price tag. There was also much hand-wringing about the process that Jeff was running for the NBCU divestiture. If his goal was to maximize the price, he would have—should have—conducted a full-blown auction for the business. He would have—should have—made a big public announcement about the fact that GE was selling the business and wanted to hear from all interested parties. But he did not do that. From the start, he seemed focused on getting a deal done with Comcast, and Comcast alone.

He knew from years of being pestered by Brian Roberts that Comcast not only wanted the business but had the financial wherewithal to get a deal done. Jeff also seemed locked into the idea of continuing to own a large minority stake in the business, most likely to get what bankers on Wall Street call "schmuck insurance," protection against having sold a business too cheaply at a time of the seller's own financial distress. GE was plenty stressed at the time of the NBCU sale, so to the extent that Jeff believed Comcast "snookered" him, as he would later admit, the pain would have been far worse for him and GE had GE sold 100 percent of the business in December 2009, rather than just 51 percent. In *Hot Seat*, Jeff wrote, in passing, that there *were* other bidders for NBCU, although the evidence was slim that any of them had a serious shot at buying the company. He mentioned News Corp, which owned Fox, and Time Warner, which owned CNN and TBS. One other potential bidder for NBCU (that Jeff did not cite) was a combination of John Malone's Liberty Media, which was then the largest shareholder in DirecTV, and Discovery Media, which was also partially owned by Malone and of which David Zaslav, the former senior NBC executive, was the CEO. Malone and Zaslav wanted in on the NBCU deal. They called Jimmy Lee and told him they could bid more for NBCU than Comcast could. Zaslav had been talking regularly to Jack, who was "puking on this fucking thing" and thought it was "the worst fucking deal ever." Word quickly got back to Brian Roberts that Malone and Zaslav were interested in NBCU.

Roberts was pissed. "You talk to anyone else," he told Jeff, "I'm going pencils down. Good luck." When word of Roberts's comment leaked into the market, other potential bidders were incredulous. "Fucking pencils down?" said one. "Who sells a media company without putting it up for auction? But Jeff was so scared that GE was going to be cratered and he wouldn't be able to get this deal and the whole company would go down that he didn't even bid it out. He just closed the deal with Brian, and he didn't let any of us in." (In September, after news of the deal negotiations leaked, Comcast and GE signed a standstill agreement, effectively preventing other bidders from joining the sale process.)

Jeff's Harvard Business School classmate Steve Burke, the Comcast executive who ran NBCU after Comcast acquired it, said that as Comcast had been the cover bid for Universal when GE bought it a few years earlier, Comcast knew the business well. He remembered that Jeff wanted the deal to have a sticker price of $30 billion, or about nine times NBCU's $3.3 billion of EBITDA. To get there, Comcast agreed to contribute its cable assets, at a valuation of $7.25 billion, "even though we both know they're worth $3 billion," Burke said. Comcast also agreed to put in $6.5 billion of cash, for $13.75 billion of total consideration. According to Burke, Comcast valued the deal with GE for NBCU at $26 billion—that's how Comcast's roughly $13.75 billion got it 51 percent—and there were another few billion dollars' worth of tax assets and cash left in the company that allowed Jeff to say it was a $30 billion deal. "We got them to agree to that," Burke recalled. Burke said Jeff made it clear to Comcast that the deal structure was like a hedge for GE, the aforementioned "schmuck insurance." Recalled Burke, about Jeff's thinking, "If I stay in fifty-fifty, I'm still in and it won't be embarrassing, and I'm getting a good price. And then we had a great mechanism to, over time, buy the other fifty percent with the cash that was generated from the business. So in theory, we're getting an asset for $7 billion of cash and throwing in a bunch of junk."

In early December 2009, GE and Comcast announced the deal. "They're selling NBC to a company called KableTown, with a *K*," joked Alec Baldwin's character, Jack Donaghy, on NBC's long-running sitcom *30 Rock*, a spoof of sorts about GE's ownership of NBCU. As designed, the headline valuation for NBCU was $30 billion, according to *The New York Times*. It took some eighteen months for the deal to close, awaiting regulatory approval. "It was a pretty brutal experience," one person involved with the deal said of the approval process. Steve Burke and Mike Angelakis, the Comcast CFO, used the interregnum to travel to New York from Philadelphia at least once a week to learn the business, evaluate the personnel, and figure out how to integrate the business into Comcast and away from GE to the extent possible.

As he learned more and more about the business, Steve Burke decided he wanted to run NBCU. That meant dispensing with Jeff Zucker, whom Jeff had appointed to take over from Bob Wright. "Zucker was the wrong guy for the job," Burke told me. "Jeff is very talented at saying, 'That font should be larger' or 'The red there' and 'You shouldn't have four people, you should only have two.' But he was not an executive." He recalled how one day between the signing and the closing of the deal—when Zucker was still running NBCU—there was a major snowstorm on the East Coast. Burke was in his office at around 8:45 a.m. No one else was on the executive floor at 30 Rock except for Zucker.

Burke walked down to Zucker's office to say hello. Zucker was standing right next to the TV screen on the wall. There were twelve separate boxes on the screen, feeds from the various NBC-owned affiliate screens around the country, including New York, Philadelphia, and Washington. Zucker was standing right up against the screen. Burke asked him what he was doing. "Well, I normally don't let anybody preempt the *Today* show," Zucker told him, "but this storm is so bad, I think local audiences should want to know about the storm. So I've let Washington, Philadelphia, and New York preempt, and only two of the three are preempting, and it's pissing me off." Burke couldn't believe the extent to which Zucker was a micromanager.

By then, Burke had decided he was going to run the company and that Zucker was going to have to go. "Shit, that's not what I would be doing this morning," he told Zucker. He walked back down the hallway to his temporary office and thought to himself, "What the fuck is he doing? He's the CEO of the company and he's acting like the head of news." Burke thought Zucker should never have had the job in the first place. "GE really brought out the worst in him," he said. "Because GE would go to him and say, 'Cut $100 million by the end of quarter.' You'd have a Christmas movie and they'd say, 'Well, you can't market the movie. Move it to January.' And they'd say, 'Well, it's a Christmas movie.'" (Through a spokesperson, Zucker declined to comment.)

Burke told Roberts, who was fine with Burke's decision to replace Zucker and to run NBCU. He asked Burke to tell Jeff. "Well, it's kinda your job to tell Immelt," Burke told Roberts. But Roberts insisted that Burke tell Jeff. He said, "No, no, no—go tell him." Burke called Jeff and then went up to Fairfield to have dinner with him. He told Jeff he wanted to fire Zucker. "Why would you wanna do that?" he asked Burke.

"Well, you know, it's interesting," he responded, "and I love these businesses."

"Well, you're gonna be in the *New York Post*," Jeff replied.

"Yeah, I'll be in the *New York Post* every once in a while, but if we can do this, it'll be a great adventure," Burke responded.

"Okay, let me tell Zucker," Jeff said. Burke agreed that Jeff should tell Zucker. Jeff said he would tell Zucker in June, about six months or so before the deal was to close. Burke's idea was to have his team in place and issuing directives a few months prior to the close, so that the plans could be implemented shortly after closing. "I'm up here every day, meeting with people and seeing Zucker, and he doesn't say anything," Burke recalled. "And so I wait, I wait, I wait, I wait." Around the middle of August, Burke was sitting in Zucker's office. "When are you gonna tell me?" Zucker said to Burke. "Just put me out of my misery. Are you keeping me or not?"

"Well, Jeff wants to talk to you," Burke said. "Has Jeff talked to you?"

"No, Jeff's never talked to me about it," Zucker responded. "If Jeff were going to talk to me, what would he say?"

"He would say that I'm going to fire you," Burke said.

"Well, Jeff's never talked to me," Zucker responded. Zucker called up Jeff and said, "Steve Burke just fired me." Burke recalled that Jeff "gets pissed off at me" because "it was all about him and his image." He added, "But [Jeff] loved Zucker. He thought he was great and thought he had done no wrong." (Zucker quickly got hired to run CNN, which he did successfully until 2022, when he was summarily fired, or allowed to retire, as they say, in part for having a consensual, undisclosed affair with a woman who reported directly to him.)

When the deal finally closed, and Comcast owned 51 percent of NBCU, Burke took over. "We do the deal," he remembered, "and it was a very scary time. And we didn't think we stole the company. We went to our board and said, 'Okay, here's our five-year plan,' and we get a decent ROI [return on investment]." Burke quickly concluded that NBC had been terribly managed under GE and the culture was abysmal. At his first big management meeting, in Orlando, he invited the company's two hundred top managers. "We have twenty-five different businesses," he said, "but we're in ten categories of business. We're in broadcast entertainment, we're in sports, we're in theme parks, we're in film, and let's look at how much money we make in each of these ten businesses and see if we're in the first quartile, second quartile, third quartile, or fourth quartile." Of the ten categories of business, Burke discovered that eight of the ten were in either the third or the fourth quartile of performance. "Raise your hand if you were in the bottom half of your class in high school or college," Burke asked them. Nobody raised his or her hand. "This is just unacceptable!" he said.

Recalled Burke, "The room was stunned, because everybody thought everything was fine. And you'd go to people and you'd say, 'NBC's been in fourth place in prime time for seven years in a row.' And they'd say, 'But we're only one hit away from being in second or first.' And then you'd say, 'My God, Universal Studios made $700 million at the peak. It's down to making $150 million a year.' They'd say, 'Oh, yeah, yeah—but that's gonna turn around.'"

He quickly ascertained, he said, that the incentive at NBCU wasn't to make money. "The game here was to keep GE, to keep Connecticut, happy," he said. "So the game then became telling a story, as opposed to running businesses, and the incentives were to do that. And then GE—and I fault Immelt on this—they never had the intellectual curiosity or the drive to really understand the businesses." As one example, Burke cited the way GE managed the local NBC news affiliates. Under Comcast's management, the local stations made $650 million, up from $150 million under GE. "We invested in local news and we put guys in place who knew what they were doing," he said, "and GE never took the time to do that."

Burke said Comcast cared about one thing: making money. The focus was on increasing EBITDA. Period. When GE and Comcast shook hands on the deal, NBCU was making $3.4 billion of annual EBITDA. In 2018, NBCU made about $8.2 billion in EBITDA; in 2019, it made $8.8 billion in EBITDA. "We could definitely sell this company for over a 100 billion, and we bought it for 26 billion," he said. "We've done a lot right. NBC's been number one for five years in prime time. Universal Studios has had four of the best years in a hundred years. . . . It was a complete cultural change, and we had to get everybody focused on—you know, I have all these mantras—'Think like an owner, not a renter.' Well, GE encouraged people to think like a renter, because they would move people around so quickly and they would do all this mumbo-jumbo."

Eighteen months or so after the Comcast and GE formed the joint venture, Angelakis went to see Sherin up in Fairfield. He told Sherin that the partnership was working fine, NBCU's financial performance had improved, and GE had a "stranded asset," in the sense that it was a minority position in a business that was no longer strategic and was being run by others. Angelakis proposed that Comcast buy GE's 49 percent five years early.

Instead of the seven years as initially agreed, Comcast wanted to immediately buy the rest of NBCU that it didn't own. Comcast also wanted to buy the 1.4 million square feet of office space in 30 Rock that NBCU used as its corporate headquarters. (By then, Comcast had already put its name atop 30 Rock, as it was contractually allowed to do once the first part of the deal closed, just as GE had done twenty-five years earlier.) "Put a number up," Sherin told Angelakis. "We're always open to talking." After about a month or two of negotiations between the two CFOs, with occasional input from Jeff and Brian Roberts, the two sides reached a deal. Comcast agreed to pay GE about $18 billion for its remaining 49 percent stake in the business, including the real estate in 30 Rock. In about eighteen months, the valuation of NBCU had increased 20 percent to $36 billion, from $30 billion. A new check for $18 billion went to GE from Comcast. "We actually created a lot of value for them in a relatively short time," said one Comcast executive. "They had a stranded asset that they had no control over, and all of a sudden, someone walks in the door and says, 'I'll give you this amount of capital.' That's a pretty good day. Did they leave money on the table? Listen, who knows? It's easy to have twenty-twenty hindsight and say they left a lot of money on the table. But we've invested a lot of capital in the business. We've changed the strategic direction of the business.... We've had some pleasant surprises around parks. We've really invested in parks, and that's done extremely well for the company." (That is, until 2020, when the pandemic caused the EBITDA in parks to plummet to a loss of $541 million, from $2.5 billion of EBITDA in 2019. In fact, by 2021, largely because of the pandemic and brutal competition, NBCU's EBITDA had slipped to $5.7 billion.)

One GE executive told me that had NBCU not been sold to Comcast, GE would have had to take it public in some way, giving the company its own currency to make the kinds of acquisitions that characterize the industry. Incredibly, he said, "there's no way we had the capital to do the deals" in a competitive situation, and suddenly, to win, GE would have needed to raise billions.

Mighty GE did not have the capital, or access to it? That was a new constraint. In any event, he continued, he thought the Comcast deal was a good one for GE "at the time," while noting that since then, "everyone else criticized it up and down. We bought high and sold low. I can't win that argument." He argued that Comcast was a better owner of NBCU than GE was. "They were able to take the media assets, the content, both at the network and the cable channels, and lever it for better pricing with all the other providers, ten times better than any deal we ever cut with Comcast themselves," he said. "They had better information and they did a better job of running it. There's no question about it for me."

But the fact that Comcast had increased the value of NBCU substantially—to around $100 billion, from the initial headline valuation of $30 billion—irked Jeff. Around the time Comcast bought GE's remaining stake in the company, Jeff was getting increasingly peeved, particularly at Steve Burke. One day, Burke got a call from Jimmy Lee at JPMorgan Chase. "Immelt is pissed at you," Lee told Burke. "He hears you're bad-mouthing him."

"Jimmy, I give you my word," he told Lee. "I literally have never mentioned the guy's name. I've never mentioned Zucker's name."

"Well, he's really pissed," Lee responded.

Burke was confused. "My job is to paint reality for people, but I have literally never used the words 'Immelt' or 'Zucker' since I've been here," he told me. "Why would you? You'd look like such a shit if you come in and trash your predecessors."

He called Jeff. Jeff didn't return the call. Two weeks later, Lee called Burke again. "I saw him again," he said of Jeff. "That's all he's talking about is how pissed off he is at you and Brian for bad-mouthing him."

At a black-tie event soon thereafter, with about thirty people in a small room, Burke spied Jeff. "He's there and he can't get away from me," Burke recalled of his Harvard Business School classmate. "I walk up and I said, 'Jeff, I gotta come see you. I know you're mad at me. I wanna talk about it.'"

"Okay, fine," Jeff told him.

Jeff's office in 30 Rock was on the fifty-third floor. Burke's office was a floor below. He went up one floor to see Jeff. "I sit down, and we start with pleasantries, and then I said, 'Jeff, I just wanna put it out there, I think you think I'm bad-mouthing you and I'm not.'"

"I know you are," Jeff told Burke. "I have people inside the building, I have people who know what's going on, and I know you're bad-mouthing GE, me, and Zucker."

"I've never mentioned your name, and I give you my word on that," he told the GE CEO. "If you're upset that I say the businesses are not doing well, that's different. I've said that, and I'm going to continue to say that until the businesses are doing well, but I would never make it personal."

Burke said Jeff wouldn't let him "off the mat."

At a certain point, Burke recalled that Jeff said, "Look, maybe I made a mistake. I thought you and Brian would say nice things about me and about the company, and I could've sold to Rupert; I don't really like the guy. I could've sold to [Jeff] Bewkes [the CEO of Time Warner]; I don't really like the guy. I thought I was selling to guys who would be good to me and be good to GE."

"Well, jeez, I'm sorry you feel that way," Burke told Jeff.

When he got back in the elevator for the short trip back to his office, he thought to himself, "The guy just told me that all he gives a shit about is how we talk about him, and he didn't run an auction. What the hell?" He went on, "And then I was like, 'Fuck him,' you know? And I never bad-mouthed him, but at that point I was like, 'I can't win with this guy. All he wants us to do is just embellish his image.'"

He talked over the situation with Brian Roberts, who told him that it wasn't great to have "the CEO of one of the most important companies in the country pissed off at you."

"I don't know what I can do!" Burke told his boss. "He's pissed off at me for crazy reasons."

Burke told me, "It did feel scary being on the bad side of GE. No one would've ever guessed that all this stuff happened. But we saw firsthand how these guys ran stuff, and they screwed this company up like you can't believe and then sold it for a third of what it was worth and then got pissed off because they did it and we kind of exposed them. As soon as we came in, we started to turn everything around, and I think he was kind of embarrassed, and he should be."

WITH NBCU OFF HIS HANDS, JEFF'S NEXT TASK WAS TO MAKE SURE TIM Geithner, Chris Dodd, and Barney Frank—the three men crafting the Dodd-Frank legislation—didn't write the bill in a way that would force GE to spin off or divest GE Capital before Jeff was ready to do it. And he was definitely not ready to do that. He told me that, fortunately, by September 2009 "cooler heads prevailed" and "people had gone to Geithner and said, 'Look, there's no reason to kill GE just for all this stuff, so let's not. They're the ones that would be hurt the worst by all this.'"

By then, Jeff said, he knew that GE and GE Capital would be named a SIFI—short for "systemically important financial institution"—since it had availed itself of so much government funding and that, as a result, it would come under heavy regulation by the Federal Reserve and the Treasury. Somehow, Jim Himes, the freshman Democratic congressman who represented Fairfield County, Connecticut, came to GE's rescue. Working together with Barney Frank, the Massachusetts congressman leading the charge on the reregulation of Wall Street, Himes had the language in what became the Dodd-Frank law changed so that GE Capital would not have to be divested from GE. Himes's insight was to create an intermediate holding company between GE and GE Capital that would keep the Federal Reserve from regulating GE as a whole. "And neither the Fed nor GE wanted that," he told me.

GE set up the intermediate holding company. "Boom, there's no risk that the Fed is going to understand turbines and airplane engines," Himes continued. "That was a big success." At one point, Himes and Frank had a colloquy on the floor of the House of Representatives about the change in the language that would benefit GE, nearly alone. Himes recalled how "all of a sudden," GE was "in a world of hurt" because of the financial crisis and the intermediate holding company "was a nice resolution of that problem. . . . It wasn't curing cancer, but it was important to a whole lot of lawyers and to predictability."

Getting the language changed in the Dodd-Frank law was another major victory for Jeff, and another one that few knew about or fully appreciated. Jeff said he never debriefed with either Himes or Frank to figure out how they made this sausage. Recalled Jeff: "I think Barney would be smart enough to say, 'Look, we want Dodd-Frank to happen, and there's no reason to destroy things along the way, destroy GE. That will make it harder to pass a bill.'" But there was no avoiding becoming a SIFI, not once GE availed itself of the Debt Guarantee Program. There was nothing Himes could do to help there. "GE not being a SIFI was really never on the table," the congressman said. "They were absolutely massive. They knew right out of the box that they were going to get designated. And in fact, they were. And I never tried to—and I don't think politically I ever would have tried to—convince somebody that they shouldn't be a SIFI, because they quite clearly were systemically important. They were doing every kind of finance, and while they were in no way, shape, or form involved in the nightmare of mortgages and securitizations"—actually, through WMC, GE Capital was—"they were pretty clearly systemically significant."

Jeff said he never fought the July 2013 SIFI designation—it would have been pointless—but he hated every minute of it.

*Chapter Twenty-Three*

# POWER MAN

As his tenth anniversary as CEO of GE approached, Jeff embarked on a number of new initiatives to modernize GE, many of which the GE "antibodies" (as he called the entrenched forces at the company) rejected. He also was looking for a transformational acquisition in one of GE's core sectors that would help to dilute GE Capital's outsize effect on the balance sheet, which ironically had increased further in the wake of the sale of NBCU. During different eras, different GE CEOs have tried to change the management culture. Ralph Cordiner had done it in the 1950s, Jack had done it in the 1990s, and in November 2010, with the blessing of the GE board, Jeff began the Global Growth Organization.

"In order to break into new markets (and maximize existing ones), we needed to make our big, broad business more local, giving in-country executives more power to act," Jeff explained. In essence, Jeff's idea was to appoint GE executives as country managers around the world, in the 185 countries where GE did business, and empower them to make decisions without his approval or the approval of the various leaders of GE's business units. In effect, Jeff created a CEO of

Mexico, a CEO of Iraq, a CEO of Thailand, and gave them the power to cut deals with customers on the sale of GE products on their own authority, without being hindered by, say, the head of the locomotive business, or the jet engine business, or the power business. The problem was that the GGO was quickly viewed within GE as another example of Jeff "not listening" to his most senior executives before deciding to commit the company to a new, expensive global initiative. Many believed GE was already global in its mindset and felt that the GGO added a layer of unwieldy bureaucracy to what was already a plenty complicated company.

Jeff's only demand was that the country CEOs figure out a way to grow revenues. "We didn't conjure the GGO out of nothing," Jeff remembered. In 2009—the year before he rolled out the initiative globally—Jeff asked John Flannery, by then a longtime GE veteran, who had started at GE Capital in September 1987 financing leveraged buyouts, to become the CEO of GE India. Flannery, the son of a Connecticut bank president and a graduate of Fairfield University and Wharton, had worked his way up at GE through GE Capital, where he had positions running GE Capital's equity investment portfolio and positions in Argentina and Japan.

Jeff gave Flannery full control of GE's business in India. He could hire and fire at will. Every GE employee in India reported to him. He had his own profit-and-loss statement and could make decisions, such as how capital should be spent or whether to open a new factory or to sell new products, on his own authority, without input or approval from Fairfield. He just had to perform. The longtime unachieved goal was to get revenue in India up to $1 billion a year. "Flannery delivered," Jeff reported. Sales in India increased, as did the sourcing of GE products made in India that could be sold elsewhere—to around $5 billion, from $500 million before Jeff created GE India. Jeff deemed the GE India experience a success and, based on it, decided to roll out GGO worldwide.

He asked John Rice, another longtime GE executive, who was then a vice-chairman after stints in variety of GE businesses, including running the power business, to move to Hong Kong to run GGO. Jeff described Rice as "widely liked and respected," traits he believed important for an initiative that was "bound to break glass and ruffle feathers." He also cited another important attribute of Rice's for selecting him to head GGO: he wasn't a candidate to succeed Jeff, since they were the same age.

Jeff spent $700 million on the new initiative, "funded by the CEO's office," he explained, and he hoped the business unit leaders would view it as a "free resource" that would "drive their growth." But he also suspected there would be resistance from them because by giving the "last word" to country leaders, such as Flannery and Rice, the GGO "would diminish their power" and therefore be a "bitter pill for them to swallow." He told his direct reports, "From now on, when you fight with our guy in Riyadh, sometimes he's going to win."

Jeff counted on John Rice to serve as an authoritarian bridge between the country executives and the business unit executives. It was an assignment destined to require great diplomacy. He hoped the top GE executives wouldn't find their power threatened and instead would appreciate how the initiative would benefit the whole company. But many of his direct reports saw GGO as another of Jeff's flaws as a leader. "What became apparent to me over time," one former GE executive told me about Jeff, "was he never really grew from the national sales manager that he was at Plastics. That was his core competency. It's what he always talks about on the sales side. So the full P&L capability that you've got to have to be the successful CEO—that Jack had, Cote had—where they understood how their initiatives or actions, everything they did, came back, connected to the numbers, the value creation, [Jeff] couldn't connect the dots on that stuff. It's shocking. I don't know how people didn't see it. But he was national sales manager, parading, masquerading as a CEO."

The wives of the GE executives seemed to pick up this flaw in Jeff especially quickly. At the 2013 Christmas party Jeff had at his house in Fairfield, he told one corporate spouse how happy she must be to no longer be overseas, after a stint with her husband and their family. They all enjoyed themselves immensely, but Jeff kept insisting that they must be happy to be back in the United States, where it was safer for them. "Everyone knows it sucked," he told her. On the car ride home, she was incredulous. "That guy is a nut job," she said. "He doesn't listen to anything. It's unbelievable." Suzy Welch had a similar experience at a party Jack was hosting at his Boston mansion, at 40 Beacon Street. Jeff had just been named CEO. Jack had started dating Suzy. Most of the party was on the rooftop, with people mingling and drinking. At the end of the evening, Suzy told Jack, about Jeff, "What an asshole. That guy just talked at me the whole time. He didn't listen to one thing I said. He didn't react to one thing I said. He doesn't listen to anybody." Jack was upset. "I just didn't understand the person," he'd tell people about Jeff.

Jeff brooked no dissent about the GGO, and it cost John Krenicki, a GE vice-chairman who was also the CEO of GE's energy business, his job. GE recruited Krenicki in 1984 directly from the University of Connecticut, where he had studied engineering. He was recruited into a two-year rotational assignment as a sales engineer in the power business. He worked in a field sales office in Florham Park, New Jersey, supporting customers such as ExxonMobil in their oil and gas refineries in the state. "It was a very tough period in that business," he recalled. "I was bored to tears." At a random New York charity dinner, to which he and a group of other young GE trainees had been invited, he met a GE executive who worked in the plastics division. A month later, he was working in a sales job in Plastics, "when Plastics probably felt a lot like Apple today," he said. He moved around with the plastics business seven times, from Syracuse to Avon, Connecticut, and eventually to Pittsfield, the Plastics headquarters. His boss's boss was Jeff Immelt.

"I knew Jeff when he was a driving a Ford Taurus," Krenicki said. Jeff was not yet a GE officer. But along with many others, Krenicki felt that Jeff had "been anointed" coming to GE out of Harvard Business School "and was on Jack's radar." He liked him. Plastics was "in growth mode," he remembered. "The business was doing well." They called on customers together. "I wouldn't say I was his best buddy by any stretch, but I had a very good relationship with him," Krenicki said. When Jack asked Jeff to go to the appliance business, he was replaced by Dave Calhoun. Krenicki worked for Calhoun for a year. He was then asked to work in the silicones business, which was part of the plastics group. "It was a killing field," he recalled. "People just struggled. It was tough."

After five years in Silicones, he returned to the larger plastics business. GE also paid for him to get an MBA from Purdue University, which he did remotely at night and then by showing up in Indiana for two weeks every other month. He believed he was in the "pile" of talented young GE executives at the business level, not at the corporate level. "I was so far below an officer of the company," he said. In the mid-1990s, he got his first chance to run a business—the European division of Silicones, "a $125 million P&L in a business that most leaders were failing in," Krenicki said. He initially worked for Dave Cote, then Tom Tiller, who was running Silicones at that time. "They went through vice presidents like you go through socks in that business, because it was a complicated business," he said. "It was never big enough to where it could move the needle for the company. I think Jack used it to test a lot of officers, to see what they were made of." Before Jack moved Cote to Appliances, he and Jack had discussed that to be competitive in Silicones, GE needed to invest around $500 million in a new manufacturing plant—a risky bet, especially for a business that was generating only $150 million in annual revenue.

Twice a year, Jack would travel to Europe to administer performance reviews. At his 1995 performance review with Jack, at Le Bristol

Hotel in Paris—the first time Krenicki met Jack—Jack told him he was nixing the idea of the new silicone manufacturing plant. On the other hand, Jack asked him, what would he do—could he do—with $50 million, if Jack were to make that capital available? Jack knew the business, of course, because he had started in Plastics and run it for years. Krenicki told him exactly what he would do with $50 million, right down to the line-item detail. Jack loved it. He invited Krenicki to lunch at Le Bristol. "Candor was the basis of a lifelong friendship," he said. At the time, GE was the number-four player in European silicones. Rather than spend the bulk of the $50 million, Krenicki came up with the idea of combining GE's silicone business in Europe with Bayer's to become the number-two silicone business in Europe. GE invested $100 million to get its ownership stake to 51 percent. "It was a home-run deal," he said. From that moment forward, Krenicki was on Jack's radar.

In 1999, Jack made him a GE officer and asked him to run the North American lighting business, based in Cleveland. Revenues were $2.5 billion and it was profitable. He worked for Mike Zafirovski, who had replaced Calhoun, who had replaced McNerney. The succession battle was in full swing. He was just generally thrilled with how things were going for him at GE. Jack wrote him letters explaining that he promoted him to officer because at Silicones he had "made something out of nothing." He was happy in Cleveland. But soon enough, he was asked to show up in Fairfield. John Opie, a GE vice-chairman, asked him to run the once highly controversial synthetic-diamond business, which was also a division of Plastics. He accepted and moved to Columbus. Two months later, Gary Rogers, who was running Plastics, asked him to return to Europe to run the business there. "My wife was pissed that we were being asked to move back to Europe again," he recalled. Krenicki turned it down. "Gary, three tours of duty in Europe in like five years," he told Rogers. "I've got to say no to this one. I can't do it." Two weeks later, Jack called him. "You're working for me," he told Krenicki.

Jack wanted him to run the locomotives business. He replaced John Rice, who had replaced Calhoun, who had replaced Nardelli. Krenicki couldn't believe how quickly his career had advanced inside GE after his first meeting with Jack five years earlier at Le Bristol. "That was the greatness," he said. "It was absolutely amazing. You got your own game. You're working for the CEO. Wow." But it was a down cycle for the business. "I was always the guy who always had a down cycle," he said, "and had a job that was never the job with a lot of momentum." At one point during a meeting in Fairfield, Jack kicked everyone but Dammerman and Conaty out of the room and asked Krenicki which of the three finalists for Jack's job he would choose. He said that he had worked for one only of the three—Jeff—and while he knew the other two, he didn't feel comfortable opining on them. "That was the honest truth," he said, adding he couldn't believe that, at thirty-seven years old, he was being asked to weigh in on the succession process. "My guess is he asked a hundred thousand people," Krenicki joked.

In 2003, Jeff named Krenicki head of Plastics. He turned it around. In the 2003 annual report, Jeff referred to Krenicki as "one of the company's best young leaders." In 2005, Jeff promoted Krenicki again to run GE's energy business, which was three times the size of the plastics business. In 2008, Jeff named him a GE vice-chairman, in addition to continuing to run the energy business. Under Krenicki's leadership, GE's energy business doubled in size and profitability, generating $50 billion in revenue and nearly a third of GE's profits. At one point, in 2011, the publicity-shy Krenicki allowed Beth Kowitt, a *Fortune* reporter, to accompany him on a business trip to Africa. She asked him whether it was his ambition to become CEO of GE. He'd been there twenty-seven years and was clever enough to evade the question. "It's a complicated job," he said of running the energy business. "It's a great portfolio, so I don't look beyond that."

But suddenly, thanks to Jeff's GGO initiative, Krenicki was out.

Jeff explained that Krenicki was one of the few senior GE executives who was not on board with GGO. Krenicki had been Flannery's boss during 2009 when Jeff sent him to India to test out GGO. Some eighteen months later, Jeff called his vice-chairmen together, along with other senior GE executives, in a conference room to decide whether to spend an incremental $750 million to roll out GGO on a global basis, given the success of GE India under Flannery. It was a meeting to prepare to get board approval for the GGO initiative. According to someone in the room that day, Krenicki shared with Jeff his displeasure with GGO. "Jeff, this is the end of the company," Krenicki said. "These businesses are complicated. They're big. They're global. And you're going to put people on top of the businesses who do not have the domain knowledge of those businesses to make these calls." Krenicki believed GGO would dilute materially the deep technological and commercial expertise that had been built up in individual businesses about GE's sophisticated customers, such as ExxonMobil and Tokyo Electric, and the nuances and subtleties of the relationships and risk-management processes that the GE executives in the business units had developed over long periods of time. Suddenly, with GGO, a local GE executive in, say, Egypt, could overrule the business leader who was accountable for the bottom-line results. "This is the stupidest idea," someone recalled Krenicki telling Jeff. "Dumb as hell."

Jeff was pissed. "[Krenicki] was an incredible operator," Jeff wrote in his book, "self-disciplined and highly organized and I respected him immensely." He wrote about how he had known Krenicki for thirty years and promoted him three times, including to vice-chairman in 2008. But, Jeff continued, his ire growing, "there was one key area in which Krenicki fell short: when he dug in on something, it seemed to me he would refuse to collaborate with others unless he got his way." He wanted his vice-chairmen to be "big-picture thinkers" who could not only do their day jobs but also help him think "horizontally" across the company. "These people took good ideas and made them better (while deconstructing and discouraging bad ideas)," Jeff continued. On the heels of the financial crisis, Jeff was "wounded," he wrote, and needed his advisers "more than ever." Krenicki "never made that list," he continued. On the topic of GGO, Krenicki "refused to suit up," Jeff remembered, defaulting to a sports metaphor. "And he wasn't quiet about it. His feeling was that he knew his business best and he shouldn't have to suffer the input of anyone else."

Jeff grew impatient with Krenicki, someone with a legitimate chance to succeed him. "We fought over everything," Jeff said. Jeff complained to him about how GE was "falling behind" in gas-turbine technology. But when Jeff raised it with him, "he made it clear I should butt out." He wanted Krenicki to help him develop a talented young executive—Lorenzo Simonelli, who became head of GE's oil and gas business and then the CEO of Baker Hughes after it was spun out of GE—but Jeff said Krenicki wouldn't do it. "Years before, I'd considered him a great candidate to succeed me," Jeff explained. "But the way he behaved as he rose within the ranks, he took himself out of the running." The final straw for Krenicki came when the energy business missed its numbers in 2011. It was a tough market for GE's power business, but Jeff was unsympathetic. "Why do I put up with all the guff he gives me," Jeff wondered, "when he doesn't hit the numbers either?" This assertion was a bit disingenuous on Jeff's part. Explained one GE power executive, power was "a long-cycle business," and both Jeff and Krenicki knew 2011 would be tough but that 2012 would be much better, and it was. "In fact, it was the all-time record the business ever had, and he knew that, too," the executive said.

But that didn't matter. Jeff had made up his mind about Krenicki. In February 2012, two months after the glowing *Fortune* article appeared— which Jeff likely did not appreciate either—Jeff told Krenicki to take his time but to find a new job. Jeff recalled that Krenicki looked "shocked" by the news. "John," he told him, "I made you vice chairman because I wanted to keep you in the company. But as part of that, it would have been great if you'd helped more broadly." (Krenicki declined to be interviewed about his departure from GE.) In *Hot Seat*, Jeff defended his decision to fire Krenicki. There were "many in the company who second-guessed my decision" about Krenicki, he recalled. "His skills might have helped us over time, especially as we hit tougher energy markets. But I never looked back." Jeff told me about Krenicki, "He never would do anything to make my life easier. Now, when you are hitting your numbers, that's okay. Like [Gary] Wendt didn't make Jack's life easier, but he always hit his numbers. John didn't do anything to make my life easier, and when he started to miss his numbers, I said, 'Life's too hard.'" But Jack told me that Jeff got rid of Krenicki because he threatened Jeff. "Everyone was saying he was good," Jack said about Krenicki. "Jeff didn't want anybody around there that had any stature and could fight him. I wanted every fucking smart guy I could get my hands on."

In many ways, Jeff's firing of Krenicki was illustrative of the power struggles that happen regularly at corporations and businesses around the world. The CEO expects loyalty, of course, and is entitled to get it. A business is like a crew boat: everyone must be rowing in sync in the same direction to get to the destination ahead of the competition. That was probably especially true when trying to implement highly disruptive, and expensive, new global initiatives. But one test of leadership often becomes how much dissent a leader can tolerate, or encourage, before jettisoning those who are sowing the seeds of resentment. Jack's tolerance for "wallowing"—mixing it up intellectually with all comers before reaching a decision—seemed to be infinitely higher than Jeff's. And that impatience or intolerance or need for obeisance had cost Krenicki his job, and perhaps GE a valuable future leader. "I didn't expect our leaders necessarily to be loyal to me," Jeff continued. "I did, however, insist that they be loyal to our collective mission." But as one GE executive recalled about the two men, "It's absolutely true that John and Jeff were oil and water."

After Krenicki got back from Davos in January 2012, the discussions between him and Jeff about his departure began in earnest. They wanted to try to make his sudden removal seem as seamless as possible, as if it were no big deal. Observed Jeff, "Unlike in Jack Welch's day, when all hangings were public hangings"—an exaggeration at best—"during my tenure we tried to let people we'd fired maintain their dignity." Krenicki came up with the idea of presenting his departure from GE as a cost-saving move, a bureaucratic delayering that would save GE a bunch of money. Jeff latched on to it. He decided to have Krenicki's three deputies—among them Steve Bolze, who ran Power and Water—report directly to him, as of the July 2012 announcement. "We were just removing a layer," Jeff recalled, "saving $1 billion in the process." That's what he said publicly. "What I didn't say publicly was that for years, this particular layer"—Krenicki—"had made every problem inside the company harder." The *Fortune* article didn't help, nor did its subheadline, "John Krenicki is the most important executive at GE's most important business. Could he run the whole company someday?" Jeff was also said to be miffed about another article, in an industry magazine, where Krenicki supposedly said something about how he thought he could run GE better than Jeff.

Ed Galanek, who was one of Jeff's closest security guards, flew around the world with him, and knew him as well as anyone inside the company, remembered hearing from Gary Sheffer, GE's communications chief, when the industry magazine story came out. "What a fucking scumbag Krenicki is to fucking say that," Sheffer said to Galanek. Galanek admired Jeff's restraint. "Jeff doesn't kill him," he said to me. "I'd have killed him. That week he would have been dead to me. Instead, Jeff kept him around for a year and then broke that business up into pieces. So though Krenicki broke Jeff's heart, Jeff never told Krenicki in a year's time, while he was there, going down to the end. He just was not great to him, but he never fucking told him, 'You fucked me. That was a scumbag move. That was disloyal.'" He said Jack would have handled Krenicki very differently. "Do you think Jack would have let somebody that runs a business of his say that he could do a better job than Jack and have the job the next day?" he asked rhetorically.

Galanek said Jeff prized loyalty almost as much as any other quality in his senior team. "That word in his life—'loyal'—has fucked him like nobody's business and has always been his concern," he said. "He was always concerned about loyalty." He said Jeff always prided himself on being loyal to Jack, even when Jack bad-mouthed Jeff, including so publicly on CNBC after the first-quarter-2008 earnings miss. "Jeff felt he was extremely loyal to Jack," Galanek continued, "and that he wanted to prove that he could do good. But he wasn't gonna be Jack. Then when Jack motherfuckered him [on CNBC], Jeff would never answer back when people would say, 'What do you think? Jack Welch just said this about you.' Because you know what? He had to be loyal to Jack. General Electric is one of the greatest American companies ever. It's the company of Thomas Edison, for Christ sakes. And that a brilliant guy—Jeff is brilliant—a brilliant guy that put a million hours of work in and tried his ass off to do everything he could, yet things still failed. And he couldn't stop it. . . . The whole Krenicki story puts it on a plate for you, because he was a guy being disloyal. Jeff did not know how to handle disloyalty. He wanted loyalty, and loved it if he got it, but didn't know how to handle disloyalty."

As many other departing senior GE executives had before him, Krenicki turned to Jack for advice about what to do next. Jeff had suggested to Krenicki that he could run GE Capital. But that was a nonstarter for a variety of reasons: While Krenicki had been on the internal GE Capital board—so he could weigh in on the occasional energy financing—he knew little, if anything, about financial services. Nor was he even remotely the right person to lead GE Capital into the new era of government regulation that was unfolding between Washington and Wall Street. Krenicki was an industrial guy. Not a finance guy. Jeff probably knew Krenicki would turn the offer down. And he did.

Krenicki called Jack. "John, what took you so long?" Jack told him. "I've been waiting for this call." Jack offered him the typical entrée to the top headhunters, Gerry Roche and Tom Neff. He also offered to introduce him to Don Gogel, the CEO of Clayton Dubilier & Rice, the New York private-equity firm where Jack had been a senior adviser for a decade. "Don and I would love for you to come work for us at CD&R," Jack said. And that's what Krenicki did. He's been an operating partner at CD&R since he left GE in 2012 and is currently vice-chairman of the firm. He signed a three-year noncompete agreement—meaning he wouldn't work for a rival of GE's energy business—with GE and received a severance package that paid him $12.9 million—$89,000 a month for ten years plus a cash bonus of at least $2.9 million.

If the loss of Krenicki was an unintended consequence of Jeff's GGO initiative, Jeff could live with it. He had no regrets. "While it was controversial within the company," he remembered about GGO, "there's no question that it paid off." When he became CEO, he noted, 70 percent of GE's customers were in the United States. By the end of his tenure, in 2017, GE was operating in 180 countries—out of 195—around the world and 70 percent of its revenue came from outside the United States. He cited 26 countries where GE's revenue was more than $1 billion. He took a GGO victory lap.

GGO was not Jeff's only big initiative, of course. He'd also made some aggressive forays into beefing up GE's digital business. He launched Predix, a proprietary cloud-based software that "enabled industrial-scale computing" both inside GE and beyond. Jeff reported that Predix logged "every aspect of a machine's life—from initial assembly to repairs to replacement parts." Jeff often boasted that Predix could monitor performance "while a jet engine was in the air" or could "tilt turbine blades to increase power output while they were spinning." GE bought an office building in San Ramon, California, on the outskirts of Silicon Valley, gutted it to make it hipster friendly, and began a hiring spree of some three hundred expensive Silicon Valley engineers, about one third of the way to its goal of having a thousand engineers in Silicon Valley. He hired William "Bill" Ruh, a former Cisco Systems executive, to run GE Digital, GE's Silicon Valley soft-ware and data analytics center. Jeff staked Ruh with some $200 million early on, and GE earmarked another $100 million of equity to invest in promising Silicon Valley start-ups.

In September 2011, Jeff lent his prestige as the CEO of GE to the Obama administration after the president asked him to lead his Council on Jobs and Competitiveness. Although there certainly was a tradition of GE CEOs serving and advising presidents of the United States—"Electric" Charlie Wilson during World War II and Reg Jones advising Jimmy Carter—what was unusual about Jeff's decision was that he was a Republican and Obama was a Democrat, and a progressive-leaning one at that, who had repeatedly during his first term criticized big banks and big business. He agreed to take the role in part because he wanted to do what he could to replace the jobs lost as a result of the financial crisis—300,000 new jobs a month would be needed to get the country back to the level of employment prior to the crisis—and because he wanted to repay with his time and energy the help the federal government had provided to GE during the financial crisis. "This was the least I could do to pay some of that back," he recalled. When Jeff told his parents, back in Cincinnati—whom he described to *60 Minutes* as "real right-wingers," and his father as someone who watched five or six hours of Fox News a day—that Obama had asked him to head up the new jobs council, his mother said to him, "Well, you said 'no' of course, didn't you?"

JEFF ALSO WANTED TO CHALLENGE HIMSELF PHYSICALLY. IN THE SUM-mer of 2011, he decided to climb Mount Kilimanjaro, in Tanzania, the tallest mountain—at 19,340 feet—on the African continent. While a former college athlete who remained active—playing golf, mostly, or putting in an hour a day on the treadmill—Jeff was not a mountain climber. The impetus for the trip came from his daughter, Sarah, who had graduated the year before from Hamilton College. Of course, Jeff loved his daughter, his only child, but had been a somewhat neglectful father, given the demands of his job running GE. He wanted to make it up to her. He told Sarah that after she graduated from Hamilton, he would do with her whatever she wanted. He figured she'd want to go to Hawaii or something like that. "Instead," Jeff remembered, "she said, 'Let's climb Kilimanjaro.'" He continued, "I kept waiting for this idea to dissipate, but she stuck to her guns. To set the scene for you, we're a family that has never even gone camping. The closest I have come to nature had been a golf game."

Suddenly, the fact that the CEO of GE and his daughter were planning to climb the highest mountain in Africa required a mobilization of both logistics and security—all of which was to be paid for by Jeff, not GE. "Of course," Jeff once explained, "we took the fat-cat special—we had twenty Sherpas and stuff like that—but you still have to make it to nineteen thousand five hundred feet." The Sherpas carried their possessions, set up their individual tents, and cooked their meals. Joining the two Immelts on the expedition was one of Sarah's closest childhood friends.

Jeff's trip to Kilimanjaro coincided roughly with a changing of the guard in his GE security detail, the men and women who traveled with him everywhere and who kept him on schedule and safe. Such protection, including the requisite fleet of private jets, had by the twenty-first century become de rigueur among a wide swath of corporate America and the Wall Street crowd. GE, of course, had been among the earliest adopters of the private-jet fleet to ferry around corporate executives quickly, efficiently, securely, and discreetly. At the time of the trip to Kilimanjaro, Ed Galanek was taking over Jeff's security detail from his predecessor, Bob Connolly. Galanek had been working for a private security company that had been hired by NBC to guard *Saturday Night Live*, in Rockefeller Center. He provided security for GE board members meeting at 30 Rock. He once protected Jeff on a walk from 30 Rock to the Waldorf Astoria for an event. They got to talking about sports and *Saturday Night Live*. They hit it off. He protected Jeff at another event in New Jersey. Soon, Galanek traveled with Jeff to Abu Dhabi and then Kuala Lumpur. He was still working at *SNL* and chatting with Jeff about NBC and sports. After Galanek had been working for Jeff as part of the private security company for about a year, Connolly, who was looking to retire, asked him if he wanted to work at GE full time as part of Jeff's security detail.

At GE, Galanek's job was to "set up all logistical moves" for Jeff. Anytime Jeff traveled, either domestically or internationally, Galanek's job was to arrange the cars and the planes for the trip. He worked closely with GE's Corporate Air Team, or CAT, to arrange the flights and the times of travel, making sure the jets and pilots were available. He said that Jeff's schedule was so packed on a daily basis, "going to so many places in a short amount of time," that it fell to Galanek to make it all work logistically. "As anybody that does close protection, dignitary protection–type stuff knows, the job isn't when you're working at the moment," Galanek explained. "The job is done ahead of time. Everything is planned in advance."

Over time, his job also evolved to include being Jeff's unofficial gatekeeper on overseas trips. He would meet with the GE country leaders and review the agenda for when Jeff was there. He would eliminate the time set aside for thirty-minute prep sessions and that sort of thing. "I'd look at their itinerary," he explained, "and go, 'No. He doesn't need a half-an-hour debrief on this. He's a real smart guy. You sent them a playbook already. He'll want to punch you in the head if you fucking want to talk to him for a half an hour about a subject he knows already.' I was like, 'You've got to build in time for this guy to go take a shit.'" Later in Galanek's tenure with Jeff, when Jeff discovered that Galanek had been his traveling gatekeeper, he was surprised.

He didn't know Galanek had been doing that for him. "What the fuck?" Galanek told him. "How do you think this has been happening? Don't you look at your itinerary? How do you think it's been knocked down to five minutes? Somebody's got to do that." Jeff started laughing.

About a month into working full time for Jeff, Connolly told Galanek about the Kilimanjaro trip. "He's gonna be climbing Kilimanjaro," Connolly said, "and he's gonna need security." Galanek said he would coordinate with the GE executives in Kenya, who he believed would also cover the situation in Tanzania. He explained to the GE executive that on the appointed day and time, he needed to be in Tanzania to meet Jeff's jet. "Just make sure he gets there," Galanek told him, "and be around in case some shit happens to him." About six weeks before Jeff and Sarah's trip to Kilimanjaro, Jeff was in San Francisco. Galanek was with him. "Hey, Ed," Jeff said, "I don't know if you know. I'm going to be climbing Kilimanjaro on my vacation. The board says I have to have security." Galanek knew about the trip, of course, and told Jeff he had arranged with the GE country manager in Kenya to meet him in Tanzania and to be helpful. "No, no, no," Jeff told him. "I need security for the climb."

"Jesus, I don't know if Michael"—the GE executive in Kenya—"is a fucking mountain climber," Galanek replied.

"No, no, no," Jeff said. "I want you to go."

Galanek was stunned. "That's not me," he told Jeff. "I'm not an outdoorsman. I was born and raised in the Bronx. My idea of water sports is turning on the hydrant. So not me."

"No, no, no," Jeff insisted. "We really want you to go. Sarah loves you." Galanek knew Jeff's daughter from providing security at 30 Rock. She had worked as an assistant for Lorne Michaels, the longtime *SNL* executive producer, and as an intern for Jimmy Fallon at *The Tonight Show*. "Nobody had any idea she was Jeff Immelt's daughter," Galanek said. "There she is walking down the fucking hallway, carrying two cases of water, as opposed to Jimmy Buffett's fucking niece, who is trying to get on camera."

"No, no, but you're so funny and we want you to go," Jeff reiterated to him.

"I think it's really unfair to kill me, just to have a fucking court jester to go up a mountain," Galanek told Jeff. "I don't understand your reason. Why don't you and Sarah just get tomatoes and throw it at me one day and we'll call it even?"

"No, no, no," Jeff repeated. "We want you to go. Come on, come on."

After Galanek had deposited Jeff safely in his hotel in San Francisco, he called Connolly to see if Jeff was just joking around or whether he would actually have to climb Kilimanjaro. "I think you have to go," Connolly said. "I don't think it's an option. Let me check into it." Connolly called him back and said that he had to go. "Fuck," Galanek responded.

Galanek's vacation schedule at GE mirrored Jeff's. He could take vacation when Jeff did. He was planning to take off the week that Jeff was climbing Kilimanjaro. His wife had arranged for them to go on a cruise with their daughters. "I would rather you punch me in the face repeatedly than go on a fucking cruise ship, but this is something she likes," he said. But since he was having to climb Kilimanjaro with Jeff and his daughter, he told her he couldn't go on the cruise. His wife responded by telling him she was going to throw his "shit" on the lawn if he didn't go on the cruise. "Look, it's not something I'm looking forward to, but I've got to do it," he told her.

A few days before he had to leave for the trip, Galanek took one of his daughters, Mary, and her friend Christine to meet Will Ferrell, who was at 30 Rock for a show. Galanek had known Ferrell for years. They were milling around waiting to meet Ferrell when suddenly Galanek heard a booming voice. It was Ferrell: "He can't fucking climb Kilimanjaro." Ferrell walked around the corner, saw Galanek, and said, "You can't fucking climb Kilimanjaro."

"Will," he replied, "I'd like you to meet my daughter Mary and her friend Christine."

Ferrell, who is around six feet four inches tall, bent over, hands on his knees, and looked into Mary's eyes and said, "Your dad can't fucking climb Kilimanjaro." Everyone was hysterical. "Ed, are you crazy?" Ferrell continued. "How are you climbing Kilimanjaro?"

"I started working out," Galanek replied. "I should be fine. Mary, tell Mr. Ferrell how Daddy's been working out."

"Well," she said, "Dad doesn't drive to the bagel store. He walks there now."

"And does Daddy buy bagels?" Ferrell asked.

"Why the fuck would I go to the bagel store if I'm not gonna buy a bagel?" Galanek replied. "I eat a bagel on the way back when I'm walking."

It turned out that walking back and forth to the bagel store instead of driving was insufficient training for Galanek. The hike was planned for Sunday to Friday, giving the group time to acclimate. "We're seeing bodies get carried off, going past us, going down, which was fucking something in itself," Galanek said. He got extremely ill. He suffered from severe dehydration. He could not urinate for nearly two days. "I had the fucking shakes," he recalled. He lost eighteen pounds in five days. "I'm hurting at this time," Galanek continued. "[Jeff's] eating like he's sitting down in a fucking restaurant in Dallas. He looks over at me and he goes, 'So how you doing?'"

"How am I doing?" Galanek replied. "How the fuck do you think I'm doing? I haven't pissed in a day and a half. I keep feeling like I'm gonna shit my pants. All I can eat is fucking banana chips and sip on a fucking cup of tea. How am I doing? My wife is on a fucking cruise ship. I'm figuring she's having shrimp scampi or a fucking shrimp cocktail. And I'm hanging out with you on a mountain, fucking dying. You fuck." That night Jeff came by Galanek's tent and woke him up inadvertently. "I'm sorry if I woke you up," Jeff told him, "but I woke up in the middle of the night and I couldn't stop laughing, thinking about you with your wife on the cruise ship."

"You fucking prick," Galanek said.

Still, he couldn't quite get over the turn of events. "I was laying on that mountain and I said, 'You know, I do have a fucking interesting life. I'm on fucking Mount Kilimanjaro with the chairman of the board of General Electric. What the fuck?' It was like, this is some fucking crazy life." Galanek didn't make it to the top. One of the guides said he would pay extra attention to him to get him to the top along with the other three. He decided to stay put at 16,800 feet while the others went to the summit. "Listen, this shit was not on my bucket list," he told the guide. "This is about Jeff and his daughter doing this. I'm here to make sure if something happens to him, I'm here to take care of him. If something fucking happens to me, what are we doing here? Why am I here?" He went down, which was also a struggle. "The thing that ended up killing me was when we walked down the last day," he said. "It took six days to get up to that point. It took one day to come down. But the going down, that pounding into your thigh like every step you go down . . . like what your feet were stepping into. And who the fuck does that six hours a day?"

The descent was difficult for Jeff, too. "The altitude didn't bother me," he explained to Michael Useem, a professor at Wharton. "I'm a guy who does roughly an hour on the treadmill every day. Getting to the summit wasn't nearly as hard as I thought it was going to be. But then you start coming down, and I must have fallen on my butt five times in the first ten minutes. The guides have this expression in Swahili that's 'pole, pole,' which means, 'slowly, slowly.'" Jeff said that climbing the tallest mountain in Africa taught him about the importance of persistence. "As a CEO, you are running a marathon," he said. "Don't let the organization backslide. Set new goals. Every day has to be better than the day before. This notion of 'puli, puli' has to do with resilience, persistence, and sticking with your vision and your goals.

That helps you when you're at seventeen thousand feet on the way up, and it helps you when you're running a company. Essentially anything you want to do that is meaningful in life must be done over time. If you want to change big institutions, you've got to have incredible persistence and constancy of purpose. That's what I learned."

For all the time Galanek spent with Jeff flying around the world, they never socialized. They never had a drink together until after they all made it off Kilimanjaro. Then Galanek had a beer and Jeff had a glass of wine. "When we came down, our bodies were broken," Galanek said. "It might have been the greatest-tasting beer I ever had in my life when I got down there, and he was sipping a glass of wine. I asked him, 'Why wouldn't you want a fucking beer now, man? What's with the fancy-ass wineglass?'"

"I've got to tell you," Jeff said. "I haven't had a beer since I was in college."

Jeff couldn't wait to get back to work, Galanek recalled. Being out of touch with GE had been difficult for him. The first thing he did was call his wife and tell her that both he and Sarah were well. Then he called John Rice, one of his vice-chairmen. "I'm still in charge," Jeff told Rice. "What's going on?"

STILL, GE'S LONG-RUNNING PROBLEMS CONTINUED TO HOUND HIM. Would the persistence he supposedly learned climbing Kilimanjaro help him cure what ailed GE? Jeff's appointment as Obama's "job czar" resulted in the fast-circulating rumor on Wall Street that maybe Jeff was looking for a graceful exit. NBCUniversal had been sold to Comcast at a significant discount, many observers believed, to its actual value. And GE was no longer the most admired company on earth or the one where most business students wanted to work. And of course, GE's stock was then trading around $16 per share, less than half of the $40 price when Jeff took over from Jack.

As Jeff had promised he would reduce to 30 percent, from 45 percent, the percentage of GE's earnings that came from GE Capital, he had to take some steps toward beefing up GE's other lines of business. There were three ways for Jeff to get GE Capital down to his stated goal of 30 percent of GE's earnings: shrink GE's financial business, grow GE's industrial businesses, or a try to do both things at once. Jeff decided to try to do both things at once. He liked to talk about GE having a "power island"—in effect, offering customers a one-stop shop for all their power-generating needs (gas turbine, boiler, steam turbine, and generators)—and he believed that buying a bunch of power-generation and transmission businesses from Alstom, the big French industrial company, might be a major step in getting GE's stock price out of the doldrums. The deal would not be simple. It would rival in size and complexity GE's Universal acquisition, also from a French company.

In January 2014, Patrick Kron, the CEO of Alstom, called Steve Bolze, the head of GE's power business following Krenicki's departure, to broach the idea of GE buying most of Alstom. John Flannery, then Jeff's business development leader, and Rob Duffy, a business development executive in the power business, followed up the next day with Kron's business development leader to gauge Alstom's seriousness about trying to do a deal. It turned out Alstom was genuine about selling and was solicitous of GE. Kron also called Jeff to formally discuss the idea. "GE Power was the most successful competitor in the sector by almost any measure," Jeff recalled. The combination would have merged the first-largest—GE—and fourth-largest—Alstom—power-generation and transmission companies in the world and put a significant distance between GE and its two rivals, Siemens in Germany and Mitsubishi in Japan. Kron "put the company in play," Jeff said. "We were ready. And we went for it."

Jeff believed he had these two competitors "on the run," especially after GE's launch of the "H," the industry's largest and most fuel-efficient turbine. Thanks to GGO, Jeff also saw that GE was winning more than its fair share of new contracts from overseas customers "that had previously gone to our rivals." Alstom also had a big high-speed rail and signaling business, which presumably would have fit well with GE's locomotive business, but Jeff didn't want it. Jeff was interested only in Alstom's power and transmission segments. In the past, Kron had been unwilling to sell those parts of Alstom. But he had changed his mind, apparently. "We had looked at Alstom over the years," Jeff told me. "Everybody in your industry, you kind of know." In 1999, GE had bought a portion of Alstom's turbine-manufacturing business.

For years, GE would just sell, say, a gas turbine and then drop-ship it to a customer and let the customer take it from there. Jeff's idea for the "power island" was for GE to sell customers everything they needed to generate electricity, including construction management and servicing. The executives inside the power businesses did not want to be in the construction-management business. It was low margin, and if you screwed up, you could end up owing a ton of money. But Jeff had a different view. "If you're already in that neighborhood, why not pick up some extra business?" was Jeff's thinking, according to one GE executive. But there was also more risk in becoming more vertically integrated. "In order to keep getting to that trough, you had to do more and more complicated stuff and risky stuff," explained the GE executive. "There's a broad trend in the industry for a long time that business is getting harder for OEMs [original equipment manufacturers], forcing them to do more aggressive things."

Based on the call between Kron and Jeff, a meeting was set up for Paris in February for them as well as one of Kron's deputies, plus Flannery, Bolze, and Jeff Bornstein, GE's new CFO. Jeff was on his way to Sochi, Russia, for the Winter Olympics and also had a few other stops on the trip for other possible M&A deals. The dinner was at the luxurious Le Bristol Hotel in Paris. "We all came away intrigued," Jeff recalled. Alstom also had an offshore wind business that would be a nice complement to the wind business that GE had bought from Enron years earlier and that was still thriving. Kron also made it clear that he wanted an answer from GE quickly. One way or another, he was determined to sell the business, and if GE wasn't going to buy it, then Kron figured Siemens or Mitsubishi would.

Jeff made the decision to buy Alstom seem like a no-brainer. But others inside GE wondered about the wisdom of the Alstom deal—albeit in retrospect. "Your energy industrial business is highly complex, technologically intense, and your customers are in emerging markets that can't pay you, and it's shrinking," one of them told me. "That's your business." He couldn't get over the fact that Jeff seemed determined to buy Alstom and to dilute GE Capital, the earnings machine, in one fell swoop. It also seemed as though Jeff had forgotten the headaches of trying to close a deal with the European Union that Jack had experienced when he walked away from the Honeywell acquisition a decade earlier. Another former GE executive said Jeff's decision to buy Alstom was just another one of his strategic mistakes. "From the moment he came in, he basically was the anti-Jack," he said. "Every decision he made, if all you did was say, 'What would Jack do, and what's the opposite of what Jack would do?' you could pretty much, with a ninety-eight-percent confidence level, predict what this guy would do. There must have been some deep-seated self-loathing or resentment, because most of those moves were the wrong moves."

The Alstom deal progressed quickly after the February meeting in Paris. It was a big deal—GE's largest industrial deal—and it cut across GE's business divisions—70 percent of what Alstom was selling would go into the power business that Bolze was running, while the remaining 30 percent was a power-grid business, which was then part of a different GE division, energy management. That meant that Jeff would be intimately involved, as would both Flannery and Bolze.

Bolze was the all-American boy. He grew up in Bethesda, Maryland. His father was an antitrust litigator at a Washington, DC, law firm. After graduating in 1981 from Walt Whitman High School, he was off to Duke, where, in 1985, he received his degree in electrical engineering. During his last two summers at Duke, Bolze worked at the NASA Goddard Space Flight Center, in Greenbelt, Maryland. After Duke, he joined Westinghouse Electric's two-year rotational training program. He was a program manager in the communications department, near the Baltimore airport, and designed chipsets for radar systems. He also worked in quality control in a semiconductor lab. He decided he wanted to be in management and enrolled in the two-year MBA program at the University of Michigan.

After his first year at Michigan, he worked for GE Lighting, in Cleveland "of all places," he said. That's also where he met his future wife, who was in the MBA program at Northwestern. They married after they both graduated. "Neither of us went to GE" after getting the MBA, he said. Bolze spent the next four years at Corporate Decisions, a small consulting firm in Boston that was later bought by Mercer. In 1993, Bolze joined GE in its corporate business development group, working for Gary Reiner, Jack's M&A guy at the time, in Fairfield. Reiner hired about ten ex-consultants a year. Bolze was one of them. "The goal was to help the businesses with lean production, quality, procurement, and special projects," he said. "It's like all internal consulting. They dispatch you to a business. You work with the business. I worked with Power Systems, Locomotives, Plastics, and then for the last year I did mergers and acquisitions." As in his consulting days, he'd fly out of Fairfield on Monday and stay at the operating division for the rest of the week, returning home on Friday. He spent most of his time on projects for Bob Nardelli, first in Locomotives, then at Power Systems.

In 1995, Bolze went to work for Nardelli in the power systems business in Schenectady as his strategic planner. He moved with his family to Albany, from Connecticut. He said his first years working for

Nardelli were some of the "darkest years" in the power business. "When Nardelli came in, in late 1995, the business that year was supposed to make $700 million and they had a major miss," he said. "They made around $285 million. It was a big hit." He rotated through the steam turbines business and then through the business of providing services to the power customers. Two years later, Bolze got his own business to run, a joint venture with Harris Corporation, based in Melbourne, Florida. He grew the business from $50 million in revenue to $400 million. He had about five hundred people working for him. Toward the end of his tenure working on the joint venture, GE moved him to Atlanta with his family. He continued to run the joint venture and other businesses in his group from Atlanta. After about seven years in Power Systems—and when he was about 75 percent done with building a new house—GE moved him to Milwaukee to work in GE's health care division. He was in charge of the functional-imaging business, which included PET/CT scanners and nuclear medicine equipment.

By that time, Jeff had moved out of Healthcare to become the CEO. Bolze was working for Joe Hogan, who took over running Healthcare after Jeff moved to Fairfield. A few months after his family moved to Milwaukee from Atlanta, GE announced the $9 billion acquisition of Amersham. Jeff and Hogan wanted Bolze to lead the complex integration of Amersham into GE. To do that, he had to move to London. He remembered having to tell his wife the news. "She literally is taking the kids around to school," he said. "We'd been there a couple of months. Milwaukee is a very friendly place. It's really fine, cold, but fine. I said, 'Well you're not going to believe this one,' as she's picking out window shades and stuff. 'I'm going to London.'" He spent the next year and a half in London, integrating the Amersham deal alongside Bill Castell, the former Amersham CEO whom Jeff had tapped to run all of GE Healthcare, the headquarters of which was moved to London—

the first GE business to be headquartered outside of the United States. He worked on the integration for another seven months.

With the school year over in Milwaukee, his wife and children were set to move to London. They had picked out a house and a school for his kids. In August, he got a call: GE wanted him to move to Paris to run the international side of the health care business. "Literally that day my wife had just [had] her third day of moving all of our shit into a house in downtown London—a rental flat or whatever," he re- called. "I went home that night and I couldn't tell her. It was Thursday night and I couldn't tell her." He told her on Friday, with the weekend still ahead. "It's August second," she told him. "You have one month to get the kids in school in Paris." He said that was "next to impossible," and if he couldn't, his kids would miss the school year. "Sure enough, we dumped that place [in London] and went to Paris," he said. "We were there in a month." His office was in the old Thomson headquar- ters, outside Paris near Versailles.

After a year, as his oldest child was getting ready for high school— having been in five different schools in five years—Bolze's wife and kids moved back to Hingham, Massachusetts, a southern suburb of Bos- ton. Bolze stayed in Paris for another year, continuing to run most of the international health care business. One Friday night around ten o'clock, Bolze got a call from Bill Conaty, the longtime GE HR execu- tive, who told him that Jeff wanted Bolze to return to Schenectady to run the power-generation equipment division of the power business, reporting to John Krenicki. The guy running the business had just quit. He talked to Jeff the next morning. "I knew the business and the team a little bit," Bolze said. "I'd not been on the equipment side. I'd been on the services side mostly. Jeff just said, 'Well, it's probably one of the toughest jobs in the company.' I'm like, 'Great.'" He kept taking the jobs being offered because each one was bigger and more impor- tant than the one before. He didn't love the constant moving and the disruption. But he was a loyal GE man. "If that's where the company needs me, then fine," he said.

Bolze was put in charge of both GE's natural gas and steam turbine businesses, as well as GE's wind turbine business. A natural gas–fired turbine is akin to the inside of a jet engine enlarged ten times, turned on its side, and bolted to the floor. Fuel—in this case natural gas—is fed into the turbine, which spins a generator, which generates electricity. A steam turbine is larger than a natural gas turbine. But the idea is similar, except that a steam turbine is fueled by burning coal, which creates the steam that turns the generator that produces electricity. The electricity is then added to the grid. Some parts of the world have abundant supplies of coal and burn that resource; other parts of the world don't have coal and burn natural gas. "It's about energy independence," Bolze explained. "You don't want to be dependent on some other country for all your power needs." At that time, he said, GE's power turbines generated about 25 percent of the world's electricity and about 45 percent of the U.S. electricity. A typical turbine, as long as it is properly maintained and serviced, can last more than twenty-five years. GE also did a big business in servicing and maintaining the turbines, often selling long-term service contracts alongside the turbines, "great service contracts," he said. At the time, there were five major competitors in the world (in descending order of market share): GE, Siemens, Mitsubishi, Alstom in France, and Ansaldo in Italy. "Think about it this way," Bolze continued. "One point two billion people in the world still don't have any electricity. None. So seventy percent of the new power in the world over the next ten to twenty years is going to be in developing markets. There's not going to be so much new power in the U.S. and Europe, outside of some select projects and renewables. You're selling a lot of new gas power into Iraq. I went to Iraq multiple times."

Bolze settled into Schenectady. At one point, Jeff wanted him to move back to Milwaukee to take the number two job in Healthcare.

Another time, Jeff asked him about moving to Erie, Pennsylvania, to run the locomotive business. Each time, he discussed the idea with Jeff. "Do you really need me to do this?" he asked. "Unless you really need me to do this, I'd rather stay with what I'm doing." He stayed in upstate New York. "At that point I'd moved so many times, that was kind of like okay," he said. He was happy with his career trajectory to that point. "They're bigger roles," he said. "They're more senior roles. They're bigger operating jobs."

Flannery and Bolze led the "due diligence" on the Alstom deal—meaning that they were responsible for the daily nitty-gritty details of understanding Alstom's business and putting together the presentation materials for the GE board of directors as it decided whether to green-light the deal or not. Flannery and Bolze were living in a hotel in Paris and trekking back and forth to an attorney's office, since confidentiality mandated that the GE team not go into Alstom's offices. They had access to some confidential information about Alstom, but they were not permitted to examine individual customer contracts or legal disputes. One GE executive recalled, "In hindsight I'm sure Alstom was like, 'Hey, this isn't the prettiest baby here, so let's keep them as far away as we can.' But at the time we were having conversations of 'We'd been competing with this company every day for the last fifty years' and 'I know that project, I know that project. We lost that deal to them. We see these guys every minute basically.' It wasn't like a pig in a poke." Flannery would be on the phone with Jeff many times a day, keeping him abreast of how things were going. Jeff started getting up at 4:00 a.m. to be on European time. Bolze and his team were on board for the deal, at least initially. Jeff viewed the Alstom deal as a good way to test Flannery and Bolze, both as a team working together to get the deal done and as individuals, as part of the still-unannounced race to succeed him. "While succession wasn't yet top of mind," Jeff remembered, "I knew both Bolze and Flannery could be contenders to replace me, and this was one way to prepare them."

The three men briefed the GE board on the deal in March 2014. "We viewed the Alstom acquisition as a simple transaction that would grow our share of the world's existing turbines by 50%—a huge long-term advantage," Jeff recalled. He also believed the acquisition would help GE boost its large gas-turbine technology, which trailed behind the two large competitors. There were things Jeff didn't like about the deal: For one, the political risk of doing such a large deal in France involving one of the country's most prominent companies and one that was an essential piece of its infrastructure. A national champion, in other words. He also worried that the regulatory approval process at the EU could make what Jack experienced with the Honeywell deal look like an elementary school curriculum. And Alstom also had a business building power plants that "we didn't understand," Jeff remarked. The more work Flannery and Bolze did on Alstom, the more Jeff believed the team became convinced that the deal would add to GE's earnings. The board was positively disposed and gave its preliminary approval.

Ironically, at Jeff's urging, the GE board was also considering two other large transactions at the same time the Alstom deal came around: an oil and gas company and another power deal. "Jeff still has the hots for making more oil and gas deals," explained one GE executive, who thought more capital allocated to that industry at the top of the market was a mistake.

At the board meeting, Lorenzo Simonelli, the head of GE's oil and gas division, made a presentation about the oil and gas deal. Bolze made a presentation about Alstom. "There was a board discussion," remembered someone in the room. "After a lot of debate back and forth, a lot of directors were saying, 'Hey, like, let's do both.'" But cooler heads prevailed when it was clear that GE didn't necessarily have the resources to do both deals. So Jeff and GE board decided to go full bore on Alstom. To help get the deal done, GE hired bankers at Lazard and Credit Suisse for their European connections, along with

Blair Effron at Centerview Partners, who was a close adviser to Jeff, helping him think about "capital allocation" at the corporate level.

By GE's 2014 annual shareholder meeting, in Chicago on April 23, Jeff recalled that the "economics" of the deal had been agreed. That was almost true. Two thorny but important issues remained unresolved with Alstom: the final purchase price and the amount of a breakup fee if GE walked away from the deal. To resolve these matters, Kron and his team flew to Chicago for secret meetings with Jeff and his team. They stayed in a different Chicago hotel from the one where the GE executives were meeting with their shareholders.

GE and Alstom were something like $500 million to $1 billion apart on purchase price. Several of the GE executives were trying to hold the line on the price, or maybe go up just a little bit to get the deal done. "These guys"—Alstom—"clearly want to do this deal," one explained, before adding, "Now, in hindsight, we know exactly *why* they wanted to do this deal." The amount of the breakup fee, which should have been a relatively minor point in a $14 billion deal, took on more importance than normal. It became embroiled in last-minute negotiations as a trade-off between purchase price for the businesses GE was buying from Alstom, how much GE would pay Alstom if it walked from the deal, and what the threshold for a walkaway would be. Alstom didn't want there to be a breakup fee. It did not want GE to be able to back out under any circumstances. In hindsight, this, too, should have raised concerns on the GE side of the table.

Jeff decided to negotiate these final two deal points—price and breakup fee—himself with Kron alone, in Chicago. "Nobody else there," recalled one GE executive. "The kind of scenario we were trying to avoid." He agreed to go up on price, to $13.5 billion, and to increase the breakup fee to $700 million. His colleagues were a little miffed. It was more than they wanted him to agree to pay. But on the other hand, it wasn't that big a difference in the context of a multibillion-dollar deal and in the context of GE's market value. "That's kind of Jeff's perspective," explained one GE executive, "and it's not grossly inaccurate. On another level, it's like, 'Hey, if we'd just waited another week, we could have spent $400 million less.' It's kind of important." GE's only exit from the Alstom deal would be if the European regulators forced GE to sell more than 10 percent of what it was acquiring from Alstom. In other words, if the EU tried to take a couple of holes out of the golf course GE wanted to buy. If the regulators wanted GE to sell 10 percent of Alstom or less, GE would have to close. If it was more than 10 percent, then GE could get out of the deal by paying the $700 million breakup fee. In time, that would become a fateful decision.

A day after the major terms of the deal were agreed between Kron and Jeff during their meeting in Chicago, word of the deal leaked to *Bloomberg*, which reported that GE would be buying most, but not all, of Alstom for about $13 billion. The news sent Alstom's shares soaring 17 percent and brought a truthful but misleading and swift denial from Alstom. "In response to recent speculation in the economic press, Alstom is not informed of any potential public tender offer for the shares of the company," Alstom told *The New York Times*, which was, of course, true but irrelevant. GE was not buying Alstom in toto, so there would never have been a tender offer contemplated for Alstom stock, since that would have resulted in GE owning the whole company, which it did not want.

One big problem with the leak of the deal to *Bloomberg* was that Kron had not fully informed the French government that a deal with GE for 70 percent of Alstom's business was in process. Kron was on a flight back to Paris from Chicago when the leak occurred. The French government immediately summoned him to explain what he was doing. "He tried to play a very close game that he lost at the last second," explained one deal participant about Kron. He had told the GE executives he intended to tell the French government, "Hey, this is getting serious," the deal participant said, "because he'd gotten to a price agreement and a breakup fee." But the leak foiled Kron's plan. "He has an absolute atomic bomb on his hands when he lands," this person recalled, "and the immediate reaction from the government and everybody is an uproar. Partly over the content and partly over being blindsided by Kron. That triggers a whole other wave of the government trying to either stop it from happening or finding some other deal that can preserve the crown jewel of the French power industry. Siemens shows up and Mitsubishi shows up."

Jeff quickly moved into playing a statesman role. "Which he does well," a GE executive explained. In Jeff's telling, as he was trying to make the deal with Kron stick, he flew to Paris, on a Sunday, to try to meet with François Hollande, the French president. But, Jeff explained, Hollande rebuffed him, sending in his stead the deputy secretary general at the Élysée, Emmanuel Macron. "Macron asked us not to sign the deal for a few days," Jeff recalled. "I wanted to tell him to jump in the Seine."

Clara Gaymard, GE's France CEO, urged Jeff to keep calm and carry on. Two days later, on April 28, Jeff met with Hollande. "His manner was formal, and it was clear he was under a lot of pressure from his opponents on the Left," Jeff said. He added, in a statement, "The dialogue was open, friendly and productive. It was important to hear in person President Hollande's perspective and to discuss our plans, our successful track record of investing in France, and our long-term commitment to the country. We understand and value his perspective, and we are committed to work together."

The leak forced Jeff to announce the deal a week later, on April 30, earlier than he otherwise would have. For the meeting with Wall Street analysts to review the deal, by his side Jeff had Bolze, Flannery, and Bornstein—the three men who would soon be under serious consideration to succeed him. In addition to how great the deal was and would be, both financially and operationally, for GE, Jeff's main message about the Alstom deal seemed to be that it would further reduce the importance of GE Capital to the overall GE equation, and much faster than Jeff had previously advertised. The Alstom deal would make the ratio between GE Capital and GE closer to 25 percent/75 percent when the deal closed in 2015.

Right from the start of the meeting, Jeff sought to realign the analysts' thinking about GE Capital. "This transaction reinforces our position as the most competitive infrastructure company, with a specialty finance division," he said. Suddenly, Jeff was referring to GE Capital as a "specialty finance division"? This was an utter misrepresentation of the fact that the Federal Reserve had declared it a SIFI and that the business was, in 2013, still the single largest generator of GE's segment profits, at $8.3 billion. (The next closest was Power and Water, at nearly $5 billion in profit.) The analysts pushed backed on Jeff's disingenuous use of the term "specialty finance division."

The remainder of the press conference was devoted to deal terms: the $13.5 billion that GE would pay for Alstom using $9.5 billion of its cash and $4 billion in new borrowings, which GE pledged to pay down over five years. The price tag represented a multiple of 7.9, the businesses trailing twelve-month EBITDA, and after fully realizing the projected $1.2 billion in cost "synergies," the EBITDA multiple would be reduced to 4.6. The internal rate of return of the deal was in the "midteens," and it would be accretive to GE's earnings by between 4 cents and 6 cents in the first year. The deal also came with $3.4 billion in cash that would be put in a lockbox for GE's benefit, as of April 1. (The cash made the "headline" value of the deal $17 billion.)

Operationally, both Jeff and Bolze talked up the deal, of course. Bolze highlighted ten operating benefits of the deal, including Alstom's big servicing revenues and $38 billion of backlog, of which most was services. "We like Alstom's fit with our strategy to serve the world's power needs," Bolze said. "Electricity demand is expected to grow 50% by 2030 and 65% of that power comes from centralized generation with additional investment in grid infrastructure required."

What Jeff tried to portray as a fait accompli with the Alstom deal on April 30 was pretty much anything but accomplished. First came the political blowup in Paris. Arnaud Montebourg, France's economic minister, accused Kron of a breach of "national ethics" and wrote to Jeff, "We have been surprised to learn that General Electric and Alstom had engaged in advanced discussions... without any prior interactions with French government authorities." He also invited Siemens, GE's chief rival in the power business, into the process of trying to disrupt GE's deal for Alstom's assets by encouraging it to make its own bid.

From there the deal just continued to get more and more complicated. "The shit hit the fan" was the way Jeff described it to me in a memo about the Alstom transaction. There was the seemingly out-of-control French politics, with Socialists who were eager to protect the country's supposed prized national assets. There was the competing bid that Montebourg had encouraged Siemens to make, which it was seriously considering, perhaps joining with another rival, Mitsubishi. And all this noise was before the European Union got into the picture, with its persnickety regulators taking their pounds of flesh.

In May 2014, the French parliament issued a surprise decree allowing the economic minister to block foreign takeovers of French companies deemed to be in "strategic" industries, a move viewed as a direct shot across GE's bow. Jeff described this move as stopping the GE-Alstom "deal in its tracks" and beginning "sixty days of limbo." A few days later, Montebourg said GE should make changes to its deal if it still wanted to try to do it. "Make us a new proposal because this one doesn't work," Montebourg said he told Bolze. "We were stuck at the starting gate," Jeff said.

On May 22, GE agreed to keep its offer for Alstom open until June 23, to continue the negotiations about the deal with the French government. Five days later, Jeff addressed the French National Assembly, its parliament. He made the argument that GE had been a longtime investor in France. He started speaking in broken French and then switched to English. "This industrial project is really about our long-term commitment to invest and grow these businesses in France," he said. "This will have benefits over many decades for employees, for small and medium-size suppliers that serve these industries and for the people of France." He added, "We have been a loyal partner to France for decades. . . . We are the company of Thomas Edison." While Jeff insisted to me there was no "deal heat" in trying to win Alstom, he also said, "We also decided to put the entire clout of GE behind winning the deal." The next day, Jeff went to the Élysée Palace to meet with Hollande. A now-infamous picture showed Jeff and his teammates, Flannery, Bolze, and Gaymard, leaving the building after meeting with the French president.

On June 15, Siemens and Mitsubishi made a complex joint bid designed to thwart GE's bid for Alstom. It was pretty rare for an interloper to try to upset one of GE's acquisitions. Unlike GE's offer, which was to buy the power and transmission assets from Alstom, leaving it with its high-speed-rail business, the Siemens/Mitsubishi bid was a way to keep Alstom intact, with the Germans and the Japanese injecting capital—and getting ownership—in the pieces they wanted. Siemens offered to buy Alstom's gas turbine business for €3.9 billion in cash. Mitsubishi offered to buy 10 percent of Alstom from Bouygues, its largest shareholder, and inject €3.1 billion in cash into Alstom.

Four days later, Jeff announced that GE had restructured its offer for Alstom, without increasing the deal's original value, and that it had been negotiating with Alstom's management. It was plenty complicated. A new fifty-fifty joint venture was created with the two companies' electrical grid assets. GE would also buy 50 percent of Alstom's wind and renewable energy business in France. The companies also agreed to create a fifty-fifty joint venture in the nuclear power business in France, with the French government having the deciding vote, to make sure the French continued to control that business. GE also agreed to sell its locomotive-signaling business to Alstom for $825 million. Alstom agreed to invest $3.5 billion in the joint ventures at a price equal to a higher multiple of EBITDA than GE agreed to pay for Alstom's power businesses. Finally, GE agreed to create one thousand new jobs in France over three years. GE, of course, would also get Alstom's gas turbines business and most of its steam turbine business.

In the end, on June 21, GE prevailed over the joint Siemens/Mitsubishi bid. The Alstom board of directors chose GE. "This is good for France, GE and Alstom," Jeff said. The French government claimed victory, too. "This is a big political, economic and, above all, industrial victory," said a spokeswoman for Montebourg. "We haven't given up on the creation of European champions in either energy or transport."

The French government also agreed to buy a 20 percent stake in Alstom for nearly $3 billion, about $500 million more than the then-total trading price of Alstom. "The deal for Alstom's electricity business would reduce GE's reliance on its banking arm and help Mr. Immelt deliver on a pledge to refocus the conglomerate around its industrial operations," *The Wall Street Journal* reported. And then the *Journal* added, about GE's industrial businesses, "which investors value more," without providing evidence that that was, in fact, true. The paper also made it clear that Jeff had put the Alstom deal on his back and hauled it over the finish line. "It was almost like a Final Four bracket," Jeff told Charlie Rose in June 2015. "We got to Paris [and] we had to compete with the German government, the Japanese government, the French government, all at the same time. We prevailed because we had been in France a long time. And the people we had on the ground wanted us. They wanted GE to be successful. And we were extremely hard for the French government to think about: well-known American company, great French citizen, a better French citizen than most French companies in good times and bad." Whether Jeff's victory was genuine or Pyrrhic remained to be seen.

GE still had to get the deal through the European Union regulators—never an easy thing—and still had to integrate the two businesses. Jeff named three GE executives to head up the four-hundred-person integration effort, but neither Bolze nor Flannery was among them. (Flannery was named CEO of GE Healthcare in October 2014.) Bolze and Jeff butted heads on who should lead the integration team. (They had also butted heads on the price GE should pay for Alstom.) Jeff selected Mark Hutchinson, an ex–GE Capital executive in China, to lead the effort. "I thought he'd have the political savvy we'd need in Brussels," Jeff explained.

THE ALSTOM ACQUISITION HAD BEEN THE FIRST SHOT IN A ONE-TWO combination that would finally allow Jeff to remake GE into a company whose profitability was less dependent on GE Capital and more reliant on power, medical, jet engines, and other traditional businesses. The second shot would involve greatly reducing GE Capital itself, which had long been a thorn in his side, even more so after it had become a SIFI during the 2008 financial crisis. In order for Jeff to convince Sheila Bair at the FDIC that GE should be allowed to participate in the Debt Guarantee Program, he and his team had effectively had to argue that GE Capital was a "systemically important financial institution," or SIFI. But being a SIFI and having GE Capital regulated by the Federal Reserve meant there was an immense amount of time, energy, and resources that went into complying with the Fed.

To help GE Capital navigate these challenging regulatory waters, Jeff hired Dave Nason, a former assistant secretary of Treasury for financial institutions and a senior member of Hank Paulson's team during the financial crisis. Nason was the son of a UPS driver and "wicked smart," according to Jeff, as well as "unpretentious" and having "immense common sense." GE was fortunate to hire Nason, who was working at the Promontory Group, a consulting firm, after leaving the Treasury. But Nason found out quickly how "unprepared"—Jeff's word—GE Capital executives were for the more stringent regulatory regime. When representatives of the Federal Reserve Bank of New York called Nason, early in his tenure, to set up a meeting with GE Capital executives to begin the regulatory process, he sent the proposed dates for the meeting to the top GE Capital executives. But the GE executives told Nason that the proposed dates didn't work because they conflicted with internal meetings. Jeff recalled that Nason "had to explain that when the government says, 'This is when we can be there,' there is only one correct response: 'Works great for us.'"

For a massive financial services company that had been essentially unregulated for decades to suddenly find itself subject to nearly daily oversight by the Federal Reserve, one of the most powerful financial institutions in the world, was not only a huge culture shock but also required a sea change of behavior in the GE corporate offices in southern Connecticut. The Fed wanted GE Capital to have more equity in the company and less leverage. That made it harder for the company to achieve its historical 25 percent, or higher, return on investment. The Fed wanted reams of data, on a regular basis, that GE Capital had a difficult time providing because the often-siloed divisions inside GE Capital didn't have uniform, networked computer systems. "All of us chafed under the weight of their processes," Jeff said. He quickly grew frustrated not only by the Fed's intrusion but also by the shocking annual cost—around $2 billion—of regulatory compliance, given the increased head count, capital, and IT infrastructure required. "We added two thousand people," Jeff told me. "We would do capital plans that were six hundred pages long. . . . They really had first lien on all your cash."

By early 2015, GE had added more than five thousand people to grapple with the new regulatory regime. To satisfy the Fed, the GE board of directors expanded to eighteen members, from thirteen. Jeff added board members such as James Rohr, the former CEO of PNC Financial Services Group, Inc., a big Pittsburgh-based bank; Mary Schapiro, the former chair of the SEC, as well as the Commodities

Futures Trading Commission and FINRA, Wall Street's self-regulatory organization; and Francisco D'Souza, the CEO of Cognizant Technology Solutions. It fell to Nason to explain to his new colleagues to think of the Fed and its representatives, who started crawling all over GE Capital, in the same way that someone would view a police officer. The difficult message was: "When a cop pulls you over, you don't tell him he's stupid. When an examiner tells you that you need to do something different, you don't tell them they're idiots. You just do it. That is your new reality." It wasn't an easy adjustment. Not for Jeff Immelt. Not for the GE board of directors. And there would be no rewards for compliance. Just requests for more compliance.

Jeff especially hated the Federal Reserve regulators showing up at meetings of the GE board of directors, which they did regularly. They would request meetings with directors without management. They demanded reams and reams of data. And there was no denying their requests. At one board meeting, the Fed's chief GE Capital regulator, Caroline Frawley, gave a "withering presentation," listing GE's "flaws one after another." Jeff "wigged out" and told her, at the meeting, "You're trashing us for no reason." After she left the room, board member Geoff Beattie, a Canadian private-equity investor, upbraided the CEO. "You are out of line because GE Capital is out of line," Beattie told Jeff. "This is not going to work the way you seem to think it will. We need to work with Frawley, not against her." Jeff vowed to be more accommodating and diplomatic. "I was chastened, and I shaped up," he recalled.

But he was also more determined than ever to break the covalent bonds that had developed between GE and the federal government regulators. He wanted out of GE Capital. It was driving Jeff nuts that GE Capital had to hire *five thousand* new full-time employees to work with the regulators, at an annual cost of "nearly $1 billion." Other costs brought that annual expense closer to $2 billion. But only a part of the cost of being a SIFI was financial and bureaucratic; there was also the psychological toll it was taking on Jeff and his regime. "It was extremely heavy, I mean extremely heavy," he told me. "I wasn't ready for it. We weren't ready for it. I think we fought it for a while, but then I think it took us a little bit of time before we realized, 'Hey, look, it just is what it is.'"

In June 2013, just before the SIFI designation became official, Mike Neal, who had been running GE Capital since Jeff decapitated Denis Nayden, retired from GE after thirty-four years. (Neal had previously, in 2007, turned down an offer to lead Citigroup.) Keith Sherin, who was GE's CFO, replaced Neal as the CEO of GE Capital. Jeff Bornstein, who had been GE Capital's CFO, replaced Sherin as GE's CFO. The old switcheroo. Beyond the important personnel changes that the SIFI designation catalyzed was what Jeff called the "mindshare" that GE Capital and its new regulators consumed. He said that dealing with GE Capital being a SIFI was taking 80 percent of his time. "It was really running two different companies," he told me. "You would have a board meeting and eighty percent of the board meeting would be GE Capital, and then you'd have fifteen minutes at the end to say, 'Oh, by the way, we're going to invest $2 billion on a new jet engine. What do you think?' Or we're building fifteen factories in China. The board would be like, *groan*. It was really like running two different companies, and it started to be just untenable. . . . You had a massive effort going underway. It sucked all the oxygen out of the room. I was always worried about the Fed's ability to say, 'Hey, you can't invest in a next-generation engine technology because we worry about your cash.' So there were just a lot of reasons why it was not fit for purpose."

As bad as that was from Jeff's perspective, worse was the fact that all twenty-five of GE's largest investors told him, "Hey, look, we don't really care about GE Capital at all." They wanted GE to jettison GE Capital one way or the other. Jeff had come around to their thinking. "With regulators involved, and with our cost of funds shooting dramatically up, I knew we had to pivot," he recalled. He started to think seriously about the once unthinkable: getting rid of GE Capital. "It wasn't the first time we'd considered this," he continued, "but now it felt more urgent. The bottom line was that we weren't being sufficiently compensated for the risk we were taking."

If your remit in life, as CEO of GE, was to "allocate capital," the idea that you would continue to invest that precious capital into a business where you were no longer being compensated for the risks being taken quickly became untenable. And when that business was newly burdened by a high-powered and determined regulator, then it was just a matter of time before Jeff would ask his direct reports to figure out a way to exit the business, despite the billions of dollars in profits that GE Capital had contributed to GE since Jack had decided to arbitrage GE's AAA credit rating into an earnings juggernaut. It wouldn't be an easy assignment. "Whenever we'd looked at dismantling GE Capital in the past," Jeff recalled, "the tax implications alone—$20 billion or more—had appeared prohibitive. Could we figure it out? We had to try." One of Keith Sherin's first tasks as the new CEO of GE Capital was to analyze whether GE Capital could be spun off to GE shareholders; that was dubbed Project Beacon. The idea behind another approach, Project Hubble, was to break up GE Capital and sell it off in pieces. A third possibility was to "stay the course" and keep GE Capital intact. Each choice was pretty much unbearable. If you got rid of GE Capital—one way or another—how would GE replace those earnings? If you kept it, how could investors and Wall Street analysts ever again assign a meaningful valuation to those earnings? The problem for Jeff remained a lackluster stock price. It was down roughly 40 percent during his time as CEO, while the S&P 500 stock index was up 50 percent. Worse, the stocks of GE's industrial competitors had been on a tear. United Technologies' stock was up 200 percent; Honeywell's stock was up 125 percent. Something had to give.

Sherin recognized the magnitude of the problem from the outset. "From the minute we were in financial distress in 2008 up till even today, there isn't a single investor that didn't want GE Capital to be smaller, and we wanted GE Capital to be smaller," he explained. "We were uncompetitive from that minute on. We couldn't borrow at the rates we were. We couldn't do the short-term commercial-paper leverage. We were uncompetitive." Jeff tapped Sherin to lead the internal effort to try to figure out what to do with GE Capital. After Thanksgiving weekend in 2014, he put together a group inside GE Capital to analyze the idea of selling off GE Capital in pieces. That was Project Hubble, named after the Hubble Space Telescope. "Everyone loves the shoot-the-moon audacity of Project Hubble," Jeff recalled. "The name stuck." Previous efforts to try to seriously consider selling the parts of GE Capital had run into the problem of creating enormous tax liabilities, in part because the logic of the deal was to sell off GE Capital's overseas businesses first and then to repatriate the proceeds.

Sherin put together a Traveling Wilburys–worthy lineup of GE Capital executives—Mike Gosk, head of tax, Daniel Colao, head of financial planning, and Aris Kekedjian, head of M&A—and told them to find a more tax-efficient solution. They recruited more GE Capital experts, including Mike Schlessinger, a longtime GE lawyer, to GE Capital's new home in Norwalk, Connecticut, to try to come up with great ideas. The team worked intensely and secretly through December. There were two eureka moments. First, Schlessinger had the bright idea to sell GE Capital's U.S. assets first, while continuing to operate the overseas assets, which were mostly comprised of aviation leasing. Selling off businesses in the United States would save on taxes and allow GE to make the argument to the Fed that it was no longer a SIFI. The second eureka moment came when the Project Hubble team's analysis showed that if they could sell GE Capital's assets at book value, investors should reward the company with a higher stock price, given that it seemed like the market was valuing GE Capital's assets at 70 percent of book value. This would be one of the most complicated restructurings in the history of corporate finance. Clever financial buyers were in a position to pick GE's pocket, knowing that it was a forced seller. There were likely to be billions in impairment charges, due to GE having to sell assets before they achieved their expected value. There were likely to be other restructuring charges—due to layoffs and office closures—and, of course, tax leakage, primarily the $6 billion in taxes that would be paid to repatriate GE cash sitting overseas. Bondholders would have to be satisfied, too. That could mean paying billions for consents or having GE guarantee the GE Capital debt. On the other hand, GE would not have to put additional capital into GE Capital during the asset-sale process, and GE would get the proceeds of the asset sales. As important, a slimmed-down GE Capital would probably be freed of the SIFI designation relatively quickly (which finally happened in June 2016). The question hanging over the whole analysis, like a sword of Damocles, was, How would GE replace the earnings that GE Capital had long provided to GE?

The March 6, 2015, board of directors meeting was devoted to Project Hubble. After Sherin and the GE Capital team presented the divestiture plan, Jeff told the board that he was going to have to "float the idea" with the Fed, the SIFI overlords, and GE Capital's bondholders, the last of which, all things considered, might have been the easiest of the three tasks, given that GE was going to have no choice but to guarantee the GE Capital bonds. "Rumors may spread," Jeff told the board. "I would never want you to hear something this important from anyone other than me." He asked for the board's approval to have the needed conversations. "The board gave us the greenlight," he recalled.

Jeff, Sherin, and Bornstein, the new GE CFO, sprang into action. There was no time to waste. They had to manage the difficult balancing act of gauging, as discreetly as possible, both whether there was a viable market for the assets and how the GE stock would react to the news, while also keeping the whole thing under wraps. First stop on the tour was to see Jamie Dimon, Jeff's Harvard Business School classmate, and Jimmy Lee, at JPMorgan Chase. Could the bank sell the

U.S. pieces of GE Capital? No self-respecting banker would answer that question in anything but the affirmative, and of course Dimon and Lee said the market would leap at the opportunity to buy the assets. The dramatic public announcement of Jeff's decision to sell off GE Capital was set for April 10.

That gave him and his team about a month to get done all the important tasks. Jeff was determined to couple the April 10 presentation with a sealed deal to sell a chunk of GE Capital assets, to show the market that GE knew what it was doing and that buyers would be interested in the assets that were for sale. Jeff made the decision to approach Jon Gray, the head of real estate at the Blackstone Group, to see if Blackstone was interested in GE Capital's real estate assets. Blackstone could have an exclusive, Jeff told Gray, provided the firm "moved fast and paid us well," Jeff recalled. Gray said he would try. The Project Hubble team also kept the Obama administration in the loop about the coming announcement.

On April 9, Jeff had lunch with Larry Fink, the powerful founder and CEO of BlackRock, the huge investment management firm, a large GE shareholder. Getting Fink on board was akin to what Jeff had done a month earlier by seeking Jamie Dimon's support for the plan. After Jeff walked Fink through the proposed divestiture, "he looked stunned," the GE CEO recalled. "If I hadn't already known how unprecedented this restructuring would be, Fink's facial expression would have confirmed it." He asked Fink for his support: "If you get a chance in the coming weeks to talk this up—even if you say, 'This wasn't completely stupid'—I'd appreciate it."

Jeff still had a long day, and night, ahead of him leading up to the big announcement. After the market closed, *The Wall Street Journal* broke the news that GE Capital was close to selling $26 billion of its real estate assets to Blackstone, which would take GE Capital's equity stakes in buildings across the globe, and a portion of its loans to various real estate projects as well. The *Journal* reported that Wells Fargo, one of the largest lenders to commercial real estate, would also take a portion of the loans. The article suggested that the sale of the real estate assets, if it were to happen, would be a piece of a larger plan to sell off all of GE Capital. The paper conveyed the situation accurately. "GE's top executives have come to see the bulk of the company's $500 billion lending operation as severable," the *Journal* reported. It took the rest of the night to get an agreement with Blackstone and Wells Fargo. At 6:30 a.m., "everyone signed off on what we felt was a fair price," Jeff recalled. Blackstone and Wells Fargo had agreed to buy $26.5 billion worth of equity stakes and loans to a variety of commercial real estate properties worldwide—one of the largest real estate deals ever. That news would be a welcome part of the overall package announcing GE's massive strategic pivot.

Two hours later, Jeff and his team assembled in a Fairfield conference room to share the news with investors. On the one hand, the divestiture of GE Capital assets, first in the U.S. and then abroad, was straightforward: Jeff was committing to getting out of the financial services business for all intents and purposes in order to return GE to its industrial and technological roots. It made for a charming narrative and, at first, investors applauded the move, with the stock closing at $28.51 that day, up 11 percent, the biggest one-day percentage gain in six years. The details were complex. When all was said and done, GE would be 90 percent an industrial company and 10 percent a finance company, since a smaller GE Capital would retain its aircraft leasing business, as well the businesses that financed GE's energy and health care customers.

In short, with the help of bankers at JPMorgan Chase, who would lead the charge on the planned divestitures, GE Capital would be returning to being a business that helped GE's customers finance their purchases of GE products. Jeff and his team explained that, as part of the plan, GE would repatriate $30 billion in cash from overseas (net of $6 billion in taxes; no wonder he wanted Paulson's help) and would take $16 billion in after-tax charges related to the write-down of good-will and assets that might be sold below their carrying value.

Crucially, as part of the deal, GE agreed to guarantee all the trad-able GE Capital debt and its commercial paper, thus eliminating nearly completely any concerns of GE Capital debtholders who might have worried that they would be stuck owning paper in a company without any assets. Instead, GE would pay off those debts, in effect making that debt instantly more valuable than it had been before. GE Capital's debtholders would be happy. The ratings agencies also seemed to ap-prove of Jeff's decision: S&P affirmed GE's credit rating, while Moody's dropped it one notch. GE was still able to boast that it was among the highest-rated industrial companies.

After applauding the GE executives for the deal, the research ana-lysts were understandably focused on what the loss of GE Capital's earnings would mean for GE's future earnings. Jeff's emphatic short answer was that the company's earnings per share would be essen-tially unchanged from the previous guidance he had shared about the company's future performance through 2018. A key part of the al-chemy was using the cash proceeds of the asset sales plus the repatri-ated cash to buy back as many as two billion GE shares—reducing the shares outstanding to around eight billion, from ten billion—and re-ducing the denominator in the earnings-per-share calculation by 20 percent.

Jeff Sprague at Vertical Research pressed Jeff about the 2018 earn-ings estimate. "You characterize 2018 as being in line with your plan but I don't think we know exactly what your plan is," he said. He then did some mental gymnastics and figured that Jeff was predicting that GE would earn $1.80 a share in 2018, between the robust industrial businesses and the slimmed-down GE Capital.

Jeff disagreed with Sprague's calculus. "I think you missed the buy-back and you missed Alstom," Jeff replied. "You missed a lot of stuff that is in there. It is substantially higher than that."

But Sprague said it wasn't "apparent" to him.

Again Jeff pushed back. "I think it is, actually," he said. "I think it actually is. So I think it is substantially higher than that." And that's how Jeff more or less committed to Wall Street that GE would earn $2 a share in 2018. "We'll still give you $2 a share," one GE executive recalled promising to investors as part of the Project Hubble presentation. "That was premised on the industrial businesses. It was premised on the numerator and denominator. It was premised on the industrial businesses delivering a numerator of earnings, and it was premised on the GE Capital Hubble buyback reducing the denominator to provide earnings-per-share neutrality. That was the premise, and everybody was on board, all the management. I don't remember a single person saying, 'I think this is bullshit.' Everybody said yes."

That promise of $2 a share would be a fateful one for Jeff. "When we did Hubble," Jeff explained to me, "you basically take half the earnings and you throw it up in the air and say, 'I don't exactly know where this is going to land, but we're going to do the best we can.' We didn't want to make the company uninvestable, so we needed to give it a guidepost." That calculus turned out to be a combination of the $35 billion in cash from the sale of most of GE Capital and the ongoing 8 percent growth in GE's industrial businesses. "That is a number that's above $2 a share, right?" Jeff continued. "That's how we built it, and we felt like we needed to offer something. It's not every day that you announce the complete wholesale exit of a huge business that, at least before the financial crisis, had been an earnings driver. We needed some context for the stock." Jeff then added that both the board and the management had fully vetted the $2-a-share analysis. "It wasn't like I woke up and thought, 'Two dollars is a nice round number. Let's go for that.'"

Summing up Jeff's latest moves, *The Economist* offered grudging praise, noting that it had taken him thirteen years "to escape" Jack's "legacy" and to "steer the industrial colossus in a new direction." The magazine traced Jeff's difficulty in grappling with GE's businesses in the post–September 11, post-Enron, post–Great Financial Crisis era. The Alstom deal, the magazine's thesis went, would be Jeff's chance to transform GE in his image, having finally shed—or expressed a desire to shed—most of what he considered the liabilities Jack left behind for him—insurance, NBCU, and a bloated GE Capital funded in the short-term-credit markets. He was making a big bet on power, and as his new colleagues at Alstom might say, *les jeux sont faits*.

# IF YOU STRIKE
# AT THE KING

In September 2014, about three months after the public announce-ment of the Alstom deal—but still more than a year before the deal closed—Steve Bolze, the head of the power business, which would end up managing 70 percent of the Alstom assets, sought to clarify with Susan Peters, the head of HR, where he stood in the succession pro-cess. To him, the process had been masked in secrecy. He didn't know for sure if he was even being considered. His name had been men-tioned on CNBC as a possible successor. And since he had been run-ning one of GE's biggest businesses and reporting directly to Jeff, he figured he would be, or should be, under consideration. But it wasn't something that was discussed directly or openly. That was frustrating. He wanted to have a discussion with Peters to try to get some insight into the succession process and whether he was in it. After all, he was fifty-one years old and was wondering, as was quite natural, if he had a legitimate chance to succeed Jeff, since both Jeff and Jack had been in their forties when they were selected as GE's CEO. Bolze and Peters had dinner together in Darien, Connecticut. All she would tell him then was that he was one of six possible candidates.

Jeff had a different, less charitable, view of Bolze's gambit. "When Jeff steps down, I want to be CEO" was how Jeff understood what Bolze had said to Peters at their dinner. "Unless I get a clear indication that I am getting his job, I'm going to leave." Jeff was appalled. Bolze, he explained, lacked "succession etiquette." He added that Peters was "surprised less" by the "bluntness of his ambition" and more by how "little he understood the process." At dinner, Peters told Bolze that the GE board would choose Jeff's successor, and it wasn't going to be anytime soon, so be patient. "Not because she thought he was the right person for the job (she didn't)," Jeff continued, "but because she knew some on the GE board thought highly of Bolze, and she wanted him in the running so they could consider him. As for me, I thought Bolze was just naïve and that he would grow out of it."

Whatever respect, or affection, Jeff might once have had for Bolze was lost after what Jeff considered a brash power play with Peters at dinner, although it seems innocent and logical enough. He noted that Bolze had "initiated the Alstom acquisition" and "seemingly relished the early negotiations." But then, he continued, he believed Bolze lost interest in the deal. "It wasn't that Bolze failed to communicate the details of the deal to our board and executive team," Jeff allowed. "It simply seemed he wasn't committed to executing on those details." As the Alstom deal progressed through its lengthy regulatory approval process—a crippling eighteen months between signing and closing—Jeff believed something changed about Bolze. "I had the feeling Bolze was more maître d' than chef," Jeff explained, "terrific at social graces, but less comfortable when asked about the ingredients that went into the main course." He continued that he thought Bolze "lacked what I can only describe as a CEO's intuition." As it did with the other possible candidates for Jeff's job, GE got Bolze a coach and worked with him on strategy. "But it wasn't enough," Jeff concluded. "A leader stops leading."

It seemed that the simmering feud between Jeff and Bolze, one of his direct reports, had become personal. It was yet another example of Jeff not being able to get along collegially with his direct reports, despite his ongoing belief that he was nurturing them and encouraging them. This was a bizarre example of cognitive dissonance on Jeff's part. Jeff lost faith in Bolze, and in particular in his ability to deliver the Alstom numbers the deal team had promised Wall Street in June 2014. There also seemed to be a growing disagreement between the two men about whether GE should close the Alstom deal at all or instead figure out a way to invoke the provisions in the contract that would allow GE to pay the $700 million breakup fee and walk away and be done with it. That would be challenging from a legal point of view because the EU negotiators, mindful of the 10 percent threshold, seemed to be gearing their requirements for approving the deal toward divestitures that could be valued as being close to the edge of the provision that would let GE walk away but not over it. Whether to abandon the Alstom deal or not was a major source of contention among executives inside GE. In *Hot Seat*, Jeff sought to derail the idea that walking away from Alstom was a realistic possibility. "We had walked away from deals before," he wrote, citing Westinghouse Electric in 2005, Dow Jones in 2006, and Abbot Diagnostics in 2007. He didn't have to mention that Jack had walked away from Honeywell in 2001. "With Alstom, though, we decided to go forward," he continued. He explained that the board reviewed the deal no fewer than a dozen times, with the GE Power and Energy Management teams making the presentations, "so there were plenty of chances to ask questions." KPMG, the outside auditor to which GE paid $110 million annually, reported its findings about Alstom. Jeff made the additional point that there were four board visits—without him around—to meet with GE Power executives during the eighteen months it took to close the Alstom deal. During those visits, his argument went, GE Power executives could have raised their objections to the Alstom deal had they felt passionately enough that it was a mistake. That seemed particularly unrealistic, especially for someone like Bolze, who was hoping to succeed Jeff as CEO.

Jeff explained that Bolze, along with Brackett Denniston and Mark Hutchinson, the head of the integration team, conducted the final negotiations with Alstom and the EU. That might have been technically true, but this was GE: Jeff was the final arbiter, always. During the summer of 2015, it was becoming clear that to grant their approval, both the EU and U.S. Justice Department were going to demand that GE sell Alstom's Florida-based subsidiary, Power Systems Manufacturing, or PSM, which was one of GE's main competitors in the sale of aftermarket parts and service for GE gas turbines in the United States and Europe. If GE owned one of its main competitors in this business, the regulators argued, customers with GE gas turbines would suffer. GE executives were not surprised by the regulators' demand. They would have liked to own PSM, of course, but figured it would be "a big ask," according to one GE executive. At the time, GE's aftermarket and servicing revenue from its customers was around $750 million. PSM's worldwide revenues were around $250 million, $90 million of which came from providing parts and service to GE's U.S. customers. PSM also had the capability to service the gas turbines of GE's competitors, Siemens and Mitsubishi. GE wanted to be able to continue to do that work.

A two-month row with the EU resulted over whether GE could continue to service Siemens's gas turbines. Siemens had around three hundred of these turbines installed at its customers, and at first, the EU wanted GE to give up the rights to service them. "We said, 'No way in hell,'" recalled one GE executive. "'We make all of our money servicing gas turbines.'" GE had some leverage. Had the EU insisted on this demand, GE would have had a colorable claim that 10 percent of the value of the Alstom deal had been lost, and it could have walked away, after paying the breakup fee. The negotiations continued.

Once the regulators had demanded that GE sell PSM, some executives in the power business didn't like the decision and were trying to figure out what to do. There was a debate. Some wanted to pull the plug on the deal (if they could) because the "integration" of the two businesses was going to be "monumental" and was going to be "too much of a pain in the ass," explained one executive. Others wanted to push forward. "There's only two deals, three deals you could do in this industry in the next twenty years," their logic went. "We can't buy Siemens. We can't buy MHI. So let's go." There was also a fair amount of "value erosion" during the course of the eighteen months it took to close the deal: there was litigation; there were customers leaving; there were problems with project reserves; and with the sale of PSM, there was a diminution of servicing revenues. "The question becomes then, 'Okay, how much value have you eroded and do you take a breakup fee—because that'll fight out in court for a couple years—or do you try and negotiate a purchase-price reduction?" recalled one GE executive.

Within GE Power there was a faction that thought a purchase price adjustment in the billions of dollars was appropriate. At one point, Bolze went to see Jeff to discuss what to do. They did not see eye to eye. He shared with Jeff the idea that a massive price reduction was in order, at least from his perspective as the person about to be charged with running 70 percent of the Alstom assets. Bolze told Jeff he thought GE should cut the Alstom purchase price by "3 or 4 billion" dollars "or we should not do this deal." Jeff disagreed: "Steve, I hear you, but, you know, stand tall. I need you to stand tall. We're going forward." Jeff's reaction to what Bolze was telling him was, alas, consistent with the way he had been handling news he didn't want to hear. "With Jeff, if you bring him bad news, he doesn't like it," explained a former GE executive. "And by the way, he sometimes won't call you back."

In any event, Jeff continued to be unhappy with Bolze. Despite the ongoing internal debate about the wisdom of closing the Alstom deal, GE reached an agreement with the EU about the sale of PSM and was also able to negotiate a reduction of the overall purchase price by about $300 million. At the same time, and without much fanfare, GE raised dramatically the dollar amount of the synergies it expected from the Alstom deal to $3 billion, from the initial $1.2 billion. What the board was seeing was both a price cut and a synergy increase. In that scenario, why not close the Alstom deal, its proponents argued. "Look," Jeff told his colleagues, "at this point, the best thing we can do is close the deal and start executing. Let me take the blame for how much we paid." In his book, he laid into Bolze in a way rarely seen in a corporate memoir. "I wasn't seeing the leadership I was hoping from Bolze," he wrote. "At one point, he came to me, worried about Alstom's diminishing unit sales. . . . I was dumbfounded. We had been proceeding to close a deal that Bolze had sold to our board based on the numbers and projections that he and his team had calculated." He asked Steve, "Do you believe in this deal or not?" According to Jeff, Bolze responded that he did; he just wanted a meaningful price reduction. Jeff interpreted Bolze's concerns as being a problem with Bolze, not a problem with the Alstom deal. "I thought the Alstom deal made all kinds of sense," he continued. "I still do. But at this point, I was beginning to understand just how serious a problem we had at the top of GE Power." He noted he was hearing "rumblings" that Bolze had to memorize data before his presentations to Jeff and the board because "he didn't have command of the most basic material" and was busy "cramming, last minute" as if "for an exam." He also blamed Lynn Calpeter, the Power CFO, who, he wrote "seemed to have checked out" and who visited France only once during the two years the deal was being nurtured, negotiated, and closed. "Three integration leaders told me that if, when I retired, Bolze was named CEO, they would sell all their GE stock," Jeff continued. "That wasn't a good sign. More than once I told Bolze, 'This was your idea. Own it.' As the CEO of Power, he was responsible for assessing the strategic impacts of the transaction. Instead, he kept complaining about how the numbers weren't as good as they'd once been."

In late summer 2015, Bolze wanted to have another visit with

604 – POWER FAILURE

Susan Peters to try to get his mind around the succession process, if she would be willing to share. Of course, he was then a year older and still concerned that he might be aging out of the chance to be GE's CEO. Jeff interpreted Bolze's move as yet another attempt at angling for Jeff's job. This time, Jeff recalled, "he was *really* going to leave GE." Bolze told Peters, "I have decided. Because if the company doesn't see me as the right person to succeed Jeff, then that says something." At their dinner in Southport, Connecticut, Peters tried to calm Bolze. "Steve," she told him, "as I told you a year ago, nobody is going to declare you the CEO-in-waiting when succession still could be a couple of years out. And again, it's the board that makes the decision." Bolze proposed an interim step that would make clear that he was next in line. "You could make me GE's COO [chief operating officer]," he posited. He understood that being the COO would be unusual—GE had never had a COO—nor would it be a guarantee he would get Jeff's job. In any event, that idea flopped, too. "Here's what I recommend you do," Susan Peters told him. "I want you to think about this over [the August] vacation. Really think about this. And if you feel the same way when you come back after the August break, then you should talk to Jeff." He asked Peters how many candidates there were now. She told them the field of six had been narrowed to four. He asked her what his chances were. She said 25 percent. He thought to himself, "Thanks. I can do the math, too." What made the situation for Bolze more excruciating was that there was no one inside GE, other than Peters, he could talk to about the succession process. He knew it was happening. But he couldn't seek advice from anyone. And he wasn't getting any younger.

According to Jeff, in September 2015, Bolze "beelined" to his office and "asked to be made COO." Jeff recalled that at that moment the four possible candidates to succeed him as CEO were Bolze, Flannery, Bornstein, and Lorenzo Simonelli, the head of GE's oil and gas business. "Bolze's entitled self-assurance astonished me," Jeff wrote, "as it had astonished Peters." Jeff's loss of confidence in Bolze was now complete. "I felt Bolze had to go," he continued. "He was not operationally sharp and he didn't seem to be thinking ahead. And I was disappointed that at this critical time, he appeared to be putting his own interests ahead of the team's." He decided Bolze "could be putting the [power] business at risk." (At the time, as opposed to what appeared years later in *Hot Seat*, Jeff wrote in Bolze's 2015 annual performance review that Bolze had had "on the whole a good year . . . lots of strong execution" and that he had done "a really good job on Alstom" and that his bonus would be in line with his annual target incentives. In his 2016 performance review, Jeff wrote about Bolze that "you really connect," that he did "the little things to rally the team" and was "very collaborative . . . does the right thing for the Company." Go figure.)

One top GE executive said he was "incredulous" that Bolze had asked to become COO, and that such a request was "most un-GE-like." Another executive told me, "Steve was an ambitious guy and sometimes his ambition got ahead of him. . . . Steve got out ahead of himself and made a couple mistakes. I think I'll just leave it at that. Steve made some mistakes."

But Bolze had some enthusiastic supporters, too. Jeff went to speak with Jack Brennan, the lead director on the GE board and the former CEO and chairman of the Vanguard Group, a large investment company. Jeff told Brennan that although he knew some board members favored Bolze in the succession process, he didn't think it was "fair" or was "merited" to make Bolze COO, in effect putting him before the others in the succession race.

Jeff explained that had it been a decade earlier, he would have fired Bolze right then and there, but since Jeff knew that his time as GE's leader was coming to an end relatively soon, he deferred to the board to make the decision about Bolze. "I had problems with Bolze," Jeff continued, "but I wanted the compensation committee to get its chance to weigh in." The committee ratified Jeff's thinking: Bolze would not be named COO and would have to wait to see how the succession process played out. Brennan subsequently called Bolze and told him that the succession process would continue to play out as planned—no COO appointment for him—that the board's powerful Management Development and Compensation Committee preferred that Bolze stay and participate in the succession process and that, if he wanted, some of the MDCC members would be happy to speak with him. (For his part, Brennan refused multiple requests to be interviewed about Bolze or anything else having to do with his tenure as the lead director on the GE board of directors. He has also, incredibly, expunged his role as the lead director from his résumé. For that matter, most GE board members I asked to be interviewed for this book steadfastly ignored my requests, a sign of true cowardice.)

Then, at the September meeting of the board of directors, things went nearly completely off the rails. After Bolze gave what Jeff described as an "enthusiastic" presentation about the Alstom deal, Douglas A. "Sandy" Warner, a longtime GE board member, looked at Jeff, smiled, and said, "I think we have our next chairman." Jeff freaked out. He described Warner, who had once been chairman and CEO of J.P. Morgan & Co., as "a forceful guy," and he was worried that he would try to force Bolze through the board as the next GE CEO. Jeff couldn't allow that. "I also suspected that Warner, who we'd passed over for lead director in favor of Brennan, was anxious to see me leave," Jeff recalled.

Warner *was* angling to get rid of Jeff. "Sandy was the only guy really carping at Jeff, and Sandy was poking around," explained one GE board member. Jeff thought Warner had "become a corrosive influence." The irony was that once upon a time Warner had been Jeff's biggest booster, especially during the time he hoped to become the lead director on the board. Ken Langone recalled the time that Warner came out of a board meeting and told Langone, gushing, "You know, Jeff is better than Jack."

Langone couldn't believe it. "You know what, Sandy?" he replied. "Why don't you give it some time before you break that gate? Asshole."

In any event, those days were over. Time for Jeff to eliminate Warner. With the support of GE's governance committee, Jeff asked Warner not to stand for reelection to the board in April 2016, just as he had years before eliminated Langone. "He was angry," Jeff conceded about Warner, "and wrote a letter to the board questioning how long I should be allowed to remain as CEO."

At some point, Warner asked for there to be an executive session of the board—without Jeff there—and told the other independent board members he believed the time had come to make a change at the top of the company. It was time for Jeff to go and for Bolze to replace him. Even though it was an executive session, word quickly got back to Jeff that Warner wanted him gone. Jeff then turned the tables on Warner. Warner was out. Warner was "the most likely to push back, the most likely to ask questions, the most likely to say, 'Hang on a second,' the least likely to be pushed around like in your committee report," explained one GE board members. "From my vantage point, Jeff was like, 'This guy, if I'm looking for a rubber stamp, he's not one. He's kind of got my head in his crosshairs.'"

For his part, Warner wrote to me that he was not comfortable talking about GE. "The facts are pretty much all in the public domain," he wrote in an email. "The mistakes have been noted. A lot of the damage was done after I left. I don't think anything useful comes from me piling on with another opinion or perspective. Litigation"—presumably referring to ongoing shareholder lawsuits and a recently concluded SEC investigation—"will clarify some of what happened and who was responsible." In a subsequent email responding to what Jeff wrote in *Hot Seat*, Warner wrote, "First I heard about Jeff's concern for my support of Bolze was when I read his book. Nothing to be gained, however, by my prolonging this very sad story. I am going to keep my head down."

Some board members found Jeff's defenestration of Warner to be deeply troubling, and it made Jeff's already precarious position with the board more so. Deadpanned Langone, "I love watching rats fucking chew on each other's tails." All of which goes to prove the point that Warren Buffett made in his 2019 Berkshire Hathaway shareholder letter about what CEOs too often want in their board members: "When seeking directors, CEOs don't look for pit bulls. It's the cocker spaniel that gets taken home."

Bolze found out that Warner had lost his GE board seat because of his outspoken support for Bolze only after he had a chance encounter with Warner on a commercial flight. They sat next to each other. Warner was somewhat circumspect and diplomatic but made it clear that he had supported Bolze and it had cost him his seat on the GE board. Until that moment, years later, Bolze had no idea why Warner did not stand for reelection in April 2016. "I felt there needed to be a change in leadership," Warner told Bolze. "You know, you were a person that I felt good about."

ON NOVEMBER 2, 2015, GE FINALLY CLOSED THE ALSTOM DEAL, AFTER eighteen months. Under Bolze's leadership, Alstom's energy business was combined with GE's power and water businesses. With some sixty-five thousand employees and annual revenue of $30 billion, the combined business was a behemoth. Jim Cramer of CNBC called the acquisition "brilliant" and predicted that GE had "a hammerlock" on the world's migration to cleaner power plants. GE's stock rose 5 percent for the day. But the victory for Jeff, however Pyrrhic, was short-lived.

A week after the Alstom deal closed, according to Jeff, on November 9, he and Susan Peters were in Greenville, South Carolina, visiting a Power production facility. Just as their visit was ending, as part of his annual review, Bolze was again thinking about leaving GE. "Jeff and I had agreed to meet at the end of the day," Bolze told me. "And he said, 'Hey, tell me what you want to do. Because if you're definitely leaving, I have to put the balls in motion." Bolze told Jeff he wanted to resign. It was déjà vu all over again, all over again. "You obviously don't see my worth," Jeff said he told him. "You don't see that I should be your guy, and, therefore, I'm going to go." Jeff was fine with Bolze's decision. "Honestly, I was exhausted by Bolze at this point; to me, his behavior had become self-serving to the point of outrageous," Jeff explained. On the plane back to Fairfield, Jeff and Peters "strategized" about who could replace Bolze, and they quickly settled on John Rice, who had once run GE's energy business and who was then head of GGO. They called Rice and asked him to take Bolze's job. With some reluctance, Rice agreed. Jeff then called Brennan, the board chairman, and told him "Bolze was out; Rice was in."

But according to Jeff, Bolze was still not going to go quietly, if at all. "He called several members of the board to discuss his resignation," Jeff continued, including Jim Cash, the longtime GE board member and Harvard business school professor who came to Jack's aid during the machinations with Suzy and the *Harvard Business Review*; Robert Lane, the CEO of Deere & Company; and Andrea Jung, the CEO of Avon. Bolze was just taking Brennan up on his offer to speak with members of the MDCC, but Jeff believed Bolze still wanted to be CEO and was making his case directly to the board. "In conversation after conversation," Jeff went on, "he pled his case—he was underappreciated." In the long annals of GE history, such a thing had never happened, if indeed that was what happened. "Highly unusual" was how one GE executive put it to me.

On November 10, at nine o'clock at night, according to Jeff, Bolze called Brennan to say he had changed his mind and wanted to stay at GE. The next day, Brennan called Jeff, told him that Bolze had changed his mind again, and asked Jeff to grant Bolze's wish. They spent some thirty minutes on the phone discussing the situation, which was increasingly complicated and made more so by Jeff's decision to whack Warner. He had to be careful. "I told him I would sleep on it," Jeff explained. Peters told Bolze that Jeff and the board would "think about" letting him stay and politely told him that he had "probably just screwed himself" in the succession process. He said he understood but told her that a couple of board members had told him that unless he has something better to do, he might as well stick around and see what happened. He might end up as CEO of GE after all. He didn't want to regret, when he was ninety years old, that he had left the game in top of the ninth when he could have been the hero in the bottom of the ninth. Some thought Bolze thought he still had a chance to become GE's CEO, even though he didn't. "Steve was in the mix," explained one GE board member, "but Steve was a puppet. He wasn't really in the mix. He was just there because everybody knew when he didn't get CEO, he would leave."

After his phone calls with the three members of the MDCC, Bolze and his wife talked over what to do, whether to follow through on his threat to quit or to stick it out. He had told Jeff he was quitting. Bolze now was rethinking that decision, yet again. They decided he should stay, that he had been too rash. He called Jeff back and reiterated that we wanted to stay. But now Jeff was in a tough position because he had told the board Bolze was leaving. It was a big mess. Bolze told Jeff he would never bring up the topic of succession again. He would put his head down and just do his job. Either way, stay or go, Bolze had learned in real time Ralph Waldo Emerson's adage about coups: "When you strike at a king, you must kill him."

In the end, Jeff decided to let Bolze stay. Turnover at the top of GE's biggest business right after closing one of its largest acquisitions would have made investors nervous, which was the last thing Jeff wanted to do given that the GE stock seemed finally to be moving in the right direction. Additionally, having Bolze in the succession process seemed to be something some members of the board wanted. Jeff figured half the board at that point thought Bolze would be his successor. And the board wanted to run the succession process, a distinct change from what Jack did. Jack ran succession back in the day. "He was the King and the Pope and the Holy Roman Emperor," Jeff once said about Jack. Jeff decided to respect the wishes of his lead director. "I didn't want to be the imperial CEO," he recalled. Bolze could stay, running the newly expanded power business. "I made what I now see as a terrible mistake," Jeff continued, "maybe the worst mistake I ever made: I listened to [the board] and acquiesced. In doing so, I violated one of the essential rules that I lived by: the company always comes first, ahead of the individual. Leaving Bolze in his job is something I'll always regret."

## Chapter Twenty-Five

# A FOX GUARDING
# THE HENHOUSE

About a month before GE closed the Alstom deal, Jeff made another fateful decision that would have serious implications for both GE's future and his tenure as CEO. He got into business with Trian Partners, a hedge fund founded by Nelson Peltz and Peter May that was well known for being aggressive, ruthless, and relentless in getting its way. Ironically, Jeff invited Trian to become a top-ten GE investor in the hope that it would be a vote of confidence for his two huge strategic moves in 2015: getting out of GE Capital and completing the Alstom deal. Adding to the irony was the little-known fact that Jeff had known Ed Garden, Peltz's partner and son-in-law, for years and was a Dartmouth classmate of Garden's brother, Tom. While at Dartmouth, Jeff was a visitor to the Gardens' home—a three-family house in Melrose, Massachusetts. "Tom and I were good friends," Jeff told me, "and that's how I got to know Ed," who went to Andover and to Harvard. Jeff had known Ed Garden since he was an upper, or junior, at Andover. Jeff used to joke that he would visit the Gardens' "dump" in Melrose. The first time Ed Garden met Jeff was early on a Saturday night at a fraternity party at Dartmouth, where Jeff was "holding court" with six or seven guys around him. "He had command of the room," Garden said. "I remember my brother at one point telling me, 'This guy is going to be the most successful guy in my class.' Over the years, that sort of turned out to be true."

Garden's firm was one of the most powerful pioneers of a new breed of activist investors. Peltz and May started investing in the 1980s and soon gained a reputation as "turnaround experts," buying troubled companies, fixing them up, and then selling them for a tidy profit. They invested their own cash in a money-losing conglomerate, Triangle Industries. They turned it into one of the largest packaging companies in the world, American National Can. They made a fortune when they sold it to Pechiney, a French conglomerate, in 1988 for nearly $4 billion, including the assumption of $2.6 billion in debt. The two partners' stake in American National Can was worth nearly $800 million, a tidy sum back then (and still). They were just getting revved up. They bought into DWG, a troubled holding company that owned Arby's, and renamed it Triarc, which then bought Snapple from Quaker Oats, turned it around, and sold it to Cadbury Schweppes for another fortune.

Garden joined the business in 2003 with the idea of bringing in outside capital into the firm to do more and bigger deals. That effort became known as Trian, with a similar mandate as Triarc: take stakes in public companies, help turn around the operations and management, if needed, and create value for themselves and other shareholders. Trian made a series of high-profile investments in companies such as Wendy's, Heinz, DuPont, P&G, PepsiCo, and Mondelez, among others. Over time, Trian gained a reputation as savvy, supportive investors as long as the operations, financials, and stock price were improving. If not, they were not shy about insisting on changes to management and to strategy, and if they didn't get their way discreetly, they were happy to make their fights public in order to get board seats and then to make their argument for change.

They wanted to invest in great companies where management might have lost the plot operationally. They wanted to bring the ownership mentality back into the boardrooms of publicly traded companies. "What Nelson and Peter had really demonstrated over a long period of time was corporate turnaround skills," Garden explained to me. In other words, they were a serious, no-nonsense, and determined duo.

Peltz had been on Jeff's radar since 2007, when Jeff reached out to him to learn more about how he thought about running businesses, solving problems, turning them around, and getting the best results. "He started spending some time with Nelson and picking his brain," Garden said. In *Hot Seat*, Jeff explained that he had read, in *Fortune*, about Peltz's proxy fight at Heinz and how the Heinz CEO ended up appreciating Peltz's input on the Heinz board. "I liked the sound of that," Jeff remembered, "so I reached out to take his temperature about possibly joining GE's board. . . . I knew Peltz's hard-nosed reputation, but I figured if I invited him onto the board (instead of having him force me to accede at the end of a bayonet) he might be in my corner."

They discussed how to reduce complexity in big organizations, how to structure compensation packages to get the desired results, and how to ensure accountability. Their conversations, which occurred regularly over the years, were philosophical, not about Trian investing in GE. After the financial crisis hit and GE was fighting for its life, Jeff floated the idea of Peltz joining the GE board. Ken Langone, for one, thought Jeff was just taunting Peltz by suggesting he wanted him on the board. When he called Langone to tell him the news, Langone was skeptical. He didn't think Jeff wanted to mix it up "with fucking gunslingers," like Peltz or Carl Icahn, another hard-boiled activist investor. Jeff kept telling Peltz he was going to put him on the GE board. But month after month it didn't happen. A few months later, Langone and his wife invited Ralph Larsen and his wife to La Grenouille, in Midtown Manhattan, for dinner. "The girls are on a banquette, and Ralph and I are on chairs looking into the banquette," he said. "I'm not above a little mischief."

"Jesus," Langone said. "Ralph, I hear you're going to put Peltz on the board."

"What?" Larsen said.

"Yeah, you're going to put Peltz on the board," Langone said.

"What are you talking about?" Larsen said.

"Ralph, it's very simple," Langone said. "Nelson Peltz is going to be put on the GE board."

"I don't know anything about this," Larsen, the lead GE director, said.

"Ralph, pardon me," Langone continued. "All I know is I've heard— I won't tell you where I got it—I heard. All I know is I've heard."

"Well, it can't be," Larsen replied. "I'd know about it. I'm the lead director."

"Well, I assume you would," Langone said.

He looked at me and said about Jeff, "He was having these talks with Peltz and never told Larsen."

Then came Jeff's invitation to Peltz to speak to GE's leaders in Crotonville. In 2014, after GE became a SIFI, Jeff invited Peltz and Garden to visit with him in Fairfield to give him their advice. By then, Trian had become an investor in the Bank of New York and had learned firsthand what it meant to regulated by the Federal Reserve. "Just doing it as a friend," Garden recalled. There was still no thought of Trian investing in GE. Garden continued, "We said, 'Look, if it were us, we would be getting out of GE Capital because the environment has changed so drastically and it's so hard for you to compete now. We would be getting out of GE Capital and making GE a pure-play industrial business. For what it's worth, it's free advice. If we owned a hundred percent of the company, it's what we would do.'" Jeff listened intently.

A few weeks after Jeff announced Project Hubble, in April 2015,

Jeff got a call from Joe Perella, the famous Wall Street banker who, after an impressive career at Morgan Stanley, First Boston, and Wasserstein Perella, had joined with Peter Weinberg, part of the Goldman Sachs royal family, to create Perella Weinberg Partners. "Okay, Jeff," Perella said. "You thought you were out of the deep end, but you're not." He wanted Jeff to know that four or five activist investors had focused on GE now that it had decided to sell off GE Capital.

Jeff wasn't naïve about the possibility that an activist investor might pounce on GE but always believed he or she would have to be "pretty dense" to buy into GE as long as GE Capital was a SIFI. Perella agreed. "None of them would touch you with a ten-foot pole when you were a SIFI," Perella told Jeff. "But now that you're on your way off that list, here they come." He cited ValueAct Capital and Trian, in particular, as two funds that he was aware of that had bought some GE stock and were pondering what to do. "Look," Perella continued, "your stock has underperformed. You're a conglomerate. You've cleaned up, in theory, what they saw as a toxic mess. Now these guys are going to come in and say, 'Fire the CEO, break up the company, blah, blah, blah.' I just wanted you to be aware."

Shortly after he got off the phone with Perella, Peltz called Jeff to congratulate him on his decision to pivot out of GE Capital, essentially taking Trian's advice. "I remember our meeting," Peltz told Jeff. "I think that's a smart move on your part." Jeff told Peltz the many good things that were underway at GE and that he wanted to figure out a way for Trian to become a "big shareholder" in GE. "We'd love to have you in the stock," he told Peltz. By then, he said, GE had done enough analysis to determine that Peltz was not a "bomb thrower" but rather would be "more reasonable" and "constructivist."

Jeff had enlisted Bornstein, his CFO, and Matt Cribbins, the head of investor relations, to figure out which activist investor might be better to have than others. Jeff figured there was "an extremely high likelihood" that an activist would swoop in, and he hoped that the devil he knew would be better than the one he didn't know. "We're big, but we're not so big," he told me. "These guys with one percent, they can cause trouble." He said that, perhaps "naively," he thought Trian could provide the market with a vote of confidence for Jeff and the "new GE," similarly to how Warren Buffett had helped the perception of GE in the depths of the financial crisis. "There's a lot of people that haven't bought the stock for a long time who've sat out during this process," Jeff explained. "This could be a catalyst for the stock. . . . We had been through an exceptionally difficult six- or seven-year time period where we were just uninvestable for lots of people out there." Blair Effron, one of Jeff's most trusted banking advisers, suggested inviting Peltz into GE. Jeff agreed that Effron could be right. He also believed that the best activist investors could be helpful and insightful. He was hoping for the best from Trian. "If we do well, will it get ugly?" Jeff told Effron. "No. And if we don't do well, will it get ugly? Yes. But why shouldn't it? If I'm fucking up, why shouldn't somebody beat me up?" (For their parts, neither Peltz nor Garden recalled that Trian had been *invited* to be a large corporate shareholder. Usually, it worked the other way around, where Trian stealthily bought stock in a company it believed was undervalued or could use a good goosing from it and then, when it crossed the threshold for making a public ownership filing, its presence became known. And Jeff said he knew Trian was already in the GE stock by then.)

In the end, Trian bought a $2.5 billion stake in GE, its largest single investment ever. If GE did what Jeff promised it would do—$2 a share in earnings by 2018—he hoped that Trian would be quiescent and supportive of him and the rest of GE management. In retrospect, perhaps Trian should have been wary of Jeff's invite, or whatever it was, and may have miscalculated given the unusual combination that Jeff was literally a friend of the family and also a CEO looking to burnish his reputation and his legacy, figuring the Trian imprimatur would help ratify Project Hubble as well as the Alstom acquisition. "Jeff was always looking for validation," explained one senior GE executive. "The whole Hubble [project] and all that stuff is going to take three or four years to unfold that whole process, and so in the meantime the stock's not doing much. A little bump when Hubble's announced, but it's not doing anything. The dogs are not wolfing down the dog food. I'm sure he was like, 'If I could get somebody to endorse this, if I could get the investors behind me here, that would be a good thing. Even if the investor happens to be a jail bomb.' And who's better validator than Nelson Peltz, right?" Recalled Garden, "I remember saying to Immelt: 'Understand that we're going to hold you to the commitments you're making to us, and if you don't deliver, despite the relationship'— I remember using the words—'we'll hold you accountable.' It was an awkward conversation."

After accumulating its $2.5 billion position in GE through the course of the summer and early fall of 2015, on October 5 Trian announced that it had made its largest investment ever in one of the best-known companies in the world. GE's stock rose 4.5 percent, to $26.60, on the news that Trian had amassed about a 1 percent stake in the company. Trian predicted the GE stock would be trading at as much as $45 by the end of 2017 if the company delivered on its promises and said it backed the "transformation" of GE that Jeff already had started. Peltz said GE and Trian shared "much common ground" and that Trian would not be asking for a board seat. Trian released an eighty-one-page "white paper" about GE, which delineated, investment banker style, the problems GE had experienced during Jeff's fourteen-year tenure and why it believed his "new GE" plan would deliver the near doubling of the stock price that Trian predicted. "Transformation underway . . . but nobody cares," Trian declared on the cover page of the PowerPoint, reflecting Peltz's and Garden's view that investors still viewed GE as a dead-money investment. Trian's thesis was that while GE had "world-class" industrial businesses, it was being dragged down by GE bureaucracy and by GE Capital. But Jeff's decision to "pivot" changed the calculus for Trian: "Its great businesses were overwhelmed by the bad ones and the underlying defensive growth of GE's core industrial businesses was obfuscated." Trian's analysis concluded that GE "can earn at least" $2.20 a share by 2018, some 10 percent more than Jeff had promised. Trian also seemed focused—understandably—on how much money GE would be returning to its shareholders in the form of stock buybacks and dividends. Trian viewed all this as achievable based on what Jeff and the company were telling the firm's partners. It was the art of the possible.

Trian wanted no less than a stunning $110 billion returned to shareholders. Its calculus included another $22 billion in stock buybacks using new debt to fund them and another $4 billion or so in stock buybacks from GE Capital assets sales that achieved higher prices than GE expected. Seventy pages in, Trian shared its formula for how GE would get to $2.20 a share by 2018: a combination of earnings from Alstom—Trian declared its public support for the deal ("We believe the Alstom transaction creates value")—the aforementioned stock buybacks and dividends, and long-overdue operating-margin improvements (themselves a combination of cost reductions and improvements in gross margins). *Et voilà*. Trian's alchemy looked highly credible on the pages of a PowerPoint presentation. But could Jeff and his team deliver on the promise of the "new GE," an unprecedented transformation of the company? "I think they'll execute," Garden told CNBC that same day. "If they don't, then all options are on the table."

After fourteen years of trying to turn around the GE battleship, by the end of 2015, Jeff seemed to finally have it going in the right direction, or at least the direction he wanted to go in, which was some combination of an industrial and technological powerhouse. It was no longer the least bit relevant that Jack had left him the world's most valuable company to run, one that was fine-tuned to perfection. That GE no longer existed. It was Jeff's GE, and had been for years. He seemed more than satisfied with what he had wrought.

GE Capital was being unwound and would soon be a mere 10 percent of GE's profits. The SIFI designation would soon be a thing of the past. The Alstom deal had closed, and if Jeff and the Power team were to be believed, the profits and "synergies" would soon be rolling in. Trian, while not exactly Warren Buffett, had made a $2.5 billion vote of confidence in GE, seemed to ratify everything Jeff had been doing, and looked forward to getting its share of the $110 billion that it wanted Jeff to return to shareholders. And with Trian, Jeff believed he had brought a friendly face to the party, one he had known for decades and one he was hoping would not bite.

Sure, there were some loose ends. GE's oil and gas business was struggling. "Oil and gas is in a tough neighborhood now," he said. Flannery was still in the process of fixing the health care business. Alstom still had to pay off, of course. "I completely believe in the strategic rationale of the deal," Jeff told investors in December 2015, "fantastic opportunities. We see new and different ones every day. [But] it took a long time to get done and anytime you go through that the business is going to suffer."

GE's $3.3 billion deal to sell its appliance business to Electrolux, cut in September 2014, was nixed by regulators in December 2015. But Jeff was plenty optimistic that was just a blip. "We have had a bunch of people that are ready to step in and pursue this," he told investors in December 2015 at his annual state of the union, live from the *SNL* stage, which GE rented from Comcast for the occasion. "I expect it to happen relatively quickly. I can never predict exactly." He managed to joke about the unexpected regulatory denial and how he hoped the next buyer of the business would have a smoother regulatory experience. "It would have to be," he said, to laughter from the audience. "I remember when I used to work for Jack, I would call out a person and I would say, 'This person is really better than you think,' and he would say, 'He would have to be.' So the regulatory process is going to be smoother next time around probably."

From the *Saturday Night Live* stage, Jeff proclaimed himself happy and that most everything was great at GE. "I am really proud of the GE team this year," he said. "We got a lot done. We got a ton done this year. I would say essentially we completed the portfolio pivot." The next day, at a smaller breakfast meeting at 30 Rock for a group of investors, one unnamed hedge fund investor, who had been one of Jeff's harshest critics, told Matt Cribbins, "A year ago, only twenty-five percent of the people in this room would have supported Jeff. Today he is a god. How quickly things change."

THE NEXT YEAR, 2016, WOULD LIKELY BE THE CRUCIAL YEAR TO DIS-cover whether Jeff would go down in history as the god who remade GE for the future or the goat who lost control of one of the greatest American companies ever conceived. The year started with more victory laps for Jeff. On January 13, GE announced it was moving its headquarters to Boston, leaving behind its Fairfield County campus, where it had been since 1974. The move to Boston, Jeff declared, was more in keeping with the future he had staked out for the company, in industrial technology. (It was also a nod to the company's origins.) "Boston has a chip on its shoulder, like GE," Jeff wrote in his annual letter to investors. He once described his view of the city, where he got his MBA, as having "three faces." There was the "Brahmin face," Jeff said, as "the country's first city, Pilgrims and all that stuff." Then there were "the schools, which is MIT and Harvard and Tufts, Northeastern, and BC. So you have an intellectual kind of centerpiece." Then he shared his view that Boston was "the country's best sports town" with "the chip on its shoulder" and "the aspect of competitiveness, the townie in Boston." He wasn't making these observations lightly. "I see pieces of GE in each one of these three faces of Boston," he said, "the heritage. We're a third century company. We are a technology company. But we're also a company [where] we never think we're good enough. We always are trying to be more competitive. I like that." What he said he "always loved" about GE was that it was a company that "rejects arrogance" and "wants to be better."

The plan was for eight hundred GE employees to work in the new headquarters, to be built in the Seaport on Boston Harbor. GE bought a 2.5-acre parcel of land along the waterfront from Gillette, which was owned by P&G, and planned to renovate two historic buildings, once the headquarters of the New England Confectionery Co., or NECCO, and to build a new twelve-story futuristic glass-and-steel box in what used to be the NECCO parking lot. The headquarters would place a greater emphasis on "innovation," the company said. Of the eight hundred employees moving to the new headquarters, two hundred would be corporate types; the other six hundred would be "digital industrial product managers." GE intended to start moving people to Boston in the summer of 2016 to a temporary headquarters while the Seaport location was constructed. The Fairfield headquarters was sold to Sacred Heart University for $31.5 million in the fall of 2016. GE's remaining office space at 30 Rock, in New York City, was also sold. At a talk to a group of JPMorgan Chase bankers a few weeks later, Jeff explained the logic for the move. "Reg Jones . . . he lived in Greenwich and taxes in Connecticut were zero. So along with a lot of other companies, we moved the headquarters of GE to Fairfield County. . . . Today, they're all basically gone. We're kind of the last person in Fairfield County." He wanted GE to be "in the flow of ideas," and Boston fit the bill.

When Jim Himes, the U.S. congressman representing Fairfield County (and the man who helped save GE's hide during the financial crisis by appealing to Barney Frank to amend what would become the Dodd-Frank law so that it wouldn't hurt GE), got wind of the news that GE was leaving Connecticut, he and the state's governor, Daniel Malloy, went to see Jeff. "Hey, give us a chance here," they said. "We will match an offer that any other state makes for you." But it didn't work.

Even though GE didn't love the taxes in Connecticut, Himes said Boston was chosen because it was an easier location for GE to recruit the newfangled workforce it coveted.

Then, on June 28, 2016, the Financial Stability Oversight Council voted to rescind GE's designation as a SIFI, making GE the first big company to lose that label. GE Capital had been a SIFI for less than three years, but in that time Jeff had made the existential decision that GE could no longer be in the financial services business because it could no longer compete effectively or earn more than its cost of capital. The FSOC had taken notice. "GE Capital has fundamentally changed its business," the council wrote in its final determination. The FSOC noted that since 2012 GE Capital had divested $272 billion of bank and non-bank assets, the largest of which was the IPO and subsequent share exchange of Synchrony, which resulted in the divestiture of $87 billion of financial assets.

As important, GE Capital had nearly abandoned its short-term-funding model, reducing short-term borrowings to $14 billion at the end of 2015, from $98 billion at the end of 2012. Commercial-paper borrowing had been reduced to $5 billion. GE Capital's long-term debt had also been reduced, to $159 billion at the end of the first quarter of 2016, from $269 billion at the end of 2012. What's more, GE had agreed to guarantee "substantially all" of GE Capital's debt, and GE Capital had established a payable to GE to service GE Capital's debt.

What was left of GE Capital were three businesses directly tied to GE's industrial businesses: an aviation finance business, an energy finance business, and an industrial finance business. "The changes at GE Capital since [2013] have significantly reduced the potential for the negative effects of material financial distress at GE Capital, including any effects resulting from potential resolvability challenges or the company's complexity, to be transmitted through each of the three transmission channels described above and thereby to pose a threat to U.S. financial stability," the FSOC concluded. GE was finally free of the

Fed's oversight. "We had accomplished what many said was impossible," Jeff concluded. "We could focus once again on running the company."

THE PROBLEM FOR JEFF WAS THAT EVEN A GE PURGED OF GE CAPITAL remained complicated to run. His biggest headache derived from the power business. He and Bolze continued to clash, a fact that somehow was kept to a small circle of top GE executives and board members. Even Bolze's direct reports hadn't a clue. "Honestly, I thought Jeff really liked Steve," one of them told me. Jeff was quickly coming to the realization that the numbers for Power, post-Alstom, that he (and Bolze) had promised investors were unlikely to be met. And that put into serious jeopardy Jeff's repeated promise of achieving $2 a share in earnings for 2018. Without the $2 a share—to say nothing of the $2.20 a share that Trian had expected—a showdown between Trian and Jeff seemed increasingly likely. The hedge fund's fangs would soon be flashing. "There was just this unrelenting push for growth, and when the strategy wasn't working about replacing GE Capital earnings, there was just enormous, enormous pressure on everyone to keep showing growth," explained one GE executive. "That put a lot of pressure on the power business and the services business to show that they were continuing to grow earnings, especially in a declining market." He said the Power executives miscalculated the market demand for the H turbine. The product was considered superior to Siemens's similar product, but not so much better that GE would be justified in having a different view of the market. "For the last three years," he continued, "before 2018, Siemens's public statements about what they thought about the market were dramatically different than what GE was saying. GE was saying year-over-year growth and 'This is what we're going to do,' and Siemens was kind of being very negative about what the market was going to look like, and it seems like Siemens was right and we were dead wrong." He said it wasn't a question of integrity but rather "I saw a lot of bravado. I saw a lot of arrogance. I saw some misplaced arrogance."

Once again, Jeff had to confront his growing vendetta against Bolze. He explained that he began hearing from his senior executives, Bornstein and Cribbins specifically, that Power's "leadership team" had once again "checked out." "They were begging me to replace Bolze," Jeff explained. Jeff remembered that twice in 2016—in May and November—he "pleaded" with the GE board "to remove Bolze." He told the board that "we were putting a huge GE business at risk leaving him at the helm." Jeff was particularly peeved (again) at Lynn Calpeter, the CFO of the power business, for a presentation she gave in August 2016 at a GE officers' meeting. The draft of her presentation, according to Jeff, failed to explain what the power business was doing to fix its working-capital problems, an important managerial conundrum, for sure, but hardly existential. "It seemed half-assed to me," Jeff said about her presentation. "The entire company was depending on her. In my opinion, she just wasn't taking her job seriously enough." He called her up after reviewing the draft. They had known each other for around twenty-five years and had worked together in the plastics division. "Lynn, in all my time at GE, this is the crappiest presentation I've ever seen," he told her. "It's lazy. Get your shit together." (He fired Calpeter in early 2017. She could not be reached for comment.) Jeff cited the power business as the only one at GE that couldn't meet its financial goals. But once again, Jeff "backed down" from his request to fire Bolze after the GE board made clear to him that "it would look bad to remove a CEO candidate in the midst of the succession process," which, despite the public pronouncements that Jeff would be around until 2020, "was now fully underway."

BY THE END OF 2016, IT WAS INCREASINGLY CLEAR THAT GE WOULD not make the numbers it had promised investors it would make.

Somewhere between the problems in the oil and gas business and the power business, it just wasn't coming together as Jeff had planned. Trian decided it was time to ratchet up the pressure on GE. In December 2016, Jeff and Matt Cribbins choppered to New York to have lunch with a group of investors. "There was much to talk about," Jeff recalled, and little of it good. Later, Jeff flew to White Plains to meet with Peltz and Garden in a conference room at the airport. Jeff remembered that the "veneer of friendliness" between Trian and the GE executives was "still evident." Peltz, who had supported Donald Trump for president, reported to Jeff that he had been on the phone with the president-elect to discuss cabinet positions. That was the end of the small talk. "Jeff, I want you to know I love you like a son in a lot of ways," Peltz told him, "and I would never do anything to harm you, but we're at a point where we need to make a change. I want you to announce that you're going to take out $1 billion of costs. If you don't announce that, we're not going to get into a public fight with you. But we will make it known that we're selling out of GE." (For his part, Garden said he had no recollection of threatening to sell out its position in GE's stock.) Jeff told Peltz and Garden that he and Bornstein were already working on cutting costs, given the unanticipated slowdown in the power and oil-and-gas businesses. "Peltz's threat unsettled me," Jeff remembered. "For a major GE investor to walk away could have serious implications in the market." (By then, Trian, which had invested in GE at an average price of around $23 a share, had sold a third of its position when the stock reached around $32 a share during the summer of 2016. At the time of the December 2016 meeting in White Plains, GE's stock was still around $31 a share.)

Peltz also told Jeff that he wanted Garden to have a seat on the GE board. Jeff didn't like that idea; he didn't want Trian involved with picking his successor. "But the request made me aware, more than ever before, that my time as CEO was growing short," he recalled. He also acknowledged that Trian was right. "To the extent that Trian was kicking our ass, we deserved it," Jeff allowed. "We hadn't adjusted costs as quickly as we should have."

That was only the beginning of Trian's teeth baring. First, in February 2017, there was a meeting at Trian's offices, in New York, between GE CFO Jeff Bornstein, Matt Cribbins, Nelson Peltz, and Ed Garden to discuss the fact that GE was going to miss its 2016 EBIT—earnings before interest and taxes—by $1.6 billion. Garden was incensed. He told Bornstein that if the miss was going to be that large, and GE executives knew it, he was obliged to cut some of GE's already bloated $100 billion of expenses. "Other companies that we invest in we take out 10 percent of the cost regularly, and you can't find 1.6 percent?" Garden told Bornstein. "Because understand, you have to start hitting your numbers." Trian just wanted GE to deliver on what it had promised.

Garden wasn't done. "You made public commitments to us and to everyone else, and you have to start hitting your numbers," he admonished, "and if sales are weak, then as a management team you get paid to take out costs." It was made clear to Bornstein that GE management had to take the costs out of the business that Trian was demanding—or that heads would roll. Bornstein didn't disagree with the Trian partners, although he told me he wouldn't have shared with Trian the preliminary 2016 numbers—what he called "material nonpublic information"—and he doesn't recall his job being put on the line during that meeting. He was already very focused on finding ways to reduce the spending that GE was doing on Jeff's Predix software project and others of his global initiatives. "Help me, then," the GE CFO told them, as in "Help me with Jeff."

A month or so later, Garden flew up to Boston to see Bornstein on a Sunday. They met at Bornstein's townhouse, on Marlborough Street, in Boston's Back Bay neighborhood. By then, GE's 2016 results were in. At that meeting, Garden proposed a plan whereby bonuses for top GE executives would be tied to hitting operating earnings numbers and cost reductions. "And if you don't execute," Garden told him, "there will be consequences. We're going to want to go on the board," something Jeff still did not want to happen. Without prompting, Bornstein offered up to Garden, "If we don't make these numbers, I'll go." (Message received: in a March 21, 2017, filing with the SEC, GE announced that it had reached an agreement with Trian to reduce selling, general, and administrative costs by $1 billion in 2017 and another $1 billion in 2018 and tied management bonuses for those years to achieving those reductions.)

But some of the proposed cuts were irksome because they seemed to cut into GE's muscle rather than its fat. For instance, Jim Suciu, a customer executive in the power business, got the axe, causing other GE executives to scratch their heads in disbelief. "Jim was legendary as the customer-contact person at Power," one of them told me. "Knew every single utility, knew every single deal. He was just like a walking encyclopedia." He asked Bolze how Suciu's institutional knowledge would be replaced. "Are you guys out of your mind?" he wondered.

"Well, he's really expensive," Bolze supposedly responded. (For his part, Bolze had no recollection of saying this about Suciu; rather, he explained, Suciu, his friend, was looking to retire from GE after thirty-eight years.)

"Okay, but why?" the executive responded. "There's a reason we pay Jim that much. It's going to take twenty-five people to replace Jim Suciu, and nobody's going to have his knowledge of the customers or the access he's got." He was incredulous. "They package him out in a very shortsighted move because we needed to get rid of the guy who's making $5 million here because we needed to meet the cost-out number," he said.

Soon into the new year, Jeff again had to confront Bolze. Jeff was pissed this time because the power business had missed the earnings Jeff believed Bolze had committed to making and what had been told to Wall Street. The power business was supposed to make $5.4 billion;

it came in at $4.95 billion, more than a $500 million miss on the operating-profit line. Bolze had told Jeff privately, and repeatedly, that the operating income number for 2016 would be closer to $5 billion than to $5.4 billion, and it was. Jeff kicked his ass anyway and told him that the whiff in the power business was why GE had missed its earnings representation for the year. Worse, in Jeff's mind, was the fact that Bolze didn't seem to realize he had missed his 2016 numbers. When he ran into Cribbins at what was supposed to be GE's temporary headquarters, on Farnsworth Street in Boston, Bolze stopped to give Cribbins a high five. "Look at our operating income and earnings," Bolze said. "It's even better than we thought." Cribbins then went to see Bornstein. "I don't think Bolze has a clue," he said.

On a January 20 earnings call, in answer to a question from a research analyst, Jeff said, "Power didn't do as much revenue as we would have liked. It was still substantially up organically, up 5 percent organically without Alstom and up 15 percent with Alstom. So that's not bad." Bornstein agreed, "Power was light, no question." In an interview, Jeff said the problem in Power was the dysfunctionality of the Power team. They just weren't executing. "Here's one of the things that's difficult if you put yourself in my shoes," he explained. "We had this company where Aviation does pretty well, Healthcare does pretty well, Transportation does pretty well, Oil and Gas does pretty well. And Power shits the bed. The problem could be with me, right? Or the problem just could be that the people we had in place at the time just didn't do what they should have done." Jeff conceded that maybe GE overpaid for Alstom by $2 billion, quite an admission by someone who had refused to heed the advice of people like Steve Bolze to negotiate for that level of price reduction for Alstom when he had the chance. "I would take that heat," he continued. "Maybe, could be. But it wasn't like this was bizarre. . . . At no time did I think I was redlining the company by doing Alstom, or had pulled the wool over my board. . . . They were intimately involved."

Once again Jeff had "had enough" of Bolze. He told the board for the third time that Bolze had to go. "This is embarrassing," he said. Again the board said no, citing the looming succession process. The board still wanted Bolze to be seen as a candidate to succeed Jeff. Once again, Bolze had no idea that Jeff kept trying to fire him. "The world perceived Bolze to be something we believed he was not: a strong candidate to replace me," he continued. "We knew at this point that the board would not make Bolze CEO, and that when he didn't get my job, he would likely exit on his own." The reason for keeping Bolze atop the power business, Jeff asserted, was that his successor would be named "within a year" and Bolze would then leave on his own accord without "any bad press," so "why ignite a media firestorm by firing him now?"

By early 2017, *dysfunctional* did not begin to describe the relationship between Jeff and Bolze, and it was hurting GE operationally and financially. The personal conflict seemed to drive Jeff to deep introspection. It had been a brutally difficult decision for Bolze to return to be the head of GE Power after he had decided to quit and then changed his mind and asked to reinstated. Jeff took him back, yes, but he knew—and had been told by Susan Peters—that his chances of succeeding Jeff were close to zero as a result. Still, he soldiered on. But Jeff remained relentless in his attack on Bolze and the Alstom deal. "I acknowledge that the Alstom deal didn't work as we'd envisioned," he wrote in *Hot Seat.* He wondered—again—if he should have figured out a way to get out of the deal when he had the chance. That would have resulted, for sure, in a lawsuit in France. "It was a complicated transaction and we had a team in disarray and a pending CEO transition—it was just too much," he continued. "When in doubt, don't."

One of Jeff's mantras over the years has been "Truth equals facts plus context." He said he wrote *Hot Seat* to try to get at his definition of the truth. "I felt like all the context had gone away from GE over the past few years," he told Anthony Scaramucci during one 2021 interview. "Basically, we did more good things than bad things, but people were dwelling on incomplete truths." He regretted not trying to change the Alstom narrative earlier when he had the chance. "Leaders who were bullish on Alstom and thought GE could win in the market were silenced," he concluded. "People who'd failed were given too loud a voice. The result: with Alstom, GE Power was stopped at the starting line."

The power business was under increasing pressure to make its numbers. Without that business performing as Jeff believed it could, and should, the prospects for GE making $2 a share in 2018 were dimming rapidly. Like Jack in many ways, Jeff was unsympathetic when it came to his business leaders making their numbers, especially when so much was on the line for his own credibility. "He was saying, 'Oh, stop your crying,'" remembered one GE executive, when it came to concerns about making the numbers Jeff expected. "He used to make little crying symbol with his hands around his eyes. He'd do that all the time and say, 'Waah, can't make the number, waah, where's my money?' Oh, all the time. 'You guys don't want it enough. You guys aren't hungry enough.' It was almost a manhood thing."

ONE WAY TO MAKE THE NUMBERS THAT JEFF EXPECTED FROM THE power business in 2017 was to sell more of what were known as "upgrade kits," which were akin to putting a new engine in an old car and giving it a new lease on life. "It's heavy metal," explained one GE executive. "You're ordering tons of steel and you're upgrading the gas-burning section of the turbine. If somebody doesn't want to buy a new turbine, it's like, 'Hey, can you put a new engine in my car, and I'll drive it another thirty thousand miles?'"

The "upgrade kits" were very profitable for GE and relatively more affordable and profitable for GE's customers than buying a new gas-powered turbine. It was a very good business for GE, but as the cycle got longer in the tooth and demand for the "upgrade kits" was fading a bit, GE would get more lenient on terms, allowing customers to pay over years, or simply rejiggering the service contract to account for the purchase of the upgrade. Word was, around GE, that the manufacturing and the sales forecast for the new "upgrade kits" fell to two executives below Bolze: Paul McElhinney, who was in charge of the highly profitable service business in Power and who had worked in a variety of senior legal positions in GE's aviation business, and Joe Mastrangelo, who was in charge of Power's new units and supply chain. According to one knowledgeable GE executive, figuring there wasn't a prayer of making the Power numbers that Jeff wanted—and thus his promised $2 a share—without the "upgrade kits," McElhinney sought out Mastrangelo and asked him to start building a sufficient number of them to allow Power to make the numbers.

Mastrangelo was incredulous. "No way," he told McElhinney, "we don't have orders for it or any visibility." McElhinney pushed back. "Listen," he told Mastrangelo, "there's no way we can make the numbers if we don't sell this many upgrade kits, and there's no way to sell this many upgrade kits if we don't start building them now. I agree I [have to get the orders] but you have to build them now! We can't sell them if you don't build them." Building the kits without the orders was something like a $2 billion to $3 billion investment on the materials to make them. "This is crazy." Mastrangelo said, "You don't build that much without orders."

A major conflagration broke out between McElhinney and Mastrangelo. "These guys were all dysfunctional, they're all fighting with each other," explained one GE executive. "Bolze is clueless. Mastrangelo and McElhinney don't like each other. Everybody's living in different cities. It's a very fucked-up team." Part of the problem was that McElhinney had a direct pipeline to Jeff and vice versa, especially so given that Jeff had lost confidence in Bolze. Instead, he turned to McElhinney if he wanted things to happen in the power business. One person who worked in the power group remembered how difficult McElhinney could be, especially when executing on a mandate that seemed to come down from Jeff. About the request for the "upgrade kits," she said, "I just kicked myself every day. You can't believe the environment with Paul and what an incredible bully he was. And screaming. I knew what we were doing was not good. But he made it so brutal. I kick myself for not having stood up to him and [said], 'Hey, this is just ridiculous.' Everyone was like that. The whole environment was in total fear of that guy. And this guy had a direct line to Immelt, so if Immelt just called the guy and said, 'Here's what I need. . . .'" It was horrible, and I'd come home to my husband all the time and say, 'I can't believe this.'" She appealed to Bolze. "You got to help me here," she said.

"I can't really do anything about this guy," Bolze supposedly told her. (Bolze told me he did not recall saying this.)

Finally, to try to quell the infighting, Mastrangelo agreed to build the upgrade kits, spending the billions required. But then, as he had predicted, the "upgrade kits" could not be sold at any price. GE got stuck with the unsold inventory. The executives joked, with gallows humor, that maybe the thing to do was to melt it down, since it could not be sold. "That's sort of emblematic of the operation," explained the GE executive. "The core operations of the company had gone haywire, especially in the power business, trying to get somehow to some operating-profit number that fundamentally was very disconnected from what was sensible in the market."

Jeff told me he had "heard" about the "upgrade kit" fiasco. He blamed Bolze. "They had purchased two hundred fifty of them and ended up writing off a bunch of them in 2018," he said. "That was stupid, really. I would never have done that. I would never have approved it. And I understand that's what they did. I would've expected Steve Bolze to be the one that makes those purchase decisions."

IN FEBRUARY 2017, *THE WALL STREET JOURNAL* HAD A SURPRISING scoop about the GE succession process. It obviously was a leak from the

GE board. The article said there were four candidates to succeed Jeff and included their pictures: Steve Bolze, fifty-three; Jeff Bornstein, fifty-one; Lorenzo Simonelli, forty-three; and John Flannery, fifty-five. Eighteen months earlier, only Bolze had been on the list; the other three had not. The *Journal* article explained that although the time had come to contemplate a successor to Jeff, given that he had been in the job fifteen years, he had just turned sixty-one, and the stock had not moved during his tenure (despite vastly increasing profits), GE's management had "bristled" when Wall Street research analysts had asked "repeatedly" about succession at a December meeting. "The tone and answers to the question suggested the leadership has zero intention of stepping down anytime soon, in our opinion," John Inch at Deutsche Bank wrote to investors.

The *Journal* did some handicapping. It said Bolze had "risen in prominence at the company" since the 2015 deal to buy Alstom and that GE said the Alstom integration was "ahead of schedule." (In hindsight, neither assertion was true.) As for Bornstein, his "bluntness" won the praise of Wall Street analysts and the "public support" of Trian. Flannery, meanwhile, had turned around GE's health care division, giving him a real shot. Only Simonelli seemed like a long shot, in the *Journal*'s calculus, because he had been tapped to become the CEO of Baker Hughes as soon as it was spun out of GE.

The *Journal* article took the four candidates by surprise. The process was being so tightly controlled at the level of Jeff, Susan Peters, and John Brennan, the lead independent director, that the four candidates themselves had never received the confirmation they were under serious consideration until the article appeared. True, there had been site visits by groups of directors, which would suggest some sort of participation in some sort of process. But that wasn't nearly the same as being told directly by Jeff, Peters, or Brennan. "There had been no discussion of any process for a long time," one of them said. "As in zero. No discussion of 'You're on the list, you're not on the list, you're just nothing.'" The *Journal* article began to make it real for these guys for the first time. But it was hardly the way to run a GE-style succession process, or what people assumed such a process would be.

Jeff was surprised by the *Journal* article, too. But then, he told me, he was not surprised by it, because, well, nothing really surprised him anymore. "I was always kind of on a path to retiring," he explained. "The date was on the books. It was going to be, I think, September 2017. I was cool with that." In retrospect, Jeff may have been "cool" with retiring in September 2017, but that most definitely was not the plan—or certainly not his preference—at the time. The plan was to make it twenty years, just as Jack had done. That plan, though, was looking increasingly unlikely. Trian, for one, was running out of patience with Jeff Immelt.

---

ON MARCH 10, 2017, TRIAN TURNED UP THE HEAT. IN AN ARTICLE THAT appeared on the Fox Business website, Charlie Gasparino reported that Jeff was "in the hot seat with Trian's Peltz." The article didn't have a lot of specifics and clearly had come from Peltz's camp. (Jeff believed Ed Garden made the call to Gasparino.) But the hedge fund seemed to be irritated that Jeff had missed "key performance benchmarks" such as increasing earnings growth and cost cutting. "How they can't find $1 billion in cost cuts at a massive company like that is what has Trian worried," an unnamed source told Fox. Gasparino reported that "people with direct knowledge of Trian's views" said that if Jeff didn't begin to right the ship, Trian would "begin formal discussions" with the GE board to seek his ouster. In a statement Trian told Fox that it and GE "continue to work constructively together to optimize shareholder value." Jeff didn't like the sound of that comment. "Rarely has a non-answer said so much," he recalled. "By not denying that my head was on the chopping block, Trian was sending me a clear message: it was." After the article appeared, Peltz called Jeff. "I just want you to know it wasn't me," Peltz told him. Jeff wasn't buying it. "That's when you know, right?" he said.

The next day, Jeff, his top managers, and the GE board members flew to Shanghai for three days of meetings. Gasparino's article created a challenging "ambience" for Jeff with his board, although the trip had been planned as a victory lap of sorts. The main purpose of the trip was to celebrate GE's many ongoing successes in China, including a new Chinese jetliner that, Jeff said, "had more GE content than any other plane on earth." His favorite night was a dinner he hosted in Beijing for the board and fifty Chinese customers, many of whom had been trained at a GE institute in China. Jeff believed no other company on earth would have taken such a trip with its board. Only GE would have attempted it. "But this trip also included many awkward moments," he remembered.

In particular, there was the two-hour executive session that the board had without Jeff. He figured the end was near. He was madly pacing outside the room where the board was meeting. One board member at the meeting explained why the switch flipped for the board in China about Jeff, "The board was committed to succession planning, and Jeff was leaving in a natural course in 2017. When it went from natural to unnatural was the first quarter 2017, when they were $800 million off in Power, and nobody could account for it and nobody saw it coming. We know now. The business was poorly run and not making any money, and not earning a return on capital." The board member couldn't understand how Jeff, as GE's CEO, could have predicted the company's earnings and then, by the first week of April, suddenly be about a billion dollars short. "What it really signaled to people is that we were too optimistic," the board member continued, "and this was the last straw. Remember, the previous six or seven or eight quarters, everybody was just complaining that one thing after another got in the way of the company's success. The Trians of the world and everybody else was saying, 'We're fed up with this.'"

Jeff could feel the vise tightening. "I didn't need a headline to tell me my time as CEO was almost up," he continued. Susan Peters had the unenviable task of being a special envoy between the board's secret deliberations and Jeff. "This is happening," she told him. "It's happening soon. You're just going to have to suck it up." Once again Jeff was both surprised and not surprised by what was unfolding in China. "Anytime the board wanted to fire me, they could, because they'd say, 'Look, the stock has underachieved,'" he told me. "I was a big boy about that. But if you look at the end of 2016, you had a $300 billion market value. You were number seven on the *Fortune* most-admired list. You were the number one company for developing leaders. It seems weird to talk about that today, but that's where we were."

Some believe that it was in Shanghai, at the two-hour executive session, that the board concluded that Jeff had to go. And that John Flannery, the CEO of GE's health care business, should succeed him. But that was all deeply under wraps, if it was even true. Jeff said he believed that was when the board decided to start a process that would move up the timing of his departure and that Flannery had been discussed as the likely successor. In Jeff's mind, the choice had come down to Flannery and Simonelli, even though Simonelli was young and had been slated to become CEO of a publicly traded Baker Hughes. Jeff was close to Simonelli. He had come around to the view, along with the board, that the GE CEO job was no longer a twenty-year assignment, and that Simonelli could be reclaimed from Baker Hughes. He thought a combination of Simonelli as CEO and John Rice as a vice-chairman was the right way to lead GE into the future. "Lorenzo would have been better than John Flannery in retrospect by far," he said. "By far. Not even fucking close. With each guy, I did strengths and weaknesses and was very clear. All of John's weaknesses played hugely. None of his strengths played through. And Lorenzo . . . he still has the strengths and weaknesses. But his strengths have gotten better, and his weaknesses have gotten less pronounced."

But the board, at least as things seemed to be developing in China, was leaning toward Flannery.

At one point, during GE's annual shareholder meeting in Asheville, North Carolina, on April 26, Blair Effron, the founder of Centerview Partners and a close Wall Street confidant of Jeff's, pulled Flannery aside and told him, "I think you're in good shape here, buddy," in a way that implied to Flannery that he knew what was happening. Flannery told him he knew nothing about what was going on. "I think you're in good shape here," Effron repeated. Flannery was never entirely sure. Ed Galanek, Jeff's head of security, remembered that Flannery came to see him when he was head of Healthcare. "I need to talk to you," Flannery told him. "I don't know if I can take it anymore. [Jeff's] really riding me." Galanek told him, "Stay with it. He loves you. You are the chosen one. He's gonna pick you to take over for him when he goes. He's just nuts with you in Healthcare because he loves Healthcare so much."

In April 2017, as the bell was slowly tolling for him, Jeff gave an interview to an unlikely source, Mike Damiano, a young, gifted journalist writing for *Boston* magazine, in GE's new corporate home. Damiano was struck by Jeff's calm, even though, according to Nicholas Heymann, a Wall Street research analyst, "the heat in the kitchen is way up" on Jeff. Damiano noted that Jeff had spent $7.5 million on a Commonwealth Avenue townhouse and was dedicated to Boston and his makeover plan for GE—as an industrial technology and top-ten software company—for the long term. "It's this or bust," he told Damiano, who asked the right question: Could Jeff accomplish the turnaround before "Wall Street shoves him out the door"?

He also revealed that GE had bought space for its logo on the Boston Celtics game jerseys, at a cost of around $9 million a year, and that GE had pledged to donate $50 million to the Boston public schools and to fighting opioid addiction in the region. After running through Jeff's personal and professional biography and his commitment to transforming GE, the article inevitably turned to whether Jeff would survive at GE long enough to see if his vision for the company would come to fruition. Even though he was roiling inside, Jeff maintained his cool with Damiano. "Immelt's holding on for now and has given no indication he wants to step down," he wrote. Damiano asked Jeff how he would judge his tenure as GE CEO. "You're probably not going to know the impact I've had on this company for years, decades in the future," he said.

Damiano told me that GE was "furious" about his article. Shortly after it went online, Deirdre Latour, GE's newest chief communications officer, called him at eight o'clock on Easter Sunday morning to complain. For more than forty minutes, she made it clear to Damiano that she was "very mad" about the article, which had come out hours before a *Boston Globe* profile that Latour told him she vastly preferred. "She said it was bullshit," he recalled. "She said Immelt did not have a Wall Street problem, that I was spinning up this narrative out of nothing, that relations with Wall Street were great, that Immelt was sticking around, that I got the story totally wrong." She told Damiano that he had "invented" the Wall Street pressure on Jeff—he clearly had not—and that she thought he had used "a snarky tone" to characterize Jeff.

After listening to her berate him, Damiano asked Latour what she wanted and urged her to send him an email if there was something wrong that she wanted him to correct. A few hours later, her deputy followed up. But there was nothing inaccurate that needed to be changed. "There was nothing to be corrected," he said. "There was nothing to be disputed. . . . But I did interpret it as extraordinary sensitivity and defensiveness from one of the biggest companies in the world. Just the idea that the chief communications officer of General Electric would be so triggered by this story that she's berating a freelancer in Boston about this at eight o'clock in the morning was amazing to me." He was left thinking he had struck a raw nerve. "I detected in the tone of the conversation a sense of desperation or panic," he continued. "It was weeks later that this all came to a head." He was incredulous that at about the same time as he was interviewing Jeff in Boston, Jeff's head was moving toward the chopping block. Jeff gave not the slightest hint of his inner turmoil. That's a certain kind of superpower. "Nobody was indicating that 2017 would be the final calendar year that Immelt would be in charge at that time," Damiano said. "He was full speed ahead on his transformational strategy for GE, to turn it into an industrial internet behemoth, and he was nowhere close to the finish line on that."

ON MAY 13, A SATURDAY, THE EIGHTEEN MEMBERS OF THE GE BOARD of directors, seated around a horseshoe configuration of tables, interviewed the final four at the Beekman Hotel, on Nassau Street in Lower Manhattan, near Wall Street. The place was a total throwback, with a huge, wood-paneled lobby at the bottom of a glass-covered atrium. It was the kind of place favored by trendy downtown Manhattanites, not the board of one of the biggest companies on the planet. And that's why Susan Peters picked it: no one would figure out what was going on. The session began at 7:30 a.m. and ran until 4:00 p.m. Jeff was not there. The candidates were told not to bring presentations. Instead, the board members peppered them with questions for ninety minutes straight. Whereas there had been zero communication with the final four prior to the February *Wall Street Journal* article, the succession process seemed to be gearing up after the board meetings in China. Whatever happened behind closed doors in Shanghai had served as a catalyst for the board to try to figure out, and fast, who would be succeeding Jeff as GE's CEO. The interview order was Simonelli, Flannery, Bornstein, Bolze. Simonelli, Flannery, and Bornstein saw each other in the hotel lobby.

During Flannery's interview, Sébastien Bazin, the CEO of AccorHotels and a new member of the GE board, asked him a question: "What would you do to make GE sexy again? It used to have such a reputation, and what would you do to make it sexy again?" Flannery thought to himself that Bazin must be asking him about his vision for GE. According to someone in the room that day, Flannery replied, "To be honest with you . . . that's the last thing in the world we should be focused on right now. We've just got to get back to the basics of how we run the company and manage the businesses. We've been chasing sexy for a long time." He later told colleagues he thought he had blown the answer to Bazin's question for sure. His only thought was, GE had to put its "head down and get our shit together."

Later that night, Jeff called some board members to find out how the session had gone. The feedback was fascinating, as was Jeff's reporting of it. "A few board members told me that Bolze had done well," Jeff recalled, and that was that on Bolze. Bornstein "made a great impression" because he "seemed humble"—basically he told the board he didn't want the job, even though human nature dictated that he probably did, especially after he had made it to the final four. And according to Jeff, Simonelli, his preferred choice, was "brilliant, informed, decisive, strategic." The feedback on Flannery was that he "was not at his best," but Jack Brennan "didn't think that should eliminate him as a candidate." Jeff got the feedback, but none of the four did. They heard nothing back from Susan Peters, or from the GE board members, after the sessions at the Beekman Hotel.

Eleven days later, on May 24, Jeff blew himself up. He was at the annual EPG—Electrical Products Group—meeting, in Longboat Key, Florida, to give a presentation on the outlook for GE for the rest of 2017 and 2018. Two years earlier at the EPG conference, he had committed GE to making $2 a share in 2018 once GE completed the financial jujitsu of selling GE Capital and spinning off Synchrony, the largest provider of private-label credit cards in the United States, and buying back GE stock. Jeff remembered that he felt about the commitment he made to earning $2 a share in 2018 similarly to how he felt about his promise to not cut the GE dividend in 2009. "That is," he recalled, "we'd made a commitment and I wanted to honor it. I didn't want to give GE people permission to fail." Backing off the 2018 earnings number would be like raising the white flag to surrender. He had done that once in 2009 when he cut the dividend after saying he wouldn't. He wasn't going to do it again in a posh hotel off the western coast of Florida in front of a group of Wall Street research analysts.

For some deranged reason, he still believed GE would make the number he had promised to investors. He put his money where his mind was, too: in 2016 and 2017, he bought $8 million worth of GE stock. Other GE executives tried to talk Jeff off the $2-a-share ledge. It was too much of a high-wire act. But it was important to him. "He had played the company out over fifteen years, and $2 to him was an important number," one executive said, "[after] making all these strategic changes to the makeup of the company. Getting back to something that looked like $2 a share [was very important]."

There was a lot of discussion of $2 a share among the GE executives, particularly in and around the EPG conference in May. "I'm not telling you that there wasn't a scenario and a path to $2," one GE executive said, "because there was a scenario. But I thought focusing exclusively on $2 at that point in time didn't make sense. We just needed to focus on organic growth and margin expansion." Another GE executive believed when GE's cash flow was $1.7 billion negative in the first quarter of 2017—for the first time ever—that's when Jeff should have backed away from the $2 a share. "It's an elephant in the room of 'What the fuck just happened here?'" he said. But Bornstein, the CFO, could not convince Jeff to abandon the 2018 goal. He wanted Jeff to talk about a range, rather than a specific number, or better yet, leave all talk of 2018 performance to the new CEO. "Bornstein is shitting bricks," this executive continued. "Something terrible is going on. I can only imagine the discussion those guys had about 'What are we going say at EPG?' Bornstein knows for sure at this point something is very wrong." Bornstein was telling people at the Boston headquarters, "We can't do $2 a share. We can't. We can't do it. It's just not possible." But Jeff wouldn't listen. "Oh, Jeff, you're such a pessimist," he told his CFO. "We can find things. We can do things. We're not going to get there if we don't try, that's for sure." (For his part, Bornstein was adamant he never made these comments about whether GE could, or could not, earn $2 a share in 2018.)

To Jeff, the $2-a-share number remained the holy grail, the singular achievement that would ratify the many strategic decisions he had made in recent years, along with his vision and his brilliance. By tradition, the GE CEO made the last presentation at EPG. Jeff went on stage at eleven thirty in the morning. "I remember feeling nervous," he recalled. "When I took the stage, I was uncharacteristically jittery. I had the sense that the end was near for me, and I was coming to terms with that." Jeff had just arrived in Florida from the Middle East. He was tired, probably exasperated, and it showed. He seemed like he would have preferred to be anywhere else but in Longboat Key. Most of what he said was a combination of corporate pabulum and technical jargon. "Nobody likes how the stock [has] traded so far this year," he started, "but it's up to us to execute, and that's what we're committed to doing." He then spoke about the "moat" that was around GE's businesses, that GE had a "huge backlog" of business, and that "the replacement value of the GE assets is, like, $2 trillion."

Before he took questions from the Wall Street analysts, he reiterated his view about earnings in 2018. "The way I look at, in 2018, is we are going to execute on an algorithm of three percent to five percent organic growth, one hundred basis points plus of margin enhancement and eighty percent to ninety percent of free cash-flow conversion," Jeff said, repeating his mantra. "We are going to execute on that." He told the analysts that in their "underwriting case"—the case they could count on—"$2 should be at the high end of the range, and our job now is to take more cost out." He then concluded, "What I would say is three percent to five percent organic growth, one hundred–plus basis points of margin enhancement, good free cash-flow conversion. This is a strong, very strong company."

Scott Davis, a Barclays analyst who had been on Jeff's case for years, went right for the throat. "Just to confirm," Davis said, "are you reiterating that $2 is a makeable number because you're going to cut costs?" Jeff replied that "it's going to be in the range" of $2 a share and added that if he had wanted to take the $2-a-share number out of the presentation, he would have done it. Davis followed up by asking about the "elephant in the room," essentially that GE had had a tough year, financially, since the 2016 EPG meeting. Was the problem that Alstom wasn't performing as expected? Davis wondered. "Alstom is going to hit its earnings plan," Jeff answered, which was obviously untrue. He then told Davis that GE would achieve its cash-flow plan of between $12 billion and $14 billion. He said the power business needed to do a better job of getting cash out of the business by reducing inventory and accounts receivable.

His interaction with Steve Tusa, an outspoken analyst at JPMorgan Chase who had lost faith in Jeff years before, started with some light joshing. Jeff commented that Tusa had grown a beard. "I don't know," Jeff said about Tusa's beard. But the joke fell flat.

Tusa replied that he could "just drop the mic and walk out if you really want."

Jeff deadpanned, "Okay, go for it."

"I thought that was kind of funny, but guess not," Tusa said. The interaction was plenty awkward.

Tusa pressed Jeff. He seemed hung up on the idea that Jeff was arguing that the company's performance in the past few years "is what it is" and there wasn't much GE management could do about it: "These are our businesses. There's still some things we can do, but structurally, these are our end markets. They're tough to compete in."

Something seemed to snap in Jeff at that moment. All of his frustrations with running the company for more than fifteen years, facing down wave after wave of Zulu warriors, seemed to come out. "They're good end markets, three percent to five percent organic—really, three percent to five percent organic," Jeff said, knocked back on his heels. "And really, look at the last five years: thirteen percent EPS growth, four percent organic growth, three hundred forty basis points, $120 billion returned to investors, right? Crap, it's pretty good, really. And so I'm saying three percent to five percent, one hundred–plus of margin enhancement, and eighty percent to ninety percent of free cash-flow conversion. So that's a pretty good record. . . . We've doubled industrial EPS almost in five years. Let's go, right?" Jeff seemed to be coming apart at the seams. He seemed to be losing the last shred of credibility he still had with the Wall Street analyst community. He was on the defensive, not a good look for a GE CEO. He could barely muster coherent thoughts.

"Look, really, guys, look, I've been coming here a long time," he continued. "There have been three moments when I've kind of felt—I'd say, 2001, we felt a little bit frothy compared to where the company was, right, 2009, we felt a little bit underloved. Today, when I think about where the stock is compared to what the company is, it's a mismatch. That's happened three times in sixteen years, right? So look, I'm in the Fox News mode: 'We report. You decide.' Okay? This is a—so I'm not here—the last thing on earth you want is a CEO that's here bitching about their stock price. But let me tell you, I've done this for a long time. There's only two other times when I kind of felt—so look, let's end it here." He didn't quit then. He proceeded to swear—he said "shit," according to the transcript—made a comment about another analyst's gray hair, and told another he had grown "bored" with his stories.

No matter how he tried to spin it, Jeff's performance was an embarrassment. "It was a fucking disaster," one GE executive told me. "Like, unbelievably bad. Jeff was unbelievably bad. It was just embarrassing, and . . . after the presentation, Jeff got—normally he gets swarmed by people, and he kind of walks off and, like, no one goes after Jeff." He compared Jeff's zealous adherence to $2 a share to getting too far out on a tree limb "and it just broke." One research analyst who was there recalled how difficult it was to watch Jeff dissemble. "The performance was flat as hell," he emailed me, "and he seemed like he did not want to be there and was super tired."

Jeff knew it was bad, too. "Mattie, how do you think that went?" Jeff asked Matt Cribbins, his head of investor relations, in the car back to the Sarasota airport.

"That did not go well," Cribbins responded.

"No, it didn't," Jeff replied, adding that he, Cribbins, and Bornstein sat in silence for the rest of the car ride.

At the Sarasota airport, Jeff hopped on one of GE's private jets to take him to Houston, where he planned to review the integration plan for the merger of GE's oil and gas business into Baker Hughes. He was also there to be honored as the 2017 International Citizen of the Year by the World Affairs Council of Greater Houston. As he returned to the airport in Houston, Susan Peters called him and told him that Jack Brennan wanted him to call "immediately" to "discuss a timetable for my stepping down."

Jeff also got a call from Ed Garden at Trian after the EPG meeting. "Jeff, it's over," Garden told him. "It's over. People have stopped listening. People have just stopped listening. You have no credibility." Garden also again told Jeff that Trian wanted a seat on the GE board of directors. According to one GE board member, after the disastrous EPG meeting, it was "game on" to name a new GE CEO "before [the end of] the second quarter," which was just weeks away. Brennan conveyed that timetable to Jeff on the flight home. By then, their relationship had deteriorated, which was ironic given that Jeff had run off Sandy Warner, who had wanted to be board chairman, and ended up with Brennan. "Jack Brennan wanted to be the guy that ran GE," explained one GE executive. "He was the guy that ran Vanguard. He didn't like being the backup. He didn't want to be the backup quarterback. He kept rooting for Jeff to have an injury so he could go in and be the quarterback."

Jeff was distraught after getting the calls from Brennan and Garden. "I felt listless, angry and hurt," he recalled. As he told *Boston* magazine, he felt like he had finally assembled the pieces; he had wanted to return GE to its previous glory and wanted to be the CEO to oversee that transition. He didn't want to overstay his welcome but wasn't ready to leave, either. For the three days of Memorial Day weekend, after being told he was toast, Jeff moped around his Kiawah Island home—for which he had paid $7.2 million in 2006. He was in "sort of a fugue state," he remembered. He didn't tell his wife, Andy, what had happened until Monday, the holiday. "She was mad—both for me and at me," he continued. Mad for him because she did not think he deserved his fate. Mad at him because she feared the public drama surrounding his soon-to-be-public exit would drown out the focus on their daughter, Sarah, who was getting married in September. "Andy had always accommodated my 24/7 work ethic and GE's preeminence in our lives," he stated. "But enough was enough." He promised not to let the GE machinations interfere with his only child's big day. (In addition to climbing Mount Kilimanjaro with Sarah to bond with her, he also got the GE "meatball" logo tattooed on his left hip, along with the initials of his wife and daughter.)

According to one GE board member, Jeff's problems with the board began when he tried to push out Bolze. Some board members remembered that he also could not get along with Krenicki and then pushed him out. He couldn't get along with Calhoun. Now he still could not get along with Bolze. "This isn't Steve," some board members told Jeff. "This is your problem. We need to move you out and bring somebody new in, and whoever comes in should decide whether they want to move Steve out or not." It was almost that simple. "Bottom line," the board member continued, "Jeff had lost the board's confidence to run the company at that stage, and it was time for a new CEO." And then there was the matter of Jeff's adherence to the $2-a-share mantra. That rubbed some board members the wrong way, at least in hindsight. "The biggest single moment in the history of GE was using $40 billion of balance-sheet equity to buy back stock to support a $2 a share," explained one board member. "Jeff, all these guys, they should be eviscerated and chastised for taking a short-term view in a company they were building. And the reason they did it was because there was so much pressure from the Trian guys that if they didn't achieve it, they were going to get canned anyway. There you have it. Let's not water the wine. That's the story."

While Jeff was bemoaning his fate in South Carolina, the apparatus of the corporate state was already moving away from him. The process accelerated over the Memorial Day weekend. Matt Cribbins called the candidates and told them that the succession process was heating up due to the EPG meeting disaster. "This is going to happen soon," Cribbins explained. "Jeff is going to be out of here soon. The wheels are in motion. Jeff has stepped on the third rail." The GE board meeting to pick Jeff's successor was on June 9, a Friday, at the Boston Harbor Hotel, across the street from GE's temporary headquarters, which did not have a conference room.

Six days before the board meeting, Jeff asked Flannery to meet him at his office in Boston, to basically spend most of the day with him to talk about how he intended to move GE forward. Jeff asked Bornstein to meet with him the next day, Sunday. By this point both

Simonelli and Bolze had been eliminated from the succession process, a fact that Simonelli realized, since he was going to be the CEO of Baker Hughes, and Bolze, incredibly, still did not, according to Jeff and a variety of other GE executives. Remembered one GE executive, "On the weekend before the meeting, [Steve] came by my office and he goes, 'Hey, looking forward to working with you.' He was very confident." Jeff met with the final two in the conference room off his Boston office. Flannery flew in from Chicago; Bornstein was already there. The topic was: What would you do as CEO? The sessions were half questioning, half coaching. "You definitely have a shot at this," he told each of them.

When he got done with Jeff, Flannery walked past the Boston Tea Party exhibit, where the song "I'm a Yankee Doodle Dandy" was playing, and then flew back to Chicago. Nearly immediately, both Cribbins and Peters wanted to know how it went with Jeff. There was some risk with choosing Flannery in that he would be the one most likely to reverse Jeff's initiatives, and that might be a reason for Jeff not to support him. But unbeknownst to Flannery and Bornstein, Jeff's influence with the board had waned sufficiently by then that the choice of the next CEO would be the board's, not Jeff's (unlike what had happened with him; he was Jack's choice, and Jack got what he wanted).

According to former GE executives, Cribbins and Peters wanted to know how Jeff had left the conversation with Flannery.

"Nothing, really," he told them.

"Nothing?" they said, seemingly incredulous.

"No," Flannery told them. "He said, 'You definitely are a contender.'"

"Not anything else?" they pressed.

"Nope," Flannery responded.

What was supposed to happen that Saturday, six days before the board meeting, was for Jeff to *tell* Flannery that he had been chosen by the board to be Jeff's replacement. That way Flannery would have six days to prepare himself for what was coming. But Jeff didn't do it. He couldn't bring himself to tell the next CEO of GE what the board had decided.

Early in the following week, the top GE executives met in Crotonville for a long-scheduled gathering. Since the May interviews with the board in downtown Manhattan, Flannery and Bornstein had talked off and on about the succession process in a relaxed, seemingly noncompetitive way. At the cocktail party that night, Flannery and Bornstein went out to the patio, sat down, and had a forty-five-minute conversation man-to-man. Cribbins's antennae were up. After all, it was supposed to be a social event where people were mixing and gabbing, and here were the two men with the best chance to become the next CEO sitting by themselves for a long chat.

"Hey, this thing's going down soon," Bornstein told Flannery.

"I don't know, Jeff," Flannery responded. "I don't know, literally, anything."

"No—like, very soon," Bornstein said. It was clear that, by virtue of having an office in Boston, Bornstein knew more about the process than did Flannery, who was off in Chicago.

"Well, what do you mean by soon?" Flannery asked.

"This week," Bornstein replied.

"Are you shitting me?" Flannery responded.

"No," Bornstein said. "There's a board meeting Thursday and Friday."

"Oh my God!" Flannery said.

Flannery knew the decision was not going to wait until the end of 2017, but he had no idea that that the board was meeting in the next few days to make its decision. Clearly, GE's vaunted succession-planning process had broken down in a major way. But given how much more Bornstein seemed to know about what was going on than Flannery did, Flannery was quickly concluding that Bornstein would be the next CEO.

According to what Flannery later told his GE colleagues, on Thursday of that week, at the end of the day, Jeff sent Flannery an email: he wanted to know how to get in touch with Flannery after the Friday board meeting in Boston. Jeff arrived at the fateful board meeting early. "There wasn't much discussion," he recalled. Brennan had already canvassed the board "to seek consensus," Jeff continued. The board voted unanimously for John Flannery, "whose extensive international experience, humility, and willingness to make significant change worked in his favor," Jeff wrote in *Hot Seat*. "The vote was over, and so was I." Board member Susan Hockfield, the former president of MIT, thanked Jeff for his service to the company. Board member Andrea Jung hugged him. "Tears came into my eyes," he recalled. "I was overcome, and it surprised me." Even knowing what was going to happen, "the finality of it stung."

He walked out of the hotel alongside Alex Dimitrief, GE's general counsel. Jeff decided to walk back to the office with Dimitrief. It was hot out, and soon the two men started sweating through their dress shirts. They walked past the Boston Tea Party ships. Jeff felt like "a fallen general, defeated by fatigue." He called his most loyal lieutenants, David Joyce, who ran Aviation, and Beth Comstock and John Rice, the two vice-chairmen. He tried to keep it together. "We kept it short," he said. He then called Flannery, who was in Chicago in a meeting with his team to get a better understanding of how the quarter was shaping up in the health care business he was still running. His cell phone rang. He stepped out of the meeting.

"That's it," Jeff told him. "You're the guy. Congratulations, the board selected you. I think this is great. We've got to run fast here. Come back to Boston, and we're going to announce this thing on Monday morning. We've got a lot of work to do over the weekend. Bring Tracy." Flannery later recounted for his colleagues that he went back into his meeting but quickly zoned out. He couldn't quite believe what had just happened. "I always tell people it was like Charlie Brown's parents," John explained to them. "The people were talking in the meeting: 'Wah, wah, wah, wah,' 'What do you think, John?' 'Yeah, sure. Right. Good.'"

He and Tracy, his wife, flew to Boston, in a private jet, along with their dog, Mookie Betts, a black lab. They stayed at the Boston Harbor Hotel. The plan was for John to become the new CEO on August 1 and then the chairman of the board of GE on January 1, 2018. Jeff would remain CEO until August and chairman until January. In the meantime, Jeff met with other three finalists, seriatim, to tell them the board had chosen John. He recalled that Simonelli was "let down" but knew he had a sinecure in Baker Hughes. Bornstein, he said, seemed relieved as well as disappointed. Plus, he had had time to absorb the news, which Brennan had told him already. "He was aware how much GE was about to change," Jeff continued. When he told Bolze, Jeff said he was "surprised" and then said he would be leaving GE. "Then Bolze said something so uncharacteristically self-aware that I still marvel at it," Jeff remembered. "'I am sorry I made it so hard for you,' he told me." Bolze told me he was not "surprised" by the board's decision but rather he was overcome, by then, with a sense of "relief" that the open-ended and frustrating process was finally over and that he and his family could "move on." He also said Jeff got it wrong: He told Jeff he was "sorry" *the past months* had been so hard on him, given the disastrous EPG meeting, the board's succession discussions without Jeff in the room, and the abrupt decision to speed up Jeff's departure. Bolze gave a graduation speech a few days later at the Massachusetts Maritime Academy, where he told the graduates he had been a finalist for the CEO job. "I had gone for it and giving it my all," he told them. "I didn't get it. It was a tough day. That's life. However, it was an honor of my lifetime to be considered as a final candidate for this role."

The next morning, in the GE office on Farnsworth Street, Jeff told Flannery about an important decision that the board had also made the day before, without consulting the CEO-elect: not only would

Bolze be leaving on June 14—two days after the June 12 succession announcement—but also the power business he had been running would be merged into the electrical grid business, which had been run by Russell Stokes, a twenty-three-year GE veteran and one of its most senior Black executives. The big—and surprising—news was that the GE board had selected Stokes to replace Bolze without consulting Flannery.

One would think that the new CEO would have been consulted for such a material personnel decision. But Flannery wasn't. Other GE executives agreed that the decision to replace Bolze with Stokes was rash. "I liked Russell," said one senior GE executive. "Great guy. But not the guy that anyone would pick to be the CEO of Power when it's on its deathbed. You needed somebody who just reeked of credibility in the power industry at the time, and we didn't have that." Others wondered why Bolze didn't stick around for a period of transition to help Flannery learn about the power business. "How much did he know [about how] the shit's hitting the fan?" one executive wondered. "Like, 'I got to get out of here and make it look like it's a career decision and not a "my business imploded" decision.'"

That night, June 10, the Flannerys and the Immelts had dinner at Jeff's Back Bay home. On Sunday, Jeff made a bunch of phone calls to share the startling news. He had a conference call with thirty or so GE business leaders. "The tone of the call was somber," he recalled, "and no one asked a single question."

He called Bill Conaty, the retired head of HR whom Jeff had nicknamed Mr. Wolf. "Aw, shit," Conaty said.

He called Sandy Warner. "He was cordial to me," Jeff remembered.

He called his parents, who answered in typical parental fashion. "Why would you leave now?" his father asked him. "You're doing a great job." He called his brother, Steve, who was then the CEO of Hogan Lovells, an international law firm. He worried that Trian would continue to blame things on Jeff. "I fear you're not going to own the narrative on this," he told Jeff. ("As usual, my brother was right," Jeff deadpanned.) Finally, he called Jack. He wanted Jack to hear the news from him.

"Hey," Jack told him, "we both know you never caught a break." Then Jack added, according to Jeff, "Thank God the board didn't pick Bolze."

*Chapter Twenty-Six*

# EISENHOWER

On Monday morning, June 12, the press release went out announcing the news that John Flannery would be replacing Jeff as CEO, effective August 1, and then as chairman on January 1. The release also said that Bornstein would remain CFO and add the title of vice-chairman, and that he and John would work together closely. GE stated that the succession process, led by the board of directors, had been underway since 2011—an extremely misleading statement, if not an outright deception. It was hardly the kind of coronation that had accompanied Jeff's selection nearly two decades earlier, or Jack's before that. There was a scrawny PowerPoint presentation. "GE's communications team did the best it could," Jeff recalled, "but there was no way to make this transition look normal." Both the outgoing CEO and the lead director of the board praised John as the right man at the right time to lead GE. John seemed both shocked and humbled, or as much as he could seem to be in a corporate press release. "Today's announcement is the greatest honor of my career," he said. "I am privileged to have spent the last 16 years at the company working for Jeff, one of the greatest business leaders of our time. He has transformed the GE portfolio, globalized the company and created a vision for the GE of the future by positioning the company to lead in digital and additive manufacturing. In the next few months, my focus will be on listening to investors, customers and employees to determine the next steps for GE." There was also a Facebook Live meeting for employees and a meeting with Wall Street analysts to discuss the CEO change.

Left out of the press release was the news that Bolze was also leaving GE, effective nearly immediately, and that Russell Stokes was his replacement. There was also no mention that Jeff had verbally agreed that Ed Garden, at Trian, would be soon joining GE's board. Needless to say, the presence on the GE board of an aggrieved activist investor made for an ominous start to John's tenure as the eleventh GE CEO.

At the meeting with Wall Street analysts, Jeff tried to put the best spin possible on a bad situation. He called the news of the unexpected succession "exciting" and explained that, starting in 2013, he had begun planning to leave in the summer of 2017. (There was no evidence of this shared.) In other words, Jeff wanted everyone to believe that nothing about the announcement or the timing of it was unexpected or unusual. From the outset, John praised Jeff and the portfolio that he had created at GE for the future. But he also made clear that not everything was working well. "We need to address those areas with urgency and with purpose," he said. "So first, I think it's important for us to focus on both short-term and long-term execution and accountability both internally and externally. And by accountability, I mean really an environment where there's rewards for performance and consequences for underperformance. So I'll be focused on executing on the basics here: the growth of the company, the cash flow of the company, the cost structure of the company. It's important we continue to be transparent with investors and provide a balanced forward outlook."

He said he was going to undertake a "comprehensive" review of the portfolio, which he pledged would be ready in the fall. He concluded on an optimistic note. "I want GE to remain a special company that makes a difference in the world," he said. "We also need to and I know we will remain a company that delivers for shareholders. I want us to have a culture of transparency and integrity in decision-making that will continue for sure. And I want us to be objective in our thinking. But I'm convinced GE has the best team in the world today. I've seen that in my time in the company. I firmly believe our best days are ahead of us."

At their meeting with John to announce the regime change, the Wall Street research analysts offered congratulations to John, and to Jeff Bornstein, and, weirdly, to Jeff "on his retirement," although maybe they were just being diplomatic. The always-piercing Scott Davis, at Barclays, wasted little time in testing John. "John, best of luck in your new job," Davis started, "of which most of us would not want to switch places with you despite the compensation package." Davis said he'd been getting "pinged" by investors about a question he wanted to ask John directly: "How do you think in terms of balancing this whole disciplined review of the company with the reality that investors increasingly don't like conglomerates, they just don't like complexity? And maybe to you, you may see synergies. But investors, they don't."

John responded well. "I spent the first essentially two decades of my career really in investing, in private equity, underwriting debt," he said. "I have a long history of looking at things from an investor perspective via training and background." He said that GE still had competitive franchises. "There's clear leverage inside the business," he continued. "We've walked you through this before, how the company shares technology, how they share global footprint, how they share branding and things. The strong portfolio was built to codepend, if you will." In the end, though, he said the proof would be in the pudding, a mixture of cash generation, performance, cost optimization, and stock price. "I've got the background to look at it that way," he asserted.

One indicator of Jeff's humiliation was the GE stock price, which increased some 4 percent on the news of his departure, to nearly $29 a share. "Apparently," Jeff deadpanned, "investors felt my impending departure was good for the company." In the *Times*, John came off as deeply qualified to analyze GE's portfolio and its predicament and unafraid to make whatever bold changes were necessary. "A keen wine collector," the *Times* reported about him, "who often grills for guests at his Nantucket home, Mr. Flannery, who is married and has three children, is an avid reader and enjoys attending R.E.M. and Natalie Merchant concerts. He named the family's new puppy, a 12-week-old black Labrador, after Boston Red Sox outfielder Mookie Betts."

AMID ALL THE HOOPLA OF THE APPOINTMENT OF THE NEW CEO, JEFF Bornstein, the GE CFO and a newly minted GE vice-chairman, quietly put in motion a process that would have material consequences for John's tenure as CEO: he asked GE's outside actuaries to examine closely a long-forgotten liability—for health care provided to the elderly—hiding inside a small GE insurance subsidiary in Overland Park, Kansas. The problem had been festering for years in an office park in the middle of the country. "Out of sight, out of mind" seemed to be Jeff's logic for ignoring it. Its disregard intentional or not, the hidden liability became a time bomb that Jeff left behind for John to try to disarm without blowing up GE. "The thing was, like, on fire as he ran out of the building," explained one GE executive. "It's like he came in the building and lit a fire and then ran out and said, 'Hmm, the guy didn't do a very good job putting the fire out.'" As a reminder, as part of the 2004 IPO of Genworth, Jeff decided to retain a small insurance subsidiary, Union Fidelity Life Company, or UFLIC, which was in the business of providing insurance to cover the costs associated with long-term health care, nursing-home care, and home health care. In December 2005, Jeff also retained at GE something called Employers Reassurance Corporation, or ERAC, from the $7 billion sale of Employers Re to Swiss Re. ERAC was in the business of reinsuring term life insurance policies. After the Genworth IPO, UFLIC stopped writing new policies and was responsible for managing the policies already written—continuing to take in premiums, managing that money, and paying out claims.

The way GE executives tell the story, there was an "old book" and a "new book" of policies at UFLIC, the older book created at Employers Re and the newer book created by companies subsequently acquired. The old book had shown some "stress" and "had been dealt with," explained one GE executive. Up until 2016, the new book had claims that were below the "underwriting curve," or the expectations that underwriters had for the policy portfolio. But then, as the policyholders began living longer than underwriters anticipated, the payments to them to cover the costs of nursing homes and home health care began exceeding what the underwriters expected.

The old book had an older average age at issuance of insured individuals, with more known preexisting conditions. The insured people in the new book had a younger average age at issuance, and the underwriting premise was that the new book would be more profitable because there were fewer preexisting conditions. "That was the assumption for a very long period of time, including all that time of which it was performing at or better than the original underwriting," explained the GE executive, "until it didn't," as the policy holders began living longer and costing GE more money.

But things didn't turn out as expected. The top GE Capital executive reported to Bornstein that claims in the new book were starting to spike in an unexpected way. "They're looking like the claims in the old book, and we don't understand why it's doing this as precipitously as it is," he told Bornstein, who had previously been the CFO of GE

Capital. They estimated the increased cost to be in the range of $2 billion pretax. He set aside $3 billion in capital to cover it, just in case. But he had no idea what the final audit would reveal.

In June 2017, just as John was succeeding Jeff, Bornstein ordered up an external "actuarial review," a top-to-bottom reunderwriting of the entire book of policies, going back to the dates they were written. It was a big job that normally would take more than a year to complete. "I don't know how to answer the question on what the impact is," the GE CFO told them. "You're giving me a sensitivity analysis. I don't have any idea whether that's a good analysis or not." Bornstein gave the insurance team six months to get it done. "I'm not closing the year not knowing what it is," he told them. "I'm sorry."

"Well, we'll give it our best shot," they told Bornstein.

"No," he replied. "Get it fucking done by year end."

He reiterated that there was no way he was going to sign off on GE's financial statements for 2017 without knowing the extent of that growing liability. Had he not ordered that audit, or allowed the insurance team to delay doing it, and then signed off on the 2017 numbers, he could have been the target of any number of shareholder and SEC lawsuits. "Thank God I told them to go do this fucking review, because if I hadn't done that, I'd be fucking toast," Bornstein told me.

Bornstein went to see Jeff and John, who were then basically tag-teaming the CEO job. John hadn't taken over yet but was in the cockpit. Bornstein told them he had ordered the external audit. Jeff understood. "That sounds like the right thing to do," Jeff said. He didn't think it was that big a deal and, those around him claimed, he had no idea how big the liability was, or would turn out to be.

But John was aghast. Bornstein said he thought the hit to earnings could be around $2 billion and might affect the ability of GE Capital to make dividends to GE, if the regulators demanded more capital be put into the insurance subsidiary. "Whatever you say," John told him, adding that he thought GE had gotten out of the insurance business a decade earlier. Bornstein sort of rolled his eyes. "Nah, we've still got something there," he said. "It could be a couple billion." John asked Bornstein how big the problems could be. He reiterated that it could be underreserved by about $2 billion. The GE executives were incredulous: How were they ever going to explain to the world that GE suddenly had a $2 billion liability, never reserved for, popping up in a business that the company had told everyone it hadn't been in for a decade? Responded Bornstein, gruffly, "Yeah, it's not good."

The actual dynamic was more complicated. Both UFLIC and ERAC were regulated by the Kansas insurance regulators as well as by the feds, and each year there was a test of the underwriting model to make sure the assumptions were still valid and the reserves made sense. "You set up this mode," explained a GE executive. "There's all these incredible assumptions in there. How long are people going to live? When are they going to start consuming the health care? What will the price be then? What's the course of Alzheimer's, or not? And then you set up all these assumptions day one and you say, 'Okay, the net present value of my liability is $100." If the assets that UFLIC has on its balance sheet exceeded the net present value of the liabilities, then everything was cool, and no red flags would be raised. "Every year you go back and test 'Do I have a surplus with my hundred?'" he continued. "And if you do, you don't adjust anything underneath it." Everything seemed to be fine until 2016.

The net present value of the liabilities in 2016 was getting closer and closer to the value of UFLIC's assets. "They do a preliminary calculation and they're like, 'Well, we're probably going to trip this. We will probably go below a hundred here.' And if you go below a hundred, you have to go back into the model and update every single assumption and virtually all of which have gone badly against you. Literally, all of them have. Interest rates lower, life expectancy longer, consumption of health care higher, price per unit higher. Everything's a bad guy in this model." According to this person, in 2016, as the net present value of the liabilities was getting closer to the value of the assets, "They went back and said, 'Well, let's change the way we do this calculation.' And they called it something like a 'roll forward' or something. They're like, 'I've got a couple of better assumptions showing up in the last ninety days or something. If I actually extrapolate those forward, I'm still about a hundred.'"

The GE executives consulted with KPMG, GE's longtime outside auditor, about the tweaks they were making to the actuarial model. "KPMG said, 'It's okay to do that,'" one GE executive told me in disbelief. "If you've done something the same way for fifteen or twenty years, exactly the same way, then you happen to be right on the edge of this calculation and you change your calculation, it's pretty obvious what they were doing. They didn't want the [liability] to go up. They were trying to have it not go up. That looks to me like I'd say somebody somewhere was very explicitly playing around with that for sure. In the meantime, the rest of the world doesn't even think we are in the business." (At some point, you've got to figure, this kind of behavior becomes fraudulent, right? At least one class-action civil shareholder lawsuit against GE and a number of former executives is trying to get an answer to what happened. The SEC also investigated, although that investigation has concluded without charges being levied against GE.)

This executive believed that even if KPMG's advice was "technically okay, because all these assumptions have gone the wrong way, you know you're in deep shit." It would be akin to a fixed-income portfolio manager who has a portfolio full of three-year, fixed-rate bonds and who, for some reason, isn't required to mark the portfolio to market. If interest rates skyrocket, the value of the bond portfolio falls dramatically whether it's marked to market or not. You know you've suffered a devastating loss, whether you choose to acknowledge it. Or said another way, it was as if someone had cancer, but the oncologist told the patient it was just a benign cyst and hoped it would go away. "If you understand anything, you know economically there's a giant hole in that investment," the GE executive continued. "The guys here should have been the same way. Even if you don't have to make the calculation, you should be sitting at your desk saying, 'Shit, when I run this model, I'm going to need like $15 to $20 billion. Like, this thing is a disaster.' And then you wouldn't be sending money from GE Capital up to GE parent. And you wouldn't be doing your share repurchases," which cost GE tens of billions of dollars in cash that it could have used—should have used, in hindsight—to shore up its balance sheet, not speculating on its own stock.

But, the logic went, Jeff did these things in his quest to "reshape the company," make $2 a share, "get the stock to $40," and then "Jeff can go home saying it's a win. . . . Nothing would get in the way of that. No reason, no logic, no nothing could get in the way of that." Revealing the growing liability in the insurance subsidiary would have doomed Jeff's $2-a-share "gambit"—which was doomed anyway—and would have reduced greatly the chance that GE Capital could continue to dividend to GE the billions from the sale of GE Capital assets that GE used to buy back its stock. "The tar starts to bubble up from the ground," noted one GE executive. "They're looking at that and they're like, 'Holy shit! Is there a technical way we can redo our methodology of treating this so that the tar goes back under the ground for a second here?'" When KPMG said they could do it, they were relieved. "But they absolutely knew then, like, 'Hey, what's under the ground is ther-monuclear.'"

In addition to the molten lava bubbling up from Kansas City, it was also increasingly clear to the top GE executives that GE's power business was in dire straits. Whatever Jeff and Bolze had hoped to derive from the Alstom merger in terms of increased dominance in the power-generation business was not happening. Around the time of Jeff's imminent departure, the problems in the power business were becoming obvious. "I could feel it," one former GE executive told me.

To try to get a handle on *this* problem, too, Jeff Bornstein had decided to replace the Power CFO six months earlier. The Power financial executives could not give him the answers he wanted around what demand for the year looked like, what the revenue from the "upgrade kits" looked like, and what kind of payment terms GE Power was giving its customers, especially when it came to the upgrades. Some customers wanted to pay GE over three years or five years, and the Power executives were agreeing to those terms. Bornstein objected. "You've got to come up with a financing solution for that," he told them. "Because the industrial businesses are not in the business of financing upgrades." To appease the GE CFO, the Power executives started "factoring," or selling, the receivables for the upgrades to third parties, who would buy the receivables at a discount. GE would get most of the cash expected from the contract up front but at the cost of a small discount and of potentially robbing from the future, unless the pace of sales continued unabated. Factoring of the receivables didn't seem particularly alarming at the time because the profit margins on the "upgrade kits" were "enormous," making the discount that resulted from factoring "de minimis," explained one GE executive. By the middle of 2017, it was becoming increasingly clear that GE Power would not make its numbers for the year. The budget called for the sale of around fifty-five trailer-mounted power units in 2017, but by June the Power business had done only "a fraction of that," he continued. Power had also sold fewer upgrades than budgeted. Bornstein was increasingly concerned. He fought with Bolze and the rest of the Power management team about the need to reduce costs in the business. Bolze presented a cost-reduction plan of around $300 million. Bornstein wanted $1 billion of costs taken out.

"Fuck that," Bornstein told them, "it's going to be closer to a billion. You're not thinking about it right." They agreed on a reduction of $700 million of costs. Part of the problem lay with Bolze and his dysfunctional management team, and part of the problem was that GE had misread the market. "Siemens had been suffering for three or four years at that point, and we were killing them, killing Siemens," explained a GE executive. "We had a new product, the H-class turbine, and Siemens was describing a market that was way tougher than what our team was describing and, at least until that point, had been able to execute against. And then, I think with the benefit of hindsight, the gas ran out of the tank and we were looking at the same market that Siemens was looking at."

He believed that John Rice, then still head of GGO, had done "a good job generating demand" for power projects in places such as North Africa, Southeast Asia, and China, which often required IMF or World Bank funding to make them happen. "It wasn't like the Nigerians called them up and said, 'Hey, can I buy an H turbine?'" he continued. "You had to really go work this and help the project get together." Of course, these were risky deals, especially when it came to getting paid. "Doing a deal in Angola isn't doing a deal in Indiana," he said. "No two ways about that." The power business, he said, "hits the side of a mountain in a time frame we'd never expected." The numbers the business said it was going to do, and that had been shared with Jeff, John, and Bornstein in July 2017, versus what it actually produced in September and December were not "even in the [same] fucking zip codes," he said. And two days after John was named Jeff's successor, Bolze quit GE, fulfilling a promise he had made—to Jeff, anyway—a few times before. He quickly took a new job at the Blackstone Group as a senior managing director focused on infrastructure investing.

For whatever reason, Jeff didn't pull any punches in *Hot Seat* when it came to the GE colleagues with whom he had run afoul. Both Denis Nayden and John Krenicki came in for some full-throated abuse. But no one came in for more consistent condemnation than Bolze. Bolze tried to remain as discreet as possible about the criticism that Jeff heaped upon him. But after the publication of Jeff's book, he took to LinkedIn to strike back, within the parameters of someone who liked his new job at the Blackstone Group and wanted to keep it but also wanted to make some effort to defend himself. "I don't often post here," he wrote, "but I've become aware of some surprising and inaccurate statements about me and the GE Power & Water team" that Jeff "apparently included" in his book. He felt "compelled" to respond. He explained that GE's wind business had delivered "over a 6x return" on GE's original investment, that GE's "distributed power" was sold for a gain, and that GE's water business was also sold for a $2 billion gain. GE's power business, he wrote, "maintained the #1 market position in gas power globally over a decade by developing the most efficient gas turbines in the world."

He conceded that the Alstom deal did not work out as hoped, "but I have always owned my role in the deal and been content to let the actual record speak for itself. That's what leaders do." His business had followed through on "more than $1 billion in synergies" in the first year after the deal closed. "While no deal is ever perfect, I remain proud of the GE Power & Water team and its record executing in a tough and ever-changing environment." He also wrote that he was "honored" to have been considered a candidate to be GE's CEO. "I only have respect for the company and its process," he wrote, "and will leave it at that."

In July, both Flannery and Bornstein visited with the Power executives, including the post-Bolze team: Russell Stokes, the new CEO, and Paul McElhinney, the CEO of power services and the man who had insisted on the manufacturing the "upgrade kits." They peppered them with dozens of questions, slowly but surely coming to understand for the first time about the factoring of the receivables, the noncash accounting, and the drop-off in business. They came to understand that the business was in deep trouble.

At one point, the Power team was discussing with Flannery and Bornstein a $1.1 billion unsecured exposure that GE had in Angola from selling the country railway and power equipment. "We sold a bunch of power equipment in Angola, and we just took an account re-

ceivable," one person in the meeting remembered. "No cash. No security. No nothing. They're not paying it. Of course, we book the income
and the sales." The conversation turned to how there was some bilat-
eral credit agency involved and "if we ricochet off nine rocks simulta-
neously, we can get $950 million of third-party financing on this project
and pay down our receivable and we'll be down to $100 million."

Flannery was incredulous, the person recalled. He turned to
Bornstein and asked how GE ended up with a $1.1 billion account re-
ceivable from Angola. "I understand where we are today," he said.
"We've got to see what's the best deal from where we are, but how the
fuck did we do this? What process would have said, 'Let's do an unsecured loan to Angola for $1.1 billion'?" Bornstein just rolled his eyes.
The message was clear: it was another example of Jeff pushing people to make their numbers. "Hey, we'll sort this out later," Bornstein
told him.

Another GE executive believed that the same kind of obfuscation
that occurred with the long-term-health liability—fiddling with the
assumptions that went in the actuarial model to keep the problem
subordinated—also occurred in the power business in 2017, as service
contracts, in particular, were renegotiated. Once again, the accounting
entries may have somehow passed muster with KPMG but, he continued, they were misleading to investors and research analysts, as they
made it seem that accounting profits were the same as cash in the
bank, when they weren't. "At its peak, it's sort of saying, 'I've got $3 or
$4 billion of income that came from renegotiating contracts and I said
to the public, I have $3 to $4 billion of CFOA"—cash flow from operating activities—"knowing that everyone thinks that's cash flow generated in the current period from operations. . . . But in fact I have sold
one, two, three, four years of receivables and reported that sale as my
current period of cash, knowing full well that (a) an investor can't find
that number in the 10-K and (b) when presented with those two pieces
of information, this company is making $5 billion and it's generating

$5 billion cash and everything's fine, when in fact we know it's really making like $1 or $2 billion and generating no cash. There's witnesses and emails that would testify to that all day long."

The problems that were brewing in both the power business and residual insurance businesses raise profound questions about the failure of the GE board of directors—and the way Jeff managed the board—to ask the tough questions of GE's management, including Jeff, and whether the board abdicated its responsibilities in order to try to stay in Jeff's good graces. Did Jeff provide the board with the information it needed to understand the profound holes in the GE business plan? Did he give board members enough time to probe deeply and ask the difficult questions? Or was he too busy telling board members to "wrap it up," signaling them with his finger at board meetings, as one board member remembered. How could the board allow Jeff to maintain his devotion to the $2-a-share earnings projection when he must have known—as did those around him—that achieving it was a virtual impossibility?

It's not as if there weren't warning signs along the way. Bill Gross raised the specter of GE Capital's reliance on commercial paper as early as March 2002. He was right. Jim Grant raised concerns about GE and its business practices between 1998 and 2012. And he was right. Like Grant and Gross, Ravi Suria saw and articulated problems at GE with its funding and its accounting. He was right. Steve Tusa and a variety of other research analysts saw problems brewing at GE and wrote about them. They were right, too. Even Mike Damiano in his April 2017 article about Jeff was right. But there was always pushback from the GE apparatchiks to anything that deviated from the GE mantra of infallibility.

The near-death experience in 2008 should have been a serious wake-up call for the GE board of directors, in the same way that the 1993 bombing at the World Trade Center should have been a wake-up call for what occurred eight years later. Why did the GE board fail so

miserably in its fiduciary duties to GE's stakeholders—employees, creditors, shareholders—at every twist and turn in the road?

Why has nearly every board member, especially the most powerful
ones, including Jack Brennan and Sandy Warner, steadfastly refused
to engage in any effort to explain the board's responsibility for what
happened to GE? In 2001, Jeff was handed the most valuable, most
respected company in the world to manage, and in 2017 he handed off
to his successor a company badly damaged, "a company on fire," as one
executive told me. Jeff was "imperial" and "delusional," many people
explained. But the failure of the GE board to rein in Jeff's behavior and
to question Jeff's judgment more deeply, or to push back against his
imperial and delusional tendencies, together stand as one of the greatest corporate governance abdications in American history. The abject
failing of the GE board is yet another example of the old Wall Street
adage that "success has many fathers and failure is an orphan." As the
shit was hitting the fan, Jack Brennan, the lead director for the last
years of Jeff's tenure, couldn't get out of the place fast enough. And he's
never said a word about what went wrong at GE and why he did nothing to prevent it from happening.

A few weeks after Jeff was fired, *The New York Times* noted his
departure in a front-page article that proclaimed the end of the "baronial" era of corporate CEOs. There was a picture of Jeff's vacated
imperial-looking office in the Skidmore, Owings & Merrill–designed
office in Fairfield, an office that had also been Jack's. GE's executive
floor in Fairfield was known as "Carpet Land," because of the Persian
rugs that were everywhere and the hushed silence that permeated the
place. "It was so quiet, you could feel the energy drain out of you," Ann
Klee, the GE executive who oversaw the move to Boston and the development of the new headquarters there, told the *Times.*

Not only did Jeff have his own private bathroom, but so also did
his two assistants. They also had their own private pantries. There
were two helicopter pads, a shoeshine station, and many a private

dining room. There was a private twenty-eight-room hotel, known as the Guest House, that Jack authorized be built as GE's profits soared. What had been 44,000 square feet of office space for GE executives in Fairfield turned into 7,800 square feet in Boston. "It has a much more collaborative feel, and the glass replicates the transparency of working together," Klee told the *Times*. "The Fairfield campus was beautiful, but it lacked the spark you feel here. It's a different time, and we like the power and energy and creativity that comes from mixing people together."

The article noted that Jeff had not survived the era of the activist investor demanding change and a higher stock price faster than Jeff could produce one. "These people"—the GE executives—"were bigger than life, and I saw it up close," said Kevin Sharer, a former chief executive of Amgen who worked as a top aide to Jack and coached many a GE executive, including John Flannery. "They were a combination of chief executive, statesman and rock star. They were unassailable." Sharer told the *Times* that the only place that evoked the kind of power comparable to the long hallways and corner offices of GE's former Fairfield headquarters was on a nuclear-powered submarine. (Sharer was once a chief engineer on a navy submarine.) "We had the confidence, the swagger, and we felt like we had unlimited industrial potential," he said. "Could we buy RCA or NBC? Of course we could. I'm not complaining, but this is absolutely not the case today." Bill George, the former CEO of Medtronic and a professor at Harvard Business School, said John Flannery was "highly regarded" inside and outside GE but would be less likely than previous GE CEOs, he thought, to serve on a presidential task force or to tweet about Trump. "He's going to keep his head down and focus on the numbers," he said.

ON JULY 6, STEVE TUSA, THE JPMORGAN CHASE RESEARCH ANALYST, struck again, and hard. He reiterated his sell rating on GE's stock, which was then trading around $27. Tusa predicted it would fall to $22. The report was another bombshell. Tusa's view of GE mirrored in many ways Trian's views: GE was a company in trouble that needed fixing, and fast. "The GE narrative is as open and undefined as it's been in decades with the new CEO likely to set the course on four key aspects of the story" during the second half of the year, he wrote. In his report's more than 120 pages, Tusa predicted that Flannery would reset the 2018 earnings guidance closer to $1 a share, from Jeff's beloved $2 a share; would undertake a "material restructuring"; would change the "capital allocation strategy," including possibly ending Jeff's stock buyback program; and would redefine the GE portfolio, including deciding whether and when to spin off GE's aviation business or its health care businesses. Things weren't good. "We detail in this note a framework for analyzing the messaging to come over the next several months, and while we expect a fresh start, a positive, we don't see a quick or easy fix to the current predicament," he wrote.

Trian, for one, was particularly impressed by Tusa's predictions for the trouble brewing in GE's power business, which Tusa described as "the most important business in the portfolio." (Tusa's perception of the problems in the power business were so accurate that GE wondered if he had a mole inside the company and initiated a search for a leak.) He predicted that $1 per share of earnings was the "credible anchor" for the company, not the $2 to which Jeff had clung. He also believed that a cultural change was needed at GE, taking a swipe at Jeff and the era of "success theater" many believed he had fostered at GE. In February 2018, *The Wall Street Journal* published a long article about that atmosphere, arguing that Jeff "had encouraged people to act like all was well, even when it wasn't," which had "masked the rot at GE." "Bulls constantly say execution needs to be better, though we think the opposite: that execution, as defined as stretching to hit a target in any way possible, should be replaced by 'pragmatism,' where managers are not afraid to bring bad news to the C-suite," Tusa wrote.

"A product of the Welch era, we believe there is almost too much accountability at GE, to a point where bad news does not travel fast enough to senior management leading to decisions that are not perfectly informed and more often than not late. . . . From an investor perspective this 'no bad news' culture was a key reason why expectations never reset. This is a key change we will be watching for."

The July 21, 2017, earnings call was Jeff's last as CEO. John was taking over on August 1. Jeff spoke in GE jargon but left the impression that nearly everything was pretty rosy. Bornstein was a bit blunter but didn't reveal much, if anything, about the impending problems in the power business. He made a passing reference to what was going on in the legacy insurance businesses, but he obfuscated the depth of the problem. "In the fourth quarter, we will perform our annual cash flow test of our runoff insurance business," he said. "We recently have had adverse claims experience in a portion of our long-term-care portfolio, and we will assess the adequacy of our premium reserves." He said an update would be coming in the fourth quarter.

For his part, John said he was excited to be taking over on August 1 and reiterated that he was doing a top-to-bottom review of the GE portfolio and would report his findings and thoughts in November. He also said he was taking a "hard look" at GE's corporate spending, suggesting that cuts would be coming. In an answer to an analyst's question, Bornstein said that John would be rethinking how GE allocates capital, including whether the stock buyback program would continue. Jeff then piped up about capital allocation and GE's dividend. "Everybody here prioritizes the dividend at a very high level," he said. "And I just don't want anybody to ever be confused about that in the context of GE—what we do, how we do it, and whoever our CEO is, how we think about that as a context. I was here the day we cut the dividend. It was the worst day of my tenure as CEO. And the dividend is really, I think, incredibly important for our investors and for the team."

In his swan song, Jeff remembered how when he started as CEO,

GE would fax out a press release about its earnings. He was now completing his sixty-fourth analyst conference about earnings, replete with PowerPoint presentations and charts. He made a special point of thanking the GE finance team for its hard work in putting together the detailed presentations. He also thanked his two colleagues. "I am looking across the table at Jeff Bornstein," he said. "He has got a six-inch binder. He knows every fact and everything to do with the company. And I feel great about John Flannery and what he is going to do with the company going forward."

On August 1, his first day as CEO, John sent a letter to GE's employees telling them of the need to have "an intense focus on running the company well" after years of new initiatives, big deals, and portfolio pruning. He explained that he had met with more than one hundred of GE's institutional investors, whose message was "loud" and "clear." He wrote, "They understand the importance of GE in the world, but they think we are underperforming." He explained, understandably, that he was still "getting used" to being CEO and reflected on the start he got financing LBOs in a GE Capital office on Madison Avenue in New York City. For his part, Jeff wrote to GE's employees about the lessons he had learned during his sixteen years at the helm. One of them was his version of the observation often attributed to Mike Tyson, the former heavyweight boxing champion, that "everybody has a plan until they get punched in the mouth." Jeff's parallel was that "every job or decision looks easy until you are the one on the line." He also praised John as "the right person to lead GE into the future."

Despite the pleasantries, tensions between Jeff and John were running high from the start, especially as John learned more and more about the long-term health care liabilities and the ongoing problems at Power. Jeff dubbed himself the "Person to Blame." He could see Trian's fingerprints behind his untimely departure. He used to refer to Nelson Peltz as "the smiling crocodile," and the croc had shown its teeth. "They'd wanted my scalp, and they'd gotten it," he wrote. He said his relationship with John deteriorated quickly. "We need to re-underwrite the place," Jeff believed John was saying to people. "No number is too low. I want this to be a holding company, no headquarters, no corporate. I want to spin businesses out, and I want to lower the numbers. Everything's up for grabs." Jeff recalled that he tried to "counsel" John to move slowly and "just run the place for a while" before concluding that the company needed to be dismantled. "In my view," Jeff continued, "it's not enough to stand at a podium and list the company's problems. You must motivate people to fix them." He accused John of "perpetuat[ing]" the idea of Jeff's "success theater." Jeff also accused John of turning him into "his scapegoat" and claimed that people "confided" to him that John was "trashing me in front of investors and others." Jeff concluded that John "didn't want or need my input."

At a meeting with GE's top executives at Crotonville in August, John shared with GE's top executives a twenty-page PowerPoint presentation about how much cash the company was actually generating compared with how much cash people thought the company was generating. It was an eye-opener. Normally, the presentations at Crotonville were like a "puppet show," one GE executive told me, where "everybody [would] get up and say how great everything is." But this time, the executive continued, John "was like, 'Fuck that, people need to know what's going on in this company.'" He walked the group through the problems in the power business, the factoring of the receivables, the ill-fated stock buybacks, what happened to the money that GE Capital generated with asset sales, how much debt GE had, that GE's dividend had been in excess of its cash flow for three straight years. "It's just, it's like 'exhibit A' of 'What the fuck were you guys doing with this company?'" the executive told me. At the same meeting, Jeff Bornstein, the CFO, became so emotional about the need for everyone at GE to pull together as a team that he teared up. "Run the company like you own it," *The Wall Street Journal* reported he said. "Be the leaders General Electric bred you to be. You should all be accountable for every prediction made and every target missed." Then, as the tears flowed, he said, "I love this company." He was frustrated and worn out.

JEFF WAS SUPPOSED TO STAY AS GE'S CHAIRMAN UNTIL DECEMBER. But he changed his mind and left GE's board in October, a month after the GE board approved a seat for Trian's Ed Garden and a month before Garden was to join the board. "He just couldn't bear to go to a board meeting where Ed was going to be at the board meeting," said one GE board member. "He couldn't stomach that 'my friend's little brother' is going to come shit on me in a board meeting." From then on, Jeff recalled, he watched the goings-on at GE from the sidelines. "Flannery had a lot to deal with during this period," he continued, adding a minor concession, "some of which I'd helped create." He had hoped that the GE Capital "pivot" would be completed and that the Alstom deal would be "proving itself," but "that wasn't the case," and the "volatility" that Trian "created around the GE board didn't help." Although Trian had no role in picking John as Jeff's successor, Garden believed that John shared Trian's passion for putting "a bright spotlight on all the problems" and for "figuring out the reality."

Trian wanted "no more spin" and for John to begin fixing the problems. Its choice, had it been asked, would have been to find an outsider to run GE. According to someone familiar with Garden's thinking, Trian believed that "a fresh set of eyes on everything [and] new perspectives would be healthy." But the hedge fund was willing to work with John Flannery because, like Trian, he agreed that "job number one" was "to find all the snakes in the rocks" and "let the world know we're going to put a bright spotlight on it."

Jeff worried that his dismissal as CEO and then his abdication of the board chairmanship would leave him unable to control the narrative about his tenure as GE's leader since 2001. He was right. *The Wall Street Journal* was especially merciless in its criticism of his tenure at GE, and one leak after another seemed to be orchestrated to embarrass him. Mentioned in passing in an October 18 article about some of the changes John had already made or would be making at GE— delaying the construction of the new twelve-story Boston headquarters, grounding the fleet of six corporate jets, ending the company-car program that had extended to seven hundred employees, ending the annual management boondoggle in Boca Raton, closing all but two R&D centers worldwide, and demanding big cuts to corporate overhead— was the fact that Jeff had often traveled "for much of his tenure" around the world in a private jet, with a *second* jet tagging along nearby in case of mechanical failure in Jeff's primary jet. "The two jets sometimes parked far apart so they wouldn't attract attention," the *Journal* reported, "and flight crews were told to not openly discuss the empty plane."

That little squib of news initiated a media firestorm. On October 29, the *Journal* advanced the story with the news that Jeff had not informed the GE board of directors about the practice and that there had been an "internal complaint" about it that also did not make its way to the directors. The *Journal* added Mark Maremont, its jet tracking expert, to the reporting team, and he traced the various flights Jeff had made over the years with a backup plane in tow. The *Journal* documented a weeklong around-the-world trip Jeff had made in September 2016, starting in Boston and then on to Anchorage, Vladivostok, Tokyo, Singapore, Kuala Lumpur, Bangkok, Helsinki, and then back to Boston. On this trip, the two Bombardier Global Express jets took off within thirty-three minutes of each other. According to the *Journal*, the empty plane's manifest referred to a nonexistent passenger—either "Robert Jeffries" or "Jeffrey Roberts." The paper reported that the extra plane added about $250,000 to each round-the-world trip at a time when Trian, and other investors, had demanded that GE cut costs.

Obviously, the stories about the second jet were embarrassing to Jeff and to GE, and were made worse by the fact that they were easily digestible bits of corporate excess. Not unlike Jack's postretirement perk package in its way. "The average guy gets this stuff," recalled Bill Conaty, the longtime GE head of human resources. "The guy that runs the local marina said to me, 'How could any CEO be traveling around the world with two planes, Bill? Could that be true?' They don't get all this nonqualified earnings and off-balance-sheet accounting, but they get that stuff and say, 'How could a guy have such an ego to think that [was a good idea]?'" Exacerbating the scandal was the additional fact that Jeff, incredibly, denied knowing about the existence of a second jet. "This is not a practice I would have allowed," he emailed the *Journal*. He blamed GE's bureaucracy. "The Corporate Air team at GE had a practice around managing air travel that I neither instituted nor asked for," he said. "Apparently, this policy was put in place after numerous plane failures on complicated travel to difficult global locations." And GE's public relations apparatus also rose to Jeff's defense.

Conaty recalled for me the origin of the two-jet solution. Early on in Jeff's tenure as CEO, probably in 2002, Jeff and Conaty, and two other senior GE executives, were at Westchester County Airport, near Fairfield, on one of the two Boeing Business Jets that Jack had bought for GE. "Jeff was a masochist when he traveled," Conaty remembered. "We would travel and sleep on the way so that we wouldn't go to a hotel. We'd go right to the business. With these planes, you could do it." He thought it was a Saturday night and they were on their way to Brussels for some reason. The other Boeing Business Jet was disabled by a mechanical problem. They were on the one that was working, or so they thought. They started down the runway, but something went wrong before takeoff and they had to abort the flight. The mechanics went to work on the problem, they thought they had it fixed, and the plane started down the runway again. Again the flight had to be aborted. "Aborted takeoffs are not good," Conaty said. "This was two of them. . . . Now we got two planes shitted up." Jeff was pissed. "Two new planes," Conaty recalled. "Two of Jack's babies." Conaty told Jeff there was also a Challenger they could take for the five of them. It was not as comfortable but would get the job done. "Son of a bitch," Jeff said, "this would have never happened with Jack." It probably would have, Conaty assured him.

From then on, Conaty recalled, GE's corporate air division, which reported to Conaty, arranged for there to be a chartered jet available in case something went wrong with one of the big corporate-owned jets. "[Jeff] was having a plane positioned somewhere," he said. "But more often than not, it was, like, a charter that would be ready, which is okay. If you got a charter somewhere, you're not going to pay for it if you don't use it. We don't want to get stuck somewhere." He said he believed that the corporate air division arranged for the chartered jet to be available on their own initiative. "I'm convinced that the Corporate Air guys, just to cover their own ass, were making sure that they have coverage where he traveled," he continued. "He didn't ask for it, but I think he certainly knew it was there. And then that went away" after a few months.

Conaty retired in 2007. He couldn't believe the two-jet scheme was still happening years later. "Amazing," he said. He also said he thought that Jeff had to know it was happening. "For Jeff to say that he didn't know about it, I mean, you got to be Pinocchio when you say that, because when we'd pull into any FBO"—fixed-base operator, another term for a private-jet hangar—"anywhere in the world, you'd look around because we had a distinct color on the plane." There were no GE logos on the planes, but the colors were distinct and the numbers on the planes obviously were, too. "We had them painted a certain way that we'd know," he continued. "And the tail numbers. It was like 372,

373, 374. When we'd pull in, we'd say, 'Hey, who's got 373? Somebody else is here. Let's find him.' There's no fucking way that you could be making these trips and not know that the other plane [was] sitting there." He said he thought that when John Flannery dismantled the corporate air division when he took over on August 1, the pilots or the staff were peeved and took their grievances, such as they were, to the press.

According to Ed Galanek, Jeff's longtime security executive, the decision to arrange for a second plane for Jeff rested with Frank Taylor, a former air force general who was head of GE's security detail. Jeff was on a jet in Africa, on his way to South Africa, when the windshield on the jet cracked. The plane had to make an emergency landing. GE could not easily get a replacement plane, and Jeff was going to miss his first big meeting of the trip. "How the fuck do we not have this figured out yet—that I'm not stuck someplace?" Jeff wondered. He was pissed and wanted a solution. He complained to Susan Peters, his head of HR. He also called Art Cummings, a former Navy SEAL and FBI antiterrorism expert who also worked in GE's security detail. Cummings called Galanek. "Fucking Jeff is fuming mad," Cummings told him. "It'll probably get to you eventually, but try to work on it right now. So if he calls you, this is what we're doing. He was just really, really pissed at me about this plane situation." Taylor came up with the solution of the second plane. At that time, GE had eight planes. "There's only one way we can really guarantee it," Taylor told Galanek. "Let's send a second plane with him where he goes, and then it will guarantee he doesn't miss a meeting."

Over time, to save money, Galanek said GE began arranging for a second plane to be available in the region where Jeff was flying, rather than having a second empty plane start with him in the United States. "If we're in the Middle East, if we had a problem, we had already paid for a plane that would be in Dubai that could come and rescue us," he said. "If we were in Africa, we'd have something there. If we were in Southeast Asia, have something there. So that was the way we did it. . . . But going back in history, how it starts is Jeff says, 'It's got to get fixed.'"

He recalled how he used to defend Jeff in the wake of the two-plane story when people would project their fantasies about what they think happens on private jets. "My friend," he said, "let me just fucking tell you something. You probably have never worked one day as hard as Jeff Immelt worked every day of his fucking life. That's number one. Number two, there were no fucking stripper poles and fucking coke on the backup plane. There was no fucking fun times. It was all about work. And I'm going to tell you something. Go tell your fucking friends what I just told you. Because you know what? It's fun to say that guys live the opulent lifestyle, and everybody wants to say that he had fucking three girlfriends and was fucking doing coke and this and that. This is the most boring guy in the world. All he wants to do is work hard and make GE the best thing it could ever be."

When I asked Jeff about the two-jet story, he was sheepish. "Oh, God, embarrassing," he said. "Bad practice." He said he never spent any time thinking about the jets or managing Corporate Air. "It was just one of those things," he said. Asked if he knew about it, as Conaty suggested, even though he had told the *Journal* he didn't know about it, Jeff said, "Did I know it? Probably I did. Again, this is a wiener answer, right? Did I know it? I guess I did. I just never thought about it. I never really cared about it. I never really spent any time thinking about it. I took no personal gain out of it. I was traveling all the time. And the funniest thing of all is, this is, like, one of these things that I don't know whether it means I'm a good guy or a shithead. I never kissed the pilots' asses. They loved Jack, right? Because he kissed their ass. They kissed his ass. With me, I just got on and worked and flew. I went to Turkmenistan and Bangladesh and Angola and stuff like that. And so, again, I can't defend it. It was bad practice." He said he the story was leaked to embarrass him. "All Jack Brennan had to do was give them a quote to say, 'Jeff worked his ass off on behalf of the company. This doesn't speak to his character.' Two sentences would have put a fork in the story. That's all they had to do." That didn't happen. Jeff felt a bit betrayed. "I don't know what it feels like to have my wife die or my child die," he said. "But it's about that."

Another narrative that Jeff could not control was the one about "success theater," the idea that he tended to dismiss bad news in favor of some sort of perpetual optimism. Once again, his nemeses were GE people leaking to *The Wall Street Journal* in the wake of his departure. In a particularly damning article, the *Journal* cited as an example Jeff's performance at the May 2017 EPG conference—the one that got him fired—where he said GE was a "strong, very strong company" and "It's not crap. It's pretty good, really." At that time, GE's stock was trading around $28. Jeff's "success theater" was to blame, the *Journal* reported. "This culture of confidence trickled down the ranks and even affected how those gunning to succeed Mr. Immelt ran their business units," the paper explained, "with consequences that included unreachable financial targets, mistimed bets on markets and sometimes poor decisions on how to deploy cash." Jeff hated the idea that he had created a culture where people could not bring him bad news or speak their minds, although there certainly was some truth to it. The *Journal* story put him again on the defensive.

The idea that Jeff did not want to hear bad news was one that dogged him throughout his tenure as GE's CEO. It was true that Jeff was an optimistic person, perhaps to a fault. He certainly had big ideas that led to big "initiatives" for GE—GGO, the GE Store, Predix and the army of software engineers in Silicon Valley, and a nearly complete makeover of the portfolio of businesses that Jack had bequeathed him. And he rarely had doubts about these changes. As he reiterated often, sometimes a leader has to lead, even if the troops didn't agree with the direction he was leading them. But not being open to hearing news he didn't like was a criticism that really stuck in his craw. The charge really stung. "The 'success theater' thing drives him crazy," one senior

GE executive told me. "The 'success theater' thing really gets the most under his skin." He said while he thought Jeff was "extremely open in communicating," he was always focused on "the sunshine communication." He also said he believed Jeff had "Trumpian" qualities. "There's got to be some personality tests that would overlap pretty high," he said. "Artificial reality; if you say it enough times it's true. If you cross me, you're dead. If you resist, you're dead. But I do remember people saying, . . .'The guy is just kind of weird.' He's incredibly friendly and charismatic, but has no friends, basically. Like, super outgoing, but impossible to get to know."

Another GE executive shared his views of the "success theater" assertion. "If you're helping to perpetuate Jeff's reality and story, you're on his team," he said. "If you're introducing facts, bad news, reality that doesn't fit the story, his story, his view of reality, you're not on the team. That's where the narcissism and the insecurity comes from. I think he was over his head in the job. That was crystal clear. But it's the job, it's the position that is the power. It's an incredible job. People want to hear what the CEO of GE says. You can go see the president. You can do whatever you want. But that job is no different than any other job in the company, which is you've got to earn it every single day. It's not royalty. You're not the king. He assumed nobility, that his story was going to have a happy ending from the day he started. It was just focused on that, not necessarily doing what was right for shareholders, people who worked for him. It just came out every year more and more and more. . . . Jeff is very loyal, if you're perpetuating the story."

Jeff found these charges nearly unfathomable. "This is simply untrue," Jeff countered in *Hot Seat*. "I'd intentionally surrounded myself with top executives who insisted I hear when we were off course or when we couldn't achieve certain goals." He blamed John Flannery for "flog[ging]" the "success theater" claim for "too long." Of course, in his January 2015 talk in Boca, John had tossed his PowerPoint slides to focus on the reality of what he thought was going on in the medical division and how it might apply to the company as a whole; the opposite of "success theater." After Jeff's book came out, Alex Dimitrief, by then a former GE general counsel, took to LinkedIn to share that he agreed with Jeff. "What hogwash," he wrote about the idea of "success theater." "First, it's simply and demonstrably untrue." He said he shared with Jeff "plenty of news that would be unwelcome for any CEO" and he "disagreed with Jeff on plenty of occasions and viewed our back-and-forth as an essential and important part of my job. Jeff's questions and challenges led my team and me to do some of our very best work, and the same held true for other functions." He wrote that it was "an important part" of the job of business leaders at GE to "push back on Jeff" when necessary and to "challenge" him and others "constructively whenever we disagreed about something important." He witnessed many GE executives "display the courage and intellectual wherewithal to" push back on Jeff "privately and during business reviews, and such exchanges did not derail any careers. To the contrary, Jeff and GE rewarded independent thinkers who stood up in principled ways for their views and teams." He decried the revisionist history. "Shame on the people who now profess to have raised issues with Jeff or others when, in reality, they did not," he concluded.

IN EARLY OCTOBER, JOHN BEGAN PUTTING HIS OWN IMPRIMATUR ON the GE management team. First to go were Beth Comstock—no surprise, as she had been a Jeff loyalist—and John Rice, the former Power executive and head of GGO. Rice's departure was a bit more of a mystery, given the ongoing problems in the power business and the fact that he had a tremendous understanding of the business. The even bigger surprise was the departure of Jeff Bornstein, who had just been promoted to vice-chairman while retaining his CFO position. Bornstein had pledged to work closely with John to run the company. But suddenly, like both Comstock and Rice, he was gone in October.

As the GE board was assembling for its monthly meeting in October, John told them that Bornstein was leaving. Bornstein decided that with Ed Garden, from Trian, about to join the GE board, it would be a good time for him to leave. "I've got to go," he told John. "I'll own everything up until now." He told John that "somebody needs to be fucking accountable" for the disaster that was unfolding in GE's power business. Bornstein didn't run the power business, of course, but he had been caught unawares by the fact that the gas turbine market, which was supposed to be on the order of forty or fifty gigawatts annually, was actually ending up closer to twenty gigawatts. Bornstein had had no visibility into the collapse and, after giving his emotional speech in Crotonville two months earlier about accountability, he decided he had to hold himself accountable. He decided that he was the only person left who could be held accountable for the power business. Everyone else was gone. He believed he wouldn't have any credibility left inside the company if he didn't stand up.

That wasn't the only reason for Bornstein's departure. While he and John agreed on most everything—making the company smaller and more focused, cutting costs—Bornstein was increasingly frustrated by what he perceived as John's indecisiveness, having fifteen meetings on the same topic without making a decision. "What we didn't have was time," explained one GE executive. "We had to be bold." On the other hand, according to someone familiar with John's thinking, GE's third quarter was looking increasingly disastrous, including the announcement of big trouble in the power business and a liability in the long-term health care business far beyond any worst-case scenario. "We are going to announce that we have a massive liability and that our power business isn't making any money, and he's been the CFO for the entire time," this person said in hindsight. "How are we going to sit credibly in front of anybody and say, 'Okay, now our financials are okay,' even though it's the same guy? Like, there was just no way to keep him as

CFO. And in Bornstein's opinion, he was like, 'I need to get out of here. This just doesn't work.'"

On the third-quarter conference call, on October 20, Bornstein explained why he was leaving GE. He emphasized that the company's financial performance had been unacceptable. Then there were the ongoing problems with the power business. "I've talked a lot about accountability inside the company, and that sense of accountability has to start with me," he said. "We are not living up to our own standards or those of investors, and the buck stops with me." Bornstein also reported that GE Capital did not dividend any cash to GE in the quarter because the company was in the process "of performing an actuarial analysis of claims reserves in our insurance business" and until that review was completed, "we have deferred the decision to pay GE Capital dividends to GE." To replace Bornstein, John announced that Jamie Miller, the CEO of GE's transportation group and the former controller of GE, would become GE's first female CFO.

After Bornstein's resignation, John said that he was still working on a top-to-bottom review of GE's portfolio, culture, cost structure, and complexity and that he looked forward to sharing his findings on November 13. He essentially apologized for the poor third-quarter results. He called them "completely unacceptable" and said they resulted largely from "weakness in our power business." During the part of the earnings call reserved for questions, Scott Davis, the research analyst, asked John about changes at the board level, given the addition of Ed Garden. John replied that "everything was on the table," including changes to the board. He also added that a board with eighteen members was unwieldy and that the size of the board would also likely change.

In fact, Garden had already been making his views known about his fellow board members. Trian was not happy. GE's stock had fallen to around $20 a share by early November, and that meant more losses for the hedge fund. Before his first board meeting, in November,

Garden called John and told him GE had to "refresh the board" because what had occurred at GE on the previous board's watch was "the biggest failure of corporate governance in corporate history" and "we have to shine the spotlight on it." (And this call was *before* Garden became aware of the extent of the long-term health care liability that was brewing in GE's insurance subsidiaries.) John was not averse to what Garden was suggesting but was leaning toward moving more slowly, perhaps replacing one or two board members a year to avoid embarrassing anyone who had served loyally on the GE board. But Garden was emphatic. A slow "refresh" wouldn't satisfy him. Garden was determined to prevail. During his first official meeting as a member of the GE board, he told his fellow board members, most of whom he did not know, that either the entire board needed to resign immediately and he and John would pick their replacements or anyone who was on the board before 2016 needed to resign immediately and he and John would pick new board members until the board got to twelve members.

Garden also told Brennan that he wanted to be appointed to the compensation committee of the board, effective immediately. But Brennan told him that GE had a rule that a board member had to be on the board for two years before he or she could be on the compensation committee, often considered the most important board committee for obvious reasons. Garden didn't like Brennan's answer. "You mean I have to prove I can be a good boy before I can go on the compensation committee?" he said. "Well, that's going to change, I promise you." At the November board meeting, it was decided that any board member could attend any committee meeting of the board that he or she wanted to attend and that the GE board would be reduced to twelve members and that three new board members would be added at the April 2018 annual shareholder meeting.

John also had some ideas for the board in the November meeting. With GE's oil and gas business gasping for air, John suggested spinning off to GE's shareholders GE's 62.5 percent stake in Baker Hughes. That would remove the oil and gas business from GE's financial statements and make the industry's problems something for Lorenzo Simonelli and GE's shareholders to solve on their own. "The wolves are at the door," John told the board. "We have to do something dramatically different. . . . We need something radical right now."

That did not go over well. "The board basically erupted," one board member explained. The board told John, "We don't need to do anything. You're overreacting to sentiment," and "Just keep your head down." John wondered why the board was not supporting his decision. But he decided to table the conversation about Baker Hughes for the time being. He was too new—and there were too many other bombs going off—for him to get into a confrontation with his board. Garden, for one, came to his defense. "What the fuck, guys?" Garden said. "You don't get it. . . . We need to do something radical now. And you all should resign, by the way." But he was new, too, and not yet in a position to get his way. He did get some unexpected support from Susan Hockfield, the former president of MIT. "Maybe we should resign," she told her fellow board members. "I mean, let's face it, this thing blew up on our watch." (Like so many others on the GE board, Hockfield did not respond to a request for an interview.)

But the board-replenishment project ended up proceeding at a more deliberate pace. At the top of the list of potential new board members was Larry Culp, the former CEO of Danaher Corporation, a mini-GE-like conglomerate based in Washington, DC, that was formed in the 1980s by the Rales brothers, Mitch and Steven. The Rales brothers became billionaires by buying and selling a portfolio of seemingly mundane industrial companies and then squeezing more profit from them. Danaher had more or less become a publicly traded, industrial-focused leveraged-buyout firm. Culp, a graduate of Maryland's Washington College and Harvard Business School, retired from Danaher in 2015 with a net worth of more than $100 million and was doing some teaching at Harvard Business School. Garden got to know Culp in 2017 when he sought him out at Harvard after Jeff's meltdown at the May 2017 EPG meeting. At that point in 2017, Garden was still in the process of working his way onto the board—by then he had an agreement from Jeff that he would join the board—and wanted to speak to Culp to get his views about what he thought was going on at GE and how best to fix it, if possible. Garden had respected the job Culp had done at Danaher and thought Culp might one day also make a good addition to GE's board. During their visits, Culp was very curious about Trian and its approach to activist investing. He told Garden that he believed that GE was "a very important business to the world and to the country" and that it was "important to get that business fixed" and that "under the right circumstances he would consider getting involved."

With the board "refreshment" process underway, Culp's name quickly leapt to the top of the list as a desirable board member. "His performance at Danaher was revered by investors," recalled one GE executive. "Ed Garden, in particular, had Larry Culp up on a statue. But he wasn't the only one. I think that everybody we talked to, they said, 'Man, if you can get Larry Culp, that's like getting Warren Buffett, a Harvard Business School stud, an incredible record.'"

John agreed with Garden that Culp would be a great choice, especially since the small-corporate-office approach that Culp had pursued at Danaher appealed to John, who believed that GE's corporate office had grown too large and that GE needed to try to emulate Danaher. John believed Culp could be a useful adviser to him on the board as he sought to reduce GE's corporate overhead. Others, though, worried that Culp was an ambitious piranha, eager to take any opening handed to him to grab the reins at GE. Blair Effron, the Centerview banker, warned John that if "there were any hiccups," Culp "would be walking in through the front door." But John told Effron he didn't care. "The company's fucked up," John told Effron, "and we need help. I need help. This guy's the right guy." John told others, "People are warning me that I may be recruiting my successor, but I don't care."

On November 13, John unveiled the results of his deep dive into everything GE. It had been five months in the making and, given that it was held at 30 Rock, had the feel of being an "important" event. Mostly, though, it was just more of a souped-up Wall Street analyst meeting. There was news. Once and for all, John announced that Jeff's beloved promise of $2 a share for 2018 earnings was exploded. In reiterating that 2018 was a "reset" year for the company, he promised 2018 earnings at between $1 and $1.07 a share, just as Steve Tusa had predicted. There was no mention, or discussion, of what had happened to Jeff's promise.

John also explained that by April 2018, the board would be reduced to twelve members, from eighteen, and that three of the twelve board members would be new. He said that he welcomed the addition of Ed Garden to the GE board. "I love to debate," John said. "I love different ideas." And then, taking a subtle shot at Jeff and his preference for hearing only good news, he continued, "I love people coming in and saying, 'Have you looked at it this way or that way?'"

He said he had created a new board committee—the Finance and Capital Allocation Committee—and tasked it with deciding whether GE should spin off its 62.5 percent stake in Baker Hughes, which John wanted the company to do but for which he hadn't yet won the approval of his board. "When you look in many ways at the transaction we did when we merged GE Oil & Gas with Baker Hughes," he said, "part of the deliberate thinking of that transaction was to create optionality with that asset, and that's what we're evaluating right now." He also explained why, earlier in the day, he had decided to cut GE's dividend in half, to 48 cents a share, from 96 cents a share. When Jeff cut the dividend for the first time since the Great Depression, his angst was palpable. He claimed it was one of the toughest decisions of his tenure. John was more clinical. Basically, he said, GE could no longer afford to pay nearly $10 billion a year in dividends. "We've been paying a dividend in excess of our free cash flow for a number of years now," he said. "And while we had the financial wherewithal on one level to do that, we just don't think it makes sense for the company going forward." He expressed regret, of course. "We understand how important the dividend is to our shareholders, especially the people who use it for current income," he said. He also shared more details about the planned $20 billion of divestitures, including of GE's locomotive business, its "industrial solutions" business, and its—gulp—lighting business, one of GE's original businesses. He also said that any new acquisitions would be "smaller transactions, closer to home in terms of industries, less integration risk, supply chain deals, bolt-on deals." He said GE's corporate overhead would be reduced by a quarter in 2018.

While John made no mention of the ongoing audit of the long-term health care insurance business, there was plenty of discussion about the ongoing meltdown in the power business and the disappointment of the Alstom deal. He said Alstom had performed below GE's expectations.

It wasn't all gloom and doom. The first chart in John's PowerPoint presentation showed many of the incredible innovations that GE had pioneered since its creation in 1892. "GE is a company that matters to the world," he said. "For a hundred and twenty-five years, we just tackled the biggest challenges the world faces. So light, flight, health—these are the absolute underpinnings of the modern world. It's always been this mix of sparks of innovation with just sheer determination to succeed. It's been a company that does things, solves things, lines things out, sweats the details. The harder, the better; bring it on. That's been the essence of the company, affecting the world for over a hundred years." He spoke about his love of GE. "It's always been a culture of meritocracy," he said. "It's been a culture of compliance and integrity. There's a passion and love for the company inside like you can't imagine." He remained optimistic about the company's future. "The world is fundamentally going to run on the back of GE," he continued. "It's going to move on the back of GE. It's going to heal, babies are going to be born on the back of GE." But, he said, his first hundred days as CEO did not go as he thought they would.

———

IN DECEMBER, ON A COMMERCIAL FLIGHT TO SALT LAKE CITY, JOHN drafted the corporate equivalent of the Port Huron Statement, his carefully considered manifesto for how GE would survive the two unfathomable surprises he was managing through in real time—the collapse of GE's power business and the soon-to-be-revealed multibillion-dollar charge related to the long-hidden liabilities in an insurance business that Jeff should have sent out the door along with the Genworth IPO. The idea was to finally break up GE, the world's most famous and successful conglomerate. Investors were voting with their feet. GE's stock was in free fall. GE executives had proven they could no longer manage the conglomerate that had been built. He decided GE needed to do something different. It became known as the "Eisenhower" document. John told his colleagues at the time that since he was on a commercial flight—he had previously disbanded GE's fleet of private jets—and was the CEO of GE and was about to write a manifesto advocating the breakup of the company, he realized he couldn't be typing "GE" or "General Electric" on his laptop. GE became "Eisenhower." As John surveyed the state of GE's business, it felt to him like trying to crack a safe with plutonium inside: one wrong click and the whole thing blows. If the Kansas insurance commissioner told GE he wanted $15 billion immediately to cover the potential losses at UFLIC or ERAC, the jig was up. GE would be bankrupt. If GE sold the wrong division, all of GE's and GE Capital's bonds—more than $100 billion of long-term debt by then—could become due and payable, since GE had agreed to guarantee GE Capital's debt as part of its liberation from being a SIFI and there were triggers on that debt tied to asset sales. That could also lead

692 – POWER FAILURE

to a possible GE bankruptcy, since GE and the remains of GE Capital were liked roped-up mountain climbers—one wrong move and everyone goes into the crevasse together.

The Eisenhower document argued that the board needed to be "refreshed," smaller in size and with new faces, and that the corporate office needed to be much reduced and the costs pushed down into the individual business units. Trian wanted to hold the GE business leaders accountable, but there was no way to do that without the costs related to their businesses being pushed down. Too much cost was up in corporate. Larry Culp had run Danaher with a hundred people at corporate; the number of people at GE's headquarters—not publicly disclosed—was many multiples of the number Culp used to run Danaher. John agreed with Ed Garden and Trian. "Pushing all of those corporate costs down into the business units—John completely bought into this," remembered one GE board membered. "This is something I think is underappreciated even today, what's happening, taking all of those corporate costs and pushing them into the business units. For the first time, if you're running the aerospace business, you actually control your P&L"—profit-and-loss statement—"because so much cost was being allocated to you."

The message of the Eisenhower document was clear: if increasing cash flow and margins across GE's businesses, specifically in Power, could not be achieved, GE needed to be broken up. John wrote that he was "confident" that Power could be fixed—a scenario he dubbed "Plan A." But just in case it couldn't, the fallback plan, Plan B, was to break up GE. He and his team had concluded that GE's "current structure does not maximize the value of the underlying businesses, assets and technologies we possess."

Essentially, John's argument was that as stand-alone entities with their own publicly traded equities—certainly the aviation and health care businesses—would trade for higher multiples than GE did and could use that more valuable stock to make important acquisitions that GE had missed out on. In particular, GE had wanted to buy Rockwell Collins for the aviation business but could not compete effectively for it; instead it went to GE rival United Technologies. The white paper also addressed the fact that GE's depressed stock had cost its top executives real money, especially when compared with stand-alone companies with stock that had performed better. "The longer-term implication on retention is obvious," John wrote. His other concerns were ones he had stated publicly: that GE was too complex, that it had wasted something like $5 billion on GE Digital with little to show for it, and that investor appetite for conglomerates had declined dramatically in the past five to ten years. He wrote that he believed Plan B would achieve greater results for GE shareholders than Plan A. He recommended that GE "explore the option" of separating the aviation and the health care divisions into two stand-alone public companies. Power needed to be retained, he argued, until its operating performance improved.

Finally, he proposed reducing corporate overhead by another $1 billion and having the corporate suite focus only on governance, emerging-market support, research, digital, and GE's additive business. He noted that GE's corporate expense during the previous five years was a whopping $13.6 billion. "While there are significant complexities to work through—including pension, debt structures, carve out issues etc.—we believe the core idea is clear and that we should move quickly to drive the pace and control the narrative," he concluded in the document. He sent the Eisenhower manifesto to the board. But given the need for the new members of the board to be elected and then seated in April—after the annual meeting—and for them "to get up to speed," nothing happened for months on John's proposal, a crucial six months, to be exact.

In the meantime, the bad news continued unabated. On January 16, 2018, John announced that because of the "comprehensive review" of GE's "run-off" insurance businesses, GE would take a $9.5 billion pretax charge ($6.2 billion after tax) to its fourth-quarter-2017 earnings. He announced further that, with the consent of the Kansas Insurance Department, GE Capital would make capital contributions of roughly $15 billion over the subsequent seven years, with $3 billion of that injection coming in the first quarter of 2018. "If they had said, 'Give me $15 billion now,' we didn't have it," one GE board member told me. "We couldn't have done it. We would have tanked." John announced that the money would be coming from GE Capital and that, as a result, GE Capital was again suspending dividend payments up to GE. There were other harsh revelations made as well. Jamie Miller, the new CFO, said that 2017 earnings would be at the low end of the previously announced range. John said that WMC, the mortgage originator that GE Capital had foolishly bought on Jeff's watch, was in the "early days" of a discussion with the Department of Justice about what kind of fine GE would have to pay to get free of liabilities. (It ended up being $1.5 billion.)

When asked on the earnings call about his strategic thinking, John also alluded to Eisenhower in vague terms. He said, as he had in November, that GE would be a more focused company, around power, including renewables, aviation, and health care. And that he would be running the business with more "rigor" and "transparency," with the main objective of generating cash and reducing costs. He also said that he and his team were still reviewing the portfolio and that there were any number of possible structures that could achieve his ambition of improving value for shareholders, including making each business unit a separately traded business—in effect breaking up the company. "All options on the table," he said. "No sacred cows." The goal, he said, was to "make sure that these businesses can flourish in the decades ahead and that they have the right capital structures and investment resources to do that."

Once again, GE's stock ticked down on the news. It had been trading around $19 per share before the announcement of the massive insurance charge; a day afterward it was trading closer to $16 a share, a

13 percent decline in about a week. The news came as a shock. One GE board member remembered getting the call from John about the problem. "I just couldn't even believe it," he said. "The numbers started out small and kept getting bigger. We went back and looked at all the public filings, and there was nothing talking about this. My first reaction was 'Did we miss something?' . . . What I hear is that you could have solved it for, like, half a billion dollars or something" on Jeff's watch. "But that didn't happen."

Eight days later, on the fourth-quarter-2017 earnings call, Jamie Miller dropped the latest bomb—that the SEC had opened an investigation into what led GE to take the $6.2 billion after-tax charge related to the health care–insurance liabilities and into GE's practice of creating long-term service contracts in the power business. While both the aviation and the health care businesses were performing well, the power business was crashing, with operating earnings down 88 percent in the quarter, "driven by the market" and "certain execution misses."

Once again John was on the defensive. "When you look at Power, it's obviously the focus point of this whole discussion," he said. "This is a very important franchise. It's going through a very difficult period, but we still have a strong franchise here. We have 50 percent share in the high-end technology. We have got a large installed base; a third of the world's electricity. There is plenty to work with. It's clear the new unit market is soft." He also hinted again at the thoughts he had outlined to the board in Eisenhower, especially as related to Aviation, Power, and Healthcare. "I will do whatever it takes really to make sure that those businesses are positioned to flourish in the future, have the right resources, have the right investment flexibility," he said.

IN THE SIX MONTHS BETWEEN JANUARY AND JUNE 2018, JOHN AND the newly created Finance and Capital Allocation Committee began discussing Eisenhower, along with the rest of the board. The board hired Paul Taubman, a former senior banker at Morgan Stanley, who had created his own eponymous firm, to help advise it on Eisenhower. Garden and Trian were busy, too: in February, on behalf of Trian, Garden endorsed the three new nominees to the GE board. There was Leslie Seidman, the former chairman of the Financial Accounting Standards Board; Thomas Horton, an executive who had overseen the merger between US Airways and American Airlines; and, of course, Larry Culp, the former Danaher CEO whom Garden had been cultivating for months. In April, GE's shareholders elected the three new board members. Part of the arrangement with Culp coming onto the board was that he would become the lead independent director in June, replacing Jack Brennan.

Not surprisingly, there was a lot of debate about Project Eisenhower. Board members asked such questions as "Why do we have to break the company up?" and "Can't we just run it better?" But slowly but surely the board started to come around to the idea that several of GE's divisions—Aviation and Healthcare in particular—might be better as stand-alone companies outside of GE and with their own stock, which could be used to make acquisitions and reward executives for their performance. Slowly, skepticism turned into "How might we do this?" and "What's the best way to do this?" As part of the analysis, "for the first time ever," according to one GE executive, the company created a comprehensive financial model of its business. It had never been done before. To understand if the breakup contemplated by Eisenhower was possible, GE's top executives had to understand precisely where the company's debt was and what it related to, what all the credit ratings were, what the covenants were that might be breached by a breakup or spin-off.

One particularly nettlesome covenant in some of the debt indentures referred to how a breach might occur if "all or substantially all" of GE's assets were sold or spun off. What did that mean exactly? Would all of GE's debt "accelerate," or be due and payable, if the company headed down the breakup path? "This thing is like a house of cards," explained a GE executive. "You have to know when you pull one of them out what happens to the rest of the thing." It was like a massive game of Jenga: you pull out the wrong piece and the whole structure comes crashing down. John was taking his time to figure out what to do. For instance, until GE management started digging into the details of bond indentures, covenants, and tax treatments, they hadn't realized that if they decided to sell GE Capital's valuable aircraft-leasing business—one of the businesses under consideration for sale—GE would have to come up with something like $5 billion to make the sale possible, wiping out the benefit of the sale.

After Project Hubble, when GE agreed to effectively guarantee all of GE Capital's debt, "everyone was strapped to the bomb," said one executive. "I don't know how much people even understood that at the board level when they did that." There was tons of analysis and discussion about what to do. Some board members favored spinning out the aviation business; others favored spinning out the health care business. John, who had run the health care business, favored spinning it out. Garden preferred making Aviation an independent, stand-alone company. In the end, Garden told John, "Eh, you know, it's not that big a deal and I'd rather just do something. So, you know, since you want to do Healthcare, let's do Healthcare."

Then there was the board transition itself, with the three new board members and Culp becoming the lead director. "We had a board in transition while we were analyzing and sort of making quasi-existential decisions about the company," one board member said. "The new board members wanted some time." Ed Garden spent time researching the insurance problems while Culp focused on the power business. Culp made several trips to South Carolina to visit the Power manufacturing facilities. (He owned a home on Kiawah Island near both Jeff and Jeff's close friend, Peter Foss.) When Culp returned, he told people that, yes, the business had problems, but it was also "better than people think." Slowly but surely the newly reconstituted GE board, with Paul Taubman's help, came to see Eisenhower as the best path forward for the company.

Board member Geoff Beattie offered John some unsolicited advice in the form of a one-page memo where he stated up front, "We're all in this together." He wrote that John "must focus" on morale, team building, and culture. He shared his opinion that there was "too much anxiety among senior leadership" about "who is or who isn't on the team." He added, "Those on the team will take the Mountain for you. . . . Trust them to do so," and told John he shouldn't try to pick his team "by consensus" because he was the CEO, "so your words matter." By asking around about certain people he was conveying the idea that that person was "not on your team." He urged John to "make decisions" and "move forward." He believed that too many of John's decisions "are delayed in execution," which was "bad for progress" and "bad for morale." John needed to stop revisiting "decisions that have been made." He wrote that John should "lead with more heart not just more brain" and reminded him that "people inside and outside the company (including the board) are killing themselves for you." He recommended that John "recognize and appreciate that" whatever way worked best for him and said that GE's people needed to know that he knew that they were "collectively the heart and soul of GE."

He also conceded that the board "refreshment process" had become "highly charged and emotional" but said it also gave John a "fresh start" as the chairman of the reconstituted board. His last two pieces of unsolicited advice were that fixing and improving GE's problem businesses, in particular Power, was "PRIORITY ONE" and something John would be judged on; finally, Beattie told him to forget everything that happened before December 31, 2017. "It no longer exists," he wrote. "If you lead investors forward, they will follow you . . . and will do so on whatever scorecard you lay out to create value over a three-year period."

On June 26, John announced that the GE board had unanimously agreed to support his dramatic vision for the future of GE, essentially as he had outlined six months earlier in Eisenhower. GE would separate its health care business through a 20 percent IPO—with the proceeds going to GE to use to pay down debt—and then spin off the remaining 80 percent of the company to shareholders. Combined with the earlier announcement that GE's transportation business—the longtime manufacturer of train engines—would be merging with Wabtec, a competitor, and that GE would also spin off to shareholders its remaining 62.5 percent interest in Baker Hughes, this meant GE would become a company largely focused on aviation and power systems—in a return largely to its industrial roots—and on renewable energy.

John had won the support of the reconstituted GE board to end GE's conglomerate structure, replacing it with a smaller corporate structure more tightly focused around, as he said, "three highly complementary businesses poised for future growth." He planned to cut $500 million out of the costs at corporate headquarters in Boston, which would be focused solely on capital allocation, corporate governance, strategy, and talent. He also said he wanted GE to pay down $25 billion in debt by 2020, to retain $15 billion in cash on its balance sheet, and to maintain an investment-grade A credit rating. He said he expected that GE would contribute $3 billion to GE Capital in 2019—as part of the agreement with the Kansas insurance commissioner—and that it was exploring ways to reduce GE Capital's exposure to the insurance liabilities. He also announced that Culp would become the lead director, replacing Brennan, and that Culp would also become chairman of the powerful MDCC. "Larry's track record on strategy development and execution, capital allocation and talent make him well suited to take on this role," John said. "I appreciate his clarity, transparency and business-first philosophy, and I believe his leadership will be invaluable to GE as we enter our next chapter." Some worried that while John appreciated what Culp had accomplished at

Danaher and was hoping to benefit from his wisdom, he had made a tactical error, especially since it was no secret around GE that Culp was Trian's guy and that he was hugely ambitious. Ed Garden "loves, loves, loves [Larry Culp], like, mega man crush," said another GE board member. In any event, Nelson Peltz sent John a note. "Hey, it's great to see the coming to fruition of all the dialogues you guys have been having," he wrote.

Although the market reacted positively to what John announced, the irony was that on the same day that he announced his intention to break up the GE conglomerate, GE's stock was dropped from being a component of the Dow Jones Industrial Average. GE was an original member of the Dow average when Charles Dow created it in 1896 and it had been a continuous member of the Dow since 1907. It was replaced in the average by the drugstore chain Walgreens Boots Alliance. By that time, GE had lost 80 percent of its value from the all-time high value it had achieved under Jack. On CNBC that day, David Novak, the former CEO of Yum! Brands, put it bluntly. GE, he said, "has lost its luster. There's nobody in business that thinks about GE today and says, 'Wow, what a company!'"

John's presentation to Wall Street research analysts about breaking up the company had the aroma of consensus, but there was no minimizing the shock waves it sent through the GE establishment. After 126 years, he was calling for the breakup of the most venerable and highly respected conglomerate in the history of the United States! It was an extraordinarily radical and divisive moment. The employees in Boston were upset, since John's plan would eliminate most of them. The more traditional members of the GE board—the board members who were not Larry Culp or Ed Garden—were also having trouble stomaching the proposal to end GE's reign as the most successful conglomerate ever. Did the board just unanimously support the CEO's radical plan to bust up the company? It did, of course, but the repercussions of what it took for John to get the board's unanimous support were still reverberating. *The Economist* noted that while the "road ahead" for GE remained "rocky," if John's "bold plan succeeds," he will "have moulded a humbler but fitter GE that may yet endure another century."

A few weeks later, on July 20, John reported GE's second-quarter earnings. The problems in Power were metastasizing. Nearly every component of the business reported lower financial performance: Orders were down 26 percent from a year earlier, with equipment orders for gas power systems down a stunning 78 percent. Operating profit was down to $421 million. In answer to an analyst's question, John said that fixing the power business was his top management priority. "We are working that intensely, a total sense of urgency," he said. "The market is challenging, but we need to work through that. It's going to be a multiyear fix, I think, with some volatility. This is not something that's going to move straight line quarter to quarter." The problem, he explained, was that the power market had collapsed 50 percent in the previous two years, but obviously, the demand for electricity was not disappearing. For instance, in 2017, GE had sold 107 gas turbines; John believed that number would be around 50 in 2018. The challenge was to get a better handle on what the size of the market was going to be and then design a cost structure around it. Already, John said, some $550 million of cost had been removed from the business in 2018, with another $450 million to come. "The industry is not going away; the short-term cycle is severe and we've got to manage through that," he said. "But there's an asset worth maintaining and preserving and expanding the value here." He was optimistic the company was on the right track. "A lot has happened in twelve months," he said. "As we stand today, I'd just say we look forward and say the path is clear."

On September 19, the operational difficulties in the power business took an unexpected turn. In a note on LinkedIn—of all places, yet again—Russell Stokes, the head of GE's power business, dropped a bombshell, in the penultimate paragraph of a ten-paragraph post: that "an oxidation issue" in GE's new HA gas turbines—the *A* stood for air-cooled—could affect the "lifespan of a single blade component." He wrote that GE had "identified a fix" and was working "proactively" with its customers. "Obviously, this was a frustrating development, for us, as well as for our customers," he wrote. He added that the "minor adjustments" didn't diminish the HA's "record-setting" position in the market. He explained, in his windup before dropping the news, that "innovation is hard. But it's also the lifeblood of what we do. In our business, one rarely arrives at a significant breakthrough without taking on seemingly insurmountable obstacles that might make others give up."

In a note to his clients, Steve Tusa, the JPMorgan Chase analyst, revealed that the problem had resulted in a broken turbine blade at an Exelon power plant in Texas. Tusa's note, which claimed GE was "seeking to play down the issue," was sufficiently material to Exelon that two power plants had to be closed temporarily. His revelation made Tusa once again the focus of an intense investigation at GE, including among the management and board ranks, of whether someone inside GE was leaking the information to him. Actually, he had been talking to GE's customers. The news of the turbine-blade problem sent GE's shares down 3 percent, to $12.49 a share. Tusa lowered his target for the stock to $10.

## Chapter Twenty-Seven

# THE LAST EMPEROR

With the operational challenges mounting and the stock price collapsing, the newly reconstituted GE board began—at long last—to assert itself. This was a board with Ed Garden in the driver's seat and Larry Culp as the lead director. Without John having a clue (and mostly in executive sessions), the board began to contemplate the once-unthinkable decision of firing John—after a brief fifteen-month tenure—and replacing him with Culp, who would be the first outsider to lead the company in its 126-year history.

What happened next was nothing less than a "palace coup," orchestrated by Garden and Culp. As it became clear that what was needed was an operational fix in the power business rather than a financial-engineering fix, one board member remembered that the question became: Who is better to navigate that, John Flannery or Larry Culp? "By the way," the board member continued, "Larry was not lobbying for the job," but "when you have Larry Culp sitting there . . ."

Garden had made no secret of wanting to break from GE's long-standing tradition of promoting from within when it came time to replace Jeff Immelt. He had long felt GE needed an outsider at the helm, someone who saw things from a new perspective. That "new perspective" was one of things that Trian had liked about having Ed Breen replace Ellen Kullman at DuPont when Trian had engineered that coup in 2015. But Garden wasn't on the board when the decision was made to fire Jeff and to choose John as his successor. Now Garden was inside the inner sanctum of GE's boardroom. He began to bring his fellow board members around to the idea that John needed to be replaced with Culp.

It wasn't only the perception that Culp had more of the operating experience for addressing what was then ailing GE. In addition, Trian was suffering, along with hundreds of thousands of GE employees, retirees, and shareholders, as the GE stock had dropped to nearly $10 a share, from $25 a share when John took over. Few executives could survive that kind of destruction of value, especially with an activist investor with $2.5 billion of skin in the game camped out on the board of directors. That's the kind of moment when activist investors seek a "catalyst" for change and then demand it.

Trian's mission was to facilitate what it called "positive change" at GE with its board seat. "As April and May and June and July—as those months went by, it became clear that maybe this was beyond John's scope of expertise," explained the board member. "Everyone felt that John had been given a bad hand. He didn't know about long-term care. He had no idea how bad Power was. He didn't know any of those things.... We didn't want to penalize John. But we needed to make the change for the company."

During the summer of 2018, the GE board met regularly to discuss John's fate and whether Culp should replace him. In addition to the plunging stock price and the ongoing operational difficulties in Power, there were claims that John was "indecisive" and that his personnel choices—of Jamie Miller to replace Bornstein as CFO, of Michael Holston to replace Alex Dimitrief as general counsel, and of Raghu

Krishnamoorthy to replace Susan Peters as head of human resources—weren't the best. There was also a complaint that John "panicked people" because he continued to speak about all the problems and then floated the idea, in Eisenhower, that the best way to solve GE's problems was to pull the plug on GE as a revered conglomerate. That freaked people out. "We were getting feedback from people within the organization saying, 'You've got to do something. He's lost the organization,'" explained the board member. An example the board member gave of John's "indecision" was his unwillingness to pull the trigger on the sale of GE's biopharma business, which GE acquired in the Amersham deal and whose sale John and the board had been talking about since early 2018. Danaher, Culp's old stomping ground, had been calling and wanted to buy the business. "We knew we could sell it for a big price," the board member said. Ed Garden wanted to sell it to Danaher. But John disagreed. Having run Healthcare, he thought the prospects for the business were especially bright and didn't want GE, or its shareholders, to lose the upside. Then there was the matter of the impairment of goodwill—the recognition that the business was no longer worth what GE valued it on its financial statements—in the power business because of how poorly it was performing and because it was becoming increasingly clear that GE had badly overpaid for Alstom. The staggering number that needed to be written off—even if it was just accounting, not cash—was looking like it was coming in in excess of $20 billion.

The confluence of events proved too overwhelming. There had to be fallout from stock price collapsing to $10, from $25, a 60 percent decline in fifteen months. Who gets held responsible for a team's poor performance? As a practical matter, you can't fire the team, even if the problems on the team predated the manager's appointment and the manager has laid out a comprehensive strategy for dealing with the team's ongoing problems. That left the manager, regardless of whether he had been given enough time to execute on his turnaround plan, as the one to take the fall. And the ambitious Larry Culp, Trian's man, was sitting right there, anxious to be helpful if asked.

The question became when and how to pull the trigger. On Wednesday, September 26, John held an impromptu telephonic board meeting to confirm that the impairment of goodwill in the power business could exceed $20 billion. John's message to the board: "We have an ongoing disaster in Power." There was not much discussion, other than that the news was terrible. "Yeah, no shit," John replied. Then, without John and without his knowledge even though he was CEO and the chairman of the board, the GE board had telephonic meetings on both Thursday and Friday nights where it finalized the decision to have Larry Culp replace John. The board also decided that, for the first time in GE history, it would hand its new CEO an employment contract, giving him four years to see what he could do. The meetings hadn't been properly noticed, per the GE bylaws. But apparently, that didn't matter to anyone on the board.

A coup was underway.

On Saturday, September 30, as John relayed to colleagues at the time, Culp emailed John asking him if he could meet John at the Boston headquarters on Farnsworth Street.

"Can we talk?" Culp asked.

John wondered if he wanted to speak by phone instead.

Culp told him he was in Boston.

That was a little odd, John thought. Culp said he would meet John in the office. John wondered why Culp was in town. It was weird and ominous. "We've decided to make a change," Culp told him matter-of-factly when the two men were face to face. He said that he was taking over as CEO, effective Monday, and that Tom Horton, the former chairman and CEO of American Airlines, who had also joined the board in April, would be the new lead director. John was not only out as the GE CEO but was also kicked off the board, effective immediately. (How that happened was also a mystery. The board can fire the

CEO, but only the shareholders can remove a sitting board member at the next annual meeting. "In the abstract these"—the failure to properly notice a board meeting and the removal of a board member before his or her term ends—"would both make for fascinating final exam questions for my corporate law class," one attorney close to GE told me.) In any event, on Monday, October 1, in a press release, GE announced John had resigned, even though there's no evidence that he did. As part of the early-morning announcement that Larry had replaced John, GE announced that it was reducing its 2018 earnings guidance and that it would likely be taking an impairment charge of $23 billion related to goodwill in the power business.

GE had its revolution. John was in a state of shock. His family was in shock. There were plenty of tears. People wondered if John had done something wrong that resulted in this unexpected fate. But of course, he hadn't. The last thing he expected was to be fired from GE after barely fifteen months, especially after he had identified various problems, was in the process of fixing them, and had just laid out a board-approved restructuring plan that he was in the process of implementing. But as an at-will employee, there was nothing he could do. If the board wanted to fire him, it could. And he had no contract that would guarantee him severance or any kind of payout. He was left with his supplemental pension—more or less in the range of $1.5 million a year, as compared with Jeff's $5 million a year—and whatever the board deigned to give him for severance. He was fired at noon on Saturday, left the office an hour later, and never came back. He went to his home on Beacon Hill to see Tracy. Together they walked their two black labs and tried to figure out what had just happened and why.

One board member told me that after John got the news, he was "gracious and a gentleman." He had been dealt a bad hand. "I'm rooting for you guys," he told the board member. "I've got a lot of stock. I want it to all work out for you guys. I'm rooting for you, and anything I can do to help, and let's stay in touch."

One former GE executive emailed John, "At least your wife and your dog and I still love you."

John replied, "Well, I'm not too sure about my wife and dog if the stock goes below $10."

Was it a coup? "That's one version of it," said one longtime GE executive. Was there another version of events? "The other version would be [the board] felt like they wanted to make a change," the executive continued. "[and that] if you're asking me right now would I rather have Flannery or Larry as the jockey, I'm going to take Larry."

Still, the fact remains that somehow Culp emerged with a fully negotiated and complex contract by Monday, October 1. Was it negotiated between the time Culp told John he was out on Saturday morning and the time Culp was announced as the new CEO, at 7:07 on Monday morning? It's possible. Or had Culp been negotiating the contract for days, or possibly weeks, once the board had decided that John had to go? That's also possible.

There's little question the decision to replace John with Culp had been in the works for weeks, if not months. In September, Culp was having dinner with a friend at a restaurant in Georgetown and was explaining that he could not attend an upcoming meeting of the board of trustees of his alma mater, Washington College, because he was "about to get a big new job at GE," where his "role would expand."

There's also the matter of Culp's ruthless PR man Declan Kelly. Kelly, then still at Teneo, the powerful public relations and investment firm, appeared to know about the changing of the guard at GE before the Culp-Flannery meeting. One board member told me that Kelly was on the phone during the crucial board meeting where the board decided to replace John with Culp, and that Kelly led the discussion about Culp's contract and compensation after Culp dropped off the call. (Kelly hung up the phone when I called him to discuss his role. And GE declined to answer my questions about how Culp came to oust John and take over as CEO.) "I'm a lifelong student of Thomas Jefferson," Culp told Wall Street analysts in April 2014 when he announced his retirement from Danaher to spend time with his family and go fishing, "and have always been struck by the wisdom of what he's written about the benefits of revolution, every 20 years or so."

There were, no surprise, many postmortems about what happened to John and why. "John is a detail guy," one GE executive explained. "He felt that because he personally owned things as CEO, he needed to personally work through them and understand them and make the decisions." That slowed down decision making. "John wanted the job to slow down so that he could do the work that he needed to do, he felt, to do it the right way," he continued. "In hindsight, it took too long on some things. A lot of the decisions that John made were of the sort that if you look at them in hindsight, it would have been better just to rip off the Band-Aid." He said when the November 2017 investor presentation occurred, "people were expecting pearls of wisdom," and instead many felt like "Ugh, there's nothing new here." Given the lack "of earthshattering news, at least from the perspective of investors," he said, "it would have been better not to build it up by having them put it off for three months."

He also believed that John didn't have the right team around him. He recalled how, when John was preparing for the November 2017 presentation, John didn't have the help he needed. "He had an average, at best, public relations person," he continued. "He had an average, at best, institutional investor person—at best. We had hired [PR specialist] Joele Frank, who was supposed to be a superstar. She was nowhere to be seen. I don't remember ever seeing her on-site. You think GE is in the fight of its life, for its survival. The question I ask myself—I asked it at the time and I ask it even more now—is, Where the hell was Jamie Dimon? Or Lloyd Blankfein? Where were the big guns? John needed to get the superstars in to help. It was all hands on deck and we had a *McHale's Navy*. It was unbelievable." He believed that, whether or not it was actually Jamie Dimon or actually Lloyd Blankfein, John should have surrounded himself with people like that as the shit was hitting the fan. "There's a reason why whenever you're in trouble you bring in the A team," he said. "I mean, we were GE. I just don't understand it. I just don't understand. When we were in trouble during the financial crisis, we had Rodge Cohen. He moved into Fairfield. We had the senior people from Goldman Sachs move into Fairfield to help us get through it. Yet in 2017 . . . I didn't see anybody."

FROM THE START, IN OCTOBER 2018, LARRY CULP LEFT NO DOUBT THAT his regime at GE would be a clean break from his eleven predecessors. Not only was he the first outsider to lead the company, but he was also the first one to get the GE board of directors to give him a contract. And what a contract it was! Culp's four-year contract paid him an annual salary of $2.5 million, plus a target bonus equal to 150 percent of his base salary, or another $3.75 million. He was also given an annual stock grant valued at $15 million. In addition to Culp's roughly $21 million in annual compensation for each of his four years, he received a stunning one-time "inducement award" of between 2.5 million and 7.5 million GE shares, with the shares coming to him in 2.5 million–share chunks after the GE stock increased by a specific amount. GE was in a difficult spot—having defenestrated two CEOs in fifteen months—and Culp used his leverage to extract a potentially very lucrative contract. Exactly how many of these GE shares Culp would ultimately receive depended on the increase in the GE stock price during the tenure of his four-year contract, ending September 30, 2022.

When Culp took over, the GE stock was trading around $12.40 a share, having increased roughly 16 percent on the news that John was out and he was in. If GE's stock increased 50 percent to $18.60 per share, Culp would get 2.5 million GE shares, worth about $47 million. If the stock increased 150 percent, he'd get 7.5 million GE shares, worth about $233 million. There would be no payout for Culp if the stock did not increase at least 50 percent. It was a heavily incentive-based contract, rewarding Culp hugely for increasing the dormant GE stock price. "It's the best performance-oriented contract I've ever seen," CNBC's Jim Cramer said at the time. "And I don't think anyone's seen a better one."

During his first earnings call, on October 30, Culp wasted little time in putting the past behind him. In this he was taking the advice of Declan Kelly, his private PR adviser. "Declan didn't fuck around when he came in," recalled one GE executive. "He was telling Larry, 'You disassociate yourself from Jeff Immelt. You disassociate yourself from John Flannery. You start every press conference or every media thing by saying, "I don't talk to any of those guys. I don't want to talk to any of them. Look at what they did."' He was brutal about legacy GE people. . . . He was kind of a prick for telling him to do that, but it was effective."

On that first call, Culp confirmed that GE would take a shocking pretax impairment charge of $22 billion related to Power and that it would reduce the quarterly dividend to 1 cent a share, from 12 cents a share, saving GE another roughly $4 billion in cash a year and wiping out the last vestige of the idea that GE's dividend was sacrosanct. He also was consolidating the power division into two distinct units and consolidating its headquarters. A few weeks later, Culp brought back John Rice out of retirement to help him run the power businesses. He made other personnel changes, too, including firing Jamie Miller as CFO.

He also announced that part of GE Digital would be sold off to Silver Lake Partners, a large private-equity firm, and that the remainder of the business would be separated into its own wholly owned subsidiary and run as a separate company. This, too, had been one of John's strategic initiatives. But Culp was getting the credit for it. As part of the transaction, GE announced that Bill Ruh, whom Jeff had recruited with great fanfare from Cisco years earlier to be in charge of

GE's battalion of software engineers in Silicon Valley, would be leaving the company. GE would not become a top-ten software company after all.

Culp was just getting revved up. In February 2019, he agreed to sell the biopharma division of Healthcare—a move that Trian wanted to happen and John had resisted—to Culp's former company, Danaher, for $21.4 billion, $21 billion of which was in cash. Culp used the cash to pay down some of GE's debt of around $115 billion. He also—at least superficially—hewed to the strategic plan that John had laid out in Eisenhower and again in the June 2018 board meeting. In December 2018, GE filed confidential documents with the SEC for an IPO of GE Healthcare, with the intention of the public offering coming in June 2019. If Jeff Immelt had merely "burned furniture" to try to keep GE afloat, Culp went into bonfire mode. He sold off GE's stake in Baker Hughes. As of May 2021, GE still owned 26 percent of the company, down from 62.5 percent, having raised several billion dollars from the stock sales. Culp also sold most of what GE owned of Wabtec, the transportation company with which GE merged its train manufacturing business under John's watch. In March 2021, Culp sold GECAS— GE Capital's ballyhooed aircraft-leasing business—for $30 billion in cash and stock to AerCap, a competitor, generating $24 billion in cash for GE and a 46 percent stake in the combined company. GE said it would use the proceeds of the sale to pay down debt and that, once the deal was completed, it would no longer report GE Capital's financial performance separately. In effect, GE Capital was no more.

---

IN FACT, OTHER THAN SELLING BIOPHARMA TO DANAHER—THE MARket value of which exceeds GE's by more than $100 billion these days— as of this writing, Culp had done nothing particularly special or innovative at GE. His other strategic moves were pretty much taken from the Eisenhower playbook. In July 2021, he introduced one innovation: he initiated an eight-for-one reverse stock split, so that the GE stock would no longer trade in the teens. But it was just a cosmetic maneuver. Despite these Band-Aid measures, as of June 2022, GE's stock is down 32 percent since Culp took over in October 2018.

Culp managed to take care of himself, at least financially. When Equilar, the executive compensation expert, published its annual list, in April 2021, of the CEOs of public companies with the highest compensation for 2020, there was a surprising name at the top of the list: Larry Culp. As it was for many companies, 2020 was a rough year for GE. Its dominant jet engine business fell off a cliff with the onset of the coronavirus pandemic, as travel screeched to a halt. Demand to buy big power plants decreased, too. According to GE's filings with the SEC, GE's aviation revenues fell by one third in 2020 to $22 billion, from nearly $33 billion the year before. Revenue in the power business fell 5.4 percent to $17.6 billion, from $18.6 billion. GE's operating income for 2020 fell 78 percent, to $2.1 billion, from $9.5 billion. It was also a rough year for GE's stock price. As the stock prices of other big industrial and technology companies soared, GE's fell 10 percent, to $10.80 at the end of 2020, from $12 at the beginning of the year. The stock was as low at $5.50 a share in mid-May.

To account for the challenges posed to the company by the pandemic, Culp and the GE board decided to start making changes to his 2018 compensation package. First, in April 2020, Culp voluntarily agreed to forgo the balance of his 2020 salary, leaving him with the $653,000 he already had been paid. He also voluntarily decided to give up his 2020 *cash* bonus. But he did receive his $15 million annual *stock* bonus. That gave Culp some bragging rights that he was taking a compensation hit along with others. Not for long.

On August 18, 2020, the GE board amended Culp's 2018 contract. First, the board extended his contract through at least August 2024, or up to and through August 2025, assuming everyone agrees. As part of the contract extension, Culp voluntarily relinquished the "inducement award" granted at the time of his hiring. And then the board replaced it with a new "Leadership Performance Share Award," which, helpfully, according to GE's 2020 proxy statement, was "intended to provide Mr. Culp with the incentive to continue to provide services to GE during this extended employment term" and "as it became clear . . . that the GE transformation would take longer than previously contemplated." It was worth, potentially, some $233 million to Culp if the GE stock price increased by 150 percent.

As part of his contract extension, the board recut Culp's deal upon which his original stock award was based—a decision made, according to GE, by the independent directors of the board at the recommendation of the independent compensation committee and one with which Culp "was not involved." But several former GE board members told me it was absurd to think that Culp had nothing to do with recutting his contract and that of course he was intimately involved with the process, as would not be the slightest bit unusual for such a material change to an executive's contract. Of course, Culp was involved in recutting his compensation package in August 2020. (Culp repeatedly declined my requests to be interviewed for this book.)

What had been a baseline of about $12.40 per share—meaning that his inducement award would not kick in until GE's stock price was 50 percent higher than $12.40, or $18.60—was drastically lowered to $6.67 per share, making it much more likely that Culp would get the big stock payout. And in fact, by mid-2021, his recut contract meant he would receive about two thirds of the shares, as long as he remained at GE. GE pegged the value of the revised incentive stock award for Culp at around $57.1 million in its proxy statement.

The timing for Culp was particularly fortuitous. Along with the market as a whole, GE's stock increased nearly 100 percent after the board agreed to recut his deal, although the stock has moved little, if at all, beyond where it was when Culp took over. It had merely rebounded off its pandemic lows, giving Culp a windfall. Culp was getting even richer than he already was, while GE's shareholders got basically nothing. The first award of about 4.65 million shares went to Culp in December 2020 after the GE stock averaged about $10 for thirty days. The second batch of about 4.65 million shares was his to keep, too (assuming he stayed at the company until at least 2024). He still had the potential for one last award of roughly 4.65 million shares if the GE stock hit about $16.68 and stayed there for thirty days (or $133.44 per share, following the eight-to-one reverse stock split, which also reduced his stock award to 1.1625 million shares). Incredibly, had the GE board not given Culp the August 2020 repricing deal, his original 2018 stock grant would still be out of the money. Instead, at the moment, anyway, Culp is about $110 million or so richer.

One longtime GE insider told me Culp's compensation package was outrageous. He blamed the board for caving into Culp's demands. "[The] board should be shot," he said, adding that the GE board was "scared of Culp" because "he has made it clear if he doesn't get what he wants all the time, he will quit." (In addition to Culp, in September 2020, GE's new CFO, Carolina Dybeck Happe, also received a "leadership" grant, although it was not nearly as large as Culp's. Happe has the potential to receive a maximum of about 1.6 million GE shares—now 200,000 shares after the reverse stock split—worth around $20 million at the time of this writing.) The GE insider was not alone in criticizing the GE board's compensation practices. In May 2021, GE's long-suffering shareholders voted overwhelmingly—some 57.7 percent—for a nonbinding resolution that took issue with the way the board compensated Culp as a result of the pandemic.

But the GE board continued to stand by its man, and his compensation package. In a statement about the shareholder referendum, a spokesperson wrote to me on behalf of the GE board, "We value the views of our shareholders, and we will gather additional feedback from them as we take into consideration the outcome of the vote and evaluate our executive compensation program moving forward. Over the last year, the Board has taken steps to align GE's compensation practices with the interests of shareholders, including the importance of maintaining a strong leadership team to guide our multi-year transformation. The GE Board is confident that the continued leadership of Larry Culp and his team are unquestionably in the long-term benefit of all shareholders." (In March 2022, the GE board took away $10 million of Culp's annual $15 million stock grant, leaving him with a total compensation package for 2022 of $11 million. The board claimed it had listened to its shareholders' objections about Culp's compensation.)

Trian was also sticking with him. On CNBC, in September 2019, Nelson Peltz said Trian made a mistake when it sold only a third of its stake in GE when the stock increased to $32 a share, from the $23 a share that the hedge fund had paid. "We should have sold three thirds of our position," he said. "We make a mistake about once every twenty-five years. So you guys can relax; for another twenty-two years everything's going to be cool." He also said he remained a big fan of Larry Culp. "I think Larry Culp is a star," he continued. "I think he knows how to run a business. I think he knows how to deal with these issues."

At GE's annual meeting in May 2021, lead GE director Tom Horton defended the board's decision to recut Culp's stock award in August 2020. "The board believed it was in GE's best interest and our responsibility as the board to secure Larry so he can continue to drive GE's transformation," he said. "If the maximum number of shares are earned in 2024, it will mean all shareholders will have benefited." And if the stock hits the roughly $16.68 threshold ($133.44 under the reverse-split regime)—still below his original 2018 payout levels—his August 2020 gambit will be worth about $233 million to him. Nice work if you can get it.

With GE's stock around $70 a share these days—meaning it's down some 30 percent since Culp's coup—it's clear that Larry Culp is the only winner, so far, in all of the turmoil that has been created at GE since he took over on October 1, 2018. He's been paid roughly $21 million a year since then—aside from his modest give-ups in 2020 and 2022—plus he'll get his 1.1625 million additional GE shares as long as he is still around Boston at the end of 2024, or 2025, as is currently the intention. As for GE shareholders, they continue to suffer.

## Chapter Twenty-Eight

# WHO LOST GE?

Reginald Jones, Jack's predecessor, died on December 30, 2003, at his home in Greenwich. He was eighty-six years old. He had kept to himself in his final years, rarely speaking with either of his two successors. (He's not even mentioned in *Hot Seat*.) He left behind no memoir and kept his name out of the papers, particularly when it came to GE. But he did share his dying words with his wife, Grace, which she conveyed to the Reverend Robert Naylor, who shared them with the three hundred or so people who attended his memorial service at the Second Congregational Church in Greenwich in early January 2004. "Leadership requires ethics, morals, and values," he told Grace, before smiling and taking his final flight.

Nearly eighteen years after Jones's death, on November 9, 2021, Larry Culp put GE out of its misery. He pulled the plug on the company after 129 years. It had once been the most admired and most valuable company in the world. But by November 2021, those days seemed long gone. In an early-morning press release, Culp announced that GE would be broken up into three separately traded public companies: a health care company, to be spun off in early 2023, with GE retaining a

19.9 percent stake; an energy company that would comprise GE's power, renewable energy, and digital monitoring businesses, to be spun off in 2024 (and renamed, absurdly, GE Vernova); and a jet engine business that would keep the GE name, and that Culp said he would continue to run. (In June 2022, Culp named himself CEO of GE Aviation while remaining CEO of GE.)

The company also announced that it would be selling its minority equity stakes in Baker Hughes, the oil and gas business it had spun off years earlier under the leadership of Lorenzo Simenelli, and in Aer-Cap, the Dublin-based aircraft-leasing business that had a few months earlier completed the purchase of GECAS, GE Capital's hugely successful aircraft-leasing business and one of the last remaining pieces of GE's once-formidable finance company. The proceeds of the sale of these equity stakes would be used to pay down GE's debt. Without giving his predecessor, John Flannery, the slightest nod, he was essentially announcing his slightly tweaked version of Eisenhower, the breakup plan that John had conceived four years earlier sitting in the first-class section of a cross-country commercial flight, perhaps itself a fitting metaphor for the company's stunning decline. Whether Culp's—and the GE board's—decision to flip the switch on GE was an act of Jones's definition of leadership or an act of cowardice and expediency remains to be seen. In his commentary, Culp made no reference to anything resembling ethics, morals, and values. Instead, it was a total capitulation.

In the meantime, we are left to wonder: What the hell happened? Had Jack set Jeff up to fail by leaving him a company that was too hot to handle? Had he hyped up GE so far and so fast that a mere mortal could not possibly follow him into the corner office in Fairfield and expect to be successful? Should Jack have checked his ego sufficiently with Mario Monti, at the EU, for GE to complete the Honeywell acquisition, even if it was only a fifteen-hole golf course? Wouldn't even a slimmed-down Honeywell have provided to GE the industrial ballast the company desperately needed at the time, to counter the perception that GE had become nothing more than a bank with some attached manufacturing businesses?

Did Jack set Jeff up for failure? Did Jack choose the wrong guy to succeed him? Or did Jeff play poorly the royal flush that Jack believed he had bequeathed his successor? Or did the world change materially for GE after September 11, four days after Jeff took over from Jeff? Or did the change come for Jeff and GE in the wake of the Sarbanes-Oxley disclosure law, making it increasingly difficult for GE to play the kinds of accounting shenanigans in which Jack seemed to revel, quarter after quarter? Why did Jeff fail to heed the powerful warnings he received from the likes of Bill Gross, Jim Grant, and Ravi Suria about the dangers lurking at GE Capital? Why didn't Jeff—a product of a superior Ivy League education—take the time to understand GE Capital fully, including the risks in the way it funded itself and in its gargantuan asset portfolio? After all, GE Capital was generating more than 50 percent of GE's earnings, meriting more than just a superficial understanding of what was going on there.

Why was Jeff such a lousy dealmaker, overpaying for Alstom while selling NBCU on the cheap? Why did he send Genworth out the door without the two insurance subsidiaries that would later nearly blow up GE? Why did Jeff freak out after GE Capital became a SIFI and then decide the business should no longer be part of GE, before selling much of it for discounted prices—at a time, after the 2008 financial crisis, when big banks have never made more money? Why did he think it was a good idea to bring Trian—a highly aggressive activist investor—under the GE tent? Was its support of Project Hubble worth the Sturm und Drang that the hedge fund later caused Jeff, John, and the staid GE board? And why, for the love of God, did Jeff stick to his desperate $2-a-share mantra for 2018 when he had to know there was no chance of GE achieving such results? Why, why, why? One reductive answer can be found in what Dave Calhoun, the CEO of Boeing, has been telling people about what happened at GE: "The GE board let Jack make all the decisions and he generally made the right ones; it let Jeff make all the decisions, too, and he generally made all the wrong ones."

STEVE BOLZE MAY HAVE REPEATEDLY PISSED OFF JEFF AND MAY HAVE miscalculated miserably his chances of becoming CEO. He was a GE lifer who had come within a heartbeat or two of achieving his ambition. But his political ham-handedness does not diminish the wisdom of his insights into what went wrong at GE after he had some time to reflect on it. In an interview at his Park Avenue office at the Blackstone Group, where he was a senior managing director from 2017 to 2021 at the firm's infrastructure fund before becoming a senior adviser to the fund, Bolze told me that he could sum up some of what he learned at GE as the "three *C*s"—capital allocation, corporate governance, and culture. He had thoughts to share on each. He said that starting with Jack, and increasingly under Jeff, too much capital was allocated to GE Capital and to M&A deals that didn't pan out. GE Capital was a good business that got too big. He said Jeff "leaned into" real estate at GE Capital, and "that bit him in 2007." He also said Jeff overpaid for acquisitions. "We would get into auctions, or not, and we would pay big bucks," he said. "Any banker would tell you that." He said Jeff did some good deals, such as the acquisition of Enron's wind business and Jenbacher's distributed power business. But, he said, "the majority of GE's acquisitions were overpaid for." He said too much money was spent building out GGO and GE's software business in Silicon Valley. "GE put some money in stock buybacks at probably the wrong levels," he continued. "And the whole time along, you've got Blair Effron, one of the top guys from Centerview, advising the board on capital allocation, at least for the last four or five years. But hey, we didn't get some things right on capital allocation."

As for corporate governance, he continued, "As an up-and-coming leader in GE, you are not trained on corporate governance. When you go to Crotonville, you are trained to be a leader . . . but corporate governance—and Warren Buffett says this, you've seen it a hundred times—putting somebody in the CEO job when they haven't trained [in] corporate governance tends not to go well. You've got to get CEO succession right. That's the number one function in corporate governance—CEO succession. And top leadership succession. And I'm not saying that Jack got it all right. Maybe he picked the wrong person. But there was even a more systematic process eighteen years ago than what I went through, and I have talked to some of my friends who have gone through the succession-planning process at fairly large companies, like Fortune 50. And that's not what I experienced during my succession process at GE." He also believed the GE board had become too big in the wake of GE Capital becoming a SIFI, and that GE had way too much debt.

He said GE's culture had changed during his time at the company. As an electrical engineer, he said, GE was the place to be. "Everyone wanted to work for Jack Welch," he said. "You came to GE because of the culture." He remembered the "four *E*s" that Jack always talked about: edge, execution, energy, and energize. And the "three *S*s": simplicity, speed, and self-confidence. "Jack was notorious for 'You gotta make your numbers and deliver,'" he said. He told me he presented to Jack three times in his career. "Everybody dives under the table when it's your time to present, and basically, it was like a do-or-die experience," he continued. "But that was the culture, and Jack listened to you and challenged you." He said Jack advised him not to take one job he was offered, and he listened to Jack. "He was also very much a people person," he continued. "He was tough. I would say he didn't bat a thousand on his people choices, but he probably batted at least seventy-five to eighty percent. He maybe stayed an extra year or two too long with Honeywell, but overall, not bad."

He wanted to talk about what happened to GE's culture. "Culture takes years to change," he said. "It doesn't happen overnight." What had been the four *E*s and the three *S*s turned into Jeff's relentless push for growth. But he worried there wasn't enough focus on profitable growth. "Growth without profitability is kind of irrelevant after ten years," he said. As have others, he cited the problem he perceived as Jeff's unwillingness to accept constructive criticism from other executives. He talked to his friends, including Dave Calhoun, who worked with him at Blackstone before becoming CEO of Boeing; John Krenicki at Clayton Dubilier; and Gary Reiner at General Atlantic. "Jack was smart," he said. "Jack was people oriented, Jack was tough, but you could talk back to him. You had to have your shit together. Why did Krenicki leave? Why did Calhoun leave? Why did Gary leave? Because they butted heads with Jeff. I'm not sure Jeff likes constructive conflict. He doesn't like it. And that's tough."

He said he worked directly with Jeff for five years and he "butted heads" with him on a variety of issues from the Alstom purchase price and the Alstom integration team to Jeff's decision to take Vic Abate, one of Bolze's key lieutenants at Power, who was leading the Power equipment business at the start of the Alstom integration effort, and make him GE's chief technology officer. He also clashed repeatedly with Jeff on the succession process, obviously. "These were maybe three or four months apart," he continued. "These weren't everyday occurrences." He said he thought GE's vice-chairmen Calhoun and Krenicki disagreed with Jeff about the company's earnings forecasts—that Jeff was always too optimistic. He wondered why John Rice, when he was a vice-chairman, didn't push back harder on Jeff's beloved $2-a-share mantra, given his vast GE experience globally. "You're not directly responsible for that," he said, "but you're a vice-chairman. You sit in every board meeting. You sit in on most committee meetings." He said he thought Jeff was "a master manipulator." He said he "definitely" got "caught up" in the opportunity to be the next GE CEO. "My dad even said, 'Well, hey, you'll be talking about this when you're eighty.'"

THEN THERE WAS THE NOVEL VIEW SHARED BY ONE LONGTIME WALL Street research analyst, who told me that Jeff's tenure as CEO was another example of the hubris that comes with an MBA in general, and the Harvard MBA in particular. "Jeff Immelt is exhibit A for the failure of a Harvard MBA to perform well as a CEO," he said. "You could even argue . . . that many of the 'skills' taught at Harvard Business School (constructing persuasive PowerPoint decks, presenting well, flipping an objection into a sales point) and honed at the Jack Welch–era internal training programs at GE" were exactly what made Jeff a finalist to succeed Jack. But those skills also doomed him to the failure of "success theater" instead of producing real performance. He said Jeff's decisions to get out of NBCU and to increase GE's energy exposure were the kind of choices that the Harvard MBA should prepare executives to make successfully. "But somehow Immelt didn't see that energy is fundamentally a commodity business in which technology does nothing to improve the quality of the product but just makes it possible for all producers to extract more of the commodity, driving down its global price," he continued. "It may be that the potential challenges of running an actual business are far too different and complex, given industry dynamics, technological change, and economic environments, to be successfully taught in a classroom to twenty-five-year-olds. Corporate finance, investing, and accounting all have a lot of basic knowledge that can be taught well in a classroom. But the case method doesn't seem to do a great job of teaching strategy. The truth may be that the principles of successful business decision-making are so simple they could be taught in an afternoon. My nominations are facts, focus, and integrity. Not clear what the professors would do for the rest of the year, or what Harvard could charge for that afternoon."

*⸺*

TO TRY TO GET A FURTHER ASSESSMENT OF THE GE SAGA, I TURNED TO Jim Grant, the genius behind *Grant's Interest Rate Observer* and one of the keenest observers of GE's astounding rise and unimaginable fall. "I think where GE went wrong was a loss of moral compass," he said. "The loss, the conceit, of a company that believes it can do no wrong, because after all, it is GE." He pointed to the fact that for many years, GE did not have backup lines of credit for its billions of dollars in commercial-paper borrowings. "You wouldn't fund that, because after all, you could always roll it, because after all, you're GE," he said. He marveled at GE's growing "dependence" on a financial services business "that was susceptible to all the cyclicality of credit businesses since the dawn of lending and borrowing," noting that somehow "the cyclicality of that kind of business was not front of mind in the GE C-suite." He said GE executives "in their self-assurance" believed "they could manage through any such cycle, that they were above it" and that somehow the inviolable laws of borrowing short and lending long did not apply to them. In other words, hubris. He recalled for me a conversation he once had with George David, who was the long-time CEO of United Technologies, a GE rival. GE "is essentially a financial institution," Grant said he told David, "and they're going to have to burn their furniture at some time to keep this great fire in the locomotive going, because that's the nature of financial institutions." David responded, "There's a whole lot of furniture to burn." Upon reflection, Grant told me, "There was a lot of furniture burned until the kindling caught fire, and then it all went up. It all went up in smoke, that is."

JEFF WAS RIGHT TO BE WORRIED THAT HE WOULD BE UNABLE TO CONtrol the narrative about his tenure as GE's CEO. But he wasted little time in trying to define his version of the truth, along with his version of the facts and his version of the context. (You will recall that one of his favorite aphorisms was "Truth equals facts plus context.") The paint was barely dry on his tenure as CEO when—somewhat ironically, I always thought—he took to the pages of none other than the *Harvard Business Review* to declare his time leading GE a resounding success: "How I Remade GE." His job as GE's leader, starting in 2001, was to remake a "historic and iconic company during an extremely volatile time," he wrote. And he was unequivocal about having succeeded in the task. "We didn't just persevere," he wrote. "We transformed the company. GE is well positioned to win in the future. . . . The outcomes of my decisions will play out over decades, but we never feared taking big steps to create long-term value. For the past 16 years GE has been undergoing the most consequential makeover in its history. We were a classic conglomerate. Now people are calling us a 125-year-old start-up—we're a digital industrial company that's defining the future of the internet of things." It did not take decades—in fact, it took barely a year—for this assessment of GE to seem delusional at best.

Jeff also defended his controversial decisions—at least inside the company—to create GGO and to lead GE's digital transformation, at considerable expense. "You must be profoundly convinced that the company must transform itself—that it's a matter of life or death—because when you start the play, you will immediately get pushback," he continued. Yes, well, maybe that happened because there were immensely talented people inside the company who disagreed with you and were worth heeding. He claimed he allowed people to express their opinion if they disagreed with him—one of the chief criticisms against him—but brooked no dissent when it was time to move forward. "I didn't give people an out," he wrote. He explained how in Crotonville he would tell the assembled leaders, "Guys, if we don't become the best technology company in the world, we're doomed, we're dead." When he talked about GE becoming a "digital industrial" company, he'd tell them, "There's no Plan B. There's no other way to get there. Who's coming with me? What's in your way? What do we need to be doing differently?" He wanted GE's people to be totally committed to the path he laid out for them. "Half measures are death for big companies, because people can smell lack of commitment," he continued. "When you undertake a transformation, you should be prepared to go all the way to the end. You've got to be all in. You've got to be willing to plop down money and people. You won't get there if you're a wuss."

Jeff praised a group of his former colleagues, including Beth Comstock, Jeff Bornstein, Keith Sherin, John Rice, and Bill Ruh. He made no mention of his successor, John Flannery. He made no mention of problems in the power business or to the growing liability in the long-term health care insurance business. He made only the smallest concession to his "complicated" legacy at GE.

While arguing that earnings in GE's industrial businesses had tripled during his tenure, that GE's market share had never been higher, that GE had paid out more dividends during his 16-year tenure than during the previous 110 years combined, and that the backlog, at $324 billion, was up $150 billion in the past decade, he acknowledged that the stock market had been particularly unimpressed. "Our P/E ratio has gone from 40:1 to 17:1 in the past decade," he wrote, "and the stock price has underperformed. Thus it is with transformation." He may not have mentioned John Flannery, but he wasn't shy about proclaiming what he had bequeathed him. "The company in 2017 is ready for any future," he concluded. "I'm confident that I'm handing over a company that will flourish in the 21st century. Some people at GE feel that the stock market doesn't fully appreciate what we've accomplished. But I look at it this way: Our task now is just to perform, to execute, and let the market make its own judgment." Maybe it was the lack of perspective—he had just left GE, after all—but Jeff's level of denial, delusion, and lack of self-awareness, all compressed into a drivel of corporate pabulum is, frankly, rather stunning.

In a subsequent conversation with me—years later—Jeff was far more lucid about what had happened at GE during his tenure. But he was still reluctant to blame himself. He said what was clear in retrospect—"you never see it when you're in a bubble," he said—was that GE's power business was in the middle of years of huge, artificial demand brought on by the deregulation of the power industry. "We pulled forward fifteen years of U.S. demand in gas turbines," he said. "The power business went from earning $1 billion a year to $5 billion a year in two years. . . . We would have a customer buy twenty heavy-duty gas turbines in a day, and the stock would go up. But we convinced ourselves that it wasn't a bubble, that it was going to be sustained, that people were using more electricity."

He surprised me by conceding that Jack made a mistake by walking away from the Honeywell deal. He thought GE should have closed that deal. I've often thought the same thing. Despite some of the warts that the drawn-out approval process exposed in the business, buying Honeywell would have achieved for GE many of things Jeff tried to do with the Alstom acquisition—only to fail—including bulking up GE's industrial businesses and diluting the importance of GE Capital. "If we had done better research on that, I think we would have finished the Honeywell deal, because power was going to go from $5 billion back to $1 billion in a couple of years," he said. If the "analysis" had been better, he continued, GE would have taken the cash being generated by the power business in the bubble years and used it to buy Honeywell, rather than assuming the power gravy train would continue.

Jeff's view was shared by a senior finance executive at GE whose tenure spanned both Jack's and Jeff's, especially once Honeywell's market value exceeded GE's by some 30 percent. "In hindsight," he told me, "I wish we bought Honeywell every day of the year. It would have changed so many things for me. It would have given us a big industrial base instead of growing Capital to replace the earnings of Power. I think it was a huge mistake. They had a couple of shitty businesses, but overall, the core of what it would have added to our aviation business would have been huge. And having a huge industrial add-on at that time, with all of a sudden Insurance and Power being in tough shape, I think it would have been a huge help for us and maybe avoided other things. But you can't go back." Norm Liu, who had been running GECAS, said Jack should have agreed to Monti's demand that GE spin off GECAS. "But c'est la vie," he said.

When Jeff relinquished the CEO seat on August 1, GE's stock hovered around $25 a share, down from around $40 a share when he took over from Jack. GE's market value fell by half during Jeff's tenure, including by 25 percent in 2017 alone. Still, his "realized compensation" for 2017, according to GE's 2018 proxy statement, was $30 million. He did not receive severance when he left GE, but he was entitled to his insurance benefits, his deferred compensation, and the monthly payout of his annual $5 million SUP, the coveted supplemental pension. The present value of these benefits totaled some $113 million at the time of his departure. He also left with more than $75 million of GE stock, much of which he had bought with his own money.

When Jeff called Jack in June 2017 to tell him that he had been fired and that John was the new GE CEO, Jack was polite and politic, at least according to Jeff's recounting of the conversation, and was gracious enough to suggest that Jeff never really "caught a break" after he took over as CEO. Jack could have said much worse to Jeff. But he held his tongue.

He didn't hold back with me. From the moment we first met at the Nantucket Golf Club in August 2018, he unleashed the bile and animosity he felt toward Jeff, claiming repeatedly that he had "fucked up" by choosing Jeff as his successor. Needless to say, in corporate America, it was highly unusual for one CEO to criticize so publicly his successor, especially at a company as traditional, straitlaced, and revered as GE. Rereading my six interviews with Jack, it was clear he could barely contain himself. In anticipation of our first interview, he prepared a sheaf of materials—financial statements, charts filled with numbers, chunks of Jeff's annual shareholder letters—put them in a folder, and handed it to me. His point throughout was that he had left Jeff a powerful legacy at GE, of assets, of financial performance, of market value, of people, of Wall Street and Main Street respect, and that Jeff had squandered it. It wasn't as true as Jack believed it to be. But the point was that *he* believed it to be true, and he desperately wanted me to believe it, too.

At first, he praised Jeff for keeping the company on track after September 11. "We came back fast," he said. But then Jack was critical of Jeff's early acquisitions. He pointed out that Jeff had spent $80 billion on deals. "He was the private-equity company running around buying, and then selling them two years later," he said. "Chemical business, Water, he sold them all." He mentioned Jimmy Lee at JPMorgan Chase. "Jimmy Lee took him for a ride you wouldn't believe," Jack said. "I used to call Jimmy and say, 'You fucker, what are you doing?' [Jimmy replied], 'He wants to buy them, what am I supposed to do?'" He quoted from Jeff's 2007 annual letter, which appeared in February 2008. "Immelt thought financial services would have a good year," Jack told me, before quoting from Jeff's shareholder letter: "Our financial services should do well in a year like 2008"—"That in itself will put you in jail, okay?" Jack interjected—"Pricing will improve as banks retrench. There could be $300 billion of assets available with high returns. We plan to seize the opportunities in the current turmoil and position our financial service businesses for years of profitable growth." Observed

Jack with a snicker, "Now, this is the guy who the analysts like to say was getting out of GE Capital. Read the fucking thing."

Nearly alone among all the hundreds of people I spoke with about GE, Jack thought Jeff made a mistake by getting rid of GE Capital. "It was paying dividends and all that to GE," Jack said. "But he had a rationale because it was a SIFI." He criticized Jeff's decision to use the proceeds of the sale of GE Capital to buy back GE stock. "He was buying back stock and not taking care of the debt," he said. "He got all those dividends from GE Capital. He was exiting it and he bought back the stock." Jack said Jeff didn't understand balance sheets or operations. "That's a bad combination," he continued. "He understood marketing. He was advertising like crazy." He said the most important responsibility of a CEO was delivering consistency of earnings. "Not managing earnings," he said, "but consistency of businesses to deliver. You hold back on one, you push another. Always levers, but you always have to focus on consistency. That is a fact. Because the other things didn't count. That is the heart of this bullshit about managing earnings, staying up all night fooling with the balance sheet. We didn't do that. He didn't understand it, so he took risks that were out of proportion. He didn't follow up on earnings. He didn't pay attention. He had meetings on operations and was disinterested. You've heard that, I'm sure. He'd walk out. He was dreaming and not eating. Not eating. You have to generate food to live another day to dream."

Jack blamed the GE board for *allowing* Jeff to allocate capital toward buying back stock instead of paying down debt. "It's an indictment of corporate governance," he said. "Corporate governance is a big part of this story. Where was it? These are facts, so we don't have to go with anybody's bullshit." He said that if Jeff hadn't bought back so much GE stock, "we wouldn't be having this conversation about GE now." He asked me to read Jeff's 2016 annual letter, which came out five months before he was fired. "I feel great about where we are and where we're headed," Jeff wrote. "I can't recall ever feeling such excitement about our opportunities, or such confidence in our ability to meet them." We talked about why Jeff might have written this. "That's what your book is going to be about, part of it," he said. "When you get through this, you'll know. I don't know and I can't figure it out. I'm not calling him delusional. These are all his words. This paper, this is not Jack Welch saying things about him, it's him saying them about himself." He paused briefly. "How about some tea?" he said.

He wanted to give me some advice about writing this book. "Comparing me to Jeff is not the winning angle," he said. "The winning angle is 'How did Jeff do with what he got? How did I do with what I got?' Jeff had eight years of free run with my foundation until he started fucking around. We were going beautifully. He was CEO of the year with those assets that I gave him. Or we gave him—my team gave him. He was CEO of the year. [But] a lot of people saw through him pretty quickly because 'he knew it all.' Guys would say to me initially, 'What a great pick.' Then they would start to say, 'He's full of himself.' Well, right now he doesn't have any idea what he did. He said he doesn't have any idea. He's delusional. He doesn't know what happened. He doesn't even consider it: happy-go-lucky."

He said Jeff fostered a culture of "superficial geniality," that a "game in the GE system was you went to a meeting, smiled, and then screwed the guy when you left the room. But that's different than success together.... This was a meanness. Superficial congeniality is when you say, 'Bill, nice job today,' and then [go] out in the hall and [say], 'Fucking Bill, he's a hammerhead.' I wanted candor. The best way to look at this is look at all the things John Flannery said about how he wanted a culture that was back to candor, back to straight talk, back to rigor. All the things we had. His indictment of Immelt was never direct but always about what he wanted in the company. What used to be. That is the best indictment of Immelt you will get. Flannery had it nailed."

Jack told me he didn't think John Flannery had a chance to succeed at GE. He agreed with the assessment that John gave people the impression that he was too contemplative. "The problem with John was he was 'What do you think I should do?'" Jack said. "You know? He left that question in everyone's mind. Is he just a thinker? Is he a doer? Is he going to do something? Too long. I think that lasted too long. It gave the naysayers a chance to pounce. Nothing would have been different, but I think Nelson [Peltz] pounced on that. He wasn't decisive enough in front of the crowd that was evaluating him."

After all, Jack had twenty years atop GE. Jeff had sixteen years. John had fifteen months. "He got fucked," Jack said. "John is about the most decent guy you'll ever want to meet. John can look back and say, 'Shit. How did this happen to me?' He has every right to say that. And there isn't a GE guy you talk to who thinks anything different. John got screwed, in a nutshell."

# ST. PATRICK'S CATHEDRAL

On March 1, 2020, just days before the COVID-19 pandemic took hold throughout the world, Jack died at his apartment in Manhattan. He was eighty-four years old. The cause of death was renal failure. He had been in declining health for years, which was apparent to me during our last visit together in August 2019. A week before his death, he emailed Cary Akins, his heart surgeon, "I'll never forget what you did for me."

Jack's funeral, held on March 5 at Saint Patrick's Cathedral, on Fifth Avenue, was one of the last major public events where people gathered oblivious to the gathering pandemic storm. In retrospect, it was also the end of an era, both for GE, and what it once meant to the country, and for the country itself. As with a presidential send-off, the current CEO, Larry Culp, as well as the two former CEOs, John Flannery and Jeffrey Immelt, were in attendance. There were also a slew of important former GE executives in the cathedral. Among Jack's pallbearers were the commentator and *Morning Joe* stalwart Mike Barnicle, the New England Patriots' Bill Belichick, Allied Signal's retired CEO Larry Bossidy, IAC chairman Barry Diller, billionaire Ken Langone,

Discovery CEO David Zaslav, and former NBC News chairman Andy Lack. It had the feel of an event not to be missed.

Jeff arrived in New York City for Jack's funeral from California at around three in the morning. After leaving GE, Jeff had been teaching a class at Stanford Business School about leadership. The venture capital firm he's associated with, New Enterprise Associates, also had an office in Menlo Park. On occasion, Jeff has defended Jack. For instance, after the publication in May 2022 of a book that (ridiculously) blamed Jack for destroying capitalism, Jeff took to LinkedIn. "Jack was not perfect, and some criticisms are fair," Jeff wrote. "He and I didn't always agree. But I have seen hundreds of CEOs up close in my career. Jack was pretty damn good." As he was coming back to Manhattan, he emailed his longtime friend and fellow GE alumnus Peter Foss and told him he was arriving early in the morning. He said he was going to stay at the same hotel as Foss and hoped that Foss would walk with him to Saint Patrick's to go to the funeral. Foss had grown up in Pittsfield and knew both Jack and Jeff from when they spent time in Pittsfield, working for Plastics. Jack recruited Foss to work at GE.

"[Jeff] knew that he wasn't walking into the most friendly environment personally," Foss told me, "because there were people there who were mad at him," what with GE's stock price still in the doldrums and a general sense that he was the one responsible for the tough shape the company was in. "He knew he should go, so he went," Foss continued. Jeff appreciated that Foss escorted him to the funeral, acting almost as a shield. "He's so respectful of Jack and what he learned from him. And as you well know, things weren't great between them the last few years. But he never lost his love for the guy. . . . He used to say to me, 'Geez, I hope I can develop a good relationship with him again someday.' But see, he felt badly about that." Foss recalled a visit he once made to Nantucket to see Jack. They were sitting together in Jack's hot tub and Jack was talking. "You may not agree with what Jeff's doing right now," Jack told him. "But you need to know something. You're his friend. And he doesn't have a lot of close friends. I didn't realize he was a bit of a loner. So when you're with him, you need to be his friend. He needs to relax. And you just don't beat him up on business stuff or anything. Just be his friend."

Barnicle and Langone gave the eulogies. At one point, Langone recounted a story of how unfiltered Jack could be. To wit, Jack told his longtime partner in crime, Larry Bossidy, that he was fat. Bossidy's response: "You're short. I can go on a diet. What about you?" Continued Langone, "He could motivate people to do things they didn't think they could do. Because he came from a hardscrabble background, he was passionate about merit. He believed in leaders. He believed people had a right to know exactly where they stood if they were on his team, where they were going. And if they did their part, what was in it for them." At one point, of the thirty CEOs of the companies in the Dow Jones Industrial Average, five had worked for Jack. Langone said Jack was a "wreck when he thought about those people's dreams because of what happened in the last few years, what happened with the stock."

Barnicle, Jack's longtime friend, spoke about what made Jack so special. "All of us here today knew Jack Welch in several different ways," he said. "Some of you knew and dealt with him in business, absolutely—business was a contact sport. Some of us knew him on the golf course, where the word 'gimme' was not in his vocabulary. And some knew him simply as a friend, which meant he was your friend for life. No matter what happened, through the highest of highs or the saddest of lows, he was your friend. I have always thought that perhaps maybe the best of Jack's qualities as a human being was that he was so real, so loyal, so generous, always alive, filled with energy and an eagerness to hear what you were doing: How was your life going, your business, your family?"

Barnicle spoke of Jack's extraordinary journey from being the only child of working-class parents in Salem, Massachusetts, to being the CEO of the Century, who presided over what was once the most admired and valuable company in the world. He spoke of Jack attending Boston Red Sox games at Fenway Park. "Jack was a billionaire," he continued. "He was the greatest CEO of the twentieth century. And you'd see him down there in the front row of the visitors' dugout, three or four of his friends with him, and they were all equal. Jack had no airs about him. They were all equal. He had simple rules. Show up every day to work and to life. Just show up. He was, as all of you know, pretty good at performance, and perform he did, right until the end. He stayed at the game. He didn't want to leave the batter's box. He did have game."

Barnicle recalled of Jack's final days, "Recently, my wife and I dropped in to see Jack and Suzy. My wife and Suzy were elsewhere in the apartment. Jack and I were alone in a room overlooking the treetops in Central Park. He was telling me how much he enjoyed seeing the trees and the sun and the grass each morning. Then he told me that he was sad. I couldn't help but think, 'Yeah, he was sad.' This guy we all love, this guy who inhaled life, who enjoyed every single breath, who was grounded in reality, pushed all of us to reach what was possible and keep going and reach beyond things, this guy who found a woman who he loved with every fiber of his being and held her close until the very end . . . I knew that he sensed the sadness stemmed from him thinking that he would not see another summer. He never wanted to let go. He said it again, 'I'm sad.' I said, 'Yeah, yeah.' I nodded at him and said, 'I understand.' 'I don't think you do,' he said. 'I understand you're sad,' I said. He said, 'No, no, no, not about that shit.' He looked at me and said, 'Why did they trade Mookie?'"

*Author's Note*

In May 1987, as Jack was falling further in love with GE Capital, his favorite unregulated bank, I was hired to be an associate in a GE Capital division in New York City euphemistically called Acquisition Funding Corporation. What the group did was provide financing for big leveraged buyouts, the debt-fueled financial alchemy whereby rich cats like Ray Chambers and Bill Simon could get their hands on a company such as Gibson Greetings using very little of their own cash and a whole lot of borrowed money and then make a bloody fortune. Leveraged buyouts were all the rage in the late 1980s. They were mesmerizing magic tricks. Markets were roaring. Deals were booming. Bankers and traders were making piles of money. It was a time when to get a job on Wall Street, all you had to do was fog a mirror. That probably explained why a former history major and former investigative reporter in Raleigh, North Carolina, like me, was able to get a job on Wall Street without having had the slightest bit of actual experience.

In September, after graduating from Columbia with my MBA, I started working at AFC in an office building on Madison Avenue, at the corner of Fifty-Fourth Street. It was the same building where the tony investment bank Dillon Read had its headquarters. Working at GE Capital seemed like a coveted job at the time—not quite the prestige of the Wall Street investment banks such as Goldman Sachs, First Boston, Morgan Stanley, or Dillon Read, for that matter—but a step up the Wall Street value chain from commercial banking. Plus, it wasn't lost on me that Murry Stegelmann, the valedictorian of the class of 1986 at Columbia Business School, had joined this same division of GE Capital the year before I did. If the smartest guy in the previous class had gone to GE Capital to finance leveraged buyouts and a year later it wanted me, too, how bad could it be? Who was I to turn down that opportunity? Joining me as associates and officemates in New York were Dave Moyer, a graduate of Harvard Business School, and John Flannery, a graduate of Wharton Business School at the University of Pennsylvania. Three new associates, all recent graduates of Ivy League MBA programs. I might have just satisfied all the requirements for an MBA, but I had virtually no idea what I was doing at GE.

My job financing LBOs was actually my second exposure to GE other than as someone who bought the occasional lightbulb. In 1982, during the year I spent at the Columbia University Graduate School of Journalism getting my master's degree, I got invited on an all-expense-paid, private jet–fueled day trip to visit the GE lighting center in Cleveland and the major-appliance manufacturing business in Louisville. How that came about, or why I was selected, I have no recollection. But it was informative, fun, and exciting to travel on a private jet for the first time, on GE's dime, and to see how GE manufactured lightbulbs and refrigerators. It was extremely impressive.

A little more than a month into the GE Capital job, on October 19, 1987, the Dow Jones Industrial Average fell 22.6 percent in one day—the largest single-day percentage drop in history. It didn't bother me particularly—as a former reporter, I had very little money—but quickly I saw grown men gather around the office's lone Quotron machine, the only way then to get near-real-time stock quotations. They were crying, or nearly so, as they watched their equity portfolios dissolve. It was strange, to say the least, to watch your bosses freaking out about their net worth. For my part, I was wondering what the collapse of the stock market would mean for the very important work of financing leveraged buyouts.

When the dust settled, what became apparent was the buyout business, at least at GE Capital, was going to be better than ever. All the companies these private-equity guys wanted to buy were suddenly 30 percent cheaper. If GE Capital was willing to keep lending—and it was—why not keep arbitraging the difference between GE's low cost of funds, thanks to its AAA credit rating, and what borrowers were willing to pay—high rates of interest plus equity warrants? There were risks for sure, especially if a borrower went bankrupt. But GE Capital had built a huge portfolio of different kinds of risk and was coining profits, especially from my business, financing LBOs. Our market share increased after the Dow crashed because, overnight, the investment banks and commercial banks began pulling back, out of fear that borrowers would not be able to pay back their loans. Jack loved GE Capital. "I thought it was easier than bending metal," he told me. "Fooling with money. Get bright people, find an edge. It was easier to make money. It was a home run."

In my year in New York City, I analyzed around 150 different LBOs in a variety of industries. I got close to financing the buyout of Wearever–Proctor Silex for three Israeli brothers, the Selzers, but in the end the GE Capital board turned the deal down. I worked on Norman Lear's successful deal to acquire a large group of movie theaters for $200 million. I worked with three savvy insurance investors from New Canaan, Connecticut, who bought the reinsurance business from Gen Re Corp. for $290 million. I worked for the Lehman Brothers' buyout fund on its $157 million acquisition of Chief Auto Parts. These deals were about par for the course back then, as few LBOs were more than $1 billion. I eventually figured out what I was doing.

After a year at the Acquisition Funding Corporation, someone asked me to go to GE Capital's headquarters, in Stamford, Connecticut, to work as the financial analyst for Jim Fishel, the chief credit officer. I'm still not sure how that happened, or why, either. Maybe someone thought I needed to learn a thing or two about GE Capital's extraordinary range of business lines. Or about how finance really worked. Or about how GE Capital made its billions in profit ($1.1 billion in operating profit in 1989, to be precise). It was a nine-month assignment and considered a promotion. My friend John Flannery had the job with Fishel before I did. I reverse commuted each day from the Upper West Side of Manhattan in a late-model BMW sedan that GE provided me. Fishel was the textbook definition of avuncular. He was wise. He was patient. He taught me a tremendous amount about financing inventory, aircraft leasing, manufactured-home financing, receivables finance, real estate finance, insurance, and how to properly underwrite the credit provided to finance an LBO. He was also very good at running interference with Gary Wendt, who was genuinely brilliant but also every bit as mercurial as everyone said.

In and around April 1989, my rotation with Fishel was coming to an end. I was theoretically supposed to return to New York City to the LBO finance group. But my gut told me that by April 1989, the opportunities that had existed for LBO firms after the October 1987 stock market collapse were drying up and a long, cold financing winter was emerging, a time when despite the lower valuations in the market, even GE Capital was beginning to pull back from financing LBOs. I decided to make the jump to Lazard Frères & Co., in New York, as an associate working on complex corporate restructuring assignments and advising on mergers and acquisitions. I got a second introduction to Lazard through Jon Foster, who had gone there as an associate after working with the Selzer brothers to try to buy Wearever–Proctor Silex. Had I not been at GE Capital and met Jon working on the failed deal, I never would have gone to Lazard. And then who knows what direction my career would have taken? Serendipity.

In any event, before I walked out the door of GE Capital, I will never forget the meeting that took place in Wendt's office between him and Fred Joseph, then then CEO of Drexel Burnham Lambert. Joseph had come to Stamford to see Wendt to try to convince him that GE Capital and Drexel should join forces to finance LBOs and other corporate mergers and acquisitions. The proposed alliance went nowhere— Wendt was way too smart to agree to join forces with Drexel when it was on its last legs—but it was a real watershed moment for me: if Joseph was in Stamford asking Wendt for his help, Drexel must have been in far worse economic shape than anyone knew. There was no way a healthy Drexel—the home of Mike Milken, the creator of the high-yield market—would ever come hat in hand to Stamford if things were well. Wendt told me he had no recollection of the meeting with Joseph, but he was glad to hear (from me) that he turned Joseph down flat. News of their meeting was never made public. It was clearly one of Joseph's final, failed Hail Mary attempts. In February 1990, Drexel filed for Chapter 11 bankruptcy protection, and it was eventually liquidated, proving to me how wise Gary Wendt had been a year or so earlier to tell Fred Joseph to pound sand. Once upon a time, GE Capital really did have, in addition to its billions in financial resources, a warehouse full of incredibly clever financial minds.

As GE was becoming increasingly prominent, along with its myriad accomplishments, the journalists who covered the company couldn't help but be impressed. "As I got more and more into it, I hadn't really fully appreciated what an ingenious company it was," Thomas F. O'Boyle, who covered GE for *The Wall Street Journal* in the 1990s, told me. "As I looked into what GE had invented, it was mind-boggling to me. If you talk about the American century and the invention of the American century by American ingenuity, GE is probably the most significant contributor to that of anybody in industrial America. You could maybe make an argument for Bell Labs, and you could maybe make an argument for IBM. . . . When you go through the array and vastness of what General Electric represented for the twentieth century in America, it's pretty impressive." Even what GE scientists discovered by mistake became a sensation. For instance, in 1949, GE created what became known as Silly Putty, a useless combination of boric acid and silicone oil that went on to delight a generation of American kids, who couldn't wait to press the stuff onto the comics pages of their parents' evening newspaper to capture the impression. GE and Edison were even the pioneers in developing the first electric cars. Henry Ford, the founder of the Ford Motor Company, worked for Edison in the 1890s as the chief engineer of the Edison Illuminating Company, an electricity manufacturing plant in Detroit. He was also working in his spare time on an internal combustion engine powered by gasoline, a radical departure at the time. Most "cars" at the time were powered either by steam or by lead-acid batteries. According to Michael Brian Schiffer, in *Taking Charge*, in 1896 Ford traveled to Coney Island, in Brooklyn, to attend a dinner for Edison executives. Afterward, Edison and his team were discussing electric vehicles when Ford's boss, in Detroit, pointed Ford out to Edison and told him, "There's a young fellow who's made a gas car." Edison and Ford fell into conversation, and then Edison exclaimed, "Young man, that's the thing; you have it. Keep at it! Electric cars must keep near to power stations. The storage battery is too heavy. . . . Your car is self-contained— carries its own power plant." In 1914, Ford told *The New York Times* that he and Edison were working together on "an electric automobile" thanks to Edison's "new nickel-iron batteries," manufactured by his battery company, that had an anticipated range of one hundred miles. But very few of these electric cars were produced, as gasoline became increasingly inexpensive and could keep a car's engine going for hundreds of miles. With the discovery of oil in West Texas, Bill Ford, the Ford chairman, told John Seabrook in *The New Yorker*, "gasoline became so cheap that the whole fleet converted over," adding, "I often wonder what would have happened had that discovery not occurred."

On the industrial side of its business, of course, GE housed a staggering amount of talent. It's not crazy to think of GE, and its DNA, back in the day as the technological equivalent of Apple, Google, or Microsoft today. By one estimate, GE has more than 35,000 active patents and has received hundreds of thousands of patents since 1892.

The innovations continue to this day. In 2018, GE unveiled a new family of engines that can be used to power a coming generation of supersonic jets. Dubbed "Affinity," the engines will be able to operate at altitudes as high as sixty thousand feet while also meeting stringent noise requirements. In 2019, GE announced that its GE9X was the world's most powerful commercial jet engine after reaching 134,300 pounds of thrust during ground testing at Peebles, Ohio. By the end of the year, Boeing had ordered more than seven hundred GE9X engines for the 777X, which the FAA approved for sale in September 2020.

In December 2021, a United Airlines Boeing 737 MAX 8 became the first passenger flight to use 100 percent so-called sustainable aviation fuel—made from recyclable, nonfossil fuels—in one of the aircraft's two CFM LEAP-1B engines (CFM is a joint venture between GE and Safran Aircraft Engines). The fuel can be used interchangeably with conventional jet fuel and requires no modifications to engines or airframes. In February 2022, CFM announced it was working with Airbus to test a jet engine fueled entirely by hydrogen, the most abundant element in the universe and one that produces zero carbon dioxide emissions when it burns. With luck, GE expects the engines to be on jets carrying passengers, with zero carbon emissions, by the middle of the next decade. GE is also working with NASA on the creation of an electric jet engine to power a single-aisle aircraft. There are more than twenty-five thousand GE or CFM aircraft engines in operation, powering flights that are taking off at the rate of one every two seconds.

Prior to the global pandemic, in 2019, GE's jet engine division was a $33 billion–revenue business, and immensely profitable, with nearly $7 billion of operating earnings. Not surprisingly, the pandemic hurt the business, given the decline in global travel. In 2021, revenues decreased to $21 billion, with operating profit of $2.9 billion, declines of 36 percent and 59 percent respectively. Still, the business is the envy of the world. The division received orders for $26 billion worth of jet engines in 2021, and its backlog totaled $303 billion.

To satiate my curiosity about GE's technological prowess, the GE brass allowed me to visit the company's Research Center, in Niskayuna, New York, in April 2019 (before GE decided it would not participate in any further way with the writing of this book). It was a gray upstate New York day, chilly, with few signs of spring in the air. In the lobby of the center, which is massive, a visitor is immediately confronted with the wood desk Edison used at his office at the Edison Electric Light Company, in the Edison Building at 44 Broad Street in downtown Manhattan. As part of the ongoing GE-Edison creation myth, on the desk are facsimiles of pages from Edison's notebooks— including pages that describe how he perfected the lightbulb for easy consumer use—and the self-winding stock ticker that he created in 1870 and that was used on stock exchanges for the next eighty years. Nearby is a case containing awards and citations for various GE scientists, including Ivar Giaever's Nobel Prize medal, from 1973; a Deming Prize awarded to Gerald Hahn in 1994; a Tesla Award for Manoj R. Shah in 2012, for his advancements in electromagnetic design; and an IEEE Transportation Technologies Award from 2015, for Robert D. King for his contributions to the design, optimization, and implementation of propulsion and energy-management systems for hybrid-electric vehicles. In 2019, GE's annual research and development budget was about $500 million.

I spoke to a scientist working in the field of bioelectronic medicine. He was part of a team that was developing a therapy to treat chronic diseases, such as diabetes and arthritis, using noninvasive ultrasound to target specific organs in the body that cause disease. "We went down the path of building electrodes that could go onto nerves," the fellow explained. "We quickly learned—and we had a great discovery, in around 2017—that we actually could do the same triggering noninvasively with ultrasound." The idea was to stimulate the neurons by pushing on them and getting them to fire, rather putting electrical signals into them. "And it worked," he said. The GE team had just successfully completed a study of sixty healthy people to make sure the therapy didn't cause weird side effects.

I also visited the so-called Additive Manufacturing lab. The promise here was being able to use a 3D-printing machine to get the parts you need, say, for a jet engine or for a turbine, whenever you need them, and at scale. "It's almost like *The Jetsons*," a different GE scientist told me. "You press a button and out comes the part you want, just the way you want it."

More than a century of GE's expertise in material science is brought to bear upon the process. "We leverage every one of those years to understand: How do I take a titanium alloy from this process and move it over into an additive process?" he continued.

He shared the example of GE's new Catalyst engine, a small aircraft engine used in a twelve-person private jet. GE's use of 3D printing allowed it to "penetrate" that market for the first time. "We took three hundred fifty-five parts in that engine and we consolidated them into one part," he explained. The 3D-printing process for the engine eliminated all the structural castings in the engine. The process increased the power of the engine by 10 percent while reducing its weight by 20 percent. As a parting gift, the GE scientist handed me a 3D-printed bottle opener with the GE meatball embossed on it.

On my way out, I passed through the lobby and the shrine to Edison. "I find out what the world needs, then I proceed to invent it," reads the larger-than-life quotation superimposed on the larger-than-life photograph of Edison, above his desk.

# Acknowledgments

My two years working at GE Capital in New York and in Stamford in the late 1980s gave me the sturdy foundation I needed to tackle a topic as immense as a history of the astounding rise and unimaginable fall of the General Electric Company. Researching, reporting, and writing *Power Failure* was arduous and took longer—nearly three years—than any one of my previous six books. But needless to say, the years I spent at GE Capital also provided me with invaluable insights into GE and GE Capital. They also provided me with invaluable connections to the multitude of sources who helped me, in immeasurable ways, in the writing of this book, and to whom I owe an immense debt of gratitude. People such as Jack Welch, Jeff Immelt, Gary Wendt, Denis Nayden, Dave Cote, Dave Calhoun, Steve Bolze, and John Krenicki—among many, many others—were generous with their time in sharing the stories of their careers at GE and their insights into the inner workings of what was once the most valuable and respected company in the world. To be able to talk for hours with both Jack and Jeff, in particular, and to have them reflect upon what they did and why they did it, I hope is what will set *Power Failure* apart from all the other books that have been written about GE. A huge thanks to the many people who shared their thoughts and reflections about GE with me. You may not find yourself mentioned by name, as per our agreement, but your help was most appreciated.

I also want to give special thanks to Adrian Zackheim, the founder, president, and publisher of Portfolio, and to Trish Daly, the senior editor at Portfolio, for having faith in me and in this project. John Brodie, my former editor at *Fortune*, who, on special assignment, somehow wrestled my gargantuan manuscript to the ground, also gets my profound thanks; there would be no *Power Failure* without him and his zealous, meticulous, and masterful editing. Also at Portfolio, I want to thank Niki Papadopoulos, the editor in chief of Portfolio, as well as Megan McCormack, Tara Gilbride, Margot Stamas, Kirsten Berndt, Jessica Regione, Sharon Gonzalez, Daniel Lagin, and Brian Lemus, who designed the beautiful cover (after my youngest son, Quentin, suggested the basic idea for it).

As always, it takes a supportive village and a team of dedicated villagers to pull off a project of this scope and scale. Joy Harris, my literary agent, is peerless. She has been a believer in me from the beginning, ever since I made the fateful decision to return to journalism in 2004 after my career on Wall Street came to end. (I also want to give a shout-out to Adam Reed, who works with Joy and always keeps us on track.) I want to thank especially Graydon Carter, first at *Vanity Fair* and then at *Air Mail*; Doug Stumpf, my longtime editor at *Vanity Fair*; and Jon Kelly, first at *Vanity Fair* and then at *Puck*, for never failing to support my journalistic endeavors. The three of you provide the wind in my sails.

To my friends—too many to mention, fortunately—I cannot tell you how much your company, your humor, and your wisdom have sustained and nourished me over the years, especially so during the grueling years of the dreaded pandemic and in the writing of this book. Also a quick life-saving thanks go out to a few of my favorite doctors in New

York City, Thomas Nash, Len Girardi, and Timothy Dutta. Thank you one and all, and of course to Gemma Nyack, as always.

To my immediate family, my parents, Suzanne and Paul, my brothers, Peter and Jamie, and their families, thanks for always being there for me. To my in-laws, the Futters—Ellen and Jeff and their families— how lucky am I to have been welcomed, lo these many years ago, into such a formidable and remarkable clan? To all the other Hiekens, Feldmans, Gellmans, and Cohans still out there, the dream shall never die.

And to my absolute, numero uno prides and joys, my sons, Teddy and Quentin, future daughter-in-law Annabeth "A.B." Gellman, and, of course, Roscoe, aka Dennis the Menace, words cannot express the love I have for you. As for Deb, my endlessly devoted and supportive wife of more than thirty-one years, well, somehow it just keeps getting better and better. I love you and thank you for everything.

# A Note on Sources

Once upon a time, to obtain the research and source material for a book as detailed, complex, and historical as this would require many hours seated at a dreaded microfiche machine in some dusty corner of a university library, digging through representations of one analog publication after another in search of relevant articles in newspapers, magazines, and obscure periodicals. But we live in the digital age now. And much of the material I read, or used, for *Power Failure* can be found easily enough by searching the fabulous digital archives of a variety of publications, including, among others, *The Wall Street Journal, The New York Times, Fortune, Forbes, Vanity Fair, The New Yorker, BusinessWeek, Bloomberg Businessweek, Grant's Interest Rate Observer,* the *Harvard Business Review,* and the full case-study archive of the Harvard Business School, which for various reasons loved to dine out on the comings and goings of The General Electric Company. There has also been, understandably, a plethora of Wall Street research written about GE over the years, and from time to time several of these research analysts, particularly Steve Tusa at JPMorgan Chase, became part of the story. I was happy to have these reports; they were very helpful in putting the story into the proper context at various points in time. A principal challenge for me in writing *Power Failure* was to avoid getting overwhelmed by the source material about the company, which dates back literally to its founding in 1892, and often to the years before then when the two strands of GE's DNA were slowly, but surely, becoming intertwined.

That said, I want to thank profusely Ben Kalin, a former stellar researcher and fact-checker at *Vanity Fair*. Ben compiled for me the hundreds and hundreds of magazine and newspaper articles I ended up reading and digesting about GE and its top executives. I cite many of these articles in the endnotes of *Power Failure*. Others of these articles I leave uncited in the endnotes, but it is clear, I believe, where I have used them in the narrative. I made the decision that there was no need to refer to a certain, say, *Wall Street Journal* article in the pages of *Power Failure*, and then to also add an endnote about the article, too. That would have added even more to what is already a lengthy book, especially when I believe it is perfectly clear from the context where I relied on the contents of a particular article.

Along with my voluminous periodical and documentary trove—thanks to Ben—I also conducted hundreds of interviews with current and former GE executives, many of whom have scaled even greater heights than they had already achieved at GE. I am grateful to them all for giving so generously of their time and their vivid recollections. I could not have written *Power Failure* without them. Their contributions are cited throughout the book, usually with the flourish "to me," which I'm not wild about using but I decided would be very helpful to the reader in understanding that the person made the statement during a conversation with me. Most of these conversations, I am happy to share, were on the record. Others were on background, giving the reader the interesting story or information but not revealing the precise source of it, other than to say that it was someone in a position to know.

Four of my sources, including three of my main sources—Jack Welch, Jeff Immelt, and Bob Wright—were not only generous with their time in granting me extensive interviews but also they wrote memoirs about their lives and their careers at GE. Needless to say, these books were extremely helpful to me in augmenting what they had said to me during our interviews. Where I relied on these books, you will find citations either in the narrative or in the endnotes. In any event, I recommend these books to you if you want more information about these individuals, their fascinating lives, and their careers at GE:

Jack Welch, with John A. Byrne, *Jack: Straight from the Gut* (New York: Warner Books, 2001). I also relied on the paperback edition, published two years later by Grand Central Publishing, for the amusing changes that Jack made to the original.

Jeff Immelt, with Amy Wallace, *Hot Seat: What I Learned Leading a Great American Company* (New York: Avid Reader Press, 2021).

Bob Wright, with Diane Mermigas, *The Wright Stuff: From NBC to Autism Speaks* (New York: RosettaBooks, 2016).

Beth Comstock, with Tahl Raz, *Imagine It Forward: Courage, Creativity, and the Power of Change* (New York: Currency, 2018).

While I'm mentioning books about GE that are worth your time, and that I found immensely helpful while also being entertaining, I want to be sure to mention one by the great investigative journalist Ida Tarbell. I highly recommend her excellent book about Owen Young, one of the early, great leaders of GE: *Owen D. Young: A New Type of Industrial Leader* (New York: The Macmillan Company, 1932).

It's also worth noting when I did not have access, or where I chose to avoid certain texts. Let me explain. Larry Culp, the current CEO and chairman of GE, refused my numerous requests for interviews. In fact, there was only one day when Culp's GE helped me with this book in any way. That was in April 2019, when GE acceded to my request to see with my own eyes GE's impressive facilities in both Crotonville, New York (the management training center), and Niskayuna, New York (the research laboratory). I am glad I had the opportunity to visit these places. Since then, GE has been unresponsive. Aside from the occasional criticism when I wrote about the company in *Vanity Fair*, and in particular about Larry Culp's contract and revisions to it during the pandemic, I've had no interaction with the company. As I have also noted in the narrative, most GE board members gave me the stiff arm, too, revealing an unfortunate—but hardly surprising, I guess— unwillingness to be held accountable in any way for their actions while serving as fiduciaries for GE's thousands of shareholders. John J. "Jack" Brennan, in particular, served on the GE board of directors from July 2012 to December 2018. He was the lead independent director of GE from April 2014 to April 2018. He repeatedly ignored my requests to be interviewed, even when a mutual friend tried to convince him, unsuccessfully, to speak with me. That's pretty much the way it went with most of the GE directors.

While I read—and couldn't help but be impressed by—the extraordinary journalistic display put on by Thomas Gryta and Ted Mann, the two *Wall Street Journal* reporters behind the huge December 14, 2018, article about GE, "GE Powered the American Century—Then It Burned Out," I did not read *Lights Out: Pride, Delusion, and the Fall of General Electric* (Boston: Mariner Books, 2020), the duo's book about the company. Nothing personal. Rather, I decided it would be best for me to write *Power Failure* without being influenced by their book in any way. I hope that was not a mistake, but if it was, I take full responsibility for making it. For similar reasons, I also have not read David Gelles's book about Jack Welch, *The Man Who Broke Capitalism* (New York: Simon & Schuster, 2022).

I am hopeful that, between the hundreds of endnotes to *Power Failure* and the citations included along the way in the narrative itself, the future GE historians and journalists can find all the information about GE that they need and want, just as I did.

# Notes

## PROLOGUE: SIASCONSET

3 **Like so many things on Nantucket:** For the information regarding the founding of Nantucket Golf Club, I interviewed the founder, Ed Hajim, who also shared his biography with me. More information on Hajim can be found at his website, edhajim.com, as well as in his excellent memoir, *On the Road Less Traveled*. Ed Hajim with Glenn Plaskin, *On the Road Less Traveled: An Unlikely Journey from the Orphanage to the Boardroom* (New York: Skyhorse, 2021).

5 **"And he's a fast eater":** Holly Peterson, "Jack Welch," *Talk*, December 2000/January 2001.

6 **"You son of a bitch":** Peterson, "Jack Welch."

## CHAPTER ONE: A CHILD OF TWO FATHERS

11 **Born in December 1844:** Biographical details about Charles Coffin are from Charles E. Wilson, *Charles A. Coffin (1844–1926): Pioneer Genius of the General Electric Company* (Princeton, NJ: Princeton University Press, 1946), 5–29. This monograph was delivered in Wilson's April 25, 1946, speech to the American Branch of the Newcomen Society of England at the Pierre Hotel in New York City. Wilson was a later president of GE.

12 **He "must have borne":** John T. Broderick, *Forty Years with General Electric* (Albany, NY: Fort Orange Press, 1929), 17.

13 **"Let's build a four-lighter":** Wilson, *Charles A. Coffin*, 4.

13 **"What the dickens":** Wilson, *Charles A. Coffin*, 4.

14 **"Mr. Coffin and his associates":** *New York Times*, June 15, 1926.

15 **"men of large means":** Wilson, *Charles A. Coffin*, 9.

16 **Edison had in his corner:** Much of the history of the founding of the Edison General Electric Company is taken from the May 17, 1939, testimony of Owen D. Young, then chairman of GE, before the Temporary National Economic Committee of the U.S. *Verbatim Record of the Proceedings of the Temporary National Economic Committee*, Volume 3, March 24, 1939 to May 29, 1939 (Washington, DC: The Bureau of National Affairs, 1939), 387–99. I also obtained a copy of his speech from the Owen D. Young Library at St. Lawrence University, SLU Special Collections, in Canton, New York. This version is fifteen pages long.

17 **Edison's lamp was little more:** Broderick, *Forty Years with General Electric*, 110.

17 **"had a very boyish appearance":** Frank Lewis Dyer and Thomas Commerford Martin, *Edison: His Life and Inventions* (New York and London: Harper & Brothers, 1929), 330. Much of Insull's take on Edison is from the Dyer and Martin book.

18 **installation of "underground conductors":** Dyer and Martin, *Edison*, 370.

19 **"Make it go":** "Samuel Insull," *They Made America*, PBS.org, no date, www.pbs .org/wgbh/theymadeamerica/whomade/insull_hi.html.

20 **The combination of the two companies:** Young congressional testimony.

22 **"Send for Insull":** Randall Stross, *The Wizard of Menlo Park* (New York: Three Rivers Press, 2007), 187.

24 **"It may be likened":** Broderick, *Forty Years with General Electric*, 19.

## CHAPTER TWO: TWO CRISES

29 **signed on to his plan immediately:** Charles E. Wilson, *Charles A. Coffin (1844–1926): Pioneer Genius of the General Electric Company* (Princeton, NJ: Princeton University Press, 1946), 14.

29 **"effect upon his judgment was permanent":** John T. Broderick, *Forty Years with General Electric* (Albany, NY: Fort Orange Press, 1929), 32.

30 **"so far as this was humanly possible":** Broderick, *Forty Years with General Electric*, 33.

30 **"Use cost or market value":** Wilson, *Charles A. Coffin*, 7.

32 **"We may be entertaining":** Broderick, *Forty Years with General Electric*, 22.

33 **"look like child's play":** Jean Strouse, *Morgan: American Financier* (New York: Random House, 1999), 552.

33 **"in the years that followed":** Wilson, *Charles A. Coffin*, 17–18.

34 **provide the money Coffin needed:** Owen D. Young, testimony before the Temporary National Economic Committee of the U.S. Congress, May 17, 1939, Owen D. Young Library, St. Lawrence University, SLU Special Collections, Canton, New York, 4.

37 **"then we can seek for the best":** Broderick, *Forty Years with General Electric*, 25.

## CHAPTER THREE: THE BATTLE OF THE AIRWAVES

39 **led GE with great success:** Much of the biography of Owen D. Young comes from Ida Tarbell's excellent *Owen D. Young: A New Type of Industrial Leader* (New York: Macmillan, 1932).

40 **"The course was so heavy":** Tarbell, *Owen D. Young*, 49

41 **"It was not long before":** Tarbell, *Owen D. Young*, 97.

42 **"It was an escape":** Tarbell, *Owen D. Young*, 99.

42 **"What was needed":** Tarbell, *Owen D. Young*, 119.

43 "was now embarked upon": Gleason L. Archer, *History of Radio to 1926* (New York: American Historical Society, 1938), 160.

44 "This letter produced": Archer, *History of Radio to 1926*, 161.

44 "in the history of the radio": Tarbell, *Owen D. Young*, 130.

45 "The whole picture puzzle": Archer, *History of Radio to 1926*, 165.

45 "There were no other": Archer, *History of Radio to 1926*, 165.

46 "The argument could not fail": Archer, *History of Radio to 1926*, 167.

47 "Without one or the other": Archer, *History of Radio to 1926*, 169.

47 "The Westinghouse Company": Owen D. Young, testimony before the Temporary National Economic Committee of the U.S. Congress, May 17, 1939, Owen D. Young Library, St. Lawrence University, SLU Special Collections, Canton, New York.

48 "It was not only good diplomacy": Archer, *History of Radio to 1926*, 171.

49 "All this would require time": Archer, *History of Radio to 1926*, 172.

49 "The way was now paved": Archer, *History of Radio to 1926*, 180.

50 "The idea of flinging messages": "He Climbed with Radio," *New York Times*, September 27, 1931.

53 In his congressional testimony: Young congressional testimony, 13.

## CHAPTER FOUR: THE INCREDIBLE ELECTRICAL CONSPIRACY

56 "He referred to the layers": "Atomic Energy: The Powerhouse," *Time*, January 12, 1959.

57 "When I took over": "Atomic Energy: The Powerhouse."

57 In December 1955: William B. Harris, "The Overhaul of General Electric," *Fortune*, December 1955, 110.

63 "Sure, collusion was illegal": Much of the discussion of GE's participation in the price-fixing scandal of the 1950s is based upon the excellent two-part series in *Fortune*: Richard Austin Smith, "The Incredible Electrical Conspiracy," *Fortune*, April and May 1961.

74 "We don't think": "Business: After the Great Conspiracy," *Time*, February, 24, 1961.

75 would have required GE: Felix Belair Jr., "Hearing Is Told Cordiner Knew of G.E. Price Fixing; Senators Told That Cordiner Knew of G.E. Price-Fixing," *New York Times*, April 26, 1961.

## CHAPTER FIVE: JACK FROM PLASTICS

78 "He brought home the newspapers": Jack Welch, interview with the author. Except where noted in the text, quotes from Jack Welch throughout the book are from his numerous conversations with the author.

80 "Ducking altar boy practice": Jack Welch with John A. Byrne, *Jack: Straight from the Gut* (New York: Warner Books, 2001), 12.

80 "an incredibly serious": Stratford P. Sherman, "Inside the Mind of Jack Welch," *Fortune*, March 27, 1989.

82 "No one would have accused": Christopher M. Byron, *Testosterone, Inc.: Tales of CEOs Gone Wild* (Hoboken, NJ: John Wiley & Sons, 2004), 14.

82 "Where Is It Written": Byron, *Testosterone, Inc.*, 14.

85 "there are no finite answers": Welch with Byrne, *Jack*, 18.

86 "What's a Plastics Division optimist?": Byron, *Testosterone, Inc.*, 45.

87 "was a way to compete": Peter Foss, interview with the author.

89 **"As long as I am here"**: John A. Byrne, "How Jack Welch Runs GE," *Businessweek*, June 8, 1998.

91 **"It was one of the most"**: Welch with Byrne, *Jack*, 27.

91 **"I was prepared for the worst"**: Welch with Byrne, *Jack*, 28.

91 **"Don't get me wrong"**: Welch with Byrne, *Jack*, 29.

92 **"Life appeared to be perfect"**: Welch with Byrne, *Jack*, 37.

93 **"Thank God I did it"**: Welch with Byrne, *Jack*, 38.

93 **"She was a real rock for me"**: Welch with Byrne, *Jack*, 39.

95 **according to journalist Christopher Byron**: Byron, *Testosterone, Inc.*, 55.

96 **"Billy the Kid"**: Thomas F. O'Boyle, *At Any Cost: Jack Welch, General Electric and the Pursuit of Profit* (New York: Vintage, 1999), 59.

96 **"behavior that wasn't the norm"**: Welch with Byrne, *Jack*, 43.

97 **"The workdays in Pittsfield"**: Byron, *Testosterone, Inc.*, 60.

97 **"zeroing in on his target"**: Byron, *Testosterone, Inc.*, 61.

99 **"Those who read the article"**: Welch with Byrne, *Jack*, 40.

100 **"Now I had to figure out"**: Welch with Byrne, *Jack*, 43.

100 **"The energy that radiated"**: Byron, *Testosterone, Inc.*, 63.

## CHAPTER SIX: THE BENIGN CYCLE OF POWER

102 **"I am English"**: Jack Welch with John A. Byrne, *Jack: Straight from the Gut* (New York: Warner Books, 2001), 80.

102 **"I'd never consider myself"**: Gene Smith, "New G.E. President Seen as Next Chief," *New York Times*, July 2, 1972, section F, page 5.

103 **"I've really spent more time"**: Smith, "New G.E. President Seen as Next Chief."

103 **"Reg made suggestions"**: Isadore Barmash, "America's Most Influential Jones," *New York Times Magazine*, September 16, 1979.

104 **"facing the corporation as a whole"**: Charles Fombrun, "Conversation with Reginald H. Jones and Frank Doyle," *Organizational Dynamics* 10, no. 3 (winter 1982): 43. This is a conversation between University of Pennsylvania Wharton School professor Charles Fombrun and Reginald Jones, along with his PR man, Frank Doyle.

105 **"one of our greatest successes"**: Fombrun, "Conversation with Reginald H. Jones," 44.

105 **"It was a teenage marriage"**: Donald Sherman, "Hill Climb," *Air & Space Magazine*, May 2001.

106 **"It took off for the first time"**: Tomas Kellner, "Supercharge Me: How Boeing Helped GE Reinvent Jet Travel," *GE Reports*, July 31, 2016.

107 **"In between were a lot"**: Fombrun, "Conversation with Reginald H. Jones," 45.

108 **"'the funds to do that'"**: Fombrun, "Conversation with Reginald H. Jones," 50.

108 **"causative in its approach"**: Fombrun, "Conversation with Reginald H. Jones," 61.

110 **"to know their businesses"**: Welch with Byrne, *Jack*, 42.

110 **a "strong leader"**: John A. Byrne, "How Jack Welch Runs GE," *Businessweek*, June 8, 1998.

110 **"come out of their genius"**: Holly Peterson, "Jack Welch," *Talk*, December 2000/January 2001, 110.

## CHAPTER SEVEN: REG AND JACK

111 **"chalk and cheese"**: Howard Banks, "General Electric: Going with the Winners," *Forbes*, March 26, 1984.

111 **"the king walked in"**: Many people interviewed for this book requested that their names not be used, so I attribute their statements to "a former GE executive," "a GE board member," and the like.

111 **a tailgate party:** *Bloomberg Businessweek*, March 18, 1981, 110.

112 **"This is really something"**: Christopher M. Byron, *Testosterone, Inc.: Tales of CEOs Gone Wild* (Hoboken, NJ: John Wiley & Sons, 2004), 102.

113 **"up there in Pittsfield"**: Byron, *Testosterone, Inc.*, 104.

114 **"wanted me at headquarters"**: Jack Welch with John A. Byrne, *Jack: Straight from the Gut* (New York: Warner Books, 2001), 51.

114 **"make up his mind"**: Byron, *Testosterone, Inc.*, 104.

115 **"great set of experiences"**: Welch with Byrne, *Jack*, 59.

115 **"discharged" PCB oils**: New York State Department of Environmental Conservation Division of Environmental Remediation, "Hudson River PCB Sediments OU-2 Site Upper Hudson River, New York," December 2018, section 1, page 2.

115 **"sink to the bottom"**: Peter Hellman, "For the Hudson, Bad News and Good," *New York Times Magazine*, October 24, 1976.

117 **"remobilized and deposited downstream"**: New York State Department of Environmental Conservation, "Hudson River PCB Sediments," section 1, page 2.

118 **"It was not mean"**: Janet Lowe, *Jack Welch Speaks: Wit and Wisdom from the World's Greatest Business Leader* (Hoboken, NJ: John Wiley & Sons, 2008), 214.

119 **new environmental commissioner**: "G.E.-State Pact on PCB Is Praised as Guide in Other Pollution Cases," *New York Times*, September 9, 1976.

121 **"the first few companies"**: Louis Kraar, "General Electric's Very Personal Merger," *Fortune*, August 1977. There are also transcripts of the conversations between Kraar, Littlefield, and Jones.

123 **"Don't ever forget it"**: Kraar, "General Electric's Very Personal Merger."

123 **"diversification of General Electric"**: Byron, *Testosterone, Inc.*, 110.

124 **"it made me feel better"**: Welch with Byrne, *Jack*, 61.

124 **"the Fairfield bureaucracy"**: Welch with Byrne, *Jack*, 61.

126 **"had it happen to me"**: Welch with Byrne, *Jack*, 66.

126 **"Bye-bye department head"**: Byron, *Testosterone, Inc.*, 111.

126 **Jack's favorite expressions**: Byron, *Testosterone, Inc.*, 111.

127 **"money and brains"**: Welch with Byrne, *Jack*, 233.

129 **"try to get him relaxed"**: Richard F. Vancil, "How Companies Pick New CEOs," *Fortune*, January 4, 1988.

129 **"thoughtful and smart"**: Welch with Byrne, *Jack*, 78.

130 **"ready to blow"**: Richard Vancil, *Passing the Baton* (Cambridge, MA: Harvard Business School Press, 1987).

132 **"on the verge of breaking out"**: Welch with Byrne, *Jack*, 74.

132 **"came as a complete surprise"**: Jack Egan, "GE and Cox Broadcasting Plan Merger," *Washington Post*, October 6, 1978.

133 **"It was spectacular"**: Bob Wright's biographical details come from an interview with the author and from Bob Wright, *The Wright Stuff* (New York: RosettaBooks, 2016).

134 **"at any price"**: Welch with Byrne, *Jack*, 75.

135 **"what they really thought"**: Welch with Byrne, *Jack*, 76.

136 **"reputation of the Company"**: Jack's self-assessment, dated June 2, 1980, to Reg Jones. Jack shared a copy of the document with me.

136 **"How could I resist?"**: Wright, *Wright Stuff*, 65.

136 **"All of a sudden"**: Bob Wright, interview with the author.
137 **"It was the orphan child"**: Welch with Byrne, *Jack*, 71.
139 **"our lives depended on it"**: Welch with Byrne, *Jack*, 73.
141 The **"hall gossip"**: Welch with Byrne, *Jack*, 76.
142 **"The initial pool"**: Vancil, "How Companies Pick New CEOs."
142 **"lack of self-confidence"**: Welch with Byrne, *Jack*, 77.
142 **"very deeply and intimately involved"**: Vancil, "How Companies Pick New CEOs."
143 **"a risky maverick"**: Vancil, *Passing the Baton*.
144 **"I'm not going to say it"**: Byron, *Testosterone, Inc.*, 112.

### CHAPTER EIGHT: THE BRASS RING

149 **"Bless his heart"**: Jack Welch with John A. Byrne, *Jack: Straight from the Gut* (New York: Warner Books, 2001), 85.
149 **"That sure made it easy"**: Welch with Byrne, *Jack*, 85.
150 **"most happy man"**: Thomas C. Hayes, "G.E. Names Welch, 45, Chairman," *New York Times*, December 20, 1980.
152 **"It was a terrific bash"**: Welch with Byrne, *Jack*, 88.
153 **"changing a company"**: Holly Peterson, "Jack Welch," *Talk*, December 2000/January 2001, 110.
154 **"for you and GE"**: Welch with Byrne, *Jack*, 98.
154 **"what I was looking for"**: Welch with Byrne, *Jack*, 99.
155 **"GE was interesting"**: Leslie Wayne, email to the author.
155 **"Men could not reduce"**: Welch with Byrne, *Jack*, 448.
156 **"I was probably lucky"**: Welch with Byrne, *Jack*, 107.

### CHAPTER NINE: GROWING PAINS

157 **"But all these good things"**: Ann M. Morrison, "Trying to Bring GE to Life," *Fortune*, January 25, 1982.
159 **"reward the good try"**: Howard Banks, "General Electric—Going with the Winners," *Forbes*, March 26, 1984.
160 **"Jack's like the neutron bomb"**: Susan Dentzer, "White Collar: The 'New Unemployed,'" *Newsweek*, July 26, 1982.
161 **"The early days"**: Denis Nayden, interview with the author.
161 **"I thought it was accurate"**: Bob Nelson, interview with the author.
161 **"local managers would panic"**: Christopher M. Byron, *Testosterone, Inc.: Tales of CEOs Gone Wild* (Hoboken, NJ: John Wiley & Sons, 2004), 113.
162 **"whether Welch felt the same"**: Byron, *Testosterone, Inc.*, 113.
164 **"just beyond imagination"**: Holly Peterson, "Jack Welch," *Talk*, December 2000/January 2001, 110.
165 emasculated in the **"Pit"**: Thomas F. O'Boyle, *At Any Cost: Jack Welch, General Electric and the Pursuit of Profit* (New York: Vintage, 1999), 73.
165 **"It gives off aromas"**: John A. Byrne, "How Jack Welch Runs GE," *Businessweek*, June 8, 1998.
165 **"on paper anytime, anyplace"**: Jack Welch with John A. Byrne, *Jack: Straight from the Gut* (New York: Warner Books, 2001), 110.
167 **"It wasn't very profitable"**: Welch with Byrne, *Jack*, 112.
167 **"Jack, I love it here"**: Welch with Byrne, *Jack*, 112.

## CHAPTER TEN: THE NEW JACK PACK

169 **"He was a smart guy"**: Bob Wright, interview with the author.

170 **"helluva lot more"**: Jack Welch with John A. Byrne, *Jack: Straight from the Gut* (New York: Warner Books, 2001), 120.

171 **dated back to Plastics**: Thomas F. O'Boyle, *At Any Cost: Jack Welch, General Electric and the Pursuit of Profit* (New York: Vintage, 1999), 71.

171 **"He could slice and dice"**: Welch with Byrne, *Jack,* 134.

172 **"I wanted him to lead"**: Welch with Byrne, *Jack*, 133.

172 **"I'm a little dude"**: Norman Liu, interview with the author.

173 **"get it into operations"**: Welch with Byrne, *Jack*, 135.

173 **"best résumé in America"**: Nicholas Lemann, "Protégé Power," *New Republic,* April 7, 1979.

174 **"Ben seemed a bizarre choice"**: Welch with Byrne, *Jack*, 137.

176 **McKinley was "a gentleman"**: Welch with Byrne, *Jack*, 236.

176 **"at least $800,000"**: Lee A. Daniels, "Federal Jury Charges $800,000 Fraud by G.E.," *New York Times*, March 27, 1985.

178 **"approximately 100,000 timecards"**: Daniels, "Federal Jury Charges $800,000 Fraud by G.E."

178 **implicated "higher management"**: "News Summary," *New York Times*, May 14, 1985.

181 **"nearly broke the company"**: O'Boyle, *At Any Cost*, 128.

181 **"I found that stuff thrilling"**: The story of the O'Boyle contretemps with Jack and GE comes from conversations I had with O'Boyle, plus a trove of documentary evidence, detailing it all, that he shared with me. I applaud him for his courage to come forward and share the story.

186 **"If there was a dust-up"**: Jonathan Segal, email to the author.

## CHAPTER ELEVEN: A TALE OF TWO ACQUISITIONS

191 **denied the rumor**: Isadore Barmash, "CBS-G.E. Plans to Foil Bids Reported," *New York Times*, April 4, 1985.

191 **CBS's "white knight"**: Jack Welch with John A. Byrne, *Jack: Straight from the Gut* (New York: Warner Books, 2001), 140.

192 **"None of us had a clue"**: Welch with Byrne, *Jack*, 144.

193 **"earliest private equity"**: Bob Wright, interview with the author.

194 **"out of the blue"**: Welch with Byrne, *Jack*, 141.

194 **"upsets a lot of people"**: Holly Peterson, "Jack Welch," *Talk*, December 2000/January 2001, 165.

195 **"when you meet people"**: "Merger Tango," *Time*, December 23, 1985.

196 **"could be a home run"**: Welch with Byrne, *Jack*, 142.

196 **"It was a go"**: Welch with Byrne, *Jack*, 142.

198 **"If Frederick had been"**: L. J. Davis, "Did RCA Have to Be Sold?" *New York Times Magazine*, September 20, 1987.

198 **the "good news"**: Welch with Byrne, *Jack*, 143.

198 **"important to the company's success"**: Welch with Byrne, *Jack*, 143.

199 **"We could hardly wait"**: Welch with Byrne, *Jack*, 144.

199 **"boldest move ever for the company"**: "Merger Tango."

199 **"one dynamite company"**: J. Roberts and B. K. Abrams, "Talks That Led to GE-RCA Pact Started Casually," *Wall Street Journal*, December 13, 1985.

200 **"Did RCA have to be sold?"**: Davis, "Did RCA Have to Be Sold?"

201 **"an ocean of money"**: Welch with Byrne, *Jack*, 145.

201 **"run a network"**: Welch with Byrne, *Jack*, 145.

202 **Jack's "bible" for the TV business**: Peter Petre, "A Liberal Gets Rich Yet Keeps the Faith," *Fortune*, August 31, 1987.

202 **described as "very weak"**: Welch with Byrne, *Jack*, 149.

203 **"really wants to do this"**: Welch with Byrne, *Jack*, 149.

203 **generated $100 million a year**: Welch with Byrne, *Jack*, 150.

204 **"sent a chill through the network"**: Davis, "Did RCA Have to Be Sold?"

205 **"had to act, and act fast"**: James Sterngold, "Kidder's Road to Acquisition," *New York Times*, May 5, 1986.

205 **"land of giants"**: Sterngold, "Kidder's Road to Acquisition."

206 **"quite a windfall"**: Max Chapman, interview with the author.

208 **"We thought Kidder"**: Welch with Byrne, *Jack*, 218.

209 **"For Chrissakes, Jack"**: Welch with Byrne, *Jack*, 217.

## CHAPTER TWELVE: THE BANK OF GENERAL ELECTRIC

210 **"They really consulted"**: Paul Street, interview with the author.

211 **albeit a quirky one**: Gary Wendt's fascinating biographical details come from my interview with him.

213 **"We relied on"**: Lorna Wendt, interview with the author.

216 **"I want 'em all"**: Steven Lipin and Randall Smith, "A Brutal Negotiator," *Wall Street Journal*, November 2, 1994.

216 **"grow the business"**: Christopher M. Byron, *Testosterone, Inc.: Tales of CEOs Gone Wild* (Hoboken, NJ: John Wiley & Sons, 2004), 118.

217 **"You've got to take ground"**: "The Most Imperfect Investment," *Grant's Interest Rate Observer*, November 8, 2002.

217 **"appears more suddenly"**: Ron Ashkenas, "Get Ready to Fall Off the Cliff," *Forbes*, February 14, 2012.

218 **"no due diligence"**: Jim Bunt, interview with the author.

219 **"We just bought Kidder Peabody"**: Jack Myers, interview with the author.

220 **"maybe six businesses"**: Bob Wright, interview with the author.

224 **Kidder had made "millions"**: James B. Stewart and Janet Guyon, "Damage Control," *Wall Street Journal*, June 8, 1987. By this point the details of the insider-trading scandal have been well documented in a variety of newspapers, magazines, and books, including in my *Money and Power: How Goldman Sachs Came to Rule the World* (New York: Doubleday, 2011).

226 **"decided to leave"**: Jack Welch with John A. Byrne, *Jack: Straight from the Gut* (New York: Warner Books, 2001).

227 **"Are you out of your"**: Welch with Byrne, *Jack*, 219.

## CHAPTER THIRTEEN: TOP OF THE ROCK

228 **"I was thinking to myself"**: Bob Wright, *The Wright Stuff* (New York: Rosetta-Books, 2016), 59.

228 **"They didn't like"**: Wright, *Wright Stuff*, 60.

229 **"but it was the only recourse"**: Wright, *Wright Stuff*, 61.

229 **"We were hedging"**: David Zaslav, interview with the author.

230 **"value in return"**: Wright, *Wright Stuff*, 103.

231 **"he had all the moves"**: Wright, *Wright Stuff*, 104.

231 **"I loved the idea"**: Jack Welch with John A. Byrne, *Jack: Straight from the Gut* (New York: Warner Books, 2001), 260.

231 **"I'm always watching"**: Holly Peterson, "Jack Welch," *Talk*, December 2000/January 2001, 108.

231 **"not taking off"**: Peterson, "Jack Welch," 108.

232 **"did not buy broken assets"**: Wright, *Wright Stuff*, 106.

233 **Malone wouldn't "screw us"**: Wright, *Wright Stuff*, 106.

235 **$60 million more**: Welch with Byrne, *Jack*, 260.

235 **"We weren't breaking"**: Marc Gunther, "There's No Business Like Business Show Business: How CNBC Grew from an Ugly Duckling into a Network to Make a Peacock Proud," *Fortune*, May 24, 1999.

236 **"I was an instant fan"**: Gabriel Sherman, *The Loudest Voice in the Room* (New York: Random House, 2017), 144.

236 **"Most of the shows bomb"**: Welch with Byrne, *Jack*, 261.

236 **"We got beat up"**: Welch with Byrne, *Jack*, 262.

236 **"tore the company apart"**: Peterson, "Jack Welch," 165–66.

237 **"totally one-sided deal"**: Peterson, "Jack Welch," 111.

239 **"free-spirited" founder**: Wright, *Wright Stuff*, 159.

239 **"bring our companies together"**: Wright, *Wright Stuff*, 159.

240 **"GE's conservative board"**: Wright, *Wright Stuff*, 162.

240 **"He hated it"**: Sherman, *Loudest Voice*, 161.

241 **"full of gossipy journalists"**: Sherman, *Loudest Voice*, 162.

241 **"number of other people"**: Sherman, *Loudest Voice*, 163.

242 **"that didn't happen"**: Sherman, *Loudest Voice*, 166.

243 **"This was a sad thing"**: Sherman, *Loudest Voice*, 169.

243 **In January 1996, Jack negotiated**: Sherman, *Loudest Voice*, 169.

## CHAPTER FOURTEEN: WHEELIN' AND DEALIN'

244 **"underperforming the market"**: Janet Guyon, Heard on the Street (column), *Wall Street Journal*, May 16, 1988.

246 **"winning the World Series"**: David Hilder and Randall Smith, Heard on the Street (column), *Wall Street Journal*, July 23, 1987.

247 **"thin as a dime"**: James Grant, "Hot Light on GE," *Grant's Interest Rate Observer*, September 14, 1990.

247 **"money-center banks"**: David Tice, "Cruisin' for a Bruisin'?," *Barron's*, October 22, 1990.

255 **"We did, we did"**: Gary Wendt, interview with the author.

256 **"What's the job"**: Denis Nayden, interview with the author.

257 **"a decision-making level"**: *Wall Street Journal*, June 27, 1991.

257 **"rest of my career"**: Jack Welch with John A. Byrne, *Jack: Straight from the Gut* (New York: Warner Books, 2001), 240.

259 **"crook on your payroll"**: Welch with Byrne, *Jack*, 280.

261 **"frank piece of propaganda"**: Megan Rosenfeld, "Bringing Bad Things to Light," *Washington Post*, April 23, 1992.

262 **"We haven't any intention"**: *Wall Street Journal*, June 13, 1991.

262 **"We treasure our independence"**: Welch with Byrne, *Jack*, 152.

263 **"Carolyn and I"**: Welch with Byrne, *Jack*, 146.

264 **"fabulous divorce settlement"**: Donna Bertaccini, interview with the author.

264 **net worth was $12 million**: Christopher M. Byron, *Testosterone, Inc.: Tales of CEOs Gone Wild* (Hoboken, NJ: John Wiley & Sons, 2004), 165.

264 **"wife of GE's chairman and CEO"**: Byron, *Testosterone, Inc.*, 167.

265 **"roughly $11 million worth"**: Byron, *Testosterone, Inc.*, 168.

265 **"I'm not going to marry him"**: Welch with Byrne, *Jack*, 147.

266 **"my best behavior"**: Welch with Byrne, *Jack*, 147.

266 **"until her back ached"**: Welch with Byrne, *Jack*, 147.

266 **"I had to practically sign"**: Welch with Byrne, *Jack*, 147–48.

## CHAPTER FIFTEEN: EARNINGS MANAGEMENT

270 **"It's not legal"**: Michael Siconolfi, "Bond Epic: How Kidder, a Tiger in April, Found Itself the Prey by December," *Wall Street Journal*, December 29, 1994. The summary of the Joe Jett trading scandal comes from numerous published reports at the time; an investigation into the scandal prepared by Gary Lynch, a partner at Davis Polk, and published on August 5, 1994; and Jett's book, *Black and White on Wall Street* (New York: William Morrow, 1999).

271 **"put things back together again"**: Jack Welch with John A. Byrne, *Jack: Straight from the Gut* (New York: Warner Books, 2001), 227.

271 **"in the first place"**: Holly Peterson, "Jack Welch," *Talk*, December 2000/January 2001, 165.

272 **"GE Capital's appetite"**: David Hilder, interview with the author.

272 **"You made my whole frigging year"**: Peterson, "Jack Welch."

273 **"there's no amount of money"**: Welch with Byrne, *Jack*, 228.

275 **using Six Sigma principles**: "GE Puts Emphasis on Fewer Defects; 'Six Sigma' Plan Could Become CEO's Legacy," *Cleveland Plain Dealer*, January 18, 1998.

276 **"Aircraft guys were miles ahead"**: Bill Conaty, interview with the author.

276 **was a big hit**: Welch with Byrne, *Jack*, 329.

277 **"Harry's the guy"**: Welch with Byrne, *Jack*, 329.

279 **"productivity gains and profits"**: Welch with Byrne, *Jack*, 335.

279 **"a debilitating parasite"**: Marshall Loeb, "Jack Welch Lets Fly on Budgets, Bonuses, and Buddy Boards," *Fortune*, May 29, 1995, among others.

280 **office every afternoon**: Welch with Byrne, *Jack*, 326.

280 **"We ate lots of pizza"**: Welch with Byrne, *Jack*, 326.

280 **"The vein had closed"**: Welch with Byrne, *Jack*, 326.

281 **"he had no intention"**: Linda Grant, "GE: The Envelope, Please," *Fortune*, June 26, 1995.

281 **"With my family's history"**: Welch with Byrne, *Jack*, p. 327.

281 **"Just like Jack"**: Cary Akins, interview with the author.

284 **"getting this job at my age"**: *Wall Street Journal*, May 24, 1995.

285 **"the most successful conglomerate"**: *Wall Street Journal*, May 24, 1995.

## CHAPTER SIXTEEN: THE BAKE-OFF

286 **"capable of immediately taking over"**: Linda Grant, "GE: The Envelope, Please," *Fortune*, June 26, 1995.

288 **"because of my marital problems"**: Christopher M. Byron, *Testosterone, Inc.: Tales of CEOs Gone Wild* (Hoboken, NJ: John Wiley & Sons, 2004), 223.

288 **"He never liked me"**: Gary Wendt, interview with the author.

289 **"I had achieved"**: Gary Wendt, interviewed by Lynn Sherr, *20/20*, ABC, March 14, 1997.

291 **"in a hammock"**: Joseph T. Hallinan, "For New CEO Gary Wendt, Conseco Post Cost $20 Million," *Wall Street Journal*, July 12, 2000.

292 **"drunken Japanese salesmen"**: Jeff Immelt, interview with the author.

293 **"like a slave"**: Byron, *Testosterone, Inc.*, 246.

293 **"spending the night"**: Byron, *Testosterone, Inc.*, 246.

294 **Jack's closest GE confidants**: Byron, *Testosterone, Inc.*, 247–48.

294 **"high-potential long shots"**: Jack Welch with John A. Byrne, *Jack: Straight from the Gut* (New York: Warner Books, 2001), 410.

295 **"This is your show"**: Welch with Byrne, *Jack*, 412.

296 **Wilkes-Barre and Scranton**: Bob Nardelli's biographical details are from both the Fall 2016 issue of *Western Magazine* (the alumni magazine of Nardelli's alma mater, Western Illinois University) and Pattie Sellers, "Something to Prove," *Fortune*, June 24, 2002.

298 **"working long hours"**: Matt Murray and Joann Lublin, "Denied Welch's Post at GE, Nardelli, McNerney Could Still Become CEOs," *Wall Street Journal*, November 28, 2000.

299 **"very difficult to have a surprise"**: Sellers, "Something to Prove," *Fortune*, June 24, 2002.

299 **Walter James McNerney Jr.**: Jim McNerney's biographical details are from our conversations.

308 **The third contender**: Jeff Immelt's biographical details are from our conversations as well as from an interview that Steve Immelt, Jeff's brother, gave about him that was shared with me.

313 **"always wished him Godspeed"**: Claudia H. Deutsch, "G.E.'s New Corporate Face; Jeffrey Immelt Rides a Can-Do Confidence to the Top," *New York Times*, December 1, 2000.

313 **"He actually wanted"**: Norman Liu, interview with the author.

314 **"wanted to hire the guy"**: Christopher A. Bartlett and Andrew N. McLean, "GE's Talent Machine" (Harvard Business School case study), November 3, 2006.

316 **"part of something great"**: Bartlett and McLean, "GE's Talent Machine."

317 **"boy to a man"**: Daniel Eisenberg, "Jack Who?," *Time*, September 10, 2001.

319 **"and then just nailed"**: Welch with Byrne, *Jack*, 421.

320 **"a chance to put it all together"**: Bartlett and McLean, "GE's Talent Machine."

320 **Jack's annual reviews**: Welch with Byrne, *Jack*, 457–62.

322 **New England accent**: Dave Cote's biographical details are from our conversations.

335 **GE board of directors**: The details of Ken Langone joining GE's board come from our conversations.

337 **"it came as a surprise"**: Welch with Byrne, *Jack*, 415.

338 **"None of us wanted"**: Welch with Byrne, *Jack*, 417.

338 **"without any help from us"**: Welch with Byrne, *Jack*, 417.

## CHAPTER SEVENTEEN: HABEMUS PAPAM

339 "open to the very end": Jack Welch with John A. Byrne, *Jack: Straight from the Gut* (New York: Warner Books, 2001), 421.

340 "doing fabulous things": Bill Conaty, interview with the author.

340 "down to two": Ken Langone, interview with the author.

340 "an exclamation point": Jeff Immelt, interview with the author.

341 "It probably wasn't the smartest": Welch with Byrne, *Jack*, 421.

342 "uncommonly nervous during dinner": Bob Wright, *The Wright Stuff* (New York: RosettaBooks, 2016), 216.

342 "possessed a solid mind": Wright, *Wright Stuff*, 214–17.

342 "It was clearly a choice": Bob Wright, interview with the author.

343 "all in unanimous approval": Welch with Byrne, *Jack*, 422.

343 "Immelt was Jack's candidate": Wright, *Wright Stuff*, 217.

344 "Call me with any concerns": Welch with Byrne, *Jack*, 423.

344 "unanimously and wholeheartedly": Welch with Byrne, *Jack*, 424.

345 "It was McNerney": Denis Nayden, interview with the author.

345 "got no chance": Jim Rohwer, "GE Digs Into Asia," *Fortune*, October 2, 2000.

349 "give him the great news": Welch with Byrne, *Jack*, 425.

350 a "torrential downpour": Welch with Byrne, *Jack*, 425.

350 "worst news of their careers": Welch with Byrne, *Jack*, 426.

351 "we like each other": Jim McNerney, interview with the author.

352 "heart in my throat": Patricia Sellers, "Something to Prove," *Fortune*, June 24, 2002.

353 "I felt really sad": Welch with Byrne, *Jack*, 428.

354 "we had managed the PR": Beth Comstock, *Imagine It Forward* (New York: Currency, 2018), 53.

357 "No one clapped harder": Welch with Byrne, *Jack*, 429.

358 "sitting in the front row": Welch with Byrne, *Jack*, 429.

358 "of the greatest human beings": David Margolick, "The Tao of Jack," *Vanity Fair*, October 2001.

360 "choosing my successor": Welch with Byrne, *Jack*, 358.

361 "couldn't sit on my hands": Welch with Byrne, *Jack*, 359.

361 had discussed the topic: Along with Jack's reflections on the Honeywell deal in *Jack: Straight from the Gut*, many of the details of this frantic dealmaking can be found in Matt Murray, Jeff Cole, Nikhil Deogun, and Andy Pasztor, "On Eve of Retirement, Jack Welch Decides to Stick Around GE a Bit," *Wall Street Journal*, October 23, 2000.

361 "He engineered the wedge": Dave Calhoun, interview with the author.

362 "Word began leaking": Welch with Byrne, *Jack*, 360.

363 "This is the most exciting day": Matt Murray, Philip Shishkin, Bob Davis, and Anita Raghava, "As Honeywell Deal Goes Awry for GE, Fallout May Be Global," *Wall Street Journal*, June 15, 2001.

364 "The first inkling": Welch with Byrne, *Jack*, 364.

364 "I felt good chemistry": Welch with Byrne, *Jack*, 366.

365 "More troublesome": Welch with Byrne, *Jack*, 367.

365 "stuff I like doing": Matt Murray, "GE's Jack Welch Plans to Assume Advisory Role After Retirement," *Wall Street Journal*, March 30, 2001.

366 "began to break down": Welch with Byrne, *Jack*, 367.

368 **"one Honeywell business after another"**: Welch with Byrne, *Jack*, 370.

368 **"shocked and stunned"**: Welch with Byrne, *Jack*.

368 **"doesn't work on the theory"**: "U.S. Officials Ask Europe to Reconsider GE's Bid for Honeywell as Deal Unravels," *Wall Street Journal*, June 15, 2001.

369 **"It was unacceptable"**: Welch with Byrne, *Jack*, 375.

371 **"help fuel its growth"**: Matt Murray, "Some Wonder How Long GE Can Rely on Deals for Growth," *Wall Street Journal*, July 31, 2001.

372 **"I never lose the thrill"**: Holly Peterson, "Jack Welch," *Talk*, December 2000/January 2001, 166.

372 **"'That's the one I'm going out with'"**: Peterson, "Jack Welch."


CHAPTER EIGHTEEN: ONE GOOD DAY

373 **"one good day"**: Christopher A. Bartlett and Andrew N. McLean, "GE's Talent Machine" (Harvard Business School case study), November 3, 2006.

374 **"Triple D was somewhere"**: Jeff Immelt, interview with the author.

377 **"fighting for the company"**: Transcript of Jeff Immelt in conversation with Eric Stein at JPMorgan Chase meeting, January 2016.

379 **"think about doing it"**: Jeffrey Sprague, interview with the author.

381 **"The companies that reliably deliver"**: Justin Fox, "What's So Great About GE?," *Fortune*, March 4, 2002.

385 **"He was furious"**: Gregory Zuckerman and Rachel Emma Silverman, "Why Bond Guru Gross Decided To Attack GE's Finance Practices," *Wall Street Journal*, March 22, 2002.

385 **"We were pissed off"**: Diane Brady, "The Education of Jeff Immelt," *Businessweek*, April 29, 2002.

386 **"knock them down a notch"**: Gregory Zuckerman and Rachel Emma Silverman, "Why Bond Guru Gross Decided to Take Swat at GE," *Wall Street Journal*, March 22, 2002.

386 **"They are an industrial company"**: Jim Grant, "After the Cult," *Grant's Interest Rate Observer*, May 10, 2002.

387 **"what we're doing."**: Jim Bunt, interview with the author.

388 **"It was an arbitrage"**: John Myers, interview with the author.

389 **"years of neglect"**: Jeff Immelt, *Hot Seat* (New York: Avid Reader Press, 2021), 57–58.

390 **"We never ignored them"**: Denis Nayden, interview with the author.

391 **"she was keeping ahead"**: Immelt, *Hot Seat*, 59.

391 **"ingenuity and managerial excellence"**: Bill Gross, interview with the author.

391 **"What devils lurk"**: Gretchen Morgenson, "Wait a Second: What Devils Lurk in the Details?," *New York Times*, April 14, 2002.

393 **"sprinkled in for effect"**: Bill Conaty, interview with the author.

394 **"I'll take care of mine"**: Immelt, *Hot Seat*, 60–61.

395 **"I just eliminated Nayden's position"**: Immelt, *Hot Seat*, 60–61.

395 **"There is no other"**: Matt Murray, "GE Capital Is Split Into Four Parts In Bid to Address Investor Concerns," *Wall Street Journal*, July 29, 2002.

399 he felt **"terrific"**: Jack Welch, interview by Matt Lauer, *Today*, NBC, September 11, 2001.

401 **A second novel**: Lisa DePaulo, "If You Knew Suzy . . . ," *New York*, May 6, 2002.

401 **to become a doctor**: Suzanna Andrews, "Jack Welch, the Other Woman and the Harvard Business Review," *Vanity Fair*, June 2002.

402 **"It wasn't *diamonds*"**: DePaulo, "If You Knew Suzy . . ."

402 **"ought to be more discreet"**: Andrews, "Jack Welch, the Other Woman."

403 **"the reputation of the magazine"**: DePaulo, "If You Knew Suzy..."

404 a **"romantic relationship"**: James Bandler, "Harvard Editor Faces Revolt Over Handling of Welch Story," *Wall Street Journal*, March 4, 2002.

406 **"cowed to Jack Welch"**: DePaulo, "If You Knew Suzy..."

406 **"the love of my life"**: Andrews, "Jack Welch, the Other Woman."

406 **"an insulting amount"**: Leslie Wayne and Alex Kuczynski, "Tarnished Image Places Welch in Unlikely Company," *New York Times*, September 16, 2002.

408 **valued around $330 million**: Details of Jack's and Jane's divorce are from Christopher M. Byron, *Testosterone, Inc.: Tales of CEOs Gone Wild* (Hoboken, NJ: John Wiley & Sons, 2004), 165–68.

409 **"the most appalling"**: Geraldine Fabrikant, "G.E. Expenses For Ex-Chief Cited in Filing," *New York Times*, September 6, 2002.

409 **"just beyond excess"**: Matt Murray, Joann Lublin, and Rachel Emma Silverman, "Welch's Lavish Retirement Pact Angers General Electric Investors," *Wall Street Journal*, September 9, 2002.

410 **"I really like myself"**: Katherine Ozment, "Crazy in Love," *Boston*, May 15, 2006.

410 the **"eerie consistency"**: Wayne and Kuczynski, "Tarnished Image Places Welch in Unlikely Company."

410 **"we unwound them"**: Immelt, *Hot Seat*, 59.

411 **"I've attended just one"**: Jack Welch, "My Dilemma—And How I Resolved It," *Wall Street Journal*, September 16, 2002.

412 the affair lasted **"several months"**: Matt Murray, Rachel Emma Silverman, and Carol Hymowitz, "GE's Jack Welch Meets Match in Divorce Court," *Wall Street Journal*, November 27, 2002.

413 **he had contractually agreed**: Ozment, "Crazy in Love."

415 **took a bizarre turn**: The Ravi Suria tale, as crazy as it is, came from conversations with people who know what happened but who did not want to be named, as well as Suria's presentation and a write-up about the incident in Jim Grant, "The Most Imperfect Investment," *Grant's Interest Rate Observer*, November 8, 2002.

## CHAPTER NINETEEN: DILUTING THE BLOB

419 **"There wasn't one normal day"**: Matt Murray, "GE's Immelt Renovates House That Jack Built," *Wall Street Journal*, February 6, 2003.

421 **"There was a growing awareness"**: Jeff Immelt, *Hot Seat* (New York: Avid Reader Press, 2021), 76.

421 **the effort was "a mistake"**: Immelt, *Hot Seat*, 77.

422 **"This was a Hula-Hoop"**: Immelt, *Hot Seat*, 68.

422 **"The numbers made it"**: Immelt, *Hot Seat*, 68.

422 **"at a steady clip"**: Immelt, *Hot Seat*, 69.

423 **"impossible to repair"**: Bob Wright, interview with the author.

423 **"I couldn't have been happier"**: Immelt, *Hot Seat*, 89.

424 **"We'd had our eye on it for years"**: Immelt, *Hot Seat*, 77.

425 **"I wanted to close"**: Immelt, *Hot Seat*, 79.

425 **"to save money"**: Immelt, *Hot Seat*, 79.

426 **"I did take a little shit"**: Jeff Immelt, interview with the author.

426 **"took a lot of external criticism"**: Immelt, *Hot Seat*, 80.

427 **"What they sell"**: Jim Bunt, interview with the author.

428 **"If we could make a deal"**: Immelt, *Hot Seat*, 82.

429 **"offering to help where I could"**: Immelt, *Hot Seat*, 82.

430 **"I struggled to keep control"**: Bob Wright, *The Wright Stuff* (New York: Rosetta-Books, 2016), 30.

431 **"I knew the deal was sealed"**: Wright, *Wright Stuff*, 34.

432 **"I could tell by his body language"**: Immelt, *Hot Seat*, 82.

432 **"We always believed"**: Steve Burke, interview with the author.

433 **"some artificial lump of bricks"**: Wright, *Wright Stuff*, 35.

434 **"the California guys"**: Wright, *Wright Stuff*.

434 **"It was about as good"**: Wright, *Wright Stuff*, 37.

434 **"He wanted to finish"**: Devin Leonard, "The Unlikely Mogul," *Fortune*, September 29, 2003.

435 **"I love NBC"**: Andy Lack, interview with the author.

437 **"tied to the same boat"**: Jack Welch with John A. Byrne, *Jack: Straight from the Gut* (New York: Warner Books, 2001).

437 **Jeff told *Fortune***: Leonard, "Unlikely Mogul."

437 **"NBC was the house's money"**: Jeff Immelt, interviewed by Charlie Rose, Charlie Rose, PBS, June 25, 2009.

437 **"We are toast"**: Immelt, *Hot Seat*, 84.

438 **"most durable profit centers"**: Immelt, *Hot Seat*, 83.

438 **its "biggest mess"**: Immelt, *Hot Seat*, 87.

439 **"It was clear to me"**: Immelt, *Hot Seat*, 87.

439 **"Here is a good rule"**: Immelt, *Hot Seat*, 87.

439 **"I would always be grateful"**: Immelt, *Hot Seat*, 88.

441 **GE sold three batches**: "G.E. Selling $2.8 Billion Stake in Insurer," *Bloomberg News*, February 27, 2006.

441 **"as was common"**: Immelt, *Hot Seat*, 90.

442 **"you will know"**: Carol Loomis, "GE's Awkward Exit From Insurance," *Fortune*, December 12, 2005.

443 **"and not [tell] anybody"**: Ken Langone, interview with the author.

444 **"chorus of 'no, no, no'"**: Denis Nayden, interview with the author.

444 **"in deal heat"**: Bill Conaty, interview with the author.

447 **"I wish we'd never"**: Immelt, *Hot Seat*, 121–22.

447 **"Partly that was because"**: Immelt, *Hot Seat*, 123.

447 **"an acquisition machine"**: "First International Bank of GE," *Institutional Investor*, November 12, 2004.

448 **"Jack Welch's best friend"**: Immelt, *Hot Seat*, 88.

448 **"'Sheriff of Wall Street'"**: Immelt, *Hot Seat*, 89.

449 **an opinion piece**: Ken Langone, "Let's Bring On the Jury, Mr. Spitzer," *Wall Street Journal*, June 10, 2004.

449 **"gave Enron all the money"**: "Peter Elkind, "The Fall of the House of Grasso," *Fortune*, October 18, 2004.

450 **"This fucking guy lied"**: Ken Langone, interview with the author.

452 **"has been a good friend"**: Joann S. Lublin and Kathryn Kranhold, "Financier Langone To Quit GE Board Over Role at NYSE," *Wall Street Journal*, February 8, 2005.

454 **"the right thing to do"**: Immelt, *Hot Seat*, 89.

## CHAPTER TWENTY: IMAGINATION AT WORK

458 **"someone to figure out"**: Jeff Immelt, *Hot Seat* (New York: Avid Reader Press, 2021), 97.

458 **"We didn't need rebranding exactly"**: Immelt, *Hot Seat*, 97.

459 **"a large bill"**: Bill Conaty, interview with the author.

460 **"I was proud of everyone"**: Immelt, *Hot Seat*, 98.

460 **"biggest Superfund site"**: Immelt, *Hot Seat*, 99.

461 **"Was there something"**: Immelt, *Hot Seat*, 99.

461 **"could be implemented horizontally"**: Immelt, *Hot Seat*, 100.

461 cared **"more about the planet"**: Immelt, *Hot Seat*, 100.

461 **"I wasn't interested"**: Immelt, *Hot Seat*, 101.

461 **"We're taking a stand"**: Beth Comstock, *Imagine It Forward* (New York: Currency, 2018), 116.

462 **"make our customers"**: Comstock, *Imagine It Forward*, 116.

462 **"Comstock and me"**: Immelt, *Hot Seat*, 102.

462 **"We're doing it my way"**: Comstock, *Imagine It Forward*, 117.

462 **"Jeff bravely asked"**: Comstock, *Imagine It Forward*, 117.

463 **"In the beginning"**: Amanda Griscom Little, "G.E.'s Green Gamble," *Vanity Fair*, August 2006.

463 **"Ecoimagination offered a big lift"**: Comstock, *Imagine It Forward*, 118.

463 **"We wanted to change"**: Immelt, *Hot Seat*, 106.

463 **"the change-maker's dilemma"**: Comstock, *Imagine It Forward*, 120.

## CHAPTER TWENTY-ONE: THE BLIND SPOT

465 **"were identical, identical"**: Dave Calhoun, interview with the author.

467 **"I bent myself"**: Jeff Immelt, interview with the author.

467 **"the No. 1 draft pick"**: Geoffrey Colvin, "Rising Star: David Calhoun, General Electric," *Fortune*, January 24, 2006.

468 **"He didn't even flinch"**: Bob Wright, *The Wright Stuff* (New York: RosettaBooks, 2016), 234.

469 **"what we both thought"**: Wright, *Wright Stuff*, 240.

469 **"GE's notorious bean counters"**: Wright, *Wright Stuff*, 240.

469 **"I had to fight"**: Jeff Immelt, *Hot Seat* (New York: Avid Reader Press, 2021), 84.

470 **"I was done with him"**: Immelt, *Hot Seat*, 85.

473 **"I knew virtually nothing"**: Michael Pralle, interview with the author.

480 **"making any new deals"**: Immelt, *Hot Seat*, 123.

483 **"As the world learned"**: Jim Grant, "Mythical Triple-A," *Grant's Interest Rate Observer*, November 30, 2007.

484 **"According to McKinsey"**: Immelt, *Hot Seat*, 123.

487 **"sell them at a profit"**: Immelt, *Hot Seat*, 124.

488 **"In the wake of Bear Stearn's"**: Immelt, *Hot Seat*, 124.

489 **"been no fun"**: Immelt, *Hot Seat*, 125.

489 **"Look, I'm sorry"**: Immelt, *Hot Seat*, 126.

490 **"If you had a child"**: Jeff Immelt, interview with the author.

491 **"an historic event"**: Nelson D. Schwartz and Claudia H. Deutsch, "G.E.'s Shortfall Calls Credibility Into Question," *New York Times*, April 17, 2008.

491 **"too big to manage"**: Schwartz and Deutsch, "G.E.'s Shortfall Calls Credibility Into Question."

492 **"we were doing okay"**: Financial Crisis Inquiry Commission Report, 242.

493 **"Not Your Father's GE"**: Jim Grant, "Not Your Father's GE," *Grant's Interest Rate Observer*, September 5, 2008.

494 **"This stunned me"**: Henry Paulson Jr., *On the Brink* (New York: Business Plus, 2010), 172.

494 **Jeff "happened" to be**: Immelt, *Hot Seat*, 127.

495 **"I'd known Jeff for years"**: Paulson, *On the Brink*, 227.

495 **"put out this fire"**: Paulson, *On the Brink*, 47.

495 **"he was relying"**: Immelt, *Hot Seat*, 128.

496 **"closed the investigation"**: Immelt, *Hot Seat*, 128.

496 **would have been "just colossal"**: Hank Paulson, interview with the author.

499 **"overcome by the need"**: Immelt, *Hot Seat*, 130.

500 **"Rolling commercial paper"**: Immelt, *Hot Seat*, 132.

500 **"like being granted"**: Immelt, *Hot Seat*, 134.

506 **"the gold standard"**: Rodge Cohen, interview with the author.

507 **"we'd made a mistake"**: Immelt, *Hot Seat*, 133.

508 **"If he was concerned"**: Immelt, *Hot Seat*, 135.

509 **"treacherous and unknowable"**: Immelt, *Hot Seat*, 135.

509 **"All I wanted to do"**: Immelt, *Hot Seat*, 136.

509 **"Had we launched"**: Immelt, *Hot Seat*, 136.

510 **"It was unclear"**: Immelt, *Hot Seat*, 137.

511 **"most expensive capital"**: Geoff Colvin and Katie Benner, "GE Under Siege," *Fortune*, October 15, 2008.

511 **"buying high and selling low"**: Colvin and Benner, "GE Under Siege."

512 **"an extended episode"**: Immelt, *Hot Seat*, 137.

512 **"we'd get a little break"**: Immelt, *Hot Seat*, 138.

513 **"created a circle"**: Jeff Bornstein, interview with the author.

514 **"By leaving us out"**: Immelt, *Hot Seat*, 138.

517 **"We keep everyone"**: Immelt, *Hot Seat*, 139.

517 **"I'd be difficult"**: Immelt, *Hot Seat*, 140.

517 **"disastrous for the economy"**: Immelt, *Hot Seat*, 140.

518 **"maximize their spreads"**: Sheila Bair, interview with the author.

520 **"having the stomach flu"**: Immelt, *Hot Seat*, 142.

## CHAPTER TWENTY-TWO: BURNING FURNITURE

523 **"I remember opening"**: Jeff Immelt, *Hot Seat* (New York: Avid Reader Press, 2021), 143.

524 **"That's just the way it is"**: Immelt, *Hot Seat*, 144.

525 **"because of financial services"**: Jeff Immelt, interview with the author.

525 **"No matter how dire"**: Immelt, *Hot Seat*, 144.

526 **"You're not resigning"**: Immelt, *Hot Seat*, 145.

526 **"really, really, really hard"**: Devin Leonard and Rick Clough, "How GE Exorcised the Ghost of Jack Welch to Become a 124-Year-Old Startup," *Businessweek*, March 17, 2016.

527 **"We had to seize"**: Immelt, *Hot Seat*, 146.

529 **"Finally, after 2008"**: Steve Burke, interview with the author.
530 **"worst seven months"**: Keith Sherin, interview with the author.
532 **"dispel[ling] false rumors"**: Immelt, *Hot Seat*, 146.
534 **"an island of one"**: Immelt, *Hot Seat*, 151.
535 **"In the dead of winter"**: Immelt, *Hot Seat*, 152.
547 **"neither the Fed nor GE"**: Jim Himes, interview with the author.


## CHAPTER TWENTY-THREE: POWER MAN

549 **"In order to break"**: Jeff Immelt, *Hot Seat* (New York: Avid Reader Press, 2021), 178.
550 **"We didn't conjure"**: Immelt, *Hot Seat*, 179.
550 **"Flannery delivered"**: Immelt, *Hot Seat*, 179.
551 **"funded by the CEO's office"**: Immelt, *Hot Seat*, 183.
551 **"would diminish their power"**: Immelt, *Hot Seat*, 204.
552 **"like Apple today"**: John Krenicki, interview with the author.
555 **"It's a complicated job"**: Beth Kowitt, "John Krenicki Powers Up GE," *Fortune*, December 19, 2011.
556 **was not on board**: Immelt, *Hot Seat*, 204.
557 **"he wasn't quiet"**: Immelt, *Hot Seat*, 204.
557 **"We fought over everything"**: Immelt, *Hot Seat,* 205.
558 **"I never looked back"**: Immelt, *Hot Seat*, 206.
558 **"'Life's too hard'"**: Jeff Immelt, interview with the author.
558 **"I didn't expect"**: Immelt, *Hot Seat*, 206.
559 **"removing a layer"**: Immelt, *Hot Seat*, 206.
559 **"What a fucking scumbag"**: Ed Galanek, interview with the author.
561 **"it paid off"**: Immelt, *Hot Seat*, 206.
561 **logged "every aspect"**: Immelt, *Hot Seat*, 111.
562 **"real right-wingers"**: Immelt, *Hot Seat*, 180.
563 **"a golf game"**: The narrative about Jeff Immelt's trip to Mount Kilimanjaro is from an interview he gave to Michael Useem, a professor at the University of Pennsylvania's Wharton School, on April 30, 2013, and from my conversations with Ed Galanek.
570 **"most successful competitor"**: Immelt, *Hot Seat*, 264.
570 **"put the company in play"**: Jeff Immelt, interviewed by Charlie Rose, *Charlie Rose*, PBS, June 15, 2015.
571 **"came away intrigued"**: Immelt, *Hot Seat*, 266.
573 **"of all places"**: Steve Bolze's biographical details are from his interviews with the author.
577 **"one way to prepare them"**: Immelt, *Hot Seat*, 267.
578 **"huge long-term advantage"**: Immelt, *Hot Seat*, 266.
580 **about $13 billion**: Aaron Kirchfeld, Matthew Campbell, and Jeffrey McCracken, "GE Said in Talks to Buy France's Alstom," *Bloomberg*, April 24, 2014.
581 **"jump in the Seine"**: Immelt, *Hot Seat*, 267.
581 **"His manner was formal"**: Immelt, *Hot Seat*, 268.
583 **"French government authorities"**: Matt Clinch and Katy Barnato, "French Warn GE as Siemens Plans Alstom Hijack," CNBC, April 28, 2014.
583 **"The shit hit the fan"**: Jeff Immelt, memo to the author.
583 **"We were stuck"**: Immelt, *Hot Seat*, 268.
585 **"The deal for Alstom's electricity"**: Inti Landauro, "Alstom Board Clears GE's Sweetened Offer for Power Business," *Wall Street Journal*, June 21, 2014.

586 **"have the political savvy"**: Immelt, *Hot Seat,* 270.

587 **"'Works great for us'"**: Immelt, *Hot Seat*, 235.

587 **"All of us chafed"**: Immelt, *Hot Seat*, 236.

588 **"I was chastened"**: Immelt, *Hot Seat*, 237.

588 **"nearly $1 billion"**: Immelt, *Hot Seat*, 238.

589 **"we had to pivot"**: Immelt, *Hot Seat*, 237.

590 **"risk we were taking"**: Immelt, *Hot Seat*, 239.

590 **"We had to try"**: Immelt, *Hot Seat*, 239.

590 **"From the minute"**: Keith Sherin, interview with the author.

591 **"The name stuck"**: Immelt, *Hot Seat*, 241.

592 **"Rumors may spread"**: Immelt, *Hot Seat*, 244.

593 **"paid us well"**: Immelt, *Hot Seat*, 245.

593 **"would have confirmed it"**: Immelt, *Hot Seat*, 248.

594 **"GE's top executives"**: Dana Mattioli, Eliot Brown, Ted Mann, and Joann S. Lublin, "GE Close to Selling Real-Estate Holdings," *Wall Street Journal*, April 9, 2015.

594 **"everyone signed off"**: Immelt, *Hot Seat*, 250.

596 **offered grudging praise**: "A Hard Act to Follow," *Economist*, June 27, 2014.

## CHAPTER TWENTY-FOUR: IF YOU STRIKE AT THE KING

599 **lacked "succession etiquette"**: Jeff Immelt, *Hot Seat* (New York: Avid Reader Press, 2021), 269.

599 **"It simply seemed"**: Immelt, *Hot Seat*, 268.

600 **"We had walked away"**: Immelt, *Hot Seat*, 272.

603 **"Look," Jeff told his colleagues**: Immelt, *Hot Seat*, 273.

603 **laid into Bolze**: Immelt, *Hot Seat*, 273–74.

604 **"*really* going to leave GE"**: Immelt, *Hot Seat*, 274.

605 **"as it had astonished Peters"**: Immelt, *Hot Seat*, 275.

606 **"I also suspected"**: Immelt, *Hot Seat*, 275–76.

606 **"better than Jack"**: Ken Langone, interview with the author.

607 **"to remain as CEO"**: Immelt, *Hot Seat*, 276.

607 **"the public domain"**: Douglas A. Warner, email to the author.

607 **"keep my head down"**: Warner, email to the author.

608 **"gets taken home"**: Warren Buffett, letter to shareholders of Berkshire Hathaway Inc., February 22, 2020.

608 **"the end of the day"**: Steve Bolze, interview with the author.

609 **"don't see my worth"**: Immelt, *Hot Seat*.

609 **"Bolze was out"**: Immelt, *Hot Seat*, 277.

609 **"I would sleep on it"**: Immelt, *Hot Seat*, 277–78.

611 **"the imperial CEO"**: Immelt, *Hot Seat*, 278.

## CHAPTER TWENTY-FIVE: A FOX GUARDING THE HENHOUSE

612 **"got to know Ed"**: Jeff Immelt, interview with the author.

612 **Jeff was "holding court"**: Ed Garden, interview with the author.

614 **"might be in my corner"**: Jeff Immelt, *Hot Seat* (New York: Avid Reader Press, 2021), 255.

614 **"with fucking gunslingers"**: Ken Langone, interview with the author.

616 **"wanted you to be aware"**: Immelt, *Hot Seat*, 256.

617 **"If we do well"**: Mike Damiano, "GE CEO Jeff Immelt Is All In," *Boston*, April 16, 2017.

618 **his "new GE" plan**: Trian Partners, GE white paper presentation, October 5, 2015.

619 **"I think they'll execute"**: Lewis Krauskopf, "Nelson Peltz's Trian Takes $2.5 Billion Stake in General Electric," Reuters, October 5, 2015.

622 **"in the flow of ideas"**: Transcript of Jeff Immelt in conversation with Eric Stein, at JPMorgan Chase meeting, January 2016.

624 **"We could focus"**: Immelt, *Hot Seat*, 260.

625 **"to remove Bolze"**: Immelt, *Hot Seat*, 280.

625 **"was now fully underway"**: Immelt, *Hot Seat*, 281.

626 **conference room at the airport**: Immelt, *Hot Seat*, 281.

626 **"we're selling out of GE"**: Immelt, *Hot Seat*, 282.

626 **"was growing short"**: Immelt, *Hot Seat*, 282.

630 **"had enough" of Bolze**: Immelt, *Hot Seat*, 283.

630 **"When in doubt"**: Immelt, *Hot Seat*, 283.

630 **he wrote *Hot Seat***: Immelt, *Hot Seat*, 283.

633 **a surprising scoop**: Ted Mann and Joann S. Lublin, "Wall Street Starts to Contemplate GE after Jeff Immelt," *Wall Street Journal*, February 20, 2017.

635 **"in the hot seat"**: Charlie Gasparino and Brian Schwartz, "General Electric CEO Immelt in the Hot Seat with Trian's Peltz," Fox Business, March 10, 2017.

636 **"That's when you know"**: Immelt, *Hot Seat*, 285.

636 **"But this trip"**: Immelt, *Hot Seat*, 287.

637 **"This is happening"**: Immelt, *Hot Seat*, 286.

638 **"he loves Healthcare"**: Ed Galanek, interview with the author.

638 **"kitchen is way up"**: Damiano, "GE CEO Jeff Immelt Is All In."

639 **GE was "furious"**: Mike Damiano, interview with the author.

641 **"make it sexy"**: John Flannery, interview with the author.

641 **"was not at his best"**: Immelt, *Hot Seat*, 291–92.

642 **"permission to fail"**: Immelt, *Hot Seat*, 293.

643 **"coming to terms with that"**: Immelt, *Hot Seat*, 293.

647 **"I felt listless"**: Immelt, *Hot Seat*, 294.

647 **"sort of a fugue state"**: Immelt, *Hot Seat*, 294.

651 **"There wasn't much discussion"**: Immelt, *Hot Seat*, 296.

651 **"The vote was over"**: Immelt, *Hot Seat*, 296.

651 **"the finality of it stung"**: Immelt, *Hot Seat*, 296.

652 **board had chosen John**: Immelt, *Hot Seat*, 296–97.

652 **could "move on"**: Steve Bolze, interview with the author.

653 **Jeff's Back Bay home**: Immelt, *Hot Seat*, 296–97.

## CHAPTER TWENTY-SIX: EISENHOWER

655 **"there was no way to make"**: Jeff Immelt, *Hot Seat* (New York: Avid Reader Press, 2021), 299.

658 **the *Times* reported**: Julie Creswell, "For G.E.'s John Flannery, It Was a 30-Year Trip to the Top," *New York Times*, June 12, 2017.

660 **"I'd be fucking toast"**: Jeff Bornstein, interview with the author.

666 **"I don't often post here"**: Steve Bolze, "Some Thoughts on Jeff Immelt's New Book," LinkedIn, February 22, 2021.

669 **the "baronial" era**: Nelson D. Schwartz, "The Decline of the Baronial C.E.O.," *New York Times*, June 17, 2017.

671 **Tusa's perception of the problems:** Thomas Gryta and David Benoit, "GE's Nemesis: An Eerily Prescient Bear," *Wall Street Journal*, June 17, 2019.

673 **"Person to Blame":** Immelt, *Hot Seat*, 300.

674 **"In my view":** Immelt, *Hot Seat*, 301.

675 **"I love this company":** Thomas Gryta and Ted Mann, "GE Powered the American Century—Then It Burned Out," *Wall Street Journal*, December 14, 2018.

675 **He had hoped:** Immelt, *Hot Seat*, 301.

676 **"for much of his tenure":** Thomas Gryta and Joann S. Lublin, "GE's New Chief Makes Cuts, Starting With Old Favorites," *Wall Street Journal*, October 18, 2017.

677 **"The average guy":** Bill Conaty, interview with the author.

679 **GE's security detail:** Ed Galanek, interview with the author.

680 **"Oh, God, embarrassing":** Jeff Immelt, interview with the author.

681 **the *Journal* reported:** Thomas Gryta, Joann S. Lublin, and David Benoit, "How Jeffrey Immelt's 'Success Theater' Masked the Rot at GE," *Wall Street Journal*, February 21, 2018.

682 **"This is simply untrue":** Immelt, *Hot Seat*, 279.

682 **blamed John Flannery:** Immelt, *Hot Seat*, 303.

683 **took to LinkedIn:** Alex Dimitrief, "Anonymity Is Not a License to Rewrite GE's History," LinkedIn, July 21, 2021.

701 *The Economist* **noted:** "John Flannery Gets Down to Business Restructuring General Electric," *Economist*, June 28, 2018.

702 **"make others give up":** Russell Stokes, "Making the Best Turbines Is Hard Enough," LinkedIn, September 19, 2018.

## CHAPTER TWENTY-SEVEN: THE LAST EMPEROR

709 **"always been struck":** Nick Dunehew, "Larry Culp Biography," *Brooksy*, October 8, 2020.

## CHAPTER TWENTY-EIGHT: WHO LOST GE?

721 **the "three *C*s":** Steve Bolze, interview with the author.

725 **"loss of moral compass":** Jim Grant, interview with the author.

726 **"the internet of things":** Jeff Immelt, "How I Remade GE," *Harvard Business Review*, September/October 2017.

728 **"in a bubble":** Jeff Immelt, interview with the author.

## EPILOGUE: ST. PATRICK'S CATHEDRAL

735 **"most friendly environment":** Peter Foss, interview with the author.

## AUTHOR'S NOTE

743 **"an ingenious company":** Thomas F. O'Boyle, interview with the author.

744 **According to Michael Brian Schiffer:** Michael Brian Schiffer, *Taking Charge: The Electric Automobile in America* (Collingdale, PA: Diane, 1994).

745 **"gasoline became so cheap":** John Seabrook, "America's Favorite Pickup Truck Goes Electric," *New Yorker*, January 24, 2002.

# Index